Water Resources Management in Mountain Regions

Edited by

Virendra Bahadur Singh
Department of Environmental Studies
Ram Lal Anand College
University of Delhi
New Delhi
India

Sughosh Madhav
Department of Civil Engineering
Jamia Millia Islamia
New Delhi
India

Rakesh Kumar Gupta
Professor
Department of Microbiology
Ram Lal Anand College
University of Delhi
New Delhi
India

Prerna Diwan
Professor
Department of Microbiology
Ram Lal Anand College
University of Delhi
New Delhi
India

Amit Kumar
Wadia Institute of Himalayan Geology
Dehradun
Uttarakhand
India

WILEY

Registered Offices
John Wiley & Sons, Inc., 111 River Street, Hoboken, NJ 07030, USA
John Wiley & Sons Ltd, New Era House, 8 Oldlands Way, Bognor Regis, West Sussex, PO22 9NQ, UK

For details of our global editorial offices, customer services, and more information about Wiley products visit us at www.wiley.com.

The manufacturer's authorized representative according to the EU General Product Safety Regulation is Wiley-VCH GmbH, Boschstr. 12, 69469 Weinheim, Germany, e-mail: Product_Safety@wiley.com.

Wiley also publishes its books in a variety of electronic formats and by print-on-demand. Some content that appears in standard print versions of this book may not be available in other formats.

Library of Congress Cataloging-in-Publication Data
Names: Singh, Virendra Bahadur, editor. | Gupta, Rajesh Kumar (Of Ram Lal Anand College), editor. | Diwan, Prerna, editor. | Kumar, Amit (Geologist), editor.
Title: Water resources management in mountain regions / edited by Virendra Bahadur Singh, Rajesh Kumar Gupta, Prerna Diwan, Amit Kumar.
Description: Hoboken, NJ : Wiley, 2025. | Includes index.
Identifiers: LCCN 2024038874 (print) | LCCN 2024038875 (ebook) | ISBN 9781394249589 (hardback) | ISBN 9781394249602 (adobe pdf) | ISBN 9781394249596 (epub)
Subjects: LCSH: Water resources development–Himalaya Mountains Region. | Water quality management–Himalaya Mountains Region.
Classification: LCC TC405 .W355 2025 (print) | LCC TC405 (ebook) | DDC 363.6/1095496–dc23/eng/20240923
LC record available at https://lccn.loc.gov/2024038874
LC ebook record available at https://lccn.loc.gov/2024038875

Cover Design: Wiley
Cover Image: Courtesy of Virendra Bahadur Singh

Set in 9.5/12.5pt STIX Two Text by Straive, Pondicherry, India
Printed and bound by CPI Group (UK) Ltd, Croydon, CR0 4YY

C9781394249589_150325

Dedicated to my lovely daughter Vritika and beloved son Rakshit.

Dr. Virendra Bahadur Singh

Contents

About the Editors

(1) Dr. Virendra Bahadur Singh

Dr. Virendra Bahadur Singh is presently working as an Assistant Professor in the Department of Environmental Studies, Ram Lal Anand College, University of Delhi, New Delhi, India. He has also worked as a Dr. D. S. Kothari Postdoctoral Fellow (DSKPDF) in the Department of Geology, University of Delhi, Delhi, India. He has worked in the Department of Civil Engineering, Indian Institute of Technology Delhi, New Delhi, India as National Postdoctoral Fellow (NPDF). He has obtained his doctoral degree from the School of Environmental Sciences, Jawaharlal Nehru University, New Delhi, India. He has also received a very prestigious Commonwealth Split-site Scholarship tenable at the University of Bristol, United Kingdom, awarded by the Commonwealth Scholarship Commission, UK. He has also participated as a Scientist in the Indian Arctic Expedition, fifth batch at the Indian Scientific Station "Himadri," Ny-Ålesund, Norway.

Dr. Singh's research work mainly focuses on glaciology, hydrology, hydro-meteorological correlation, meteorology, suspended sediment transport, hydrogeochemistry, water stable isotopes, mass balance, glacier retreat, biogeochemistry, water pollution, and climate change. He has worked in the fields of glaciology, hydrology, hydrogeochemistry, and water resources in the mountainous environment for the past 15 years. He has published 36 research papers in various peer-reviewed national and international journals. He has also edited three books published by international publishers such as Wiley and Elsevier. He has reviewed many research papers published by various high-impact international journals such as *Nature Communications*; *Journal of Hydrology*; *Science of the Total Environment*; *Hydrological Processes*; *CATENA*; *Environment, Development and Sustainability*; *Applied Geochemistry*; *Polar Science*; *Groundwater for Sustainable Development*; *Environmental Processes*; and *Sustainable Water Resources Management*.

(2) Dr. Sughosh Madhav

Dr. Sughosh Madhav is presently a Dr. D. S. Kothari Postdoctoral Fellow (DSKPDF) in the Department of Civil Engineering, Jamia Millia Islamia, New Delhi, India. He has obtained his master's degree from the Department of Environmental Science, Banaras Hindu University, Varanasi, India. He earned his doctorate from the School of Environmental Sciences, Jawaharlal Nehru University, New Delhi, India.

Dr. Madhav's research work mainly focuses on hydrogeochemistry, water resources, water pollution, river water, wetland, water quality, climate change, disaster management, chemical weathering, biogeochemistry, wastewater treatment, bioremediation, trace elements contamination, urban water crisis, and soil contamination. He has published 25 research papers in various national and international journals. He has also edited 12 books published by various international publishers, such as Wiley, Springer, and Elsevier, on different environmental issues.

(3) Professor Rakesh Kumar Gupta

Professor Rakesh Kumar Gupta is the Principal of Ram Lal Anand College, University of Delhi, New Delhi, India, and a Professor in the Department of Microbiology, Ram Lal Anand College, University of Delhi, New Delhi, India. He obtained his Ph.D from the National Dairy Research Institute (ICAR), Karnal, Haryana, India. He has also worked as a Postdoctoral Research Fellow at the Center for Environmental Biotechnology and the Department of Microbiology, University of Tennessee, Knoxville, TN, USA.

His research work mainly focuses on applied microbiology, water quality management, molecular biology, detection of heavy metal ions, environmental metagenomics, etc. Professor Gupta has two US Patents and two International Patent Publications. He has developed a whole cell biosensor for real-time detection of estrogen in natural water bodies. He has received various national and international prestigious fellowships such as the United Nations Development Program (UNDP) Fellowship and the National Dairy Research Institute Fellowship. Professor Gupta has associations with various professional bodies such as the Association of Microbiologists of India and the Biotechnology Research Society of India. He has published more than 50 research publications in various national and international journals and has also published eight books and book chapters.

(4) Professor Prerna Diwan

Professor Prerna Diwan is a Professor in the Department of Microbiology, Ram Lal Anand College, University of Delhi, New Delhi, India. She has obtained her Ph.D from the Department of Microbiology, University of Delhi, New Delhi, India. She has postdoctoral research experience at the Korea Research Institute of Bioscience and Biotechnology, South Korea, and the Department of Biochemistry, University of Alberta, Canada.

Professor Diwan's research focuses on water and air metagenomics, AMR, bioremediation of heavy metals, and oral microbiome studies in the field of microbiology. She has published more than 20 research papers in various high-impact national and international journals and has also published eight book chapters.

(5) Dr. Amit Kumar

Dr. Amit Kumar is a Scientist at the Wadia Institute of Himalayan Geology (WIHG), Dehradun, Uttarakhand, India, specializing in glaciology and hydrology of the Himalaya-Karakoram region. He holds a PhD in Geology from Panjab University, Chandigarh, India, and a Master's from Banaras Hindu University, Varanasi, India.

Since joining WIHG's Centre for Glaciology in 2009, he has led multiple projects sponsored by the Department of Science and Technology (DST), Government of India, on glacier monitoring and associated hazards. With over a decade of fieldwork and research, Dr. Kumar has advanced techniques in glacier monitoring and glacier hydrology, earning accolades like the Young Scientist Award from the Department of Science and Technology (DST), Government of India, in 2014. His research focuses on water resource management and mitigating glacier and climate-driven hazards, particularly flash floods, contributing to numerous peer-reviewed publications. His ongoing efforts aim to establish integrated multi-hazard early warning systems for vulnerable regions in the Himalayas.

List of Contributors

Bedour Alsabti
Water Research Centre, Kuwait Institute for Scientific Research (KISR)
Kuwait City, Kuwait

Naman Arora
Department of Hydro and Renewable Energy, Indian Institute of Technology Roorkee, Roorkee
Uttarakhand, India

Raghavendra Ashrit
National Center for Medium Range Weather Forecasting (NCMRWF)
Ministry of Earth Sciences (MoES)
Noida, India

Kainat Aziz
Department of Civil Engineering, Indian Institute of Technology, Guwahati
Assam, India

Kiran Bishwakarma
State Key Laboratory of Tibetan Plateau Earth System, Resources and Environment (TPESRE), Institute of Tibetan Plateau Research, Chinese Academy of Sciences
Beijing, China

University of Chinese Academy of Sciences
Beijing, China

Subhashree Biswal
Department of Geology, Ravenshaw University
Cuttack, India

Smitha C. Ch
Department of Computer Science and Engineering, Koneru Lakshmaiah Education Foundation
Guntur, India

Shalini Singh Chhetri
Himalayan Women Awareness and Livelihood (HIMWAL) Society
Dehradun, India

Sabarathinam Chidambaram
Water Research Centre, Kuwait Institute for Scientific Research (KISR)
Kuwait City, Kuwait

Lobzang Chorol
Department of Environmental Science and Engineering, Indian Institute of Technology (Indian School of Mines)
Dhanbad, India

Tanveer Dar
Centre for Cryosphere and Climate Change Studies, National Institute of Hydrology
Roorkee, India

Pallavi Das
Department of Environmental Science
Indira Gandhi National Tribal University
Amarkantak
Madhya Pradesh, India

Sayantan Das
School of Computer and Systems Sciences
Jawaharlal Nehru University
New Delhi, India

Mayura D. Deshmukh
Department of Geology, Shri Shivaji
Science College, Amravati
Maharashtra, India

Subash Dhakal
Rural Management and Development
Department, Government of Sikkim
Gangtok, Sikkim, India

Jaydeo Kumar Dharpure
Centre of Excellence in Disaster Mitigation
and Management, Indian Institute of
Technology Roorkee, Roorkee
Uttarakhand, India

Water Resources System Division, National
Institute of Hydrology Roorkee, Roorkee
Uttarakhand, India

Deepika Dimri
Himalayan Women Awareness and
Livelihood (HIMWAL) Society
Dehradun, India

Dwarika P. Dobhal
Centre for Glaciology, Wadia Institute of
Himalayan Geology
Dehradun, India

Enviromental Geomorphology, Wadia
Institute of Himalayan Geology
Dehradun, India

Fan Zhang
State Key Laboratory of Tibetan Plateau
Earth System, Resources and Environment
(TPESRE), Institute of Tibetan Plateau
Research, Chinese Academy of Sciences
Beijing, China

University of Chinese Academy of Sciences
Beijing, China

Vinay K. Gaddam
Department of Civil Engineering
Siddhartha Academy of Higher Education
Vijayawada, India

Vinay K. Gaddam
Department of Civil Engineering
V.R. Siddhartha Engineering College
Deemed to be University, Kanuru
Vijayawada, Andhra Pradesh, India

Nagappan Ganesh
Department of Earth Sciences
Annamalai University, Chidambaram
Tamil Nadu, India

M. Geetha Priya
CIIRC, Jyothy Institute of Technology
Bengaluru, India

Ashvani K. Gosain
Department of Civil Engineering
Indian Institute of Technology Delhi
New Delhi, India

Ajanta Goswami
Centre of Excellence in Disaster Mitigation
and Management, Indian Institute of
Technology Roorkee, Roorkee
Uttarakhand, India

Department of Earth Sciences, Indian
Institute of Technology Roorkee, Roorkee
Uttarakhand, India

Guanxing Wang
Key Laboratory of Tibetan Plateau
Land Surface Processes and Ecological
Conservation, College of Geographical
Science, Qinghai Normal University
Xining, China

Qinghai Province Key Laboratory of
Physical Geography and Environmental
Process, College of Geographical Science
Qinghai Normal University, Xining, China

Mousumi Gupta
Sikkim Manipal Institute of Technology
Sikkim Manipal University, Majitar
Sikkim, India

Sunil K. Gupta
Department of Environmental Science and Engineering, Indian Institute of Technology (Indian School of Mines)
Dhanbad, India

Sunila Hooda
Department of Microbiology, Ram Lal Anand College, University of Delhi
New Delhi, India

Sanjay Kumar Jain
Water Resources System Division, National Institute of Hydrology Roorkee, Roorkee
Uttarakhand, India

Om P. Kaptan
Department of Geology, School of Physical Sciences, Sikkim University, Gangtok
Sikkim, India

Tirumalesh Keesari
Isotope and Radiation Application Division, Bhabha Atomic Research Centre
Mumbai, India

Department of Atomic Energy
Homi Bhabha National Institute
Mumbai, India

Ashok K. Keshari
Department of Civil Engineering
Indian Institute of Technology Delhi
New Delhi, India

S.F.A. Khadri
Department of Computer Science and Engineering, Acharya Nagarjuna University Nagarjuna Nagar, Guntur
Andhra Pradesh, India

S.F.R. Khadri
Department of Civil Engineering
V.R. Siddhartha Engineering College
Deemed to be University, Kanuru
Vijayawada, Andhra Pradesh, India

Sayyad Fazal Rahman Khadri
Department of Civil Engineering
Siddhartha Academy of Higher Education
Vijayawada, India

Narayan P. Khanal
Nepal Environment and Development Consultant Pvt. Ltd.
Kathmandu, Nepal

Deen D. Khandelwal
Centre for Glaciology, Wadia Institute of Himalayan Geology
Dehradun, India

Ashita S. Khasa
Department of Microbiology, Ram Lal Anand College, University of Delhi
New Delhi, India

Vinod Khatti
Himalayan Environmental Studies and Conservation Organization (HESCO)
Dehradun, India

Liza G. Kiba
Department of Agricultural Engineering
School of Agricultural Sciences
Nagaland University, Medziphema
Nagaland, India

Anil Vishnupant Kulkarni
Divecha Center for Climate Change Indian Institute of Science, Bangalore
Karnataka, India

Ambika Kumar
Bhagalpur National College, Bhagalpur
Bihar, India

Anil Kumar
Forest Ecology and Climate Change Division, ICFRE-Himalayan Forest Research Institute, Shimla
Himachal Pradesh, India

Arun Kumar
Department of Hydro and Renewable Energy, Indian Institute of Technology Roorkee, Roorkee
Uttarakhand, India

Deepak Kumar
Department of Chemistry, Guru Jambheshwar University of Science and Technology, Hisar
Haryana, India

Devansh Kumar
Department of Microbiology, Ram Lal Anand College, University of Delhi
New Delhi, India

Pawan Kumar
Department of Natural Resources Management, Maharana Pratap Horticultural University, Karnal
Haryana, India

Priyanka Kumari
Department of Geology, School of Physical Sciences, Sikkim University, Gangtok
Sikkim, India

Chitrasen Lairenjam
Department of Agricultural Engineering
School of Agricultural Sciences, Nagaland University, Medziphema
Nagaland, India

Ambrish K. Mahajan
Department of Environmental Sciences
Central University of Himachal Pradesh
Dharamshala
Himachal Pradesh, India

Mohit Marwah
Department of Microbiology, Ram Lal Anand College, University of Delhi
New Delhi, India

Manish Mehta
Enviromental Geomorphology, Wadia Institute of Himalayan Geology
Dehradun, India

Gnana S.S.V. Mendu
Department of Computer Science and Engineering, Siddhartha Academy of Higher Education
Vijayawada, India

Anil K. Misra
Department of Geology, School of Physical Sciences, Sikkim University, Gangtok
Sikkim, India

Hemant Mohokar
Isotope and Radiation Application Division, Bhabha Atomic Research Centre
Mumbai, India

Sobhana Mummaneni
Department of Computer Science and Engineering, Siddhartha Academy of Higher Education
Vijayawada, India

Grace Nengzouzam
Department of Agricultural Engineering
School of Engineering and Technology
Nagaland University, Kohima
Nagaland, India

Jharana Nepal
State Key Laboratory of Tibetan Plateau Earth System, Resources and Environment (TPESRE), Institute of Tibetan Plateau Research, Chinese Academy of Sciences
Beijing, China

and

University of Chinese Academy of Sciences, Beijing, China

Banajarani Panda
Department of Geology
Ravenshaw University
Cuttack, India

Diksha Pant
Isotope and Radiation Application Division, Bhabha Atomic Research Centre
Mumbai, India

and

Department of Atomic Energy
Homi Bhabha National Institute
Mumbai, India

Ramesh R. Pant
Central Department of Environmental Science, Institute of Science and Technology, Tribhuvan University
Kathmandu, Nepal

Akansha Patel
Centre of Excellence in Disaster Mitigation and Management, Indian Institute of Technology Roorkee, Roorkee
Uttarakhand, India

Krity Rai
Department of Geology, School of Physical Sciences, Sikkim University, Gangtok
Sikkim, India

Deepansha Raina
Department of Microbiology, Ram Lal Anand College, University of Delhi
New Delhi, India

Prem Ranjan
Department of Agricultural Engineering
North Eastern Regional Institute of Science and Technology, Nirjuli
Arunachal Pradesh, India

Rakesh K. Ranjan
Department of Geology, School of Physical Sciences, Sikkim University, Gangtok
Sikkim, India

Gopinadh Rongali
National Center for Medium Range Weather Forecasting (NCMRWF)
Ministry of Earth Sciences (MoES)
Noida, India

Piyali Sabui
Department of Environmental Science
Indira Gandhi National Tribal University
Amarkantak, Madhya Pradesh, India

Arpan Sharma
Asian Development Research Institute
Patna, Bihar, India

Narpati Sharma
Sikkim State Council of Science and Technology, Government of Sikkim
Gangtok, Sikkim, India

Saurabh K. Sharma
School of Computer and Systems Sciences
Jawaharlal Nehru University
New Delhi, India

K. Shruti
CIIRC, Jyothy Institute of Technology
Bengaluru, India

Sunil K. Singal
Department of Hydro and Renewable Energy, Indian Institute of Technology Roorkee, Roorkee, Uttarakhand, India

Shakshee Soni
Department of Environmental Science
Indira Gandhi National Tribal University
Amarkantak, Madhya Pradesh, India

Shalini Swami
Department of Microbiology, Ram Lal Anand College, University of Delhi
New Delhi, India

Nihitha Vemulapalli
Department of Computer Science and Engineering, Siddhartha Academy of Higher Education, Vijayawada, India

Aman Verma
School of Computational and Integrative Sciences, Jawaharlal Nehru University
New Delhi, India

Arbind Kumar Verma
Department of Agricultural Engineering
School of Agricultural Sciences, Nagaland University, Medziphema, Nagaland, India

Nidhi Verma
Department of Microbiology, Ram Lal Anand College, University of Delhi
New Delhi, India

Nishchal Wanjari
Department of Geology, School of Physical Sciences, Sikkim University, Gangtok
Sikkim, India

Ayushii Yadav
Department of Environmental Science
Indira Gandhi National Tribal University
Amarkantak, Madhya Pradesh, India

Yuxuan Xiang
State Key Laboratory of Tibetan Plateau Earth System, Resources and Environment (TPESRE), Institute of Tibetan Plateau Research, Chinese Academy of Sciences
Beijing, China

and

University of Chinese Academy of Sciences
Beijing, China

Preface

Various water resources such as glaciers, lakes, rivers, and streams are found in the mountain regions which are the source of water for the downstream region of mountains. Numerous aspects are included in the management of water resources in the mountain region such as monitoring, managing, availability, planning, and importance of water resources. Nowadays, water resources management study is very important due to the high demand of water for a growing population, drinking, hydroelectric power generation, irrigation, etc. Water resources management plays a crucial role in the growth and economic development of any country. Hence, there is a need to know about the different aspects of water resources management.

This book contains the management of various water resources situated in the mountain regions. It covers various aspects of water resources management in the mountainous environment such as hydrological characteristics and hydro-meteorological correlation of water resources, the impact of climate change and anthropogenic activities on the water resources, hydrogeochemical and biogeochemical characterization of various water resources, hydrological modeling for water resources management, glacier change and mass balance in the high-altitude region of mountainous environment, use of water resources for the hydroelectric power generation, irrigation and drinking purposes, and the weather pattern in the vicinity of various water resources in the mountain regions. This book also deals with the transport of suspended sediment from various water resources, the impact of sediment transport on hydroelectric power generation, and water quality assessment of lakes and springs in the mountain regions. Disaster and flash flood caused by water resources and climate change, challenges associated with water resources management, and various indigenous techniques used for water resources management in the mountain regions is also discussed in this book. Remote sensing is a very important tool used for monitoring various water resources situated in the mountain regions and is also part of this book. Overall, this book brings a holistic knowledge about water resources management in the mountain regions.

This book is very useful for undergraduate and postgraduate students of universities and colleges, teachers, scientists, environmental engineers, and research scholars especially those working in the fields of hydrology, sediment transport, climate change, water pollution, hydrogeochemistry, flash flood, civil engineering, environmental science, and water resources management. It covers theoretical, practical, modeling, and social aspects of water resources management in the mountain regions. We have included the book chapters of various authors from different disciplines and institutions, which cover different aspects of water resources management. We are thankful to all the authors and reviewers and the publisher of this book for their support, guidance, and dedicated work in publishing an edited volume in the field of water resources management in the mountain regions.

December 2024

Virendra Bahadur Singh, Sughosh Madhav,
Rakesh Kumar Gupta, and Prerna Diwan
New Delhi, India

Amit Kumar
Dehradun, India

1

Management Challenges of Water Resources in Sikkim Mountain Regions

Anil K. Misra[1], Om P. Kaptan[1], Priyanka Kumari[1], Krity Rai[1], Rakesh K. Ranjan[1], Nishchal Wanjari[1], and Subash Dhakal[2]

[1] *Department of Geology, School of Physical Sciences, Sikkim University, Gangtok, Sikkim, India*
[2] *Rural Management and Development Department, Government of Sikkim, Gangtok, Sikkim, India*

1.1 Introduction

The Indian Himalayan Region (IHR) is a treasure trove of natural beauty and resources, but it is also a fragile ecosystem facing a myriad of challenges due to climate change. This region boasts breathtaking landscapes, from majestic glaciers to perennial rivers nourished by their meltwaters, and a rich tapestry of flora and fauna. However, the threat of climate change casts a shadow over these wonders, triggering concerns about its impact on the communities that rely on these local resources for their livelihoods. The threats posed by climate change in the Himalayas are wide-ranging and complex. They encompass a spectrum of challenges, including devastating floods, prolonged droughts, and perilous landslides, as underscored by recent research (Kuniyal et al. 2021). These climatic upheavals not only endanger human lives and property but also disrupt agricultural practices, imperil biodiversity, and erode the very foundations of food security in the region. Compounding these challenges is the rapid pace of economic growth in many parts of the northeastern Himalayan region. This growth surge has sparked an unprecedented demand for natural resources, leading to their overexploitation and drastic alterations in land use and cover. The resulting habitat fragmentation and unsustainable development practices further exacerbate the vulnerability of the region's ecosystems. Moreover, the economic boom has reshaped consumption patterns and spurred massive investments in infrastructure. While these investments promise progress, they also carry a host of unintended consequences, including adverse social and environmental impacts. Demographic shifts, with an increasing concentration of the population in urban centers, are anticipated to compound these challenges by intensifying the strain on vital resources, such as water, food, and energy. In the present chapter, management challenges and mitigation measures related to water resources of Sikkim are discussed.

Water Resources Management in Mountain Regions, First Edition. Edited by Virendra Bahadur Singh, Sughosh Madhav, Rakesh Kumar Gupta, Prerna Diwan, and Amit Kumar.

1.1.1 Sikkim State

Located within the embrace of the eastern Himalayas, Sikkim, India's second smallest state, spanning 7096 km^2, exhibits geological youth within its fold mountains. Its climatic spectrum ranges from subtropical in the south to tundra in the north. The northern expanse, covered in snow and glaciers for up to half the year, witnesses night temperatures dipping below freezing point, with the high-elevation tundra zones experiencing bone-chilling lows below −40°C. Despite this, the state maintains an average annual temperature of approximately 18°C, with peak rainfall during the monsoon months of July and August, as documented in the National Wetland Atlas (2011). Sikkim boasts a rich network of water bodies, with the Teesta and Rangeet rivers serving as major perennial sources along with high-altitude wetlands of varying sizes and shapes (Hussaina et al. 2018). Among its natural wonders, high-altitude lakes such as Gurudongmar and Cholamu punctuate the landscape, with Cholamu Lake earning distinction as India's highest lake and the source of the Teesta River. These lakes, nourished by both rivers and snowmelt, contribute to the state's ecological diversity, alongside notable water bodies like Changu (Tsomgo), Khechodpalri, Memencho, and Lampokhari (Chatterjee et al. 2010). However, Sikkim grapples with significant challenges in water resource management, owing to its rugged terrain and ecological sensitivity. The formidable mountainous topography poses formidable obstacles to essential water infrastructure development, including the construction of dams, reservoirs, and pipelines. The technical complexities and substantial costs inherent in such undertakings are further compounded by the threat of landslides and soil erosion, which not only jeopardize water quality but also compromise the structural integrity of vital water-related infrastructure.

Despite the abundance of water resources in Sikkim, rural areas located in South and West Sikkim face a significant hurdle in accessing clean and safe drinking water. Addressing this challenge requires the establishment of sustainable and dependable water supply systems, particularly in remote regions. Further, Sikkim's pursuit of energy security has led to a growing reliance on hydropower projects. However, the development of such projects must be approached cautiously, considering the environmental consequences, including the preservation of river ecosystems and ensuring adequate downstream water availability. Spring and river water pollution pose a significant threat to Sikkim's water resources, with industrial activities, agriculture, and human settlements contributing to degradation. Implementing robust monitoring and management strategies is essential to maintain water quality and mitigate the adverse impacts of pollution. The challenging weather conditions and rugged terrain in Sikkim further complicate the maintenance of water infrastructure. Regular and rigorous maintenance is indispensable to ensure the longevity and functionality of critical structures, such as dams, water pipelines, and other facilities. In tackling these multifaceted challenges, a comprehensive and integrated approach is paramount to ensure the sustainable and equitable management of Sikkim's precious water resources.

1.1.2 Climate

Sikkim's climate spans from subtropical in the south to tundra in the north, presenting diverse conditions across altitudes. Moist tropical conditions prevail up to 600 m, followed by subtropical climates up to 1500 m, transitioning to cold temperate zones up to 3000 m, beyond which alpine climates dominate. Rainfall ranges from 1250 to 5000 mm annually, with the majority occurring between April and October. Higher elevations experience significant snowfall in winter, with snow thickness reaching 30–40 m (Basu 2013). Temperature fluctuates with elevation, varying from 4°C to 30°C in the lesser Himalayan regions. Areas around 1800 m typically range between 1°C and 25°C, while elevations exceeding 4000 m see summer temperatures rise to 15°C and freezing conditions in winter, early spring, and late autumn (Basu 2013). Sikkim heavily depends on glacial meltwater for its water supply, rendering it susceptible to reduced availability and compromised quality. Erratic weather patterns, characterized by unpredictable rainfall and snowmelt, exacerbate water resource management challenges. Preserving Sikkim's rich ecosystems necessitates meticulous planning and sustainable practices to mitigate adverse environmental impacts.

1.1.2.1 Analysis of Precipitation Indices

In a recent study conducted by Dubey et al. (2022), a comprehensive analysis of spatial and temporal variations in extreme precipitation intensity indices in Sikkim was undertaken. The study used sophisticated methodologies, including Kriging for interpolation, to discern patterns and trends in precipitation across the region. The investigation focused on several key indices, notably Consecutive Dry Days (CDD), Consecutive Wet Days (CWD), PRCPTOT (total precipitation), as well as threshold indices such as R10, R20, and R30, which signify the number of days when precipitation values fall below or above specific thresholds. Additionally, percentile-based indices like R95P and R95PTOT were calculated based on a reference period spanning from 1963 to 2012.

The analysis of CDD revealed a concerning negative trend, particularly pronounced in the eastern region of Sikkim. This trend suggests a reduction in the number of days with minimal rainfall (<1 mm), indicating potential challenges regarding water resource availability in these areas. Conversely, CWD exhibited an opposing trend, with an increase in occurrences noted in the northern part of the region, implying alterations in precipitation patterns. PRCPTOT, which measures the total precipitation from wet days, displayed varied changes across Sikkim. Significant decreases were observed in the northern and western parts, suggesting potential shifts in rainfall distribution and intensity within the region (Figure 1.1).

Threshold indices highlighted spatial disparities in heavy precipitation days, with higher frequencies noted in the eastern and northern regions. Furthermore, percentile-based indices indicated significant negative trends, particularly in the northern region, suggesting a reduction in heavy precipitation events over time. Maximum single-day (RX1DAY) and five-day precipitation indices (RX5DAY) exhibited mixed trends across Sikkim, with negative trends prevalent in the northern, western, and southern regions. These trends underscore potential changes in extreme precipitation events within the region (Figure 1.2).

The Simple Daily Intensity Index (SDII), calculated as annual precipitation divided by the number of wet days (rainfall >1 mm), showed varying trends. Significant decreases were observed in the northern region, while increases were noted in the eastern part, reflecting alterations in rainfall distribution patterns across Sikkim (Figure 1.3).

It is notable that while Sikkim receives high rainfall, it primarily occurs within a short duration during the monsoon months. These findings highlight the importance of

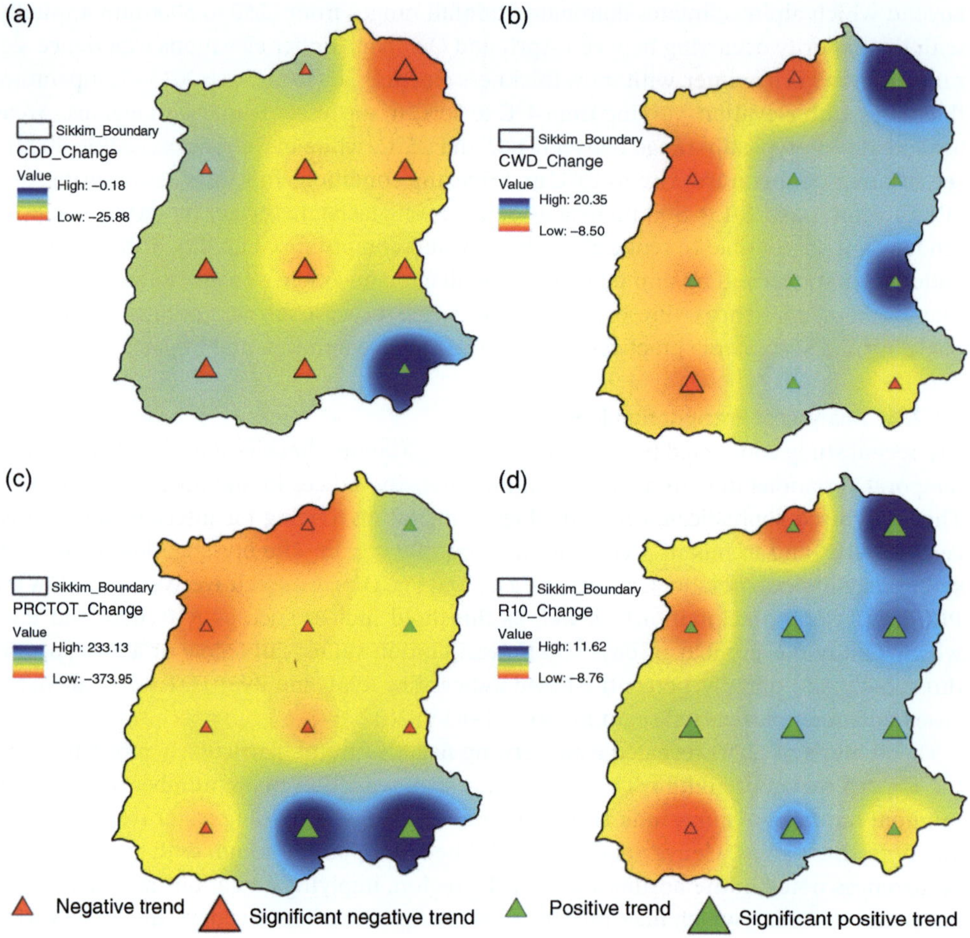

Figure 1.1 Spatial changes in extreme indices of precipitation in Sikkim state: (a) CDD, (b) CWD, (c) PRCTOT, (d) R10. *Source:* Dubey et al. 2022/Springer Nature.

understanding precipitation patterns for effective water resource management in Sikkim, with implications for agricultural practices and hydrological studies. The study emphasizes the need for proactive measures to adapt to changing precipitation patterns and ensure sustainable water resource management in the region (Dubey et al. 2022).

1.1.3 Geology of Sikkim

Sikkim is a landlocked state of India and serves as a gateway to the Eastern Himalaya. The Sikkim Himalaya has been divided into different lithotectonic units separated by various folded thrust systems. From north to south, these lithotectonic units are the Greater Himalayan sequence (GHS), the Lesser Himalayan sequence (LHS), and the Sub-Himalayan sequence (SHS), separated by the corresponding thrusts, which are the Main Central thrust (MCT), the Main Boundary thrust (MBT), and the Main Frontal thrust (MFT).

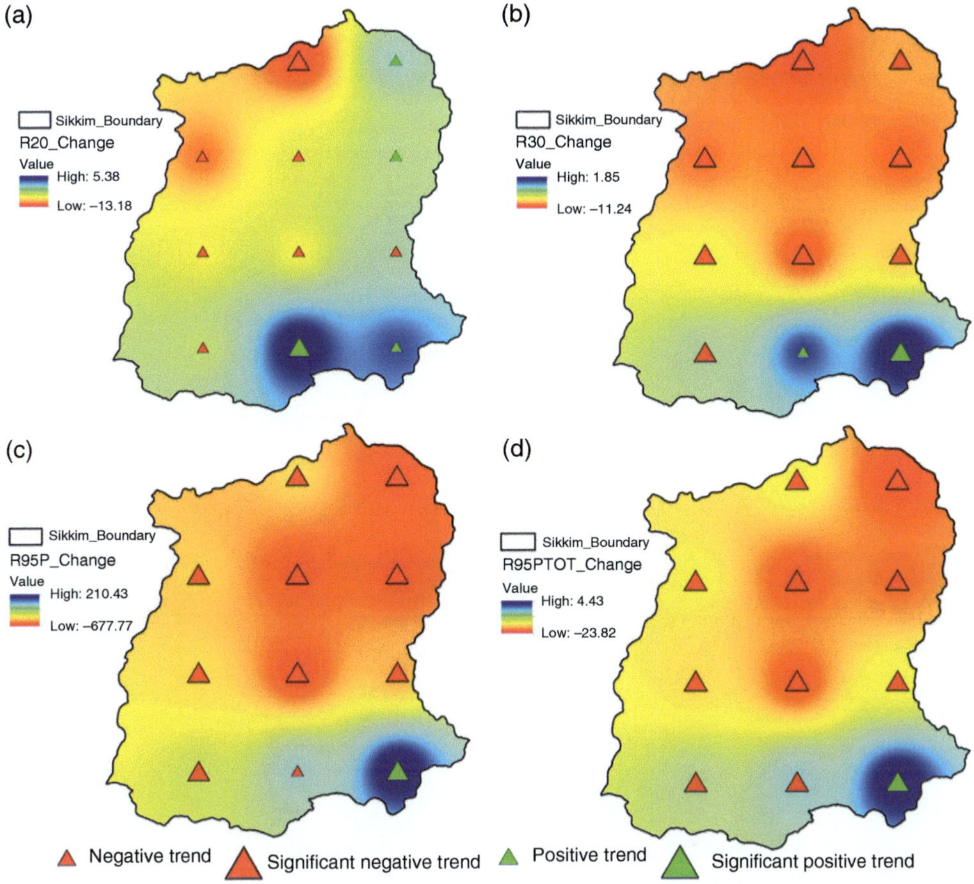

Figure 1.2 Spatial changes in extreme indices of precipitation in Sikkim state: (a) R20, (b) R30, (c) R95P, (d) R95PTOT. *Source:* Dubey et al. 2022/Springer Nature.

The main lithological units of Sikkim comprise the GHS, comprising migmatites, gneisses, high-grade schists, and amphibolites, and is generally known as Kanchenjunga gneiss or the Darjeeling Group of rocks. To the immediate south of this lies the LHS, comprising low-grade, thick metasedimentaries such as phyllites, schists, and quartzites of the Daling Group. Lingtse granite gneiss exposures are recorded along the contact of the GHS and the Daling Group of rocks (Figure 1.4). The granite gneiss is also observed intruding into the Daling Group of rocks at various places. Also, a band of garnetiferous chlorite-biotite schist is recorded at the contact of the Darjeeling and Daling Group of rocks. To the south of this granitized metasedimentary sequence, Gondwana Group of rocks comprising Rangit pebble, slate, Damuda sandstones, and shales with coal bands.

The parent rocks for the GHS and Daling Group appear to be a sequence of shale, sandstone, and limestone that got variably metamorphosed. The Buxa formation comprises quartzites and limestone/dolomite. The stromatolitic assemblage of Buxa dolomite also indicates a middle/upper Proterozoic age. It also appears that the sandstones and limestones have sufficient lateral continuity, though they were deposited in stable shelf condition.

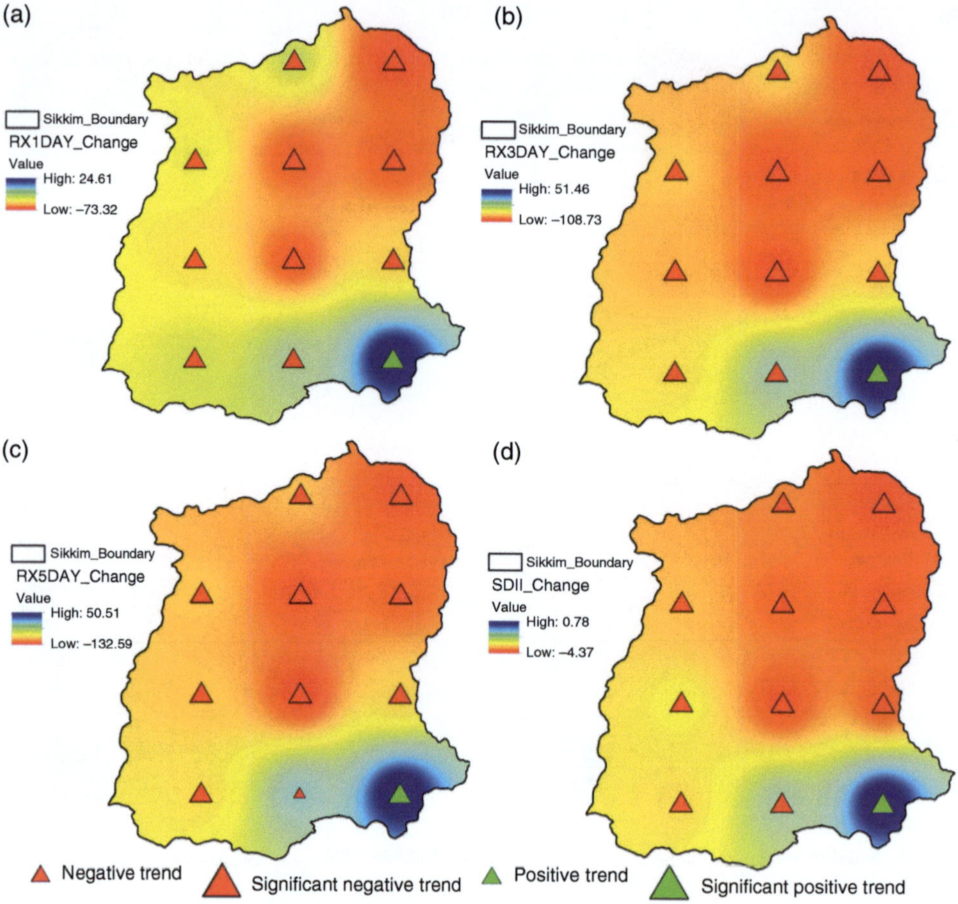

Figure 1.3 Spatial changes in extreme indices of precipitation in Sikkim state: (a) RX1DAY, (b) RX3DAY, (c) RX5DAY, (d) SDII. *Source:* Dubey et al. 2022/Springer Nature.

The SHS, found further south of Sikkim state (in the northern part of the West Bengal), comprises weakly to unmetamorphosed mollase deposits of Siwaliks. The Siwaliks are overlain by quaternary sediments in the Duar plains of West Bengal. The dip of the strata shows an entirely reverse sequence (i.e. Siwaliks overlain by Gondwanas, which in turn are overlain by the Daling Group of rocks, and then the Central Crystallines). This inversion of sequence, or overturning of beds, is explained by the concept of thrusting, which is common in Himalayas.

1.2 Water Resources in Sikkim

The Sikkim Himalaya region is witnessing a noticeable decline in glaciated areas and a shift in snow cover patterns across seasons. These changes in cryospheric resources hold profound implications for the sustainability of water resources, posing significant threats to communities and livelihoods throughout the region. The observed and projected alterations

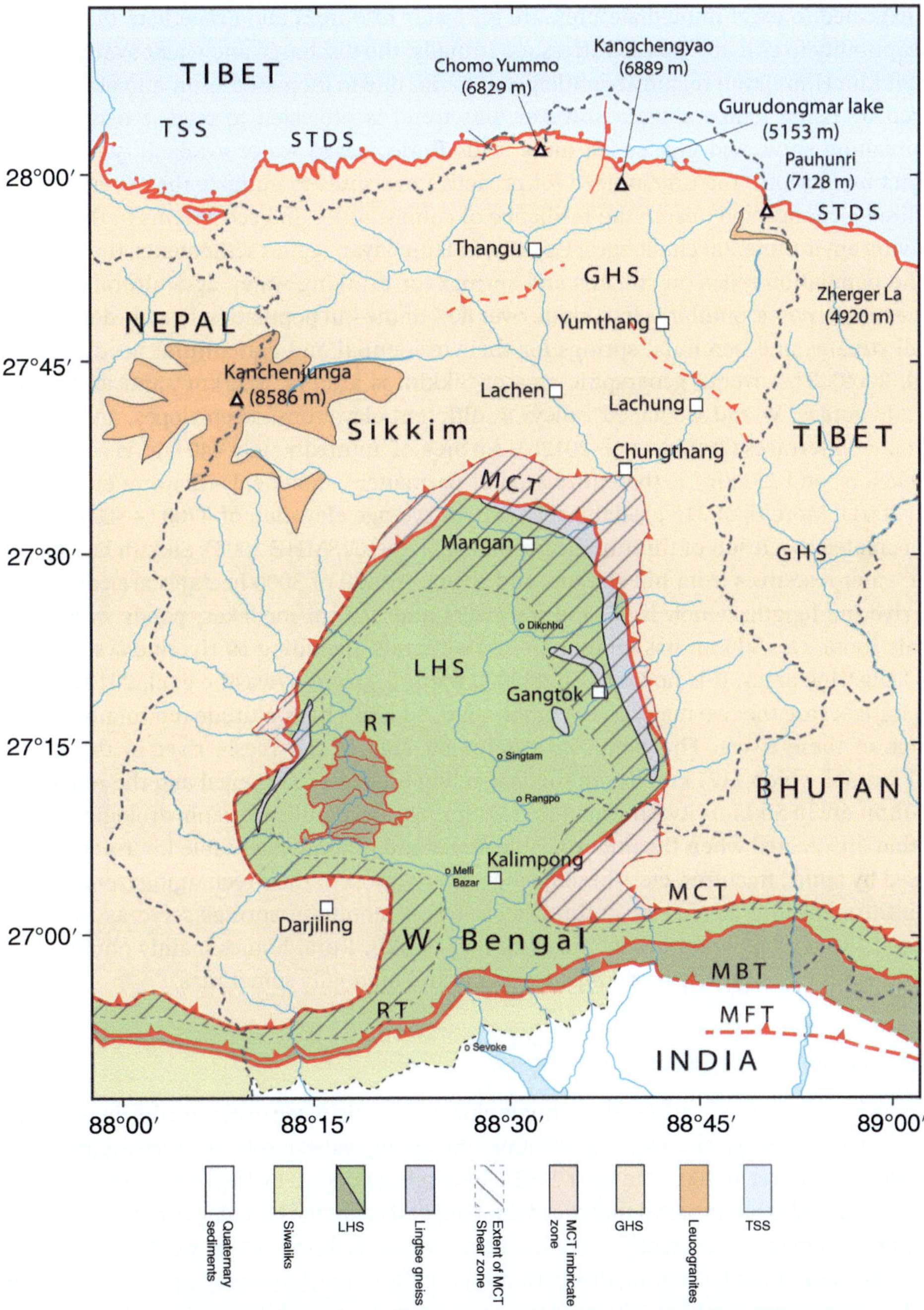

Figure 1.4 Geological map of Sikkim and the Darjeeling Himalaya. Abbreviations: TSS, Tethyan sedimentary sequence; STDS, South Tibetan Detachment System; GHS, Greater Himalayan Sequence; MCT, Main Central thrust; LHS, Lesser Himalayan Sequence; RT, Ramgarh thrust; MBT, Main Boundary thrust; MFT, Main Frontal thrust. *Source:* Modified and adopted from Kellett et al. (2014). Compiled from: Gansser, (1983), Bhattacharyya and Mitra (2009), Long et al. (2011b), Kellett et al. (2013), Mottram et al. (2014).

are expected to exert immediate pressure on water resources and exacerbate the region's susceptibility to climate-induced stressors. Initially, the discharge and water availability in the Sikkim Himalayan region are anticipated to rise due to increased snow and glacier melt driven by rising temperatures. However, this trend is projected to reverse over time as diminishing snow and glacier ice mass leads to decreased water availability. This dual impact underscores the urgent need for proactive measures to mitigate the adverse effects of climate change and ensure the resilience of communities and ecosystems in the face of evolving environmental challenges. Usually, in Himalayan region settlements, the majority of the population relies on streams and springs for drinking water, agricultural use, and domestic purposes. Similarly, in Sikkim, over 80% of the hill population heavily depends on small streams and perennial springs for their residential and agricultural needs (Tambe et al. 2009). The overall geographic area of Sikkim is around 7096 km^2, and its features include gorges, V- and U-shaped valleys at different elevations, steep slopes, and escarpments with terraces (Tambe et al. 2012b). An area of approximately 440 km^2 is covered by 84 glaciers, and 251 km^2 is the entire area of permanent snowfields (Sharma et al. 2013; SAC 2001). More than 315 glacial lakes with an average elevation of 4700 (+500) m have been created as a result of this unique geomorphology (CISMHE 2005). Sikkim has significant water resources from hill streams and rivers. A total of 3000 hectares are covered by the riverine length, which is 753 km for rivers and 147 km for lakes, ponds, reservoirs, canals/jhora, etc. Sikkim has more than 220 wetlands, including 59 riverine systems and 150 lacustrine areas. It is home to up to 2000 natural springs (Avasthe et al. 2013). Teesta and Rangeet are the two major rivers in the state. Sikkim's high-altitude mountains are the source of these rivers. The estimated catchment area of the Teesta river in the hills is 8051 km^2, of which 1121 km^2 lies in the Darjeeling hills of West Bengal and the remaining 6930 km^2 are in Sikkim (Avasthe et al. 2013). Springs are significant geohydrological markers that are created when the impermeable strata and groundwater table intersect and are caused by faults, fractures, etc. These springs receive most of their recharging from rainwater infiltration and are fed by groundwater. These mountain springs serve as the main source of water supply for urban and rural households. Rural homes mainly obtain water from springs using gravity-based piped systems and manually sometimes.

1.2.1 Water Security Issue

The rise in temperature, attributed to climate change, is anticipated to disrupt the quantity and timing of snow and glacier melt, significantly impacting water resources, particularly in the Himalayas (Arora et al. 2008; Buytaert 2012). Research conducted by Negi et al. (2018) over the past three decades has revealed that the mean temperature in the Himalayan region surpasses the global average. Concurrently, there has been a decrease in winter snowfall, accompanied by an increase in rainfall during this period. This shift in precipitation patterns underscores the dynamic nature of climate change's influence on water availability, necessitating comprehensive strategies to adapt and manage water resources effectively in the Himalayan region.

Mountain ecosystems are delicate ecosystems; even a slight variation in precipitation and temperature can affect the availability of water and other vital ecosystem services, increasing the likelihood of natural disasters such as floods, landslides, and droughts (Basu et al. 2021; Tsering et al. 2010). Natural springs are directly impacted by changes in climate conditions that cause fluctuations in the trend and pattern of rainfall. Precipitation is not distributed uniformly; it varies in form (such as snow, rain, and hail), duration, intensity, and amount, as well as geographically, seasonally, and temporally (Chatterjee et al. 2016; Hasanean and Almazroui 2015; Westra et al. 2014). Scarcity of water has emerged as a major issue in the parts of the west and south districts of Sikkim. These parts get substantially less yearly rainfall, roughly 150 cm, than other regions of the state because they are situated in the rain shadow of the Darjeeling Himalayas (Barua et al. 2012; Tambe and Arrawatia 2012). According to the Rural Management and Development Department (RMDD 2012), Government of Sikkim, the south district is extremely prone to droughts, with five of its eight blocks experiencing droughts frequently and facing serious winter water scarcity. The topography of the south district is distinguished by steep, unreachable terrain and high elevations (between 300 and 5000 m). The district's northernmost regions consist of elevated snow-covered mountains and are uninhabited. In the 500–2500 m elevation range, the central and southern regions are home to the majority of the population. Since very narrow, deep valleys are not suitable for communities, people often dwell at higher elevations. Because of this, many households find it extremely challenging to obtain water from permanent stream sources that run through the narrow and deep valleys. There is also less room for traditional water conservation methods on steep and undulating terrain (Kusre et al. 2018; Tambe et al. 2012a).

Most of the springs are either dried up or on the verge of drying; for example, S-Golai water source in Soreng and Store Dhara in Tharpu, West Sikkim; Tuk Khola near Sumbuk and Elaichibari water source near Sadam, South Sikkim; and Gari Dhara in Central Pendam and Nag Dhara in Bhanu Gram, Rongli, East Sikkim (Sharma and Shrestha 2016). The primary supply of water is springs; however, access to and availability of water are significant concerns. The region's challenging topography makes it difficult to access water. On the other hand, the primary reason for the springs' drying up during the past century has been determined to be the degradation of the catchment regions, but the change in climate is currently emerging as a new menace in the twenty-first century. Due to climate change, the majority of the surrounding springs have now become seasonal, and women now have to travel daily a greater distance to fetch water. It adversely impacted rural as well as urban water security. The issue of dying springs is becoming more and more prevalent in this area. This is owing to various human-caused factors, as well as the effects of climate change on precipitation patterns, which include increased rainfall intensity, decreased temporal spread, and a sharp decrease in winter precipitation.

Analysis of precipitation data spanning from 1951 to 2018 across four monthly scales (3, 6, 12, and 24) in Sikkim reveals trends in the Standardized Precipitation Index (SPI) and year-to-year variations. The SPI time series aids in identifying wet and dry periods, which are crucial for understanding meteorological drought, particularly affecting soil moisture and crop yield (Ajaz et al. 2019). The three-month SPI, comparing precipitation within a specific period with the same period in all included years, serves as an early indicator of meteorological drought. Both the three-month and six-month SPIs exhibit (Dubey et al. 2022) prolonged wet periods in long-term analysis (Figure 1.5a,b). However, the 12-month SPI indicates intensified drought from approximately 1992–2013, aligning with the pattern observed in the 24-month SPI. The analyses of the 12-month and 24-month SPIs suggest (Dubey et al. 2022) widespread drought conditions (Figure 1.5c,d), with maximum SPI values occurring in the years 1962 (3-month), 1962 (6-month), 1963, and 2001 (12-month), as well as 2001 (24-month).

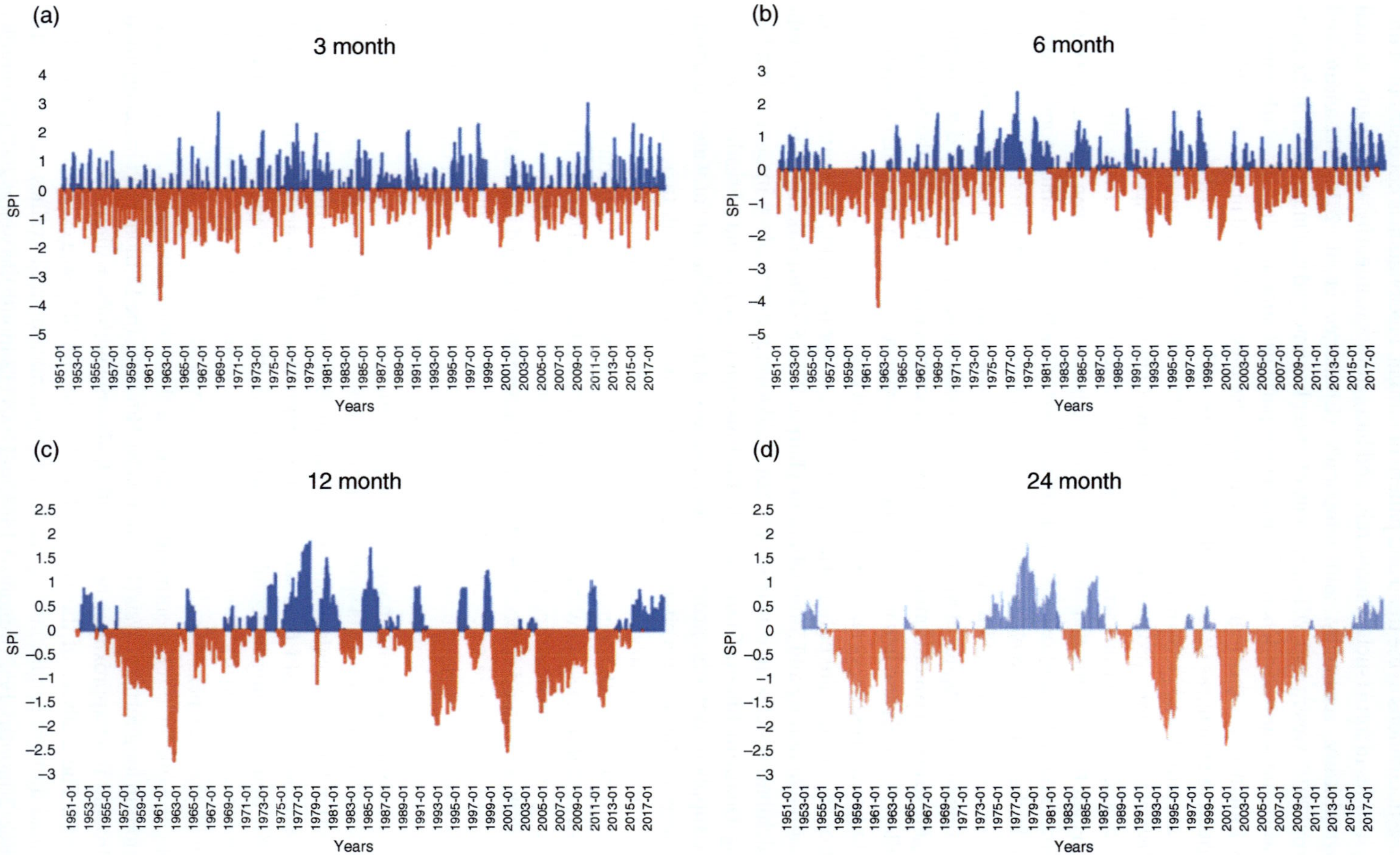

Figure 1.5 Time series of Standardized Precipitation Index (SPI) derived from Sikkim rainfall for the period 1951–2018: (a) 3 months, (b) 6 months, (c) 12 months, (d) 24 months. *Source:* Dubey et al. 2022/Springer Nature.

The most accessible and user-friendly sustainable water resource is rainwater, which has enormous potential for use in agriculture and household needs in the future. According to climate estimates, Sikkim would have a 12 mm decline in winter rainfall and an approximate 30% decrease in total precipitation by the year 2030 (SAPCC 2011). In spite of its small size, Sikkim exhibits significant variations in sensitivity, exposure, and adaptive capacity indicators within short distances. Population growth and land-use changes have escalated the demand for water per person daily across the region. However, climate change introduces uncertainty over the supply of spring water in the Himalayas. Springs, heavily influenced by rainfall patterns, are drying up gradually, posing a serious water security issue. Addressing water scarcity in hilly terrain areas like Sikkim necessitates a comprehensive approach integrating scientific insights on climate change impacts, water availability, and quality. Adaptation strategies rely on understanding current and projected climate trends, geographical constraints, and vulnerability to climate-induced disasters. Establishing water resources near populated areas and constructing rainwater harvesting infrastructure are crucial, considering the difficulty in accessing major water sources. The maintenance of water quality is essential alongside quantity management, employing both in-situ and ex-situ methods. Sustainable development requires identifying groundwater recharge zones and implementing appropriate measures like surface check dams. Initiatives such as the Dhara Vikas program focus on spring revival through runoff control and percolation enhancement. Leveraging traditional knowledge on rainwater harvesting enhances community resilience. Government initiatives like Mission Amrit Sarovar complement local efforts toward sustainable water resource management.

The seismic vulnerability of South, East, and West Sikkim adds another layer of complexity to the region's water management challenges. These areas are classified as very high to high seismic danger zones, subjecting them to frequent microtremors and small earthquakes. This seismic activity contributes to the formation of subsurface fractures, leading to water seepage in lakes. Natural lakes in region have historically enacted an important role in sustaining rural communities by recharging groundwater aquifers and supporting tourism. However, the drying up of lakes like Poison Lake (Figure 1.6) in Rinchen village, located at an altitude of 1382 m in South Sikkim, highlights the influence of subsurface

Figure 1.6 Water availability in the Poison Lake area during (a) and after (b) the rainy season in 2021. *Source:* Misra et al. 2022/Taylor and Francis/CC BY 4.0.

fractures on water resources. Geophysical studies reveal the existence of the fractures, facilitating the seepage of water from lake (Misra et al. 2022).

The phenomenon observed in Poison Lake indicates that similar processes might be affecting other lakes in the region. Given that a significant area of South and West Sikkim falls within seismic zone IV, with some areas in seismic zone V, the susceptibility to seismic activity is evident. Analysis of seismic data indicates that the epicenters of past earthquakes lie within a 100-to-150-km radius from many dried-up lakes. This correlation underscores the influence of seismic events on the hydrology of the region, emphasizing the requirement for exhaustive water management strategies that account for geological hazards.

Further, over the years, there has been unplanned urbanization in the state of Sikkim. The east district of the state is home to 79% of the state's urban population, of which 82% live in the Gangtok municipal corporation areas. Unprecedented urban growth has put enormous pressure on limited urban resources which has resulted in unplanned construction. It can also have an important impact on demand and supply of water to Gangtok city especially during the dry seasons from March to May (Acharya et al. 2012). When it comes to the capital of the state, Gangtok, the primary water supply comes from the Ratey Chu River. At 3800 m (12 500 ft) above sea level, the Ratey Chu rises from Tamze lake. Any disruption of the water supply can led to a water crisis and difficulties for the residents, as well as hotels and business owners. On 19 April 2023, due to a cloud burst near the Ratey Chu River, the pipeline that supplies drinking water to Gangtok city was damaged, disrupting the water supply and residents faced acute water shortage for one week, as shown in Figure 1.7a,b. Gangtok is a well-liked tourist destination; thus, the interruption of the water supply during the busiest time of year had a big effect on the hospitality sector.

(a) (b)

Figure 1.7 (a) and (b) Water distribution by water tankers to the students of Sikkim University at hostel in Gangtok City in 2023.

The issue of water conservation and management in mountainous regions, particularly in areas like Sikkim, is indeed a significant challenge exacerbated by various factors such as topography, climate change, and rainfall patterns. The high topography of regions like Sikkim presents challenges for water conservation. Steep terrain and deep valleys make it difficult to capture and store rainwater efficiently. The absence of a significant groundwater aquifer limits the availability of water resources for residents. This situation worsens during the dry season, leading to severe water stress problems. Further, climate change seriously affects water resources in mountainous regions. Rising temperatures, in particular, pose significant challenges. Chaudhary and Bawa (2011) highlight the negative impacts of increasing temperatures on water bodies, biodiversity, human health, and agriculture in Himalayan regions. This warming trend is consistent with findings by Gupta and Das (2006), who observed an increase in mean minimum temperature in Sikkim.

The study by Kusre et al. (2018) sheds light on the rainfall pattern in the region, indicating an average of 24 consecutive rainless days over the past three years. This pattern exacerbates water stress issues, particularly during the dry season from December to March when rainfall is scarce. The combination of limited rainfall and challenging topography amplifies the water crisis in the region. Sharma and Shrestha (2016) emphasize the relationship between temperature and spring water recharge in Sikkim. Their study suggests a powerful positive interaction, indicating that rising temperatures may delay the recharge of springs. This delayed recharge leads to lower water discharge, further exacerbating water scarcity issues during the dry season. Thus, the convergence of topographical constraints, changing climate patterns, and rainfall trends exacerbates water stress issues in mountainous regions like Sikkim.

1.2.2 Management Initiatives

The distinctive topography and geographical location of Sikkim have rendered it one of the nation's highest rainfall zones, with an extensive drainage network due to hydrometeorological conditions. However, the northeastern Himalayan region has witnessed an alarming rise in extreme warm events over the past four to five decades, coupled with a decline in extreme cold days/events and an increase in hot day events. Addressing these climate-related challenges necessitates well-planned water resource management practices designed to specific geographical requirements. The Dhara Vikas project stands as a prime example, aiming to rejuvenate and sustain springs in Sikkim's South and West districts through rainwater harvesting, geohydrology, and GIS techniques. Serving as a robust climate adaptation strategy for drought-prone areas, Dhara Vikas effectively mitigates rural water scarcity by reducing surface runoff, facilitating greater water percolation to recharge underground aquifers, and ultimately enhancing spring discharge. The initiative's implementation primarily revolves around executing a scientifically sound strategy and fostering awareness, focusing strategically on runoff control and increasing water permeation to bolster groundwater recharge. Activities in pursuit of this goal encompass the development of springsheds, strengthening the hydrological role of hill-top forests, revitalizing lakes as recharge structures, expanding minor irrigation networks for paddy cultivation, terracing sloping lands, fortifying water storage infrastructure, and nurturing para-professionals in geohydrology. Figure 1.8 shows the recharge trenches excavated under the Dhara Vikas program in Sikkim.

Figure 1.8 Various recharge trenches excavated under the Dhara Vikas program in Sikkim.

In Sikkim, the Amrit Sarovar Mission stands as another significant water management endeavor, inaugurated on 24 April 2022 in India, with a steadfast commitment to safeguarding water resources for future generations. As part of the Azadi ka Amrit Mahotsav celebrations, the mission is dedicated to the development and revitalization of 75 water bodies in every district across the nation. In Sikkim alone, a comprehensive effort was made, identifying nearly 203 sites suitable for the establishment of Amrit Sarovars, with successful completion achieved at 199 sites. The prioritization of locating these Sarovars near both major and minor settlements in Sikkim underscores the acute necessity, particularly evident during lean seasons when communities are compelled to undertake the laborious task of manually fetching water from distant springs, streams, or reservoirs.

1.3 Management Challenges

The water scarcity issue in South and West Sikkim, despite various water management schemes, indicates the requirement for more far-reaching approaches, particularly in implementing springshed and watershed management schemes. Sikkim, with its annual rainfall ranging from 1250 to 5000 mm and numerous perennial water bodies in North Sikkim, possesses the potential to address its water demand adequately. However, the challenging topography poses significant hurdles in distributing water through pipelines to different parts of the state. One major challenge highlighted is the vulnerability of water pipelines to damage caused by landslides and flash floods, particularly during the rainy season. Such incidents not only disrupt the water supply but also necessitate lengthy repair and restoration processes, lasting several weeks.

The region's geographical vulnerabilities pose severe consequences on its water infrastructure. Repairing such damages requires substantial resources and time, further exacerbating the water scarcity issue during critical periods. Consequently, addressing these challenges demands a multifaceted approach that integrates robust infrastructure planning, disaster preparedness, and innovative water management strategies designed for the unique terrain of Sikkim. The implementation of water management schemes in hilly terrain areas like Sikkim presents a unique set of challenges due to the rugged topography and specific environmental conditions. Some major challenges include:

- **Topographical Constraints**: Sikkim's mountainous landscape features steep slopes, deep valleys, and rugged terrain, making it challenging to access remote areas where water infrastructure is needed. Constructing roads and pathways to reach these locations adds to the overall complexity and cost of infrastructure projects. The narrow valleys and limited flat land available in Sikkim restrict the options for constructing large water reservoirs and treatment facilities. Laying pipelines across uneven terrain and around natural obstacles such as cliffs, rivers, and dense vegetation is another major challenging task.
- **Erosion and Sedimentation**: In hilly areas like Sikkim, erosion, landslides, and sedimentation pose significant challenges to water quality and infrastructure stability. Managing sediment-loaded runoff and erosion control are crucial components of water management schemes in such regions. Sediment runoff from eroded slopes introduces pollutants, such as silt, organic matter, and chemicals, into water bodies. These pollutants degrade the water quality, affecting its fitness for domestic use, irrigation, and aquatic life. Contaminated water can pose health risks to human populations and ecosystems, highlighting the importance of mitigating sedimentation. Further, sediment deposition in waterways also obstruct flow, leading to an increased risk of flooding and damage to infrastructure such as bridges, culverts, and dams. Sediment buildup in reservoirs and water treatment facilities reduces storage capacity and impairs the effectiveness of treatment processes, necessitating costly maintenance and repair efforts.
- **Limited Water Sources**: Sikkim is characterized by abundant rainfall, access to water sources is limited despite the apparent water availability. This is primarily owing to the existence of perennial water bodies in the limited areas of the state and absence of groundwater reservoirs, which are mostly common in flat or low-lying areas. Identifying and harnessing available water sources in hilly regions while ensuring sustainability is a difficult task. Sikkim receives abundant rainfall, but its precipitation patterns are highly variable both spatially and temporally. East and North Sikkim areas experience heavy rainfall while South and West Sikkim receive comparatively less, leading to disparities in water availability. Managing these variations and ensuring equitable distribution of water resources across the region pose significant challenges. Further, the geological characteristics of Sikkim result in limited groundwater storage capacity. Aquifers exist in some areas, but accessing them is difficult. Moreover, building infrastructure for water storage, treatment, and distribution in hilly terrains is inherently more complex and costly compared to plain areas. Constructing reservoirs, laying pipelines, and maintaining water supply networks require careful planning and engineering expertise to navigate the rugged terrain and minimize environmental impact.
- **Seasonal Variability**: Hilly regions often experience pronounced seasonal variations in precipitation, leading to periods of water shortage during dry seasons and potential flooding during monsoons. Developing efficient water distribution networks from water-rich sites to water-deficient sites via pipelines, canals, and gravity-fed systems is extremely difficult in hilly regions.
- **Ecological Sensitivity**: Sikkim state is characterized by unique and fragile ecosystems, serving as biodiversity hotspots with rich biodiversity and endemic species. These

ecosystems encompass diverse habitats such as forests, grasslands, and wetlands, supporting a wide array of plant and animal species adapted to specific environmental conditions. Water management interventions can modify natural processes such as transport of sediment, nutrient cycling, and hydrological patterns, influencing ecosystem functions and services. Alterations to natural processes can disrupt ecological balance, affect species interactions, and diminish the resilience of ecosystems to environmental stressors, exacerbating the impacts of change in climate and land-use change. Moreover, projects of water management may also inadvertently facilitate the introduction and spread of invasive species, disrupting native ecosystems and outcompeting indigenous flora and fauna. Invasive species can change habitat structure, disrupt trophic interactions, and degrade ecosystem services, posing threats to biodiversity conservation and ecosystem integrity in hilly regions.

- **Community Engagement**: In hilly regions where communities have strong ties to the land, engaging local stakeholders in the planning and execution of water management schemes is crucial for their success. Community engagement is a difficult task, as it depends on respecting the local cultural practices, land tenure systems, and traditional knowledge that must be considered to ensure project acceptance and sustainability.

Addressing these challenges requires comprehensive planning, adaptive strategies, and the participation of government agencies, NGOs, local communities, and other stakeholders. Innovative approaches that harness indigenous knowledge and leverage appropriate technologies can enhance the effectiveness of water management schemes in the hilly terrain areas of Sikkim.

1.4 Future Adaptation Strategies and Options

Future water management strategies for hilly terrain areas of Sikkim must be innovative, resilient, and sustainable to address the challenges posed by climate change, population growth, and environmental degradation. Some of the best options for future adaptation are:

- **Integrated Watershed and Spingshed Management**: Integrated watershed and springshed management involves considering ecological, social, and economic factors in water resources management. It begins with understanding the hydrological cycle, including precipitation patterns and groundwater recharge. The goal is to regulate water flow to minimize floods and droughts, achieved through structures like check dams and recharge pits. Continuous monitoring and assessment allow for adaptive strategies based on real-time data to optimize resource allocation.
- **Climate-Resilient Infrastructure**: Designing and constructing water infrastructure with climate resilience in mind, considering projected changes in precipitation patterns, temperature, and extreme weather events is very important. This includes building robust infrastructure capable of withstanding floods, landslides, and other climate-related hazards while ensuring an uninterrupted reliable water supply during periods of water scarcity, especially in South and West Sikkim areas.
- **Nature-Based Solutions**: Promoting the use of nature-based solutions such as afforestation, reforestation, soil conservation, and wetland restoration to enhance water

retention, reduce soil erosion, and improve water quality. These nature-based interventions not only contribute to water security but also furnish additional benefits such as biodiversity conservation, carbon sequestration, and recreational opportunities.

- **Smart Water Management Technologies**: Leveraging advances in technology such as remote sensing, GIS, and networking to monitor, manage, and optimize water resources in real-time. Smart water management technologies enable better decision-making, improve water efficiency, and minimize losses in water distribution systems.
- **Rainwater Harvesting and Greywater Recycling**: Promoting rainwater harvesting systems at the household, community, and institutional levels to capture and store rainwater for various uses, such as irrigation, domestic supply, and groundwater recharge. Additionally, promoting greywater recycling systems to treat and reuse wastewater for non-potable purposes reduces pressure on freshwater resources.
- **Community-Based Adaptation**: Empowering the local communities through capacity building, awareness-raising, and participatory decision-making processes to actively engage in water management and adaptation efforts is important for Sikkim. Community-based adaptation strategies leverage local knowledge, resources, and traditional practices to enhance resilience and promote sustainable water use practices.
- **Diversification of Water Sources**: Diversifying water resources by tapping into alternative water sources such as groundwater, springs, and surface water bodies to reduce reliance on single sources prone to variability and depletion. Integrating multiple water sources into water supply systems increases resilience to droughts and ensures continuous water availability.
- **Policy and Governance Reforms**: Strengthening policy frameworks, regulations, and governance mechanisms to promote sustainable water management practices, address water-related challenges, and facilitate multi-stakeholder collaboration. This includes promoting water governance principles such as equity, transparency, and accountability to ensure the effective management and allocation of water resources.
- **Capacity Building and Knowledge Sharing**: Investing in capacity building, education, and knowledge sharing initiatives to enhance the technical skills, institutional capacities, and adaptive capacity of water professionals, policymakers, and stakeholders. Building institutional resilience and fostering a culture of learning and innovation are important for executing effective adaptation strategies.

With the adoption of these future adaptation strategies and options, water management in hilly terrain areas of Sikkim can become more resilient, sustainable, and responsive to the evolving challenges of climate change and environmental degradation.

References

Acharya, S.K., Bandyopadhyay, S., and Praharaj, S. (2012). Carbon intensive urbanization, climate variability and urban vulnerabilities in hill areas: a case of Gangtok urban region, Sikkim. *Journal of International Development and Cooperation* 18 (4): 93–112.

Ajaz, A., Taghvaeian, S., Khand, K. et al. (2019). Development and evaluation of an agricultural drought index by harnessing soil moisture and weather data. *Water* 11: 1375.

Arora, M., Singh, P., Goel, N.K., and Singh, R.D. (2008). Climate variability influences on hydrological responses of a large Himalayan basin. *Water Resources Management* 10: 1461–1475.

Avasthe, R.K., Kumar, A., and Rahman, H. (2013). Potential water management technologies for sustainable agriculture in Sikkim and Darjeeling hills of India. In: *Water Management in the Hill Regions Evidences from Field Studies*, 206–231. New Delhi, Chapter 2: Bloomsbury Publishing.

Barua, A., Katyaini, S., and Mili, B. (2012). Reducing the vulnerability of the poor by building resilience: a case study from south Sikkim. In: *Climate Change in Sikkim Patterns, Impacts and Initiatives* (ed. M.L. Arrawatia and S. Tambe), 291–316. Gangtok: Information and Public Relations Department, Government of Sikkim.

Basu, S.K. (2013). *Geology of Sikkim State and Darjeeling District of West Bengal*. Bangalore: Geological Society of India.

Basu, R., Misra, G., and Sarkar, D. (2021). A remote sensing-based analysis of climate change in Sikkim supported by evidence from the field. *Journal of Mountain Science* 18 (5): 1256–1267.

Bhattacharyya, K. and Mitra, G. (2009). A new kinematic evolutionary model for the growth of a duplex - an example from the Rangit duplex, Sikkim Himalaya, India. *Gondwana Research* 16: 697–715.

Buytaert, W. (2012). Assessment of the impacts of climate change on mountain hydrology: development of a methodology through a case study in the Andes of Peru. *Mountain Research and Development* 32 (3): 385–386.

Chatterjee, A., Blom, E., Gujja, B. et al. (2010). WWF initiatives to study the impact of climate change on Himalayan high-altitude wetlands (HAWs). *Mountain Research and Development* 30 (1): 42–52. https://doi.org/10.1659/MRD-JOURNAL-D-09-00091.1.

Chatterjee, S., Khan, A., Akbari, H., and Wang, Y. (2016). Monotonic trends in spatio-temporal distribution and concentration of monsoon precipitation (1901–2002), West Bengal, India. *Atmospheric Research* 182: 54–75.

Chaudhary, P. and Bawa, K.S. (2011). Local perceptions of climate change validated by scientific evidence in the Himalayas. *Biology Letters* 7 (5): 767–770.

CISMHE (2005). *Carrying Capacity Study of Tista Basin in Sikkim*. Volume II Land Environment – Geophysical Environment. Delhi: Centre for Inter-disciplinary Studies of Mountain and Hill Environment, University of Delhi.

Dubey, S.K., Ranjan, R.K., Misra, A.K. et al. (2022). Variability of precipitation extremes and drought intensity over the Sikkim State, India, during 1950–2018. *Theoretical and Applied Climatology* 148: 1–14. https://doi.org/10.1007/s00704-022-03931-x.

Gansser, A. (1983). *Geology of the Bhutan Himalaya*, 181. Birkhauser Verlag Basel.

Gupta, R.C. and Das, M. (2006). Recent trends in temperature (cold wave condition) and rainfall over Gangtok city. *Mausam* 57 (3): 507–512.

Hasanean, H. and Almazroui, M. (2015). Rainfall: features and variations over Saudi Arabia, a review. *Climate* 3 (3): 578–626.

Hussaina, S.A., Mahara, N., Tuboia, C., and Badolaa, R. (2018). Wetlands of the Indian Himalayas: status and conservation initiatives. *Wetlands in the Himalaya: Securing Services for Livelihoods*, 33–46.

Kellett, D.A., Grujic, D., Coutand, I. et al. (2013). The South Tibetan detachment system facilitates ultra rapid cooling of granulite-facies rocks in Sikkim Himalaya. *Tectonics* 32: 252–270.

Kellett, D., Grujic, D., Mottram, C., and Mukul, M. (2014). Virtual field guide for the Darjeeling-Sikkim Himalaya, India. *Journal of the Virtual Explorer* 47: https://doi.org/10.3809/jvirtex.2014.00344.

Kuniyal, J.C., Maiti, P., Kumar, S. et al. (2021). Dayara bugyal restoration model in the alpine and subalpine region of the Central Himalaya: a step toward minimizing the impacts. *Science Reports* 11: 1–18.

Kusre, B.C., Bora, P.K., Rai, D. et al. (2018). Developing water resources management strategies for South Sikkim District, India. *Water Science and Technology: Water Supply* 18 (2): 648–659.

Long, S., McQuarrie, N., Tobgay, T. et al. (2011b). Geologic Map of Bhutan. *Journal of Maps* 7: 184–192.

Misra, A.K., Ranjan, R.K., Wanjari, N. et al. (2022). Determining fractures and their orientation for preventing in-situ seepage of water in the Poison Lake, Sikkim, India. *Journal of Taibah University for Science* 16 (1): 1225–1240.

Mottram, C.M., Argles, T.W., Harris, N.B. et al. (2014). Tectonic interleaving along the Main Central Thrust, Sikkim Himalaya. *Journal of the Geological Society 171*: 255–268.

National Wetland Atlas (2011). *Sikkim*, SAC/EPSA/ABHG/NWIA/ATLAS/36/2011. Ahmedabad, India: Space Applications Centre (ISRO), 84p.

Negi, H.S., Kanda, H.S., Shekhar, M.S., and Ganju, A. (2018). Recent wintertime climatic variability over the North West Himalayan cryosphere. *Current Science* 114 (4): 760–770.

Rural Management and Development Department (2012). *Village Development Action Plan*. Gangtok: Government of Sikkim.

SAC (2001). *Glacial Atlas of Teesta Basin (Sikkim Himalayas), Space Application Centre (ISRO), Ahmedabad and State Remote Sensing Agency*. Gangtok: Department of Science and Technology Government of Sikkim, 2001.

SAPCC (2011). Sikkim action plan on climate change (2012–2030). Government of Sikkim.

Sharma, S., Kumar, K., and Singh, K.K. (2013). Water security in the mid-elevation Himalayan watershed, east district with focus in the State of Sikkim. *Kohima Workshop* (September).

Sharma, R.K. and Shrestha, D.G. (2016). Climate perceptions of local communities validated through scientific signals in Sikkim Himalaya, India. *Environmental Monitoring and Assessment* 188: 1–11.

Tambe, S. and Arrawatia, M.L. (2012). *Climate Change Initiatives in Sikkim. Climate Change in Sikkim Patterns, Impacts and Initiatives*. Gangtok: Information and Public Relations Department, Government of Sikkim.

Tambe, S., Arrawatia, M.L., Bhutia, N.T., and Swaroop, B. (2012a). Rapid, cost-effective and high-resolution assessment of climate-related vulnerability of rural communities of Sikkim Himalaya, India. *Current Science* 101 (2): 165–173.

Tambe, S., Arrawatia, M.L., Kumar, R., Bharti, H., Shrestha, P. (2009). Conceptualizing strategies to enhance rural water security in Sikkim, Eastern Himalaya, India. Selected technical papers from the proceedings of the workshop on integrated water resource management held on 27th Nov, 2009. Central Ground Water Board, Eastern Region, Ministry of Water Resources, Government of India. Salt Lake, Kolkata.

Tambe, S., Kharel, G., Arrawatia, M.L. et al. (2012b). Reviving dying springs: Climate change adaptation experiments from the Sikkim Himalaya. *Mountain Research and Development 32* (1): 62–72.

Tsering, K., Sharma, E., Chettri, N., and Shrestha, A. (2010). Climate change vulnerability of mountain ecosystems in the eastern Himalayas – synthesis report. *International Centre for Integrated Mountain Development*, Kathmandu, Nepal (June).

Westra, S., Fowler, H.J., Evans, J.P. et al. (2014). Future changes to the intensity and frequency of short-duration extreme rainfall. *Reviews of Geophysics* 52 (3): 522–555.

Mottram, C.M., Argles, T.W., Harris, N.B., Parrish, R., Horstwood, M.S., Warren, C.J., and Gupta, S. (2014). Tectonic interleaving along the Main Central Thrust, Sikkim Himalaya. *Journal of the Geological Society, 171*, 255–268.

2

Basin-Scale Estimation of Runoff Components Using the SPHY Model

A Case Study of the Koshi River Basin, Central Himalaya

Kiran Bishwakarma[1,2], *Fan Zhang*[1,2], *Jharana Nepal*[1,2], *Guanxing Wang*[3,4], *Yuxuan Xiang*[1,2], *Ramesh R. Pant*[5], *and Narayan P. Khanal*[6]

[1] *State Key Laboratory of Tibetan Plateau Earth System, Resources and Environment (TPESRE), Institute of Tibetan Plateau Research, Chinese Academy of Sciences, Beijing, China*
[2] *University of Chinese Academy of Sciences, Beijing, China*
[3] *Key Laboratory of Tibetan Plateau Land Surface Processes and Ecological Conservation, College of Geographical Science, Qinghai Normal University, Xining, China*
[4] *Qinghai Province Key Laboratory of Physical Geography and Environmental Process, College of Geographical Science, Qinghai Normal University, Xining, China*
[5] *Central Department of Environmental Science, Institute of Science and Technology, Tribhuvan University, Kathmandu, Nepal*
[6] *Nepal Environment and Development Consultant Pvt. Ltd., Kathmandu, Nepal*

2.1 Introduction

The Himalayan region, with its higher deposition of snowpack and glacier cover, is the lifeline to more than a billion people by providing essential water resources for domestic consumption, irrigation, and economic development (Lutz et al. 2014; Yao et al. 2012). However, the water cycle of the Himalayan region is significantly affected by the cryosphere alterations due to climate change (Immerzeel et al. 2020; Wang et al. 2023; Yao et al. 2022). Being the "Water Tower of Asia" (Kang et al. 2010; Zhang et al. 2022a), the Himalayas and surrounding regions are highly sensitive to increasing temperature and shifting precipitation induced by climate change (Immerzeel et al. 2020; Yao et al. 2012), posing a severe threat to water resources and predicted to have intensified impacts on future water resources (Kraaijenbrink et al. 2021; Yang et al. 2014). Previous studies have addressed that the Himalayan region has warmed by 1.8 °C, which is comparatively higher than the global average during the last half-century (Bajracharya et al. 2023; Kang et al. 2010). Likewise, precipitation across the Himalayan region is increasing and is characterized by considerable spatiotemporal heterogeneity (Agarwal et al. 2014; Duncan et al. 2013). Therefore, the quantitative assessment of runoff processes and their variations in the alpine regions under these changing climatic patterns is crucial for understanding the factors influencing runoff processes and for effective water resource management.

Water Resources Management in Mountain Regions, First Edition. Edited by Virendra Bahadur Singh, Sughosh Madhav, Rakesh Kumar Gupta, Prerna Diwan, and Amit Kumar.

Runoff is a vital constituent of riverine hydrological systems, and the contribution of different runoff components at the catchment level is crucial to quantifying freshwater budgets (Neal et al. 2010). Major runoff components, including rainfall runoff, base flow, snowmelt runoff, and glacier melt runoff distribution over a catchment area, significantly affect water availability and seasonal variability. Changes in meteorological variables such as temperature and precipitation, significantly influence riverine hydrological systems (Gyanwali et al. 2023). Additionally, in the Himalayas, changes in snow/ice reserves also result in significant changes in total runoff.

The rainfall runoff mostly dominates the Himalayas and surrounding alpine regions. In major Himalaya Rivers having headwaters in the Tibetan Plateau, such as Upper Yellow, Upper Yangtze, Upper Mekong, Upper Salween, Upper Brahmaputra, and Upper Indus rivers, rainfall runoff contributes more than 50% of the total runoff (Lutz et al. 2014). As snow and glacial melt are major hydrologic processes and play an essential part in the hydrological cycle of the Himalayan region, climate change is posing a significant impact on melt features (Lutz et al. 2014). Therefore, the contribution of glacier melt runoff and snowmelt runoff to the total runoff cannot be ignored in the alpine region (Nepal 2016; Zhang et al. 2016). Even the basin with a glacier of less than 5%, can significantly contribute to summer runoff from ice melt. Hence, studying the hydrological regime and its runoff components, and the current climate change substantially impacts the freshwater resources systems on both the spatial and temporal scales.

Although declining glaciers initially provide abundant amounts of water by increasing glacier runoff, reduced glacier volume will eventually result in a decline in total runoff (Nie et al. 2021). Over the past century, the warming trend has intensified in the Himalayas, causing a significant reduction in snow and glacier cover (Kirkham et al. 2019; Zhang et al. 2019). The changing pattern of precipitation and melting of snow and/or ice due to increasing temperature are altering hydrological systems of this region and ultimately affecting water resources in terms of quantity and quality.

As the Himalayas are highly susceptible to climate change effects, which have significant impacts on the downstream water resources, research on these vulnerable regions has intensified in recent decades, concerning melting and runoff generating processes, water yield, and its spatiotemporal distribution. However, due to physical constraints and the scarcely distributed in-situ measurements, the ground scenarios of hydrological regimes in the Himalayas and surrounding areas are still lacking and uncertain. Therefore, hydrological modeling is useful for understanding hydrological responses to climate change and bridging existing gaps.

Several studies have been conducted in different regions using the glacial-hydrological model, Spatial Process in Hydrology (SPHY), a spatially distributed hydrological model for hydrological analysis on the river basin scale (Terink et al. 2015). This model is based on the PCRaster dynamic modeling framework and evaluates river hydrology conceptually (Terink et al. 2018). The SPHY model is widely used in hydrological research, particularly in the simulation of the total contribution of runoff components, and response to climate change (Khanal et al. 2021; Latif et al. 2020; Lutz et al. 2014; Nepal et al. 2024; Singh et al. 2021; Terink et al. 2017, 2018; Wang et al. 2023; Zhang et al. 2022b). Lutz et al. (2014) used the SPHY model to simulate the total contribution of different runoff components such as glacier melt, snowmelt, base flow, and rainfall, in the five major river basins originating from the Third Pole Region. Terink et al. (2017) used the SPHY model to assess hydrology and climate change in the hydropower development of the Tamakoshi River Basin, Nepal. It was also estimated the annual rainfall and its impact on the hydrology of a lowland catchment. Moreover, Latif et al. (2020) applied the SPHY model to differentiate the snowmelt and glacier melt contributions in the Gilgit River Basin. Khanal et al. (2021) applied this model to evaluate climate change response for the 15 major high-mountain Asian rivers. Wang et al. (2023) employed the SPHY model to know the response of the runoff process to temperature rise with glacier ratios in the southern part of the Tibetan Plateau. Zhang et al. (2022b) employed the SPHY model to estimate how runoff components in the headwater region of the Yellow River respond to climate change. More recently, Nepal et al. (2024) estimated the overall water balance and simulated different runoff components for the Chameliya River Basin, Nepal using the SPHY model.

Considering the issues of the changing hydrological patterns in the Himalayas, this study is designed for the Koshi River Basin (KRB), located in the northern and southern slopes of the Central Himalayas. The primary objectives of this study are to (i) set up the SPHY model for the KRB; (ii) simulate the total contribution of rainfall, base flow, snowmelt, and glacier melt to the total discharge; and (iii) analyze the subbasin-scale variations of runoff components.

2.2 Materials and Methods

2.2.1 Study Area

The KRB originates from the Tibetan Plateau and traverses through the Himalayan range, mid-mountains to the Terai lowland region in Nepal and India, and ultimately merges with the Ganges River System (GRS). Regarding catchment area and annual runoff, the KRB is the largest river system in Nepal and the largest tributary of the GRS, contributing the highest sediment loads to the GRS (Sinha et al. 2019). This study covers the upper and middle segments of the KRB, in-between the coordinates ranging from 27° N to 29° N and 85° 38′ E to 88° 95′ E (Figure 2.1), covering a total catchment area of ~53 000 km^2. The KRB is one of the complex river systems of the Himalayas in terms of geographical and climatic variations. It exhibits a distinct geographical division, broadly categorized into two main sections: the northern and southern slopes of the Himalayas. The KRB is also the most populated basin in Nepal, providing riverine ecosystem services to the livelihoods of these communities. This basin has significant hydropower potential and provides irrigation and drinking water facilities to ~27 million people in the downstream region. The physiographic information of the KRB, including major tributaries, is presented in Table 2.1.

The climatic system of the KRB can be characterized by moderate temperatures and significant monsoonal precipitation, largely influenced by the Indian monsoon system. However, the distribution of temperature and precipitation in the basin area exhibits a distinct spatiotemporal variation. Lying to the northern slopes of the Himalayas, the upstream region experiences a dry climate and low precipitation (~400 mm), whereas the downstream region, facing the southern slopes of the Himalayas, is dominated by the tropical climate and receives high monsoonal precipitation (~1800 mm) (Xue et al. 2017). The climate of the study area is primarily influenced by the Indian summer monsoon, with ~75% of the annual precipitation occurring from June to September. The average temperature across the basin varies from −5°C to 30°C, with distinct spatiotemporal variations (Bharati et al. 2019).

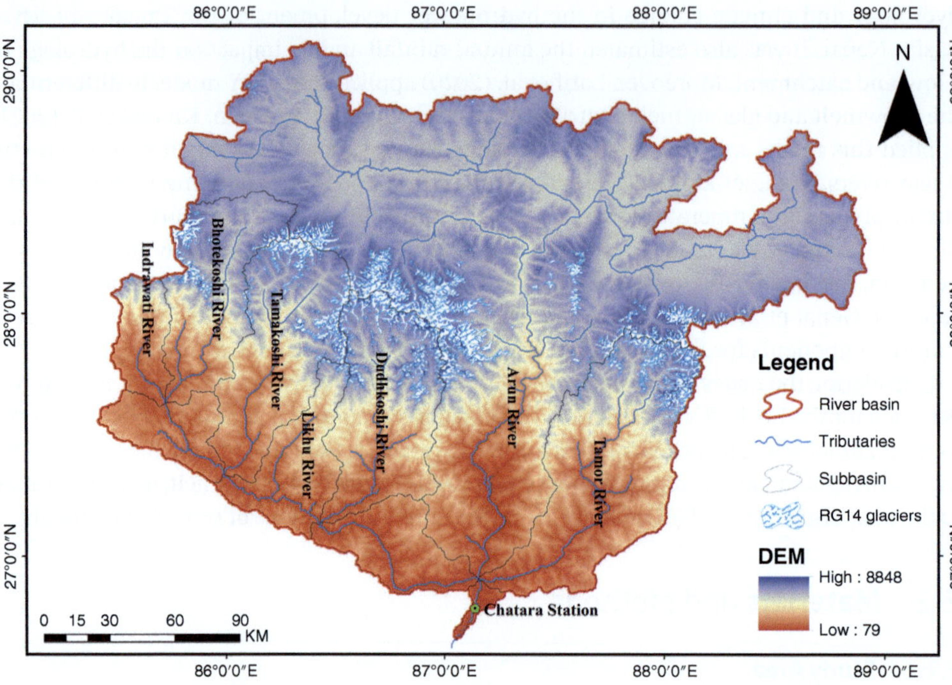

Figure 2.1 Map showing the Koshi River Basin with major tributaries.

Table 2.1 Physical characteristics of the Koshi River Basin and its major tributaries.

SN	River basin	Basin area (km²)	Glacier area (km²)	Glacier area ratio (%)	Average elevation (m)	Elevation range (m asl)
1	Indrawati	1209	16	1.32	2263	623–5854
2	Bhotekoshi	3381	390	11.52	3985	623–7974
3	Tamakoshi	4630	470	11.39	3570	472–7326
4	Dudhkoshi	4564	694	17.07	3526	342–8848
5	Likhu	3088	61	2.91	1974	342–6850
6	Arun	29524	1544	5.23	4424	187–8806
7	Tamor	6993	475	7.93	2808	187–8464
Basin average		53793	3650	6.78	2787	91–8848

2.2.2 Model Description

The SPHY model is a physically distributed hydrological modeling tool with a dynamic feature used in regional water resource management (Latif et al. 2020; Terink et al. 2015, 2018). This model is developed by integrating the most effective components of existing models such as HydroS, SWAT, HimSim, PCR-GLOBWB, etc. This model is widely applied to study climate change's impact on hydrology and irrigation management of the snow/glacier-fed river basins. It requires static data (digital elevation model, land use and land cover, glacier cover, soil characteristics, etc.) and dynamic data (daily precipitation and temperature) as input. The model produces a range of outputs, including precipitation, rainfall, snow, glacier melt, snowmelt, total runoff, rainfall runoff, base flow, glacier melt runoff, snowmelt runoff, evapotranspiration, total runoff, etc. These outputs are generated at various temporal resolutions, including daily, monthly, and annual series. In the SPHY model, the total runoff is the sum of rainfall runoff, base flow, glacier melt runoff, and snowmelt runoff.

2.2.2.1 Rainfall Runoff

The SPHY model calculates the total surface runoff using the saturated additional overland flow method, commonly known as the Hewlettian runoff method (Hewlett 1961). The topsoil layer calculates surface runoff as:

$$\mathrm{RO} = \begin{cases} \mathrm{SW}_1 - \mathrm{SW}_{1,\mathrm{sat}} & \text{if } \mathrm{SW}_1 > \mathrm{SW}_{1,\mathrm{sat}} \\ 0 & \text{if } \mathrm{SW}_1 \le \mathrm{SW}_{1,\mathrm{sat}} \end{cases} \tag{2.1}$$

where

RO: surface runoff (mm),
SW_1: water content in first soil layer (mm),
$SW_{1,sat}$: saturated content in first soil layer (mm).

In the river catchment with steep slopes and high hydraulic conductivity soil, the lateral flow becomes significant. In SPHY, lateral flow can only be employed for water greater than the field's carrying capacity. The hillslope lateral flow is computed as:

$$\mathrm{LF}_1^* = \mathrm{W}_{1,\mathrm{excfrac}} * \mathrm{v}_{\mathrm{lat},1} \tag{2.2}$$

where

LF^*_1: lateral flow (mm) from the first soil layer,
$W_{1,excfrac}$: drainable water volume,
$v_{lat,1}$: flow velocity (mmd^{-1}).

The drainable volume in SPHY is calculated as:

$$\mathrm{W}_{1,\mathrm{excfrac}} = \frac{\mathrm{W}_{1,\mathrm{exc}}}{\mathrm{SW}_{1,\mathrm{sat}} - \mathrm{SW}_{1,\mathrm{fc}}} \tag{2.3}$$

where

$SW_{1,fc}$: field capacity of the first soil layer (mm).

Based on Neitsch et al. (2011), if the catchment of interest has a duration of concentration longer than one day, lateral flow alone will drain to the main channel. This impression is also used in the SPHY model, which employs a lateral flow travel time ($TT_{lag,1}$) into the channel.

$$\mathrm{LF}_1 = \left(\mathrm{LF}_1^* + \mathrm{LF}_{1,t-1}^*\right)\left(1 - \exp\left[\frac{-1}{\mathrm{TT}_{\mathrm{lag},1}}\right]\right) \tag{2.4}$$

where

LF_l: total lateral flow on day t (mm),
$LF^*_{l,\,t-1}$: lateral flow restored on day $t-1$ (mm).

Now, the rainfall runoff (RRo) is given by

$$\text{RRo} = \text{RO} + \text{LF}_l \tag{2.5}$$

2.2.2.2 Snow Processes

In SPHY, snowfall accumulates from precipitation and contributes to runoff through snowmelt. The degree-day melt approach by Hock (2003) is applied to simulate snowmelt. The daily potential snowmelt is determined using the degree-day modeling approach as:

$$\text{A}_{\text{pot},t} = \begin{cases} T_{\text{avg},t}.\text{DDF}_\text{s} & \text{if } T_{\text{avg},t} > 0 \\ 0 & \text{if } T_{\text{avg},t} \le 0 \end{cases} \tag{2.6}$$

where

$A_{\text{pot},t}$: potential snowmelt on day t (mm),
DDF_s: calibrated degree-day factor for snowfall (mm°$\text{C}^{-1}\text{d}^{-1}$).

As the snow stored on the previous day restricts the actual snowmelt, it is further represented as:

$$\text{A}_{\text{act},t} = \min\left(\text{A}_{\text{pot},t}, \text{SS}_{t-1}\right) \tag{2.7}$$

where

$A_{\text{act},t}$: actual snowmelt on day t (mm),
SS_{t-1}: snow storage on day $t-1$ (mm).

Furthermore, when the air temperature surpasses the melting point and there is no further capacity for meltwater to freeze within the snowpack, runoff from snow (SRo (mm)) is produced and is given by

$$\text{SRo}_t = \begin{cases} \text{A}_{\text{act},t} + \text{P}_{l,t} - \Delta\text{SSW} & \text{if } T_{\text{avg},t} > 0 \\ 0 & \text{if } T_{\text{avg},t} \le 0 \end{cases} \tag{2.8}$$

where,

ΔSSW: change in meltwater stored in the snowpack (mm).

2.2.2.3 Glacier Processes

In SPHY, glaciers are melting surfaces that can entirely or partially cover a grid cell. Like snowmelt, glacier melt is also estimated using Hock's degree-day modeling approach. As the debris-free (DF) and debris-covered (DC) glaciers have different melt rates, the daily melt from DF glaciers (A_{CI}) and the daily melt from DF glaciers (A_{DC}) are given by

$$\text{A}_{\text{CI},t} = \begin{cases} T_{\text{avg},t} * \text{DDF}_{\text{CI}} * \text{F}_{\text{CI}} & \text{if } T_{\text{avg},t} > 0 \\ 0 & \text{if } T_{\text{avg},t} \le 0 \end{cases} \tag{2.9}$$

$$A_{DC,t} = \begin{cases} T_{avg,t} * DDF_{DC} * F_{DC} & \text{if } T_{avg,t} > 0 \\ 0 & \text{if } T_{avg,t} \le 0 \end{cases} \tag{2.10}$$

where

DDF_{CI}: degree-day factor for DF glaciers (mm°C^{-1}d^{-1}),
F_{CI}: fraction value for DF glaciers,
DDF_{DC}: degree-day factor for DC glaciers (mm°C^{-1}d^{-1}),
F_{DC}: fraction value for DC glaciers.

Total glacier melt (A_{GLAC}) is given by

$$A_{GLAC,t} = \left(A_{CI,t} + A_{DC,t}\right) * \text{GlacF} \tag{2.11}$$

In SPHY, a fraction of glacier melt infiltrates into the groundwater while the remaining fraction runs off. The runoff created from the glacier melt (GRo) is estimated as:

$$\text{GRo}_t = A_{GLAC,t} * \text{GlacROF} \tag{2.12}$$

where

GlacROF: calibrated glacier melt runoff factor.

2.2.2.4 Base Flow Processes

In SPHY, base flow estimation is based on two factors: the groundwater flow's steady-state response to recharge and oscillations in the water table brought on by the groundwater system's nonsteady reaction to periodic groundwater recharge (Terink et al. 2015). The base flow is calculated using

$$BF_t = \begin{cases} 0, & \text{if } SW_3 \le BF_{thresh} \\ BF_{t-1} * \exp^{-\alpha_{gw}} + \text{Gchrg}_t * \left(1 - \exp^{-\alpha_{gw}}\right), & \text{if } SW_3 > BF_{thresh} \end{cases} \tag{2.13}$$

where

BF_t: base flow on day t (mm),
BF_{t-1}: base flow on day $t-1$ (mm),
α_{gw}: base flow recession coefficient,
Gchrg_t: groundwater recharge on day t.

Overall, the total runoff (QTot) of each grid is determined by summing each of these runoff components once all the runoff components have been defined and represented as:

$$\text{QTot} = \text{RRo} + \text{SRo} + \text{GRo} + \text{BF} \tag{2.14}$$

where

RRo: total rainfall runoff (mm),
SRo: total snowmelt runoff (mm),
GRo: total glacier melt runoff (mm),
BF: total base flow (mm).

2.2.2.5 Water Balance

In the SPHY model, the basin-scale water balance is determined by computing the difference between precipitation (P) and glacier melt (GM), and subsequently subtracting the total runoff (Q) and evapotranspiration (ET). Glaciers are considered static entities that produce meltwater but are not factored into mass conservation. As a result, precipitation that falls on glacier surfaces is not included in the water balance calculations for the basin. The precipitation calculated for the subbasins in the output files only represents the non-glaciated fraction. Therefore, glacier melt is an additional input to the water balance calculations. The overall relationship is expressed by

$$\text{Water balance} = \text{P} + \text{GM} - \text{Q} - \text{ET} \tag{2.15}$$

2.2.3 Mann–Kendall Test

The Mann–Kendall (MK) test is a well-established statistical scheme used in hydrology for detecting trends in time series data (Kendall 1975; Mann 1945), providing valuable insights into changes in hydrological variables. This test is a rank-based technique for the time series data and can be computed using the following equations.

The MK statistic "*S*" can be calculated as

$$\text{S} = \sum_{i=1}^{n=1} \sum_{j=i+1}^{n} \text{sgn}\left(X_j - X_i\right) \tag{2.16}$$

X_i and X_j represent a sequence of data values, and n is the total number of data series.

$$\text{sgn}(\theta) = \begin{Bmatrix} +1, & \text{if}\,\theta > 0 \\ 0, & \text{if}\,\theta = 0 \\ -1, & \text{if}\,\theta < 0 \end{Bmatrix} \tag{2.17}$$

The variance of *S* is calculated using

$$\text{Var}(S) = \frac{n(n-1)(2n+5) - \sum_{i=1}^{m} t_i(t_i - 1)(2t_i + 5)}{18} \tag{2.18}$$

where

m: number of tied value,
t_i: number of ties for the ith value.

MK statistic "*Z*" value can be estimated using

$$\text{Z} = \begin{Bmatrix} \dfrac{S-1}{\sqrt{\text{Var}(S)}}, & \text{if}\,\text{S} > 0 \\ 0, & \text{if}\,\text{S} = 0 \\ \dfrac{S+1}{\sqrt{\text{Var}(S)}}, & \text{if}\,S < 0 \end{Bmatrix} \tag{2.19}$$

A positive and negative Z-value indicates increasing and decreasing trends, respectively. On the other hand, the absolute value of $Z \sim 0$ specifies no trend.

2.3 Results

2.3.1 Model Setup and Sensitivity Analysis

The static parameters of the SPHY model, such as elevation map (DEM), land use land cover (LULC), and soil map were obtained from HydroSHEDS, GLOBCOVER2009, and HiHydroSoil, respectively. Additionally, glacier coverage was taken from the Randolph Glacier Inventory 4.0 version (Pfeffer et al. 2014). The daily runoff data for the Chatara Station were collected from the DHM, Nepal (www.dhm.gov.np) for 2009–2019. Since the meteorological station's distribution in the KRB is very scarce, ERA5 (0.25° resolution-resampled) of precipitation and temperature data was used (https://www.ecmwf.int/en/forecasts/dataset/ecmwf-reanalysis-v5). The sensitivity analysis of the SPHY model was conducted using the global sensitivity approach, wherein one parameter value was modified at a time while keeping all other parameters constant. The most sensitive parameters in the SPHY model were recession coefficient (RC), alpha groundwater (alphaGW), clean ice-covered glaciers (DDFG), degree-day factors for debris-covered glaciers (DDFDG), degree-day factors for snow (DDFS), and delta groundwater (deltaGW). The calibrated value of RC, DDFG, DDFDG, DDFS, alphaGW, deltaGW, GlacF, and Tcrit were 0.394, 16 mm degree^{-1} day^{-1}, 8 mm degree^{-1} day^{-1}, 12 mm degree^{-1} day^{-1}, 0.50, 75, 0.90, and 3°C, respectively.

To minimize the prediction uncertainty, the calibration and verification of the model based on observed values are indispensable (Gupta et al. 2019; Nepal 2016). The major objective of the calibration and verification processes is to achieve a close correspondence between the simulated and the observed variables (Lutz et al. 2014). Statistical evaluation of simulations is also needed to quantify the reliability of the model outputs (Nepal and Shrestha 2015). The most common and widely used statistics to examine the model accuracy are the Nash-Sutcliff Efficiency (NSE), coefficient of the determinant (R^2), Percent Bias (PBIAS), etc. (Eeckman et al. 2019). Generally, NSE > 0.50, $R^2 > 0.5$, and PBIAS within ±25% indicate the satisfaction of the model simulations.

In this study, we calibrated the model for the five years from 2011 to 2015 and validated it for the subsequent four years, from 2016 to 2019, using observed daily discharge data at the Chatara hydrological station. The sensitivity of simulated results was analyzed at a daily time scale using NSE and PBIAS. The NSE and PBIAS were 0.82 and 12.57%, respectively, for the calibration period, whereas during the validation, NSE was 0.85 and PBIAS was 2.09%, displaying the simulated results that perfectly represent the real hydrological scenarios. Overall, the statistical results and hydrograph for observed and simulated daily discharge (Figure 2.2) showed that the SPHY model accurately represents the KRB and can be used for further analysis. The simulation was done for the 2010s, including two years of warm-up periods (2009–2011) to develop suitable initial conditions for groundwater and soil water storage.

2.3.2 Water Balance of the Koshi River

The basin-scale water balance of the KRB is presented in Figure 2.3a,b in both annual and monthly time scales. The average annual input from precipitation and glacier melt to total

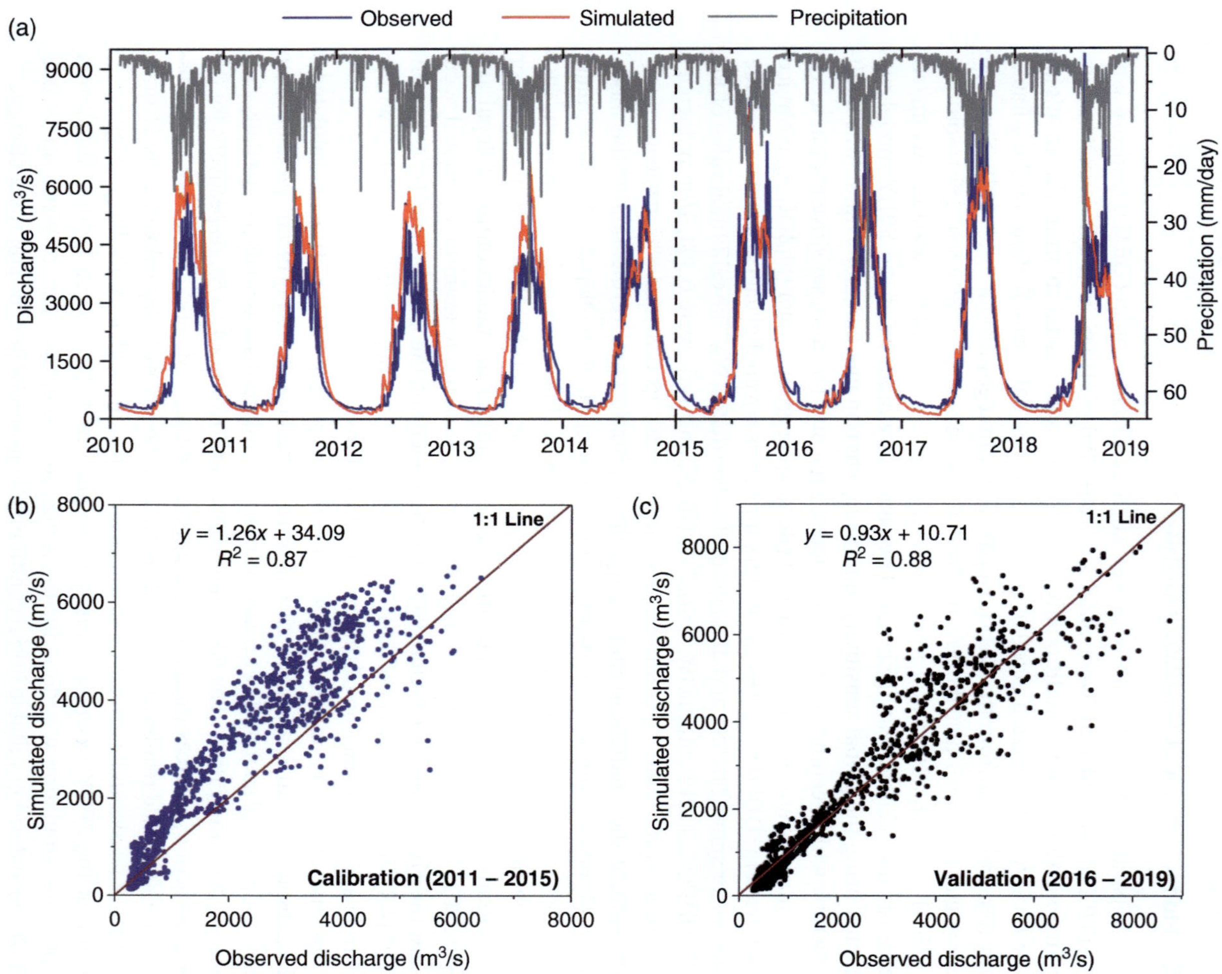

Figure 2.2 Hydrograph of the Koshi River Basin at Chatara Station: (a) line diagram showing observed and simulated discharge, including total precipitation; (b) scatter diagram during calibration period (2011–2015); (c) scatter diagram during the validation period (2016–2019).

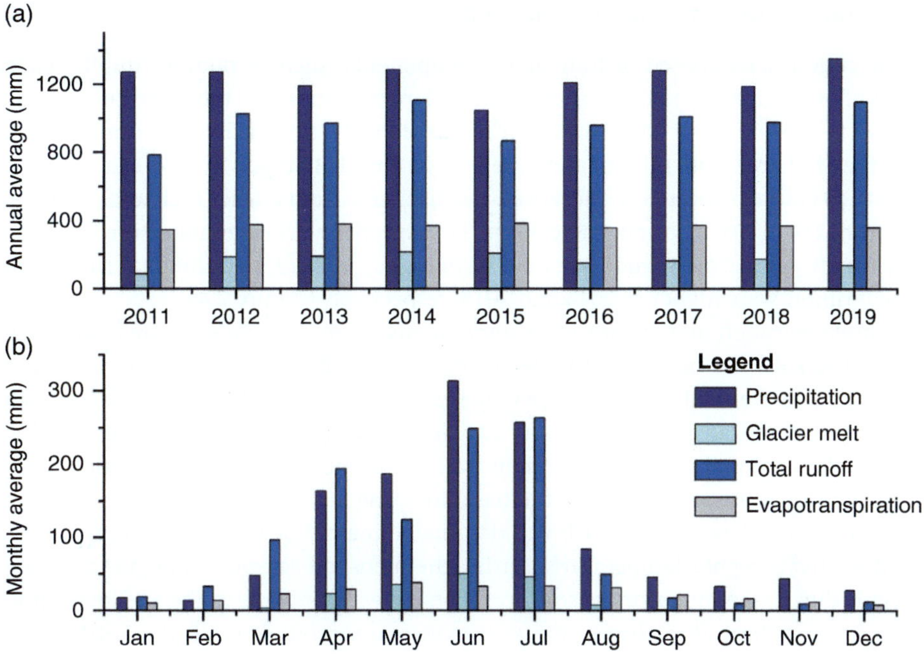

Figure 2.3 Annual (a) and monthly (b) variations of the water balance components in the Koshi River Basin.

water balance was 1233 mm (44%) and 167 mm (6%), respectively. Total basin-scale average annual output from evapotranspiration (ET) and river runoff was 367 mm (9%) and 977 mm (39%), respectively, during the 2010s. Annually, the precipitation input to total water balance ranged from 1046 to 1353 mm, highest and lowest precipitation in 2019 and 2015, respectively (Figure 2.3a). Likewise, the input from the glacier melt ranged from 88 mm to 213 mm, highest and lowest glacier melt in 2015 and 2011, respectively (Figure 2.3a). The water out in runoff and ET ranged from 783 to 1105 mm and 344 to 384 mm, respectively.

The monthly variations of water balance components are graphically presented in Figure 2.3b. The total input from precipitation in the water balance ranged from 13 to 313 mm, the highest in June and the lowest in November. The seasonal precipitation variations indicated that the Indian monsoons influenced the KRB's climate system. Likewise, the input from glacier melt ranged from 0.04 to 50 mm, with higher glacier melt in June and lower in February, indicating the influence of temperature effects. Meanwhile, the monthly variations of water output from the KRB in runoff and ET ranged from 9.57 to 263 mm and 7.99 to 37.74 mm, respectively. The higher runoff was observed in July and lower in February, showing the high influence of precipitation, whereas the higher and lower ET were found in May and December, respectively, showing the influence of higher sunlight hours during the pre-monsoon season. Overall, the variations of precipitation and discharge have a major influencing role in altering the water balance of the KRB, contributing more than 80% in total inputs and outputs, respectively.

2.3.3 Total Runoff and Runoff Components

The total runoff is composed of four major components, such as rainfall runoff, glacier melt, base flow, and snowmelt. The average annual precipitation and total discharge ranged from 1046 to 1353 mm and 1483 to 2018 m^3/s, respectively. In addition, the annual variations of the discharge from rainfall and base flow in the KRB ranged from 791 to 1240 m^3/s and 182 to 291 m^3/s, respectively. Moreover, the discharge from glacier melt and snowmelt ranged from 214 to 308 m^3/s and 198 to 306 m^3/s, respectively. The simulated precipitation, rainfall runoff, glacier melt, and base flow were higher in 2019, and the total runoff and snowmelt runoff were higher in 2014 and 2015, respectively (Figure 2.4).

The simulated runoff and major runoff components at the seasonal scale are tabulated in Table 2.2. The KRB was mostly dominated by rainfall runoff on the basin scale, contributing about 55% to the total runoff. Other major contributors were base flow, glacier melt, and snowmelt, contributing 20%, 13%, and 12%, respectively. Regarding seasonal variation, the surface flow (rainfall runoff, glacier melt runoff, and snowmelt runoff) was higher during the monsoon season. In contrast, the base flow was higher during the post-monsoon season. As discussed earlier, the KRB was dominated mainly by the Indian monsoon system and relatively higher temperatures during the monsoon season. As higher precipitation occurs during the monsoon season, considerable water storage in the shallow aquifer leads to a higher base flow during the post-monsoon season. Overall, the KRB mainly depends on rainfall runoff; nearly 70% of total runoff was generated during the monsoon season (June–September).

2.3.4 Subbasin Variations of the Runoff Component

The varying climatic conditions and availability of glacier mass across the different subbasins of the KRB are likely to influence the total discharge of rivers. The estimated average annual discharge of the Arun River (largest tributary of the KRB) was 700 m^3/s, with the highest and lowest discharge observed during monsoon (1596 m^3/s) and pre-monsoon (110 m^3/s) seasons, respectively. The annual average discharge of Tamor River, Likhu River, Dudhkoshi River, Tamakoshi River, Bhotekoshi River, and Indrawati River was 233, 60, 173, 198, 103, and 50 m^3/s, respectively. The order of tributaries based on the average annual discharge was Arun > Tamor > Tamakoshi > Dudhkoshi > Bhotekoshi > Likhu > Indrawati.

In the subbasins scale, the contribution of glacier melt ranged from 4% (Likhu subbasin) to 20% (Bhotekoshi subbasin). The higher glacier area ratio of the Bhotekoshi catchment has a significant input from glacier melt to the total discharge. The contribution of snowmelt ranged from 1% (Likhu subbasin) to 20% (Arun and Bhotekoshi subbasins). Likewise, the contribution of base flow ranged from 4% (Likhu subbasin) to 19% (Arun subbasin), and the contribution of rainfall ranged from 42% (Bhotekoshi subbasin) to 90% (Likhu subbasin). In the basin and subbasin scales, more than half of the total annual discharge was contributed by rainfall runoff in the KRB. A significantly higher contribution (>75%) of rainfall runoff was observed in the Tamor subbasin, Likhu subbasin, Tamakoshi subbasin, and Indrawati subbasin. The details of the annual average of total discharge and the tentative contribution of the contributors are presented in Table 2.3.

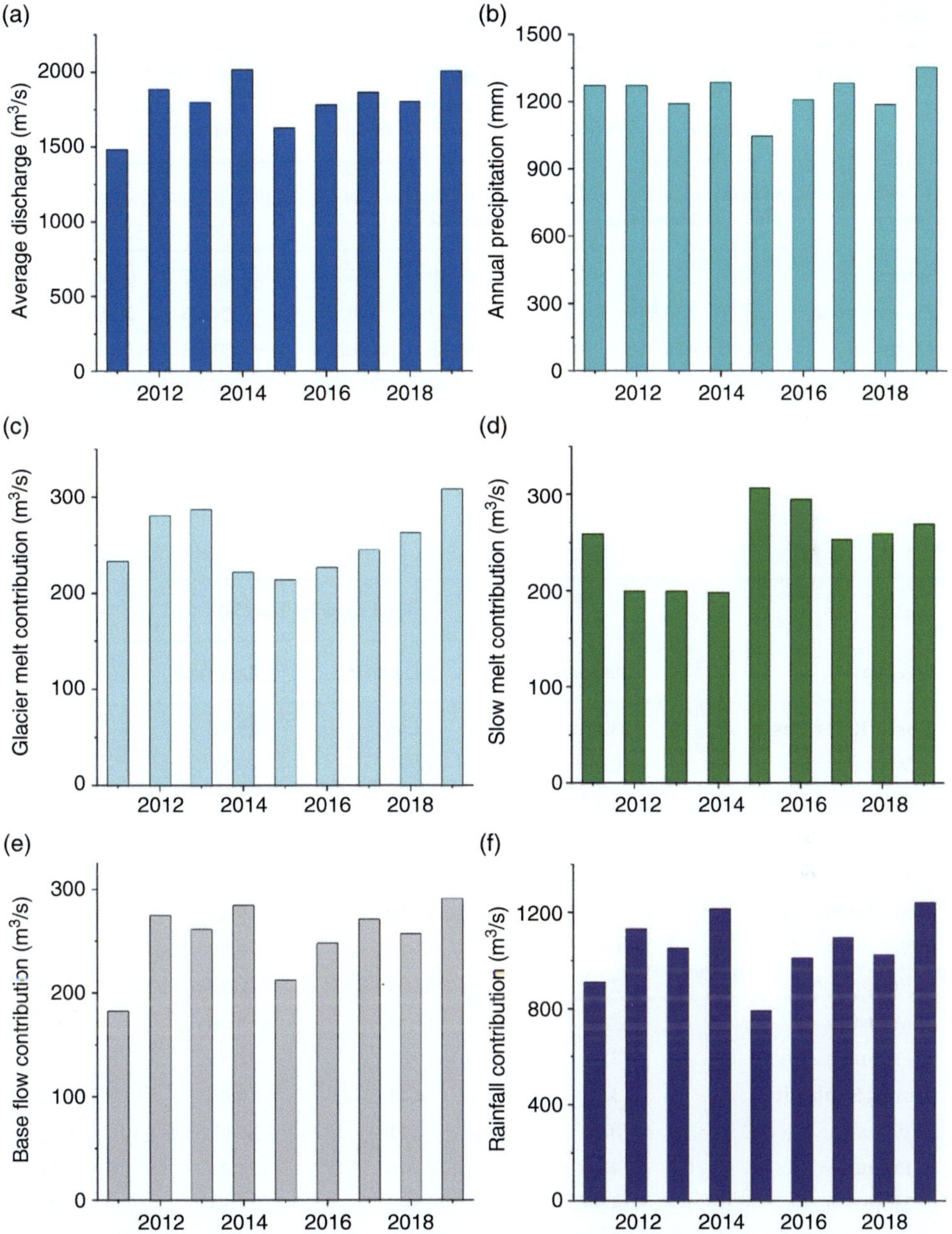

Figure 2.4 Annual variations of (a) total discharge, (b) basin scale precipitation, (c) glacier melt runoff, (d) snow melt runoff, (e) base flow, and (f) rainfall runoff in the Koshi River Basin.

Table 2.2 Seasonal variations of hydrological regimes of the Koshi River Basin.

	Runoff components (m^3/s)				Total discharge (m^3/s)	Precipitation (mm)	Glacier melt (mm)	ET (mm)
Season	**Glacier melt**	**Snowmelt**	**Base flow**	**Rainfall**				
Winter	5.97	37.34	200.36	21.39	265.05	27.67	0.12	9.49
Pre-monsoon	27.35	330.91	65.67	64.17	488.10	54.06	2.86	22.80
Monsoon	608.94	449.36	468.43	2750.83	4176.56	251.53	43.48	34.77
Post-monsoon	306.06	97.93	500.63	1348.38	2253.00	77.81	9.42	22.19
Basin average	243.11	230.68	366.72	1010.71	1851.22	1233.24	167.66	267.74

Table 2.3 Seasonal variations of hydrological regimes of the Koshi River Basin and major subbasins.

	Total contribution (%)			
River basin	**Glacier melt**	**Snowmelt**	**Base flow**	**Rainfall**
Koshi River Basin	**13.15**	**12.45**	**19.80**	**54.57**
Winter	2.34	14.17	75.84	7.65
Pre-monsoon	6.79	60.28	13.05	19.88
Monsoon	16.57	11.92	6.03	65.48
Post-monsoon	14.03	3.93	22.74	59.30
Arun Subbasin	**5.84**	**19.90**	**19.14**	**55.12**
Winter	0.07	17.79	76.19	5.96
Pre-monsoon	2.40	68.52	12.80	16.28
Monsoon	7.70	17.53	8.70	66.07
Post-monsoon	3.57	5.09	32.75	58.59
Tamor Subbasin	**3.75**	**4.21**	**10.05**	**82.42**
Winter	0.34	7.11	86.93	5.62
Pre-monsoon	4.38	38.86	21.24	35.51
Monsoon	4.65	3.95	4.15	87.25
Post-monsoon	3.72	1.53	14.20	80.55
Dudhkoshi Subbasin	**19.84**	**11.65**	**8.72**	**59.84**
Winter	0.58	41.80	48.59	9.03
Pre-monsoon	8.19	47.48	11.27	33.06
Monsoon	23.20	10.46	3.58	62.77
Post-monsoon	19.43	5.04	13.93	61.61

Table 2.3 (Continued)

River basin	Total contribution (%)			
	Glacier melt	Snowmelt	Base flow	Rainfall
Tamakoshi Subbasin	**10.02**	**5.18**	**8.17**	**76.70**
Winter	0.45	6.97	88.21	4.37
Pre-monsoon	7.77	40.28	22.20	29.74
Monsoon	11.70	5.40	2.47	80.43
Post-monsoon	9.45	1.12	12.75	76.68
Likhu Subbasin	**4.41**	**1.25**	**4.11**	**90.29**
Winter	5.09	15.11	70.43	9.37
Pre-monsoon	11.60	17.92	11.90	58.58
Monsoon	4.22	0.63	1.30	93.85
Post-monsoon	4.21	0.21	5.92	89.66
Bhotekoshi Subbasin	**20.00**	**20.23**	**8.05**	**51.60**
Winter	9.56	30.24	57.94	2.27
Pre-monsoon	24.07	58.49	7.76	9.68
Monsoon	31.78	20.27	3.19	44.76
Post-monsoon	33.55	6.99	13.86	45.61
Indrawati Subbasin	**8.50**	**6.00**	**10.00**	**75.00**
Winter	1.19	21.96	73.40	3.45
Pre-monsoon	9.87	60.45	13.80	15.88
Monsoon	9.26	1.81	3.97	84.96
Post-monsoon	7.99	0.60	17.04	74.38

The seasonal comparison of total discharge and the contribution of its contributors was carried out. This study considered the four major seasons as winter, pre-monsoon, monsoon, and post-monsoon. The results suggested a substantial seasonal variation in the contribution of major contributors to the total discharge. During the winter season, all tributaries were dominated mainly by base flow. More than 50% of the total runoff was contributed by base flow in all subbasins except for the Dudhkoshi region. Owing to Mt. Everest (the highest peak in the world) in the Dudhkoshi subbasin, rainfall and snowmelt runoff contribute more than 50% of the total runoff. During the pre-monsoon season, all the tributaries were dominated mainly by snowmelt runoff except for the Likhu subbasin. The Likhu subbasin had comparatively less glacier area ratio coverage, and rainfall was dominant in this subbasin during the pre-monsoon season. With the dominancy of the Indian monsoon, the rainfall runoff is the most profound component during both the monsoon and post-monsoon in all subbasins.

2.4 Discussion

Using the physically distributed SPHY model, this study investigated the basin-scale water balance and contribution of glacier melt runoff, snowmelt runoff, base flow, and rainfall runoff of the KRB during the 2010s.

The annual accumulated water balance and runoff components of the KRB (at Chatara Station) were analyzed using the MK trend test, as presented in Table 2.4. Overall, basin-scale total precipitation and temperature showed an increasing trend. Some existing studies observed the same results (Khadka et al. 2023; Shrestha et al. 2017) in the Nepal Himalayas. Due to the increasing trend of precipitation, total runoff, base flow, snowmelt runoff, and rainfall runoff were also observed to be increasing trend. However, a decreasing trend of glacier melt was observed, with a decrease in basin-scale glacier melt runoff. However, all trends were statistically nonsignificant ($p > 0.05$). The decreasing trend of glacier melt in the KRB also aligned with existing studies (Khadka et al. 2020; Miller et al. 2012; Thakuri and Salerno 2016). The trend analysis was conducted for data spanning only nine years, which might have some limitations, especially for detecting long-term variability. However, it can still provide valuable insights into short-term changes in hydrological regimes.

The Himalayan region faces profound challenges due to the fluctuation of total precipitation and temperature due to climate change. Estimating water balance, particularly in the Himalayan basins, is critical in assessing climate change's impacts due to the glacier melt's significant contribution. Several studies have been done in this region using the hydrological model. For instance, Bharati et al. (2012) employed the SWAT model in the KRB for the spatiotemporal variations of water balance, revealing the large temporal variability, which is higher during the rainy seasons. Yadav et al. (2022) also used the SWAT model in the Dudhkoshi River (a major tributary of the KRB). They found that about 16% of total

Table 2.4 Man–Kendall trend test and statistical summary of hydrological variables in the Koshi River Basin.

Components	Kendall Tau	*P*-value	Trend
Water balance components			
Precipitation (mm)	0.055	0.919	Increasing
Temperature (°C)	0.377	0.155	Increasing
Glacier melt (mm)	−0.11	0.761	Decreasing
Total runoff (mm)	0.277	0.358	Increasing
Evapotranspiration (mm)	−0.111	0.761	Decreasing
Runoff components			
Glacier melt runoff (mm)	−0.111	0.761	Decreasing
Snowmelt runoff (mm)	0.253	0.345	Increasing
Base flow (mm)	0.277	0.358	Increasing
Rainfall runoff (mm)	0.222	0.477	Increasing

precipitation returned as evapotranspiration with higher runoff months had a lower value of evapotranspiration and vice-versa. Devkota et al. (2024) employed the SWAT model in the Narayani River Basin and estimated that 22% of total precipitation changed to total evapotranspiration and 74% of total precipitation drained as total runoff. However, these studies had a limitation of glacier model inclusion in the model simulation, which may have caused some uncertainty in water balance analysis. Therefore, in this study, we applied the SPHY model, incorporating glacier melts to overcome existing uncertainty in the water balance of Himalayan regions.

Basin-scale estimation of runoff components is crucial for effective water resource management, flood prediction and mitigation, hydrological modeling accuracy, understanding ecological impacts, climate change impact assessment, etc. In KRB, the contribution of rainfall runoff, base flow, glacier melt runoff, and snowmelt runoff to the annual runoff was 55%, 20%, 13%, and 12%, respectively. Previously, several studies have been conducted in this river basin using different hydrological models such as SPHY, SWAT, HBV, J2000, etc. Gyanwali et al. (2023) applied the SPHY model, integrating it with the Optimal Capacity Expansion Model and explored the climate resiliency of hydropower projects in Nepal and estimated the contribution of base flow, snow flow, glacier flow, and rain flow in the KRB about 22%, 11%, 4%, and 62%, respectively. Wang et al. (2023) quantified the runoff components of the two major tributaries of KRB, i.e. Dudhkoshi and Arun River subbasins. They found that glacier runoff and snowfall runoff contributed 29% and 10%, respectively, in the Dudhkoshi, and 12% and 14% in the Arun subbasin. Budhathoki et al. (2023) used the HBV model to simulate the streamflow contribution to the water balance of the Tamakoshi River Basin (a major tributary of KRB) and estimated the glacier melt and snowmelt runoff contributed about 13% and 5%, respectively. Khadka et al. (2020) estimated contribution of major runoff components in the subbasin scale, and their results revealed that the highest contribution from snowmelt, ice melt, rain, and baseflow was found in the Arun subbasin, Sunkoshi, Likhu, and Arun subbasin, respectively. Likewise, Nepal (2016) found that more than 30% of glacier runoff contributed to the available discharge in the Dudhkoshi subbasin using J2000. Additionally, snowmelt runoff and glacier runoff contributed about 13% and 8% to river discharge in the Tamor subbasin (Gupta et al. 2019).

The spatial and seasonal variations of water balance and runoff components indicate a high sensitivity to the glacier area ratio, with notable changes in total runoff due to the alteration of glaciers (Zhang et al. 2016). Comparatively, glacier melt runoff was the dominant component of the total runoff in the Bhotekoshi and Dudhkoshi subbasins, covering 8% and 14% of the total basin area, respectively. The current scenarios of an increasing trend of meltwater runoff in the high-mountain glacier-fed river basin indicate that meltwater contribution would increase roughly until the mid-century (Huss and Hock 2018). Hence, the role of snow and glaciers in regulating hydrological processes and sustaining total runoff is paramount in the Himalayan region. Moreover, the accelerated rate of glacier melt caused by global warming and the consequent reduction in glacier area could significantly impact the hydrological regulation of downstream (Ding et al. 2019; Wang et al. 2023).

Overall, this research endeavored to assess the spatiotemporal variations of the water balance components, focusing on their relative contributions to the KRB. This study ensures that the hydrological model is one of the effective ways to investigate the runoff process. Owing to the inclusion of glacier melt, the SPHY model is one of the best models for the Himalayan river basins and can be applied to explore the runoff process in other glacier-fed river basins with effective model parameters.

2.5 Conclusion

This study estimated water balance components, total runoff, and runoff components, along with their patterns during the 2010s (2011–2019) in the KRB, using the SPHY model. For the evaluation of model efficiency, NSE, R^2, and PBIAS were employed. Additionally, the MK trend test was applied for a statistically significant trend evaluation of water balance components and runoff components. During the model calibration (2011–2014) and validation (2015–2019), the simulated data fitted with the observed data, showing the actual pattern replicated by the model. During the 2010s, the average annual input to total water balance in the KRB was 1233 mm from precipitation and 158 mm from glacier ice melt, giving a total of 367 mm evapotranspiration (ET) and 977 mm that left the basin as total river discharge. About 56 mm of remaining water was stored in the basin as a soil water content. The simulated annual average precipitation and total runoff in the KRB were about 1200 mm and 1600 m^3/s, respectively, with a significant contribution of the rainfall runoff of about 55% to the total runoff. Other major contributors were runoff from base flow, glacier melt, and snowmelt, contributing 20%, 13%, and 12%, respectively. The simulated results showed significant seasonal variations of the runoff and its components. During the winter and pre-monsoon seasons, the base flow and snowmelt contribution were dominant, respectively, whereas the rainfall runoff contribution was dominant during the monsoon and post-monsoon seasons. Except for the glacier melt and glacier melt runoff, all basin-scale water balance components, including average temperature and runoff components, were observed in a nonsignificant ($p > 0.05$) increasing trend. In summary, the simulated results indicated that water storage tends to increase during wet seasons and decrease during dry seasons, with an increasing trend of total precipitation and a decreasing trend of glacier melt. This study presents a basic estimation of total runoff and runoff components of the Himalaya transboundary river basin and recommends considering future climate scenarios to know the variations of water balance components, total runoff, and runoff components.

Acknowledgments

This research was supported by the National Natural Science Foundation of China (Grant Nos. 42125104 and 41761144078) and the Kathmandu Centre for Research and Education, CAS-TU. KB and JN were supported by the Alliance of International Science Organizations (ANSO) scholarship for international PhD students.

References

Agarwal, A., Babel, M.S., and Maskey, S. (2014). Analysis of future precipitation in the Koshi River Basin, Nepal. *Journal of Hydrology* 513: 422–434. https://doi.org/10.1016/j.jhydrol.2014.03.047.

Bajracharya, S.R., Pradhananga, S., Shrestha, A.B., and Thapa, R. (2023). Future climate and its potential impact on the spatial and temporal hydrological regime in the Koshi Basin, Nepal. *Journal of Hydrology: Regional Studies* 45: 101316. https://doi.org/10.1016/j.ejrh.2023.101316.

Bharati, L., Gurung, P., and Jayakody, P. (2012). Hydrologic characterization of the Koshi Basin and the impact of climate change. *Hydro Nepal: Journal of Water, Energy and Environment* 11: 18–22. https://doi.org/10.3126/hn.v11i1.7198.

Bharati, L., Bhattarai, U., Khadka, A., Gurung, P., Neumann, L. E., Penton, D. J., Dhaubanjar, S., & Nepal, S. (2019). From the mountains to the plains: impact of climate change on water resources in the Koshi River Basin. *IWMI Working Paper* (187). https://doi.org/10.5337/2019.205

Budhathoki, B.R., Adhikari, T.R., Shrestha, S., and Awasthi, R.P. (2023). Application of hydrological model to simulate streamflow contribution on water balance in Himalaya River Basin, Nepal. *Frontiers in Earth Science* 11: 1–11. https://doi.org/10.3389/feart.2023.1128959.

Devkota, N., Lamichhane, S., and Bhattarai, P.K. (2024). Multi-site calibration of the SWAT hydrological model and study of spatio-temporal variation of water balance components in the Narayani River Basin, central part of Nepal. *H2Open Journal* 7 (1): 114–129. https://doi.org/10.2166/h2oj.2024.084.

Ding, Y., Zhang, S., Zhao, L. et al. (2019). Global warming weakening the inherent stability of glaciers and permafrost. *Science Bulletin* 64 (4): 245–253. https://doi.org/10.1016/j.scib.2018.12.028.

Duncan, J.M.A., Biggs, E.M., Dash, J., and Atkinson, P.M. (2013). Spatio-temporal trends in precipitation and their implications for water resources management in climate-sensitive Nepal. *Applied Geography* 43: 138–146. https://doi.org/10.1016/j.apgeog.2013.06.011.

Eeckman, J., Nepal, S., Chevallier, P. et al. (2019). Comparing the ISBA and J2000 approaches for surface flows modelling at the local scale in the Everest region. *Journal of Hydrology* 569: 705–719. https://doi.org/10.1016/j.jhydrol.2018.12.022.

Gupta, A., Kayastha, R.B., Ramanathan, A., and Dimri, A.P. (2019). Comparison of hydrological regime of glacierized Marshyangdi and Tamor River Basins of Nepal. *Environmental Earth Sciences* 78 (14): 1–15. https://doi.org/10.1007/s12665-019-8443-5.

Gyanwali, K., Adhikari, P., Khanal, S. et al. (2023). Integrating glacio-hydrological and power grid models to assess the climate-resiliency of high mountain hydropower in Nepal. *Renewable and Sustainable Energy Reviews* 183: 113433. https://doi.org/10.1016/j.rser.2023.113433.

Hewlett, J. D. (1961). Soil moisture as a source of base flow from steep mountain watersheds. US Dept. Agric. Serv. 132.

Hock, R. (2003). Temperature index melt modelling in mountain areas. *Journal of Hydrology* 282 (1–4): 104–115. https://doi.org/10.1016/S0022-1694(03)00257-9.

Huss, M. and Hock, R. (2018). Global-scale hydrological response to future glacier mass loss. *Nature Climate Change* 8 (2): 135–140. https://doi.org/10.1038/s41558-017-0049-x.

Immerzeel, W.W., Lutz, A.F., Andrade, M. et al. (2020). Importance and vulnerability of the world's water towers. *Nature* 577 (7790): 364–369. https://doi.org/10.1038/s41586-019-1822-y.

Kang, S., Xu, Y., You, Q. et al. (2010). Review of climate and cryospheric change in the Tibetan Plateau. *Environmental Research Letters* 5 (1): https://doi.org/10.1088/1748-9326/5/1/015101.

Kendall, M.G. (1975). *Rank Correlation Methods*, 4e. London: Charles Griffin.

Khadka, M., Kayastha, R.R.B., and Kayastha, R.R.B. (2020). Future projection of cryospheric and hydrologic regimes in Koshi River Basin, Central Himalaya, using coupled glacier dynamics and glacio-hydrological models. *Journal of Glaciology* 66 (259): 831–845. https://doi.org/10.1017/jog.2020.51.

Khadka, N., Chen, X., Sharma, S., and Shrestha, B. (2023). Climate change and its impacts on glaciers and glacial lakes in Nepal Himalayas. *Regional Environmental Change* 23 (143): 1–14. https://doi.org/10.1007/s10113-023-02142-y.

Khanal, S., Lutz, A.F., Kraaijenbrink, P.D.A. et al. (2021). Variable 21st century climate change response for rivers in High Mountain Asia at seasonal to decadal time scales. *Water Resources Research* 57 (5): https://doi.org/10.1029/2020wr029266.

Kirkham, J.D., Koch, I., Saloranta, T.M. et al. (2019). Near real-time measurement of snow water equivalent in the Nepal Himalayas. *Frontiers in Earth Science* 7 (177): 1–18. https://doi.org/10.3389/feart.2019.00177.

Kraaijenbrink, P.D.A., Stigter, E.E., Yao, T., and Immerzeel, W.W. (2021). Climate change decisive for Asia's snow meltwater supply. *Nature Climate Change* 11: 591–597. https://doi.org/10.1038/s41558-021-01074-x.

Latif, Y., Ma, Y., Ma, W. et al. (2020). Differentiating snow and glacier melt contribution to runoff in the Gilgit River Basin via degree-day modelling approach. *Atmosphere* 11 (1023): 1–24. https://doi.org/10.3390/atmos11101023.

Lutz, A.F., Immerzeel, W.W., Shrestha, A.B., and Bierkens, M.F.P. (2014). Consistent increase in High Asia's runoff due to increasing glacier melt and precipitation. *Nature Climate Change* 4 (7): 587–592. https://doi.org/10.1038/nclimate2237.

Mann, H.B. (1945). Non-parametric test against trend. *Econometrica: Journal of the Econometric Society* 13 (3): 245–259.

Miller, J.D., Immerzeel, W.W., and Rees, G. (2012). Climate change impacts on glacier hydrology and river discharge in the Hindu Kush-Himalayas. *Mountain Research and Development* 32 (4): 461–467. https://doi.org/10.1659/MRD-JOURNAL-D-12-00027.1.

Neal, E.G., Hood, E., and Smikrud, K. (2010). Contribution of glacier runoff to freshwater discharge into the Gulf of Alaska. *Geophysical Research Letters* 37 (6): 1–5. https://doi.org/10.1029/2010GL042385.

Neitsch, S., Arnold, J., Kiniry, J., and Williams, J. (2011). Soil & water assessment tool theoretical documentation version 2009. Texas Water Resources Institute. https://doi.org/10.1016/j.scitotenv.2015.11.063

Nepal, S. (2016). Impacts of climate change on the hydrological regime of the Koshi River Basin in the Himalayan region. *Journal of Hydro-Environment Research* 10: 76–89. https://doi.org/10.1016/j.jher.2015.12.001.

Nepal, S. and Shrestha, A.B. (2015). Impact of climate change on the hydrological regime of the Indus, Ganges and Brahmaputra River Basins: a review of the literature. *International Journal of Water Resources Development* 31 (2): 201–218. https://doi.org/10.1080/07900627.2015.1030494.

Nepal, J., Pant, R.R., Shrestha, S. et al. (2024). Water balance estimation and runoff simulation of Chameliya Watershed, Nepal. *Environmental Earth Sciences* 83 (3): 117. https://doi.org/10.1007/s12665-024-11430-7.

Nie, Y., Pritchard, H.D., Liu, Q. et al. (2021). Glacial change and hydrological implications in the Himalaya and Karakoram. *Nature Reviews Earth and Environment* 2 (2): 91–106. https://doi.org/10.1038/s43017-020-00124-w.

Pfeffer, W.T., Arendt, A.A., Bliss, A. et al. (2014). The Randolph glacier inventory: a globally complete inventory of glaciers. *Journal of Glaciology* 60 (221): 537–552. https://doi.org/10.3189/2014JoG13J176.

Shrestha, A.B., Bajracharya, S.R., Sharma, A.R. et al. (2017). Observed trends and changes in daily temperature and precipitation extremes over the Koshi River Basin 1975–2010. *International Journal of Climatology* 37 (2): 1066–1083. https://doi.org/10.1002/joc.4761.

Singh, V., Jain, S.K., and Shukla, S. (2021). Glacier change and glacier runoff variation in the Himalayan Baspa River Basin. *Journal of Hydrology* 593: 125918. https://doi.org/10.1016/j.jhydrol.2020.125918.

Sinha, R., Gupta, A., Mishra, K. et al. (2019). Basin-scale hydrology and sediment dynamics of the Kosi river in the Himalayan foreland. *Journal of Hydrology* 570: 156–166. https://doi.org/10.1016/j.jhydrol.2018.12.051.

Terink, W., Lutz, A.F., Simons, G.W.H. et al. (2015). SPHY v2.0: spatial processes in hydrology. *Geoscientific Model Development* 8 (7): 2009–2034. https://doi.org/10.5194/gmd-8-2009-2015.

Terink, W., Immerzeel, W.W., Lutz, A.F. et al. (2017). Hydrological and climate change assessment for hydropower development in the Tamakoshi River Basin, Nepal. *FutureWater Report* 31: 1–93.

Terink, W., Leijnse, H., van den Eertwegh, G., and Uijlenhoet, R. (2018). Spatial resolutions in areal rainfall estimation and their impact on hydrological simulations of a lowland catchment. *Journal of Hydrology* 563: 319–335. https://doi.org/10.1016/j.jhydrol.2018.05.045.

Thakuri, S. and Salerno, F. (2016). Glacio-hydrological simulation in Dudh Koshi River Basin, Nepal examining glacier melt contribution to river flow. *International Journal of Scientific Development and Research* 1 (2): 72–78. www.ijsdr.org.

Wang, L., Zhang, F., Nepal, S. et al. (2023). Response of runoff processes to temperature rise in basins with different glacier ratios in the monsoon-influenced southern Tibetan Plateau. *Journal of Hydrology: Regional Studies* 45: 101299. https://doi.org/10.1016/j.ejrh.2022.101299.

Xue, W., Jungang, G., Yili, Z. et al. (2017). Land cover status in the Koshi River Basin, Central Himalayas. *Journal of Resources and Ecology* 8 (1): 10–19. https://doi.org/10.5814/j.issn.1674-764x.2017.01.003.

Yadav, V.K., Nema, M.K., and Khare, D. (2022). Evaluation of SWAT model for simulating the water balance components for the Dudh Koshi River Basin in Nepal. In: *Sustainability of Water Resources* (ed. B. Yadav, M.P. Mohanty, A. Pandey, et al.), 63–77. Cham: Springer 10.1007/978-3-031-13467-8.

Yang, H., Qi, J., Xu, X. et al. (2014). The regional variation in climate elasticity and climate contribution to runoff across China. *Journal of Hydrology* 517: 607–616. https://doi.org/10.1016/j.jhydrol.2014.05.062.

Yao, T., Thompson, L.G., Mosbrugger, V. et al. (2012). Third pole environment (TPE). *Environmental Development* 3: 52–64.

Yao, T., Bolch, T., Chen, D. et al. (2022). The imbalance of the Asian water tower. *Nature Reviews Earth and Environment* 3: 618–632. https://doi.org/10.1038/s43017-022-00299-4.

Zhang, Y., Luo, Y., Sun, L. et al. (2016). Using glacier area ratio to quantify effects of melt water on runoff. *Journal of Hydrology* 538: 269–277. https://doi.org/10.1016/j.jhydrol.2016.04.026.

Zhang, F., Thapa, S., Immerzeel, W. et al. (2019). Water availability on the third pole: a review. *Water Security* 7: https://doi.org/10.1016/j.wasec.2019.100033.

Zhang, F., Zeng, C., Zhang, Q., and Yao, T. (2022a). Securing water quality of the Asian water tower. *Nature Reviews Earth and Environment* 3 (10): 611–612. https://doi.org/10.1038/s43017-022-00347-z.

Zhang, T., Li, D., and Lu, X. (2022b). Response of runoff components to climate change in the source-region of the Yellow River on the Tibetan plateau. *Hydrological Processes* 36 (6): 1–16. https://doi.org/10.1002/hyp.14633.

3

Assessment and Prediction of Water Yield in the Chandra Basin, Western Himalaya, India: Using Physical Basis SWAT Model and Machine Learning

Sobhana Mummaneni[1], *Vinay K. Gaddam*[2], *Gnana S.S.V. Mendu*[1], *Nihitha Vemulapalli*[1], *Smitha C. Ch*[3], *and Sayyad Fazal Rahman Khadri*[2]

[1] *Department of Computer Science and Engineering, Siddhartha Academy of Higher Education, Vijayawada, India*
[2] *Department of Civil Engineering, Siddhartha Academy of Higher Education, Vijayawada, India*
[3] *Department of Computer Science and Engineering, Koneru Lakshmaiah Education Foundation, Guntur, India*

3.1 Introduction

Water resource management under changing climate conditions is very crucial, mainly for the Himalayas, which have limited water resources in the form of snow and glaciers and provide meltwater to the growing population for various domestic supplies and socioeconomic activities, thereby increasing water demand conditions. Effective management guarantees an adequate supply of freshwater. Also, it may allow the mitigation of water shortages by minimizing pollution risks, preserving biodiversity, increasing sustainability, supporting economic growth, and securing the well-being of present and future generations. The Chandra basin in the Western Himalayas preserves abundant water resources in the form of snow and glaciers and provides runoff to downstream communities, mainly in critical summer months. Due to the vulnerability of external factors like soot contamination (Kumar et al. 2021), climate change, and anthropogenic activities (Panwar 2020), the glaciers and snow cover in this basin have become more sensitive. Hydrological modeling is, therefore, essential to assess water for these activities. The Soil and Water Assessment Tool (SWAT) simulates runoff using factors like land use/land cover, soil, and meteorological variables, enabling precise predictions of water yield and its quality through different time periods. Several water management practitioners and decision-makers have used the SWAT model for assessing runoff. Among them, a few researchers have used the SWAT model to assess the risks associated with hydrological variability.

For example, Ayivi and Jha (2018) used the SWAT model to forecast hydrological behavior in a North Carolina watershed based on streamflow variability. This study evaluated the runoff and the model's performance by R^2 and NSE to produce the monthly streamflow, showing strong alignment with observed data. This study also explored the impact of future land use changes on water yield and runoff, projecting increases with the expansion of

Water Resources Management in Mountain Regions, First Edition. Edited by Virendra Bahadur Singh, Sughosh Madhav, Rakesh Kumar Gupta, Prerna Diwan, and Amit Kumar.

impermeable surfaces. Gaddam et al. (2018) studied the stream runoff of the Baspa River in Himachal Pradesh, India, focusing on contributions from rainfall and snow-ice melt. A streamflow simulation model was used to assess the runoff for the period 2000–2014 and validated against discharge from the Jaipee Hydropower Station. The findings indicate a strong correlation between modeled runoff and observed discharges, with 81.2% of runoff from snowmelt, 11.4% from rainfall, and 7.4% from ice melt, highlighting the impacts of climatic change on glacierized basins.

Wang et al. (2019) employed CiteSpace to analyze the evolving research on the SWAT model amidst human impacts and climate change on the water cycle. This study focused on SWAT's applications in runoff simulation, hydrological impacts, and nonpoint source pollution, suggesting enhancements for the model's adaptability to changing environmental conditions. Lee et al. (2020) used SWAT and LSTM (long short-term memory) models for runoff assessment in the Mekong River basin. A deep neural network was used to evaluate the runoff predictions using historical rainfall and runoff datasets between 1980 and 2005. This study evaluated the models' performance for the same river under RCP 4.5 and RCP 8.5 scenarios. Gupta et al. (2021) have utilized the SWAT model to simulate hydrological processes, including glacier melt and snowmelt, in the Upper Indus Basin (UIB) of the Indian Himalaya. The model was applied for the study period between 1981 and 2000 and evaluated and assessed the runoff between 2001 and 2007 for the same basin. The observations are listed on a daily and monthly basis. The model parameters are pre-calibrated, and then the SWAT-CUP SUFI-2 algorithm is used to simulate the runoff contributions from all the hydrological variables, including snow-ice melt. Mudunuru et al. (2021) introduced a swift and dependable calibration method for the SWAT analysis in the American River Watershed, consisting of 21 parameters using deep learning. This deep learning approach incorporated Convolution Neural Networks (CNNs) for analyzing streamflow. Hyperparameter tuning was used to optimize the CNN architecture and to ensemble the simulations to train, validate, and test the DL models. The DL method surpassed traditional methods like generalized likelihood uncertainty estimation (GLUE) and showed narrower ranges and higher reliability, providing more accurate predictions for both low and high streamflow compared to the GLUE model. Thapa et al. (2021) utilized daily and future climate datasets to predict snowmelt runoff using machine learning models (LSTM, GRU, RNN). A gamma test was used to identify the optimal input combinations, and hyperparameters were fine-tuned via grid search. Future temperature and precipitation scenarios were projected under RCP4.5 and RCP8.5 scenarios for simulating river discharge. Comparative analysis of the models' performance provided insights into the region's hydrological dynamics, considering both seasonal and annual changes.

Liu et al. (2021) studied deep learning applications for runoff assessment in the Hanjiang River Basin. They utilized the Conv-TALSTM model, which incorporates convolution kernels and attention mechanisms into LSTM networks. This approach was compared against traditional LSTM, TALSTM, and Conv-LSTM models, as well as the Wetspa hydrological model. The results demonstrated that Conv-TALSTM outperformed other models in high-flow event accuracy. This study highlighted that the integration of both meteorological and hydrological data as inputs significantly improved model efficiency. Mirzaei et al. (2021) evaluated the effectiveness of a novel Stacked LSTM (SLSTM) model against the physically based SWAT model for daily rainfall-runoff simulation in data-scarce regions. The study was conducted in Malaysia's Samarahan and Trusan basins; it contrasts these models' reliance on spatial data distribution, leveraging over 20 years of hydro-climate records. While SWAT excels in physically represented catchments, SLSTM, enhanced with PCA for data preprocessing, shows resilience against sparse data, highlighting the suitability of machine learning models for areas with limited hydrological measurements.

Han and Dawson (2021) employed Elastic-Net multinomial logistic regression to analyze longitudinal data from the Civic Purpose Project, involving 476 high school participants in California. Through 5000 iterations, this approach predicted the trajectory of civic purpose across political, community service, and expressive activity dimensions from Wave 1 (12th grade) to Wave 2 (two years later). Emphasizing the identification of key predictors while minimizing model complexity, the study demonstrates the efficacy of Elastic-Net regression in exploring civic engagement predictors efficiently and data-driven. Rahman et al. (2022) used the SWAT model and multilayer perceptron (MLP) to simulate the streamflow of the UIB. The SWAT model was calibrated and validated using the SWAT-CUP SUFI-2 algorithm by selecting the parameters through global sensitivity analysis. The MLP model was also considered and trained using the same input and output of SWAT by determining the optimal network architecture through a trial-and-error method. Evaluation metrics used in this study are NSE, PBIAS, R^2, and MAPE. Gaddam et al. (2022) employed a temperature index-based runoff simulation model to estimate the runoff of the Chandra River basin in the Western Himalaya. The model performance was validated against observed runoff from Sutridhaka and Chhotashigri glacial catchments, revealing notable concurrence. This study quantified snow/ice melt and rainfall contributions to the basin runoff and evaluated runoff sensitivity to project future temperature and precipitation changes. This analysis offers valuable insights into Chandra River basin hydrological dynamics, enhancing understanding of potential responses to evolving climatic conditions. Bhusal et al. (2022) compared the efficacy of the Random Forest model with the traditional Hydrological Engineering Center-Hydrologic Modeling System (HEC-HMS) in rainfall-runoff simulation, integrating it with the Hydraulic Engineering Center's River Analysis System (HEC-RAS) for hydraulic analysis. The study utilized PERSIANN-CCS for precipitation data and found both models reliable, with the machine learning approach providing a viable alternative where data scarcity limits conventional methods. The research highlights the potential of combining machine learning with physical models for enhanced hydrological predictions, suggesting areas for future research, including diverse data sources and model comparisons.

Wortmann et al. (2022) analyzed climate change impacts on glacierized river catchments in NW China's Tarim River area, vital for local agriculture. This study employed SWIM-G and WASA models, driven by outputs from eight general circulation models under three climate change scenarios, projecting glacier cover, future climate, and river discharge through the twenty-first century. The results indicate warming and increased precipitation, affecting glacier size and river flows differently across catchments. This study underscores the critical importance of sustainable water management in response to changing conditions, integrating advanced climate and hydrological models to address environmental complexities. Jayanthi et al. (2022) introduced a novel approach integrating the SWAT and Support Vector Regression to predict water levels in Pakhal Lake, Telangana, India, under climate change scenarios. The study utilized the NASA Earth Exchange Global Daily

Downscaled Projections (NEX-GDDP) dataset, with 21 Global Climate Models, and refined climate projections using the Reliability Ensemble Averaging (REA) method. Refined projections were combined with SWAT outputs to train the ν-SVR model, aiming for accurate forecasts to guide water resource management strategies effectively.

Bastiaan et al. (2022) explored the challenge of multicollinearity in linear regression models and evaluated the effectiveness of penalized regression methods, notably Ridge and Elastic-Net regression, as solutions. By analyzing a simulated dataset comprising 100 observations with 95 highly correlated independent variables, the study demonstrates that the Elastic-Net regression method markedly surpasses both ridge regression and ordinary least squares (OLS) in minimizing the mean squared error (MSE). The results emphasize the inadequacy of OLS in managing multicollinear datasets and establish the superior predictive performance of Elastic-Net regression under such conditions.

Islam et al. (2023) investigated streamflow using random forest machine learning (RFML) along with SWAT. The investigation was performed in the area of Rio Grande Headwaters for a period of 25 years between January 1991 and December 2016. The study utilized five predictors and one response variable to simulate the runoff. Evaluation metrics used to assess model accuracy are Nash–Sutcliffe efficiency, index of agreement, root mean square deviation ratio, percent bias, and coefficient of determination. Then, the results were correlated with random forest regression (RFR) to identify the influential predictor variables in the hydrological simulation affecting the streamflow. Janjić and Tadić (2023) used the SWAT model to simulate nonpoint source pollution, runoff, and assess the complex hydrological processes amid changing environmental conditions. This survey outlines SWAT's evolution since the 1990s, noting a surge in related articles, with almost 4000 published in the last decade. Emphasizing SWAT's broad application in modeling large watersheds, particularly in North America and Asia, it addresses the challenges like data requirements for accurate simulations. Despite limitations, SWAT's strengths include continuous development and the ability to simulate human and agricultural activities, highlighting its role in adaptive watershed management. Shekar et al. (2023) compared the SWAT model with several artificial intelligence (AI) models for analyzing rainfall-runoff (R–R) in the Peddavagu River Basin of the Deccan Plateau Region, India. The machine learning models include RFR, LSTM, support vector regression, multiple linear regression, k-nearest neighbor regression, artificial neural network, and extreme gradient boosting regression. The evaluation metrics R^2 and NSE were used to assess the performance of the models for calibration (1990–2005) and validation (2006–2010) periods. The study concluded that AI models, especially LSTM, showed promise for R–R simulation, suggesting their potential in sustainable water resource management. Wang et al. (2023a) investigated runoff patterns in the Hotan River Basin, employing 12 meteorologically relevant input features. Overall, eight machine learning regression models, such as support vector regression, random forest, and gradient boosting, were utilized to train, simulate, and compare the runoff for the Yurungkash and Kalakash River basins. Evaluation metrics like root mean square error, percentage bias, mean absolute error, and Nash–Sutcliffe efficiency were used to assess the performance of models. Lu et al. (2023) introduced a novel stacking ensemble learning model enhanced by an attention mechanism for daily runoff prediction on the Fuchun River Reservoir, China. This study utilized extreme gradient boosting, random forest, and adaptive boosting models, integrating these models with an attention-based meta-model to refine predictions. The methodology demonstrates significant improvements over traditional models, validating the model's efficacy in managing water resources and controlling floods through comprehensive performance metrics and comparisons.

Su et al. (2023) employed a comprehensive approach to examine the climate change impact on the hydrology of the Hulan River Basin by analyzing long-term data (1960–2020) from the Wangkui Meteorological Station, focusing on trends in temperature, precipitation, and wind speed. This study used a SWAT-distributed hydrological model, refined with snowmelt parameters and SWAT-CUP software for accurate runoff simulations. The model's effectiveness was affirmed through data from the Lanxi Hydrological Station (2008–2020), showcasing higher efficiency and correlation. This methodology underscores SWAT's applicability in cold, high-latitude areas, offering valuable insights for managing water resources. Wang et al. (2023b) introduced a novel approach for reversible data hiding (RDH) and utilized the ridge regression method to enhance the security and efficient embedding of datasets. This study distinguishes itself by prioritizing error minimization and selective emphasis on key image pixels, ensuring minimal alteration to the original image's appearance while bolstering resistance to tampering. By prioritizing crucial image areas for data embedding, this approach represents a notable advancement over prior methods, as demonstrated in experiments showcasing its superiority in reliability and effectiveness for data protection and embedding. All these studies suggest that the SWAT model performs better along with regression models for hydrological simulations. Thus, this study aims to evaluate runoff in the Chandra basin by utilizing both SWAT and regression methods.

3.2 Study Area

This study focuses on the Chandra River basin in the Western Himalayas as, shown in Figure 3.1 (Gaddam et al. 2022). The basin is located between 77° 0′ 0″E and 77° 50′ 00″E and 32° 00′ 0″N and 33° 10′ 00″N, with altitudes ranging from 2800 m to nearly 6600 m, covering an area of approximately 2440 km^2 (Tawde et al., 2019). It contains approximately 200 glaciers. This region is a popular tourist spot, particularly near the Batal area, where many tourists stay between Chandratal Lake and Chhotadhada. The Chandra River basin is one of the largest river catchments in the Western Himalayas, containing a significant ice cover (709 km^2) and draining into Tandi, where the Chandra River merges with the Bhaga River to form the Chenab River. Larger glaciers in this basin are Bara Shigri, Samudra Tapu, Chhota Shigri, and Nalk Nala. The Chandra basin is situated in a transition zone between monsoon and arid climates, receiving rainfall from the Indian summer monsoon between July and September, and from westerly disturbances between January and April during winter (Bookhagen and Burbank 2010). Despite the lack of meteorological observatories in the basin, gridded reanalysis data used by Mukherjee et al. (2018) indicate that the recent decade has been warmer, with less precipitation compared to previous decades. Similarly, Azam et al. (2014a, 2014b) observed the same pattern using data from a meteorological station at Bhuntar airport, approximately 50 km from the Chhota Shigri Glacier. For Himansh base camp and Shurti Dhaka Glacier, Pratap et al. (2019) recorded maximum

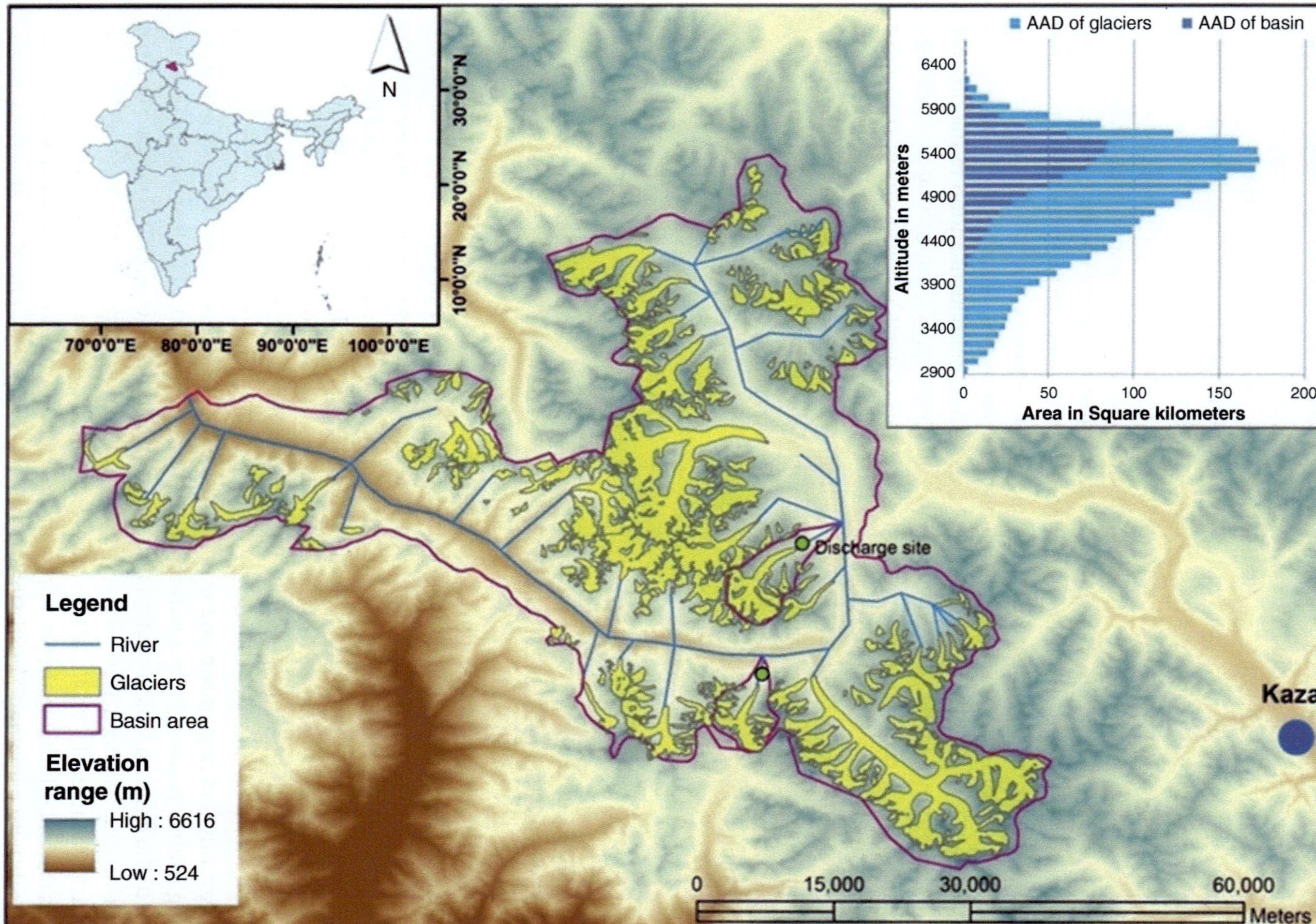

Figure 3.1 Study area: "Chandra basin" is selected in Himachal Pradesh, India, for runoff simulation. *Source:* Gaddam et al. 2022/Springer Nature.

Table 3.1 Information on spatial and temporal datasets utilized for runoff computation in the study.

Dataset category	Name of the dataset	Spatial resolution	Source
Elevation model	TanDEM-X	12 m	TerraSAR-X
Land use/ land cover	Global Artificial Land Surface Dataset	30 m	(Chen et al. 2017)
Soil	Soil category (Water, Glacier)	30 m	Harmonized World Soil Database (HWSD) (FAO & IIASA 2023)
Weather	Precipitation (mm), wind speed (m/s), minimum-maximum temperature (°C), relative humidity (%), solar radiation (W/m^2)	Daily (1969–2023)	Indian Meteorological Department (IMD) Station Name: Bhuntar (https://mausam.imd.gov.in/)

temperatures of 11.4 °C and 7.4 °C in July, and minimum temperatures of −15 °C and −15.7 °C in January, respectively, during 2016–2017. Geologically, the region is characterized by metamorphic rocks, primarily meso- to kata-zonal metamorphites, migmatites, and gneisses (Singh et al. 2019). The glaciers located here receive the precipitation from both the Indian monsoon in summer and from westerlies in winter.

The datasets utilized for assessing the runoff in the Chandra basin were acquired from diverse sources as shown in the Table 3.1.

3.3 Methodology

This study (Figure 3.2) involves the computation of water yield using the SWAT physical model for the years 1969–2023, with a detailed description of the evaluation of individual parameters required for the SWAT model, which are given in Section 3.1 (3.1.1–3.1.4). Later, the runoff from this basin under the same LU/LC conditions was calculated for the years 2024–2030, using SWAT-regression models. Sections 3.2–3.3 briefly describe the procedure for runoff evaluation by regression models.

3.3.1 SWAT Model

The SWAT is a widely used hydrological model designed to assess runoff. Initially, it used the elevation model for the delineation of stream channels, and then the outlets are defined. The watershed is also delineated by utilizing the outlet boundaries. The land use/ land cover data and soil database of the study region are obtained from Chen et al. (2017), FAO & IIASA (2023), and used to categorize the different types of land uses and soil within the watershed. The reclassified data are then overlaid onto the delineated watershed, and the Hydrological Response Units (HRUs) are defined at the microscale,

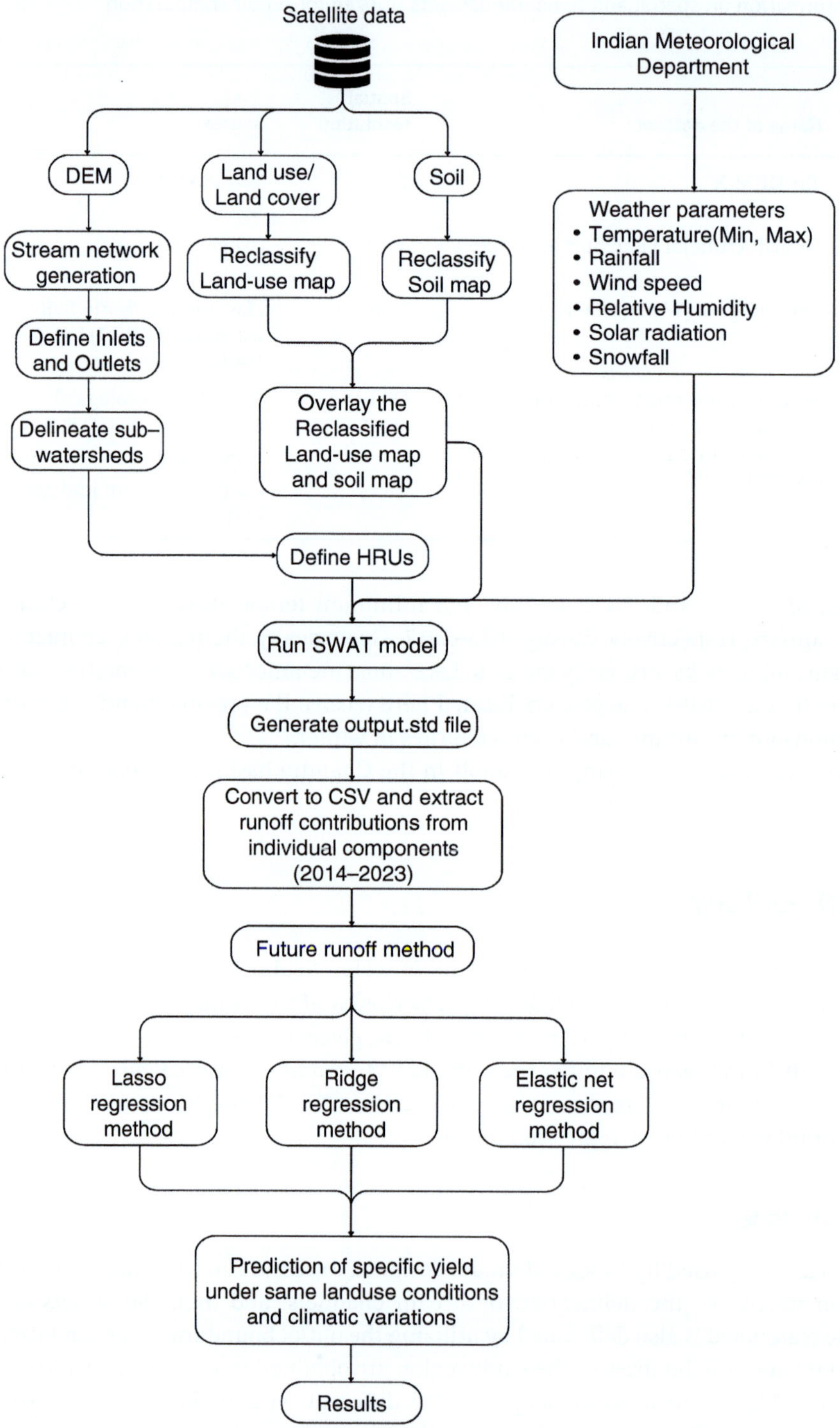

Figure 3.2 Flowchart of the methodology used in the study.

representing unique combinations of soil and land use within the HRU. HRUs represent spatial units within the watershed that have homogeneous hydrological characteristics, after integrating land use, soil type, and slope information (Neitsch et al. 2011). The weather parameters, such as precipitation, temperatures, wind speed, solar radiation, and relative humidity, are used in the model to compute the runoff, as discussed in the following sections. Initially, the SWAT setup is downloaded from (https://swat.tamu.edu/software/arcswat/), which is compatible with ArcGIS. The initial procedure involves creating a Project Geodatabase with the same name as the project folder. Additionally, Raster Storage and Parameter based Geodatabases are established in the SWAT folder. These Geodatabases look similar to the structure of the SWAT2012 database. The watershed delineation is based on the Digital Elevation Model (DEM), based on the topographical curvatures, flow length, and slope characteristics. Then, the stream channels are delineated (Neitsch et al. 2011). Outlets are thus defined and represented spatially. A total of 27 outlets, along with 27 subbasins (HRUs), are identified as per SWAT analysis. A single-step approach is employed to determine the slope to assess terrain steepness across the watershed. Subsequently, the land use/land cover and soil datasets were uploaded, reclassified, and overlaid onto the delineated watershed. 27 HRUs are defined, and weather parameters were incorporated for runoff analysis. Upon completing the SWAT model setup, the model is executed for the years 2014–2023 (Algorithm 1). The SWAT output comprises a standard output summary file (output.std), containing the necessary features for predicting water yield, as shown in Algorithm 1.

3.3.1.1 Land Use/Land Cover

Classification of land use/land cover datasets was considered as per the database in SWAT, for aligning the diverse land use categories within the standardized codes recognized by the model. This process involves uploading the land use/land cover data into the SWAT interface, along with a user-defined look-up table. The codes include aggregating or

Algorithm 1 SWAT-regression algorithm.

Input: DEM, land use/land cover, soil map, weather parameters
Output: Water yield

1: Delineate watershed
2: Reclassify land use/land cover data
3: Reclassify soil map data
4: Overlay delineated watershed with reclassified land use/land cover and soil maps
5: Define Hydrological Response Units (HRUs)
6: Input weather data into the model
7: Run the SWAT model
8: Generate the output.std file
9: Convert output.std file to CSV file and extract features
10: Split the dataset into training and testing data
11: Apply Lasso regression, Elastic-Net regression, and Ridge regression models
12: Predict water yield
13: Evaluate the performance of the models using metrics RMSE and MSE

Table 3.2 Attribute table of land use/land cover.

Area (%)	Land use SWAT	LULC description
4.26	WETN	Wetlands, non-forested
3.83	WETL	Wetlands, mixed
63.59	WETF	Wetlands, forested
28.53	WATR	Water

disaggregating the land use classes to represent the spatial heterogeneity (Neitsch et al. 2011). The attributes considered for the land use/land cover, as per SWAT, are shown in Table 3.2.

3.3.1.2 Soil Database

This process involves mapping diverse soil categories from various sources to the standardized soil classifications recognized by SWAT. It involves aggregating or disaggregating soil classes, assigning unique SWAT soil codes to each class, and updating the soil attributes such as texture, permeability, infiltration, and erosion potential (Neitsch et al. 2011). By the standard soil information library, the SWAT tool accurately simulates the rate of infiltration, percolation, runoff, and erosion processes within the watershed, thereby enhancing the reliability of hydrological predictions. The attributes considered by the soil database within the main watershed are shown in Table 3.3.

3.3.1.3 Weather Datasets

Meteorological variables, like temperature, relative humidity, rainfall, snowfall, solar radiation, and wind speed are considered (Neitsch et al. 2011) from Bhuntar observatory as input parameters for runoff simulation by the SWAT model. However, the observed data gaps were filled by the values from the weather generator table WGEN_user, comprising the monthly averaged values of the parameters. These records are obtained from the Indian Meteorological Department (IMD), Govt of India (https://swat.tamu.edu/data/india-dataset/). By utilizing the WGEN_user table, the missing values are effectively replaced, ensuring the completeness and accuracy of the weather inputs for the SWAT model. The filled weather data, spanning from 1969 to 2023, are uploaded into the SWAT model.

Table 3.3 Attribute table of soil map.

Id no.	Total area (sq km)	Area (%)	Attributes with ID
1	1372.148594	56.1858	WATER – 6997
2	1070.014006	43.8142	GLACIER – 6998

3.3.1.4 SWAT Simulation

The simulations using the SWAT model are carried out for the estimation of water yield from the basin. The study period is considered for 54 years, between 1 January 1969 and 31 December 2023, at the basin scale, and then the water yield is obtained, including (i) Soil Storage, (ii) Vel/Depth, and (iii) Log flows using Equations (3.1)–(3.5) (Neitsch et al. 2011). The outputs generated by SWAT with the Standard Output Summary File (output.std) files are selected for further analysis, to simulate the runoff in the near future from 2024 to 2030 using regression models.

$$W_{\text{yld}} = Q_{\text{surf}} + Q_{\text{gw}} + Q_{\text{lat}} - T_{\text{loss}} \tag{3.1}$$

$$Q_{\text{surf}} = \frac{\left(R_{\text{day}} - 0.2S\right)^2}{\left(R_{\text{day}} + 0.8S\right)} \tag{3.2}$$

$$S = 25.4\left(\frac{100}{CN} - 10\right) \tag{3.3}$$

$$Q_{lat} = 0.024\left(\frac{2.SW_{\text{ly,excess}}.SC.slp}{\varnothing_d.L_{\text{hill}}}\right) \tag{3.4}$$

$$Q_{gw} = \frac{8000.K_{\text{sat}}}{L_{\text{gw}}^2}.h_{\text{wtbl}} \tag{3.5}$$

where

W_{yld} = water yield (mm)
Q_{surf} = Surface runoff (mm)
Q_{gw} = groundwater flow (mm)
Q_{lat} = lateral flow (mm)
T_{loss} = transmission losses from the tributary in HRU by means of transmission through bed (mm)
R_{day} = Rainfall depth for the day (mm)
$SW_{\text{ly, excess}}$ = drainable amount of water stored in the saturated zone of hill segment per unit area (mm)
SC = saturated hydraulic conductivity (mm h^{-1})
slp = slope
$\varnothing_d$ = drainable porosity of the soil layer
L_{hill} = length of hillslope segment
S = Retention parameter
CN = soil curve number
K_{sat} = hydraulic conductivity of the aquifer (mm/day)
L_{gw} = distance from subbasin divide for groundwater system to maintain channel (m)
h_{wtbl} = water table height (m)

Table 3.4 Selected parameters.

Parameter	Description	Units
UNIT TIME	Monthly time stamp (1–12)	—
PREC	Average amount of precipitation in the month	mm
SURQ	Surface runoff for the month	mm
LATQ	Lateral flow contribution to streamflow for the month	mm
GWQ	Groundwater contribution to streamflow for the month	mm
PERCOLATE	Water percolation for the month	mm
SW	Amount of water stored in the soil profile for the month	mm
ET	Actual evapotranspiration for the month	mm
PET	Potential evapotranspiration for the month	mm
WATER YIELD	Water yield to streamflow from HRUs for the month	mm

3.3.2 Evaluation of Runoff in the Near Future

This step involves the computation of runoff using data-based machine learning processes involving the SWAT-Lasso, SWAT-Ridge, and SWAT-Elastic-Net regression methods, which are highly reliable for the decision-making process. The three methods involve the data generated using the physical based SWAT model, which are listed in Table 3.4 (Neitsch et al. 2011). The Tile Flow parameter is not considered for runoff estimation due to consistent zero observations spanning from 1969 to 2023. As a result, the modeled features during the analysis comprise the PREC, SURQ, LATQ, GWQ, PERCOLATE, SW, ET, and PET. The calibration of the model is done using the water yield from 1969 to 2015 and then validated using 2016–2023.

3.3.2.1 SWAT-Lasso Regression Method

The Lasso regression involves initializing the coefficient vector (β) with zeros or random small values (Andriopoulos and Kornaros 2023). Then, the model iteratively updates the coefficients until convergence or reaching the maximum number of iterations. During each iteration, predictions are made using the current runoff coefficients, and the residuals are computed. The gradients of the water loss concerning the coefficients are calculated by incorporating both the gradient of the mean square error and the regularization term (λ multiplied by the sign of the coefficients). The coefficients are updated using gradient descent, with a learning rate determining the step size. The process continues until the change in coefficients falls below a predefined tolerance level or the maximum number of iterations is reached. Algorithm 2 shows the working flow methodology for the SWAT-LASSO regression method.

Algorithm 2 SWAT-Lasso regression algorithm.

Input:
X: Feature matrix (m×n) containing features PREC, SURQ, LATQ, GWQ, PERCOLATE, SW, ET, PET
y: Target variable (m×1), WATER YIELD
λ: Regularization parameter
learning_rate: Step size for gradient descent
Output: Coefficients β
Initialize:
β: Coefficient vector (n×1) with zeros or random
    small values
iterations: Number of iterations for optimization
tolerance: Convergence criteria for optimization
Repeat until convergence or maximum iterations are reached:
    y_pred = $X \times \beta$
    residuals = $y -$ y_pred
    gradient=−1m×XT×residuals+λ×signβ
    $\beta = \beta -$ learning_rate × gradient
    If $||\beta_{new} - \beta_{old}|| <$ tolerance:
        Break the loop
    end loop
End

3.3.2.2 SWAT-Ridge Regression Method

The ridge regression method involves initializing the coefficient vector with zeros or small random values, setting the regularization parameter (λ), and defining the convergence criteria (Wang et al. 2023b). Iteratively, the algorithm computes predictions using the existing runoff coefficients, calculates residuals between predictions and actual values, and updates coefficient values using gradient descent. The regularization term in the gradient ensures that coefficient values are penalized to prevent overfitting. The process continues until convergence or a maximum number of iterations is reached. The resulting coefficients provide the fitted model for predicting target variables. Algorithm 3 shows the working flow methodology of the SWAT-Ridge regression method.

3.3.2.3 SWAT-Elastic-Net Regression Method

The Elastic-Net regression method involves initializing the coefficient vector with zeros or small random values, setting the L1 and L2 regularization parameters (λ1 and λ2, respectively), and defining the convergence criteria (Han and Dawson (2021)). Iteratively,

Algorithm 3 Swat-ridge regression algorithm.

Input:
X: Feature matrix (m×n) containing features PREC, SURQ, LATQ, GWQ, PERCOLATE, SW, ET, PET
y: Target variable (m×1), WATER YIELD
λ: Regularization parameter
learning_rate: Step size for gradient descent
Output: Coefficients β
Initialize:
    β: Coefficient vector (n×1) with zeros or random small values
    iterations: Number of iterations for optimization
    tolerance: Convergence criteria for optimization
Repeat until convergence or maximum iterations are reached:
    $y\_pred = X \times \beta$
    residuals $= y - y\_pred$
    gradient=−2m×XT×residuals+λ×β
    $\beta = \beta -$ learning_rate × gradient
    If $\|\beta_{new} - \beta_{old}\| <$ tolerance:
        Break the loop
    end loop
End

the algorithm computes predictions using the existing runoff coefficients, calculates residuals between predictions and actual values, and updates the coefficients using gradient descent. Both L1 and L2 penalties are applied to the gradient to enforce sparsity and control coefficient magnitudes. The process continues until convergence or a maximum number of iterations is reached, yielding the coefficients for the fitted model. Algorithm 4 shows the working flow methodology of the SWAT-Elastic-Net regression method.

3.3.3 Performance Metrics

Performance metrics elaborate the effectiveness and dependability of the regression models. In this study, the performance metrics like RMSE and MSE are employed to evaluate the model performance. RMSE assesses prediction accuracy by measuring residual spread, while the MSE offers an overall assessment of prediction quality, with lower values indicating better performance. Residual curves for individual models were also plotted to understand the effectiveness of regression models. Equations (3.6) and (3.7) illustrate the calculation of these metrics.

$$\text{RMSE} = \sqrt{\frac{1}{n}\sum_{i=1}^{n}\left(y_i - \hat{y}_i\right)^2} \tag{3.6}$$

Algorithm 4 SWAT-Elastic-Net regression algorithm.

Input:
X: Feature matrix (m x n) containing features PREC, SURQ, LATQ, GWQ, PERCOLATE, SW, ET, PET
y: Target variable (m x 1), WATER YIELD
λ1: L1 regularization parameter (for Lasso)
λ2: L2 regularization parameter (for Ridge)
learning_rate: Step size for gradient descent
Output: Coefficients β
Initialize:
    β: Coefficient vector (n x 1) with zeros or random small values
    iterations: Number of iterations for optimization
    tolerance: Convergence criteria for optimization
Repeat until convergence or maximum iterations are reached:
    y_pred $= X \times \beta$
    residuals $= y -$ y_pred
    *l1*_penalty = λ1 × sign(β)
    *l2*_penalty = 2 × λ2 × β
    gradient=−1m×XT×residuals+l1_penalty+l2_penalty
    $\beta = \beta -$ learning_rate × gradient
    If $\|\beta_{\text{new}} - \beta_{\text{old}}\| <$ tolerance:
        Break the loop
    end loop
End

Thus, we considered the above three methods to compute the runoff.

$$\text{MSE} = \frac{1}{n}\sum_{i=1}^{n}\left(y_i - \hat{y}_i\right)^2 \tag{3.7}$$

where n is the number of observations, y_i represents the observed values, and $\hat{y}_i$ represents the predicted values. The performance of the regression models is compared using the metrics RMSE and MSE.

3.3.4 Results and Observations

After running the SWAT model in ArcGIS as discussed in the methodology section with the parameters listed in Table 3.4, the runoff is assessed (Figure 3.3). The contribution of different sources to the river runoff is shown in Figure 3.3, suggests that the maximum percentage contribution is from precipitation, which is estimated at 34%, while the least

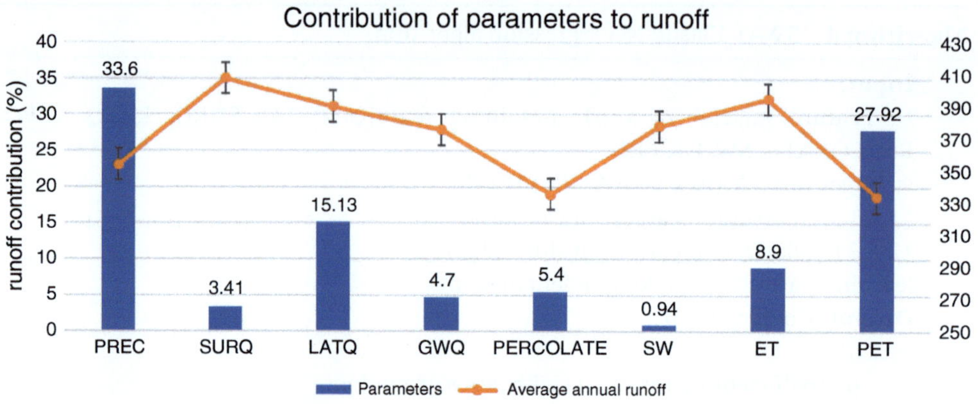

Figure 3.3 Percentage contribution of parameters to runoff and average annual runoff from 2014 to 2023 derived using the physical based SWAT model.

contribution comes from the soil water profile, which is approximately 1%. This shows the average annual runoff from 2014 to 2023 using the physical based SWAT model.

Using the results from SWAT as input, the predicted runoff for the period 2024–2030 is assessed using the regression models. Figure 3.4 shows the predicted water yield using the SWAT-regression models.

A comparative analysis of the RMSE metric for the proposed SWAT-Lasso, SWAT-Ridge, and SWAT-Elastic-Net regression models. This comprehensive evaluation facilitates the identification of unique performance attributes inherent to each regression technique and aids in discerning the most effective modeling approach. Figure 3.5 visually illustrates the comparison of the regression models in terms of RMSE and MSE. This graphical representation offers a clear visual insight into the performance differences among the SWAT-Lasso, SWAT-Ridge, and SWAT-Elastic-Net models in predicting water yield.

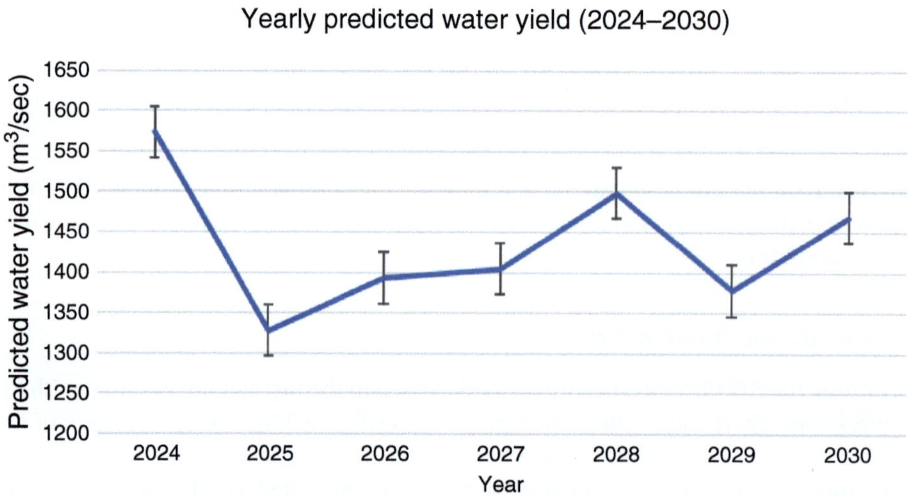

Figure 3.4 Water yield prediction.

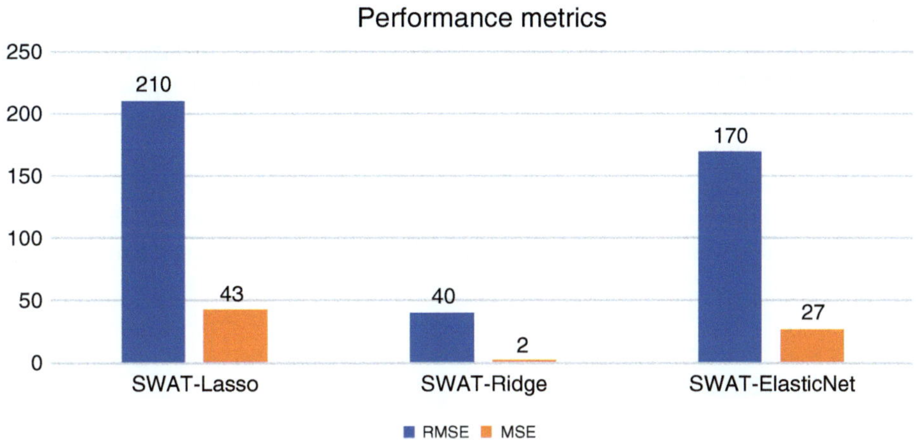

Figure 3.5 RMSE and MSE comparison.

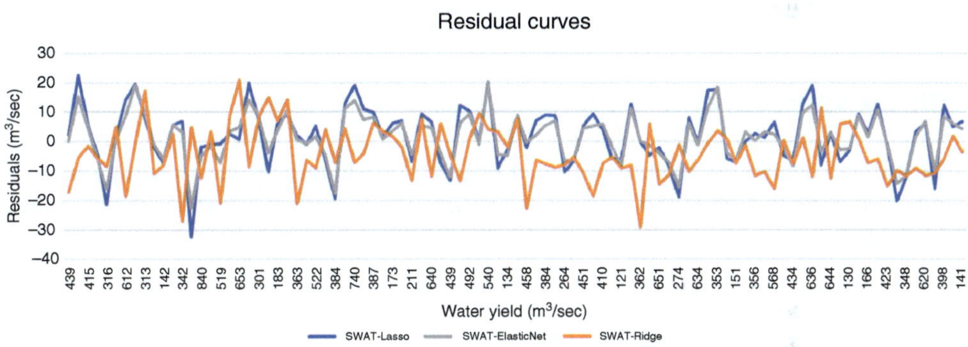

Figure 3.6 Residual curves.

As discussed, residual curves provide valuable insights into the performance and effectiveness of regression models. These curves depict the distribution and patterns of residuals, which are the differences between observed and predicted values. Hence, the residual curves of the SWAT-regression models are plotted and shown in Figure 3.6, suggesting a similar trend among the three methods.

Upon comparison, it is observed that SWAT-Ridge regression suits as the optimal model for runoff computation, exhibiting lower RMSE and MSE values compared to other SWAT-regression techniques.

3.4 Conclusion(s)

Hydrological modeling plays a crucial role in understanding water yield. Utilizing the physically based SWAT model for hydrological assessment in the Chandra basin, Himachal Pradesh, India, this study introduces the SWAT-regression framework, which integrates spatial and temporal data for water yield estimation. By employing the Lasso, Ridge, and Elastic-Net regression methods using SWAT-derived data, it is observed that the SWAT-Ridge model, with an RMSE of 4% is highly suitable for predicting water yield in a glaciated basin, suggesting to implement the same method for the assessment of runoff in data-scarce regions of the Himalayas. Future work could involve the integration of additional regression techniques and deep learning methods exploring better spatiotemporal analysis that is more suitable for water yield predictions on a large spatial scale. This helps in analyzing the water budget for sustainable water management practices in the Chandra basin and similar hydrological watersheds.

Acknowledgments

The authors would like to thank Siddhartha Academy of Higher Education for providing essential financial and infrastructure support during the study. Thanks to IMD and TERRASAR-X team for sharing the data. Special thanks for SWAT team for providing updated data on landuse landcover, soil information.

Ethical Approval and Responsibilities of Authors

The authors declare no potential conflict of interest. All authors have read, understood, and have complied with the manuscript as applicable with the statement on “Ethical responsibilities of authors.”

Consent to Participate

The manuscript does not involve any information related to human subjects.

Consent to Publish

Data used and taken from literature are properly cited.

Funding

The authors would like to thank Siddhartha Academy for seed grant received for the study on glaciers.

Competing Interests

The authors have no relevant financial or nonfinancial interests to disclose.

Author(s) Contributions

Concept and design were performed by Vinay Kumar Gaddam and Sobhana Mummaneni. Material preparation, data collection, and analysis were performed by Gnana Siva Sai Venkatesh Mendu and Nihitha Vemulapalli. The manuscript was written by the first author, and all authors contributed to the improvement of the manuscript draft in both technical and English language corrections. All authors read and approved the final manuscript.

Availability of Data and Materials

The data were obtained from TerraSAR-X, Globeland 30 Geometrics of China, FAO Harmonized World Soil Database, Indian Meteorological Department, and the information-sharing policy is as per the guidelines of Siddhartha Academy of Higher Education, Vijayawada.

References

Andriopoulos, V. and Kornaros, M. (2023). LASSO regression with multiple imputations for the selection of key variables affecting the fatty acid profile of Nannochloropsis oculata. *Marine Drugs* 21 (9): 483.

Ayivi, F. and Jha, M.K. (2018). Estimation of water balance and water yield in the Reedy Fork-Buffalo Creek Watershed in North Carolina using SWAT. *International Soil and Water Conservation Research* 6 (3): 203–213.

Azam, M.F., Wagnon, P., Vincent, C. et al. (2014a). Processes governing the mass balance of Chhota Shigri Glacier (western Himalaya, India) assessed by point-scale surface energy balance measurements. *The Cryosphere* 8 (6): 2195–2217.

Azam, M.F., Wagnon, P., Vincent, C. et al. (2014b). Reconstruction of the annual mass balance of Chhota Shigri glacier, Western Himalaya, India, since 1969. *Annals of Glaciology* 55 (66): 69–80.

Bastiaan, R.M., Salaki, D.T., and Hatidja, D. (2022). Comparing the performance of prediction model of ridge and elastic net in correlated dataset. *Operations Research: International Conference Series* 3 (1): 8–13.

Bhusal, A., Parajuli, U., Regmi, S., and Kalra, A. (2022). Application of machine learning and process-based models for rainfall-runoff simulation in Dupage River basin, Illinois. *Hydrology* 9 (7): 117.

Bookhagen, B. and Burbank, D.W. (2010). Toward a complete Himalayan hydrological budget: spatiotemporal distribution of snowmelt and rainfall and their impact on river discharge. *Journal of Geophysical Research: Earth Surface* 115 (F3): 115–128.

Chen, J., Liao, A.P., Chen, L.J. et al. (2017). Content and composition of global artificial land surface dataset at 30 m resolution (2010). *Journal of Global Change Data & Discovery* 1: 136–148.

FAO & IIASA (2023). *Harmonized World Soil Database Version 2.0*. Rome and Laxenburg: FAO & IIASA https://doi.org/10.4060/cc3823en.

Gaddam, V.K., Kulkarni, A.V., and Gupta, A.K. (2018). Assessment of snow-glacier melt and rainfall contribution to stream runoff in Baspa Basin, Indian Himalaya. *Environmental Monitoring and Assessment* 190: 1–11.

Gaddam, V.K., Myneni, T.K., Kulkarni, A.V., and Zhang, Y. (2022). Assessment of runoff in Chandra river basin of Western Himalaya using Remote Sensing and GIS Techniques. *Environmental Monitoring and Assessment* 194 (3): 145.

Gupta, C., Kulkarni, A.V., and Taloor, A.K. (2021). Streamflow modeling and contribution of snow and glacier melt runoff in glacierized upper Indus basin. *Environmental Monitoring and Assessment* 193 (11): 761.

Han, H. and Dawson, K.J. (2021). Applying Elastic-Net regression to identify the best models predicting changes in civic purpose during the emerging adulthood. *Journal of Adolescence* 93: 20–27.

Islam, K.I., Elias, E., Carroll, K.C., and Brown, C. (2023). Exploring random forest machine learning and remote sensing data for streamflow prediction: an alternative approach to a process-based hydrologic modeling in a snowmelt-driven watershed. *Remote Sensing* 15 (16): 3999.

Janjić, J. and Tadić, L. (2023). Fields of application of SWAT hydrological model – a review. *Earth* 4 (2): 331–344.

Jayanthi, S.L.S.V., Keesara, V.R., and Sridhar, V. (2022). Prediction of future lake water availability using SWAT and support vector regression (SVR). *Sustainability* 14 (12): 6974.

Kumar, A., Singh, S., Kumar, N. et al. (2021). Assessment of carbonaceous fractions in ambient aerosols at high altitude and lowland urban locations of Kumaon Province, Indian Himalaya. *SN Applied Sciences* 3: 1–14.

Lee, D., Lee, G., Kim, S., and Jung, S. (2020). Future runoff analysis in the Mekong river basin under a climate change scenario using deep learning. *Water* 12 (6): 1556.

Liu, Y., Zhang, T., Kang, A. et al. (2021). Research on runoff simulations using deep-learning methods. *Sustainability* 13 (3): 1336.

Lu, M., Hou, Q., Qin, S. et al. (2023). A stacking ensemble model of various machine learning models for daily runoff forecasting. *Water* 15 (7): 1265.

Mirzaei, M., Yu, H., Dehghani, A. et al. (2021). A novel stacked long short-term memory approach of deep learning for streamflow simulation. *Sustainability* 13 (23): 13384.

Mudunuru, M.K., Son, K., Jiang, P., and Chen, X. (2021). SWAT watershed model calibration using deep learning. arXiv preprint arXiv:2110.03097.

Mukherjee, K., Bhattacharya, A., Pieczonka, T. et al. (2018). Glacier mass budget and climate reanalysis data indicate a climatic shift around 2000 in Lahaul-Spiti, western Himalaya. *Climatic Change* 148: 219–233.

Neitsch, S.L., Arnold, J.G., Kiniry, J.R., and Williams, J.R. (2011). *Soil and Water Assessment Tool Theoretical Documentation Version 2009*. College Station, TX: Texas Water Resources Institute.

Panwar, S. (2020). Vulnerability of Himalayan springs to climate change and anthropogenic impact: a review. *Journal of Mountain Science* 17 (1): 117–132.

Pratap, B., Sharma, P., Patel, L. et al. (2019). Reconciling high glacier surface melting in summer with air temperature in the semi-arid zone of Western Himalaya. *Water* 11 (8): 1561.

Rahman, K.U., Pham, Q.B., Jadoon, K.Z. et al. (2022). Comparison of machine learning and process-based SWAT model in simulating streamflow in the Upper Indus Basin. *Applied Water Science* 12 (8): 178.

Shekar, P.R., Mathew, A., Pandey, A., and Bhosale, A. (2023). A comparison of the performance of SWAT and artificial intelligence models for monthly rainfall–runoff analysis in the Peddavagu River Basin, India. *AQUA – Water Infrastructure, Ecosystems and Society* 72 (9): 1707–1730.

Singh, Ajit T., Waliur Rahaman, Parmanand Sharma, C. M. Laluraj, Lavkush K. Patel, Bhanu Pratap, Vinay K. Gaddam, and Meloth Thamban. (2019). Moisture sources for precipitation and hydrograph components of the Sutri Dhaka Glacier Basin, Western Himalayas, *Water* 11(11), 2242. doi:https://doi.org/10.3390/w11112242

Su, Q., Dai, C., Zhang, Z. et al. (2023). Runoff simulation and climate change analysis in Hulan river basin based on SWAT model. *Water* 15 (15): 2845.

Tawde, S.A., Kulkarni, A.V., and Bala, G. (2019). An assessment of climate change impacts on glacier mass balance and geometry in the Chandra Basin, Western Himalaya for the 21st century. *Environmental Research Communications* 1 (4): 041003.

Thapa, S., Li, H., Li, B. et al. (2021). Impact of climate change on snowmelt runoff in a Himalayan basin, Nepal. *Environmental Monitoring and Assessment* 193 (7): 393.

Wang, S., Sun, M., Wang, G. et al. (2023a). Simulation and reconstruction of runoff in the high-cold mountains area based on multiple machine learning models. *Water* 15 (18): 3222.

Wang, X., Wang, X., Ma, B. et al. (2023b). High-performance reversible data hiding based on ridge regression prediction algorithm. *Signal Processing* 204: 108818.

Wang, Y., Jiang, R., Xie, J. et al. (2019). Soil and water assessment tool (SWAT) model: a systemic review. *Journal of Coastal Research* 93 (SI): 22–30.

Wortmann, M., Duethmann, D., Menz, C. et al. (2022). Projected climate change and its impacts on glaciers and water resources in the headwaters of the Tarim River, NW China/Kyrgyzstan. *Climatic Change* 171 (3): 30.

4

Assessment of Suspended Sediment Properties and Particle Load for Optimal Operation of Hydropower Plants in the Himalayan Region

Naman Arora, Arun Kumar, and Sunil K. Singal

Department of Hydro and Renewable Energy, Indian Institute of Technology Roorkee, Roorkee, Uttarakhand, India

4.1 Introduction

Mountains serve as the origin of the world's major rivers and are pivotal in geochemical circulation and the interaction between oceans and continents (Garrels et al. 1975; Whitehead et al. 2015). Rivers worldwide are estimated to transport ~20 billion tons of suspended sediment annually, with Asian rivers accounting for 80% of the total suspended sediment load (SSL) discharged into the oceans (Milliman and Syvitski 1992; Maharana et al. 2018). The Himalayas and the neighboring mountainous regions are the source of the majority of Asian rivers. The Indus, Ganges, and Brahmaputra are significant rivers for the Indian region and are influential for approximately half of the country's population (Amir Khan et al. 2018). These rivers contribute to high sediment and form the two largest submarine fans in the world due to severe erosion (Wulf et al. 2012). It is estimated that these rivers transport approximately 1.8×10^9 tons of suspended sediments, marking one of the highest sediment loads worldwide (Milliman and Syvitski 1992; Joshi et al. 2023).

There are several factors that influence high erosion in Himalayan rivers. These factors are high rainfall during the monsoon, subglacial and supraglacial activities, avalanching, and natural weathering (Hasnain and Chauhan 1993). The rivers originating from the Himalayas are fed by glaciers, which carry sediments both as suspended particles and bed loads. Weathering activities within these glacial rivers contribute more material for stream transport compared to non-glacial rivers, owing to the glaciers' active role as erosion agents (Gardner 1986). Due to the vital role glaciers play as primary geomorphic forces eroding landscapes and being key contributors of suspended sediment, the influx of suspended sediment is dependent upon the volume of meltwater discharged from the glacier (Hasnain and Chauhan 1993). Monsoonal precipitation in glaciated areas intensifies water flow, leading to accelerated runoff and enhanced erosion (Bisht et al. 2020).

The Indian Himalayan glaciers are partly covered with sediment and debris, transported to their surfaces by erosion, snow avalanches, rock falls, and debris flows (Singh et al. 2016a, 2017). As ice melts, the loose material moves readily, adding to the SSL in meltwater runoff.

Water Resources Management in Mountain Regions, First Edition. Edited by Virendra Bahadur Singh, Sughosh Madhav, Rakesh Kumar Gupta, Prerna Diwan, and Amit Kumar.

Consequently, the SSL directly correlates with the volume of meltwater discharged in any glaciated catchment (Hasnain and Thayyen 1999; Kumar et al. 2002; Haritashya et al. 2006; Kociuba and Janicki 2018; Bisht et al. 2020). The SSL from the Dokriani, Gangotri, Dunagiri, Patsio, Chhota Shigri, and Chorabari glaciers were estimated as 369, 16095, 47, 2.8, 161, and 513 tons/day, respectively (Singh et al. 2003; Haritashya et al. 2006; Srivastava et al. 2014; Singh et al. 2015; Kumar et al. 2016; Singh et al. 2016b; Kumar et al. 2018). In glacier-fed rivers, the sediment delivery is highest during the monsoon season. It is reported that the suspended sediment concentration (SSC) increases from June onward, reaching a maximum in August and subsequently start decreasing (Singh et al. 2016b; Bisht et al. 2020).

The delivery of SSC by glacier-fed rivers poses substantial implications on their material fluxes, morphology, water quality, geochemical cycling, and hydropower generation (Joshi et al. 2016). The abundance of suspended sediment in river water is a prevalent factor contributing to deteriorating water quality worldwide, making it unsuitable for direct domestic and industrial usage (Collins 1999; Grove et al. 2015; Panagopoulos and Haralambous 2020). There is a direct relationship between the concentration of heavy metals in river water and suspended sediment levels, with elevated concentrations typically occurring during monsoon events (Balogh et al. 1997; Gogoi et al. 2020). Suspended sediment holds considerable control over geomorphological and biological processes in rivers. Further, suspended sediment is harmful to the operation of hydropower plants (HPPs), as it causes sediment erosion in hydraulic machinery, thereby impacting power generation. Due to such impacts on hydropower generation and water-related processes, precise measurements of suspended sediment transports in Himalayan rivers are important (Bisht et al. 2020; Arora et al. 2022b; Gupta et al. 2023).

4.2 Hydropower in India

In India, the total installed capacity of power is 425 GW (October 2023), of which thermal power shares 56.18%, followed by renewable energy sources (42.06%) and nuclear (1.76%), as given in Table 4.1 (CEA 2023). For the assessment of hydropower potential, the Central Water Commission, Government of India, has categorized the Indian river network into six primary river systems: Indus, Brahmaputra, Ganga, Central Indian River system, East flowing rivers, and West flowing rivers. These river systems are subsequently subdivided into 49 basins. The total hydropower potential in the country is 163 GW, of which only 35% has been harnessed till now (CEA 2023). The high potential of hydropower in the country remains untapped in the Himalayan region, where hydropower stations are affected by sediment. The Himalayas have young and weak geology, releasing high sediment concentration in the river during the monsoon season. Sediment poses a great challenge for the development and operation of HPPs.

4.3 Challenges of Hydropower Development in the Himalayas

The Himalayan rivers have significant hydropower potential due to their steep gradient. These rivers contain suspended sediment, mainly comprised of sand (60–1000 μm), silt (2–60 μm), and clay (2 μm) size particles, regulated by energy conditions and availability

Table 4.1 Source-wise installed capacity of electricity in India as of October 2023 (Central Electricity Authority 2023).

S. No.	Source	Installed capacity (MW)	Share (%)
1	Coal	206 826	48.60
2	Lignite	6620	1.56
3	Gas	25 038	5.88
4	Diesel	589	0.14
5	**Thermal total (1–4)**	**239 073**	**56.18**
6	Hydro	46 850	11.01
7	Small hydro	4987	1.17
8	Wind	44 292	10.41
9	Bio-power	10 835	2.55
10	Solar	72 018	16.92
11	**Renewable energy (6 to 10)**	**178 983**	**42.06**
12	**Nuclear**	**7480**	**1.76**
13	**Grand total (5 + 11 + 12)**	**425 536**	**100.00**

(Singh et al. 2020). These rivers pose high temporal variation in SSC (Arora et al. 2021; Arora et al. 2022a). Under extreme circumstances, the SSC may reach up to ~50 000 mg/l (Thapa et al. 2015). The suspended sediment properties considerably affect various processes. In HPPs, coarser particles lead to more significant reservoir sedimentation, sediment erosion of underwater parts, and clogging of the cooling water system (Sangal et al. 2018; John et al. 2021). Sediments with a high proportion of hard minerals and angular shapes accelerate erosion (Padhy and Saini 2009; Khurana et al. 2014, 2016; Rai et al. 2017, 2021). With such characteristics, the Himalayan rivers pose considerable challenges regarding sediment management and the optimal operation of HPPs (Pradhan 2004).

There is a lot of hydropower potential in Indian Himalayan rivers, namely Alaknanda, Bhagirathi, Sutlej, Ravi, Beas, Chenab, and Jhelum. Alaknanda and Sutlej river basins have total hydropower potentials of 4081 MW and 24 566 MW, respectively. Out of the total hydropower potential, significant potential remains untapped in both river basins (HPDOE 2023; UJVNL 2023). Weak geology, natural calamities, high sediment events, and environmental and social concerns are some factors affecting hydropower development in both basins. The present chapter reports the assessment of suspended sediment properties and determination of maximum particle load for the Alaknanda and Sutlej rivers.

4.4 Impact of Sediment on HPP

Sediment in flowing water is considered undesirable, especially if the concentration is high and the mineral composition is highly abrasive for the HPPs (Rai et al. 2021; Arora et al. 2022b). Sediment has the potential to affect hydropower generation by reducing reservoir capacity and/or causing erosion to the underwater parts of generating equipment (Gummer 2018). Sedimentation of reservoirs presents a global challenge and has long posed difficulties for dam operators. As sediment-laden flows enter the reservoir at reduced velocities, sedimentation occurs, leading to a decrease in water storage capacity (Wang et al. 2005; Schleiss et al. 2016). Worldwide, it is estimated that reservoir storage capacity decreases by 0.5–1.0% annually as a result of sedimentation (White 2001). The decrease in the capacity of reservoirs could disrupt the water supply, which cannot be able to meet the demand for water storage needed for power, irrigation, water supply, and other uses, and discourages the setting up of new reservoirs (Richter and Zenz 2020; Ren et al. 2021; Richter et al. 2021).

Sediment can also impact power generation by causing erosion to the underwater parts of generating equipment, particularly during the high-flow season (Pradhan 2004; Felix et al. 2016a; Rai et al. 2019a). Sediment erosion is a function of suspended sediment properties, operating parameters, and turbine material properties (Truscott 1972; Poudel et al. 2012; IEC 62364 2019). There are several impacts of erosion, including a reduction in the efficiency of the turbine, cavitation, pressure fluctuations, and temporary shutdown of HPPs (Neopane et al. 2011; Thapa et al. 2012a; Winkler 2014). Sometimes, there is a possibility that a combined effect of erosion and cavitation may be observed in the hydraulic turbines. This effect is referred to as a synergy of erosion and cavitation (Thapa et al. 2007). Sediment erosion issues are common in HPPs located in the Himalayas, the European Alps, the Andes, and the Pacific Coast mountain ranges (Thapa et al. 2012b, 2015).

4.5 Alaknanda River Basin

The Alaknanda river originates from two glaciers, Satopanth and Bhagirathi Kharak, 13 km upstream of Badrinath, a famous Hindu shrine visited by a large number of devotees and tourists. The Saraswati river is the first tributary to meet the Alaknanda from the left, located 3 km upstream of Badrinath, in the Chamoli district of Uttarakhand state, India. The Alaknanda river then joins the Dhauliganga tributary river from the left at the Vishnuprayag confluence. The river joins with the Karamnasa Nadi, from the left, at Helang town and the Birahiganga from the left. The river thereafter flows south and joins the Nandakini tributary and Pindar tributary at Nandprayag and Karanprayag, respectively. The river then flows west and joins with the Mandakini from the right at Rudraprayag town. The river then flows southwest past Srinagar and meets the Bhagirathi at Devprayag to form the Ganga river. The total length of the river from the source up to Devprayag is 210 km. The drainage map of the study area is shown in Figure 4.1.

In the ARB, the annual rainfall varies between 1000 and 1600 mm, with 75% occurring between June and September during the monsoon season (Dasgupta 1975; Hasnain and Chauhan 1993; Chakrapani et al. 1995; Chakrapani et al. 2009; Chakrapani and Saini 2009). There are several types of rocks present in the ARB. Central crystalline, consisting of migmatized and granitic metasediments, is a dominant rock in the upper reaches. Following the main central thrust, there is a Berinag formation, consisting of a quarzitic sequence and

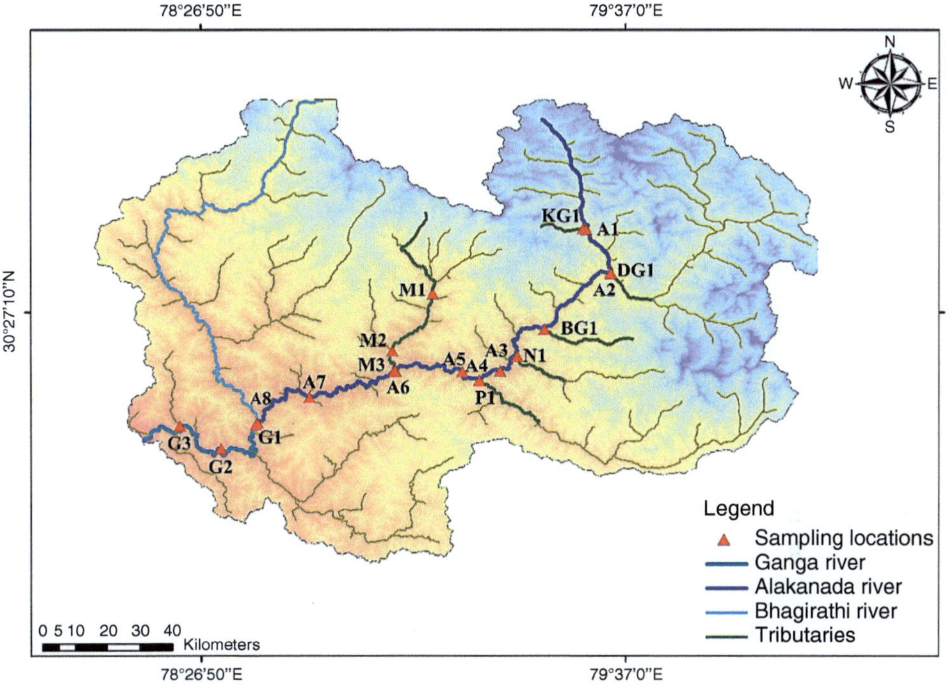

Figure 4.1 Sampling sites in ARB. A1–A8 represent sampling sites of the Alaknanda river from Pandukeshwar to Devprayag. G1–G3 represent sampling sites of the Ganga river from Devprayag to Shivpuri. KG1, DG1, BG1, N1, P1, and M1–M3 represent sampling sites in Khairo Ganga, Dhauliganga, Birahi Ganga, Nandakini, Pindar, and Mandakini tributaries, respectively. *Source:* Arora et al. (2024a)/ American society of civil engineers.

bands of crystalline limestone. There is a presence of Chandpur formation comprising micaceous phyllites and greywacke upstream of the confluence with the Bhagirathi river. There is a dominance of limestone, slates, quartzites, and shales in the tributaries of ARB. (Sarin et al. 1992; Hasnain and Chauhan 1993).

4.6 Sutlej River Basin

The Sutlej river originates from Rakashtal Lake in Tibet and serves as the largest tributary of the Indus River. The total catchment area of the river is 47 480 km^2, of which 36 448 km^2 lies within Tibet and the remainder lies in India (Jahan et al. 2023). The upper catchment area largely encompasses the entirety of the Kinnaur district, with an expansive catchment area measuring 5157 km^2 (Sharma et al. 1991). This region is predominantly defined by the towering presence of the high Himalayan range, stretching in a northwest-to-southeast direction, constituting Kinnaur's principal geographical feature (Gupta and Sah 2008; Sharma and Dhanze 2013). The elevation of the riverbed gradually decreases from 3050 meters as it traverses Tibet to approximately 1200 m upon exiting Kinnaur,

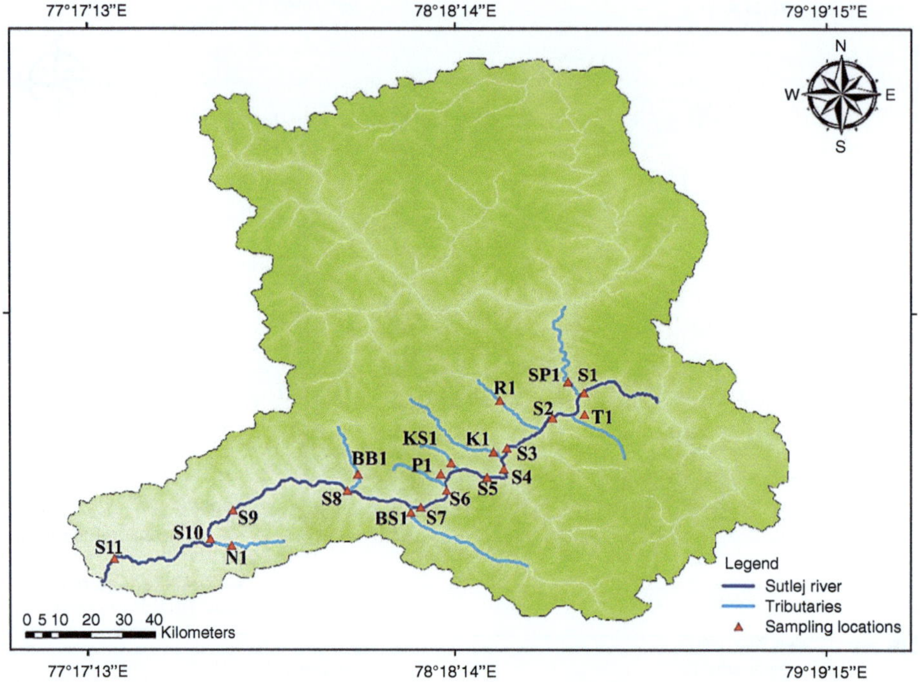

Figure 4.2 Sampling sites in SRB. S1–S11 represent sampling sites of the Sutlej river from Khab to Behna. SP1, T1, R1, K1, KS1, P1, BS1, BB1, and N1 represent sampling sites in Spiti, Titang, Ropa, Kiran, Kashang, Pangi, Baspa, Bhaba, Nogli tributaries, respectively. *Source:* Arora et al. (2024b)/ Springer Nature.

highlighting the overall high altitude of the Kinnaur basin. The drainage map of the study area is shown in Figure 4.2.

The study area is situated within the "rain shadow" zone, characterized by limited precipitation. It has been divided into two primary climatic zones based on annual precipitation levels and temperature fluctuations. These zones are subhumid temperate and semiarid to arid temperate. (Gupta et al. 1993). River runoff is primarily influenced by snow and glacier melting from the elevated, orographically protected Himalayan Crest, and it matches the magnitude of runoff from the Himalayan Front, which is predominantly influenced by monsoonal rainfall (Singh and Kumar 1997; Bookhagen and Burbank 2010). Snowfall predominantly takes place in the region from December to March. Significant seasonal rainfall typically happens between July and September, during which extreme water and sediment events are frequent (Wulf et al. 2010).

The rocks prevalent in the study area are predominantly weak, and extremely prone to mass movements. Kinnaur exhibits exposure to phyllites, mica schists, gneisses, granite, and other similar metamorphic rocks. The Spiti valley showcases a comprehensive series of Mesozoic and fossiliferous Paleozoic strata. This sequence aligns with the primary northwest-southeast trend of the Himalayas (Sharma et al. 1991).

4.7 Suspended Sediment Properties

There are four properties of suspended sediment responsible for sediment erosion in hydraulic turbines. These properties are SSC, particle size distribution (PSD), shape, and mineral composition.

4.7.1 SSC

SSC is defined as the ratio of the dry weight of sediment in the sediment-laden water to the total weight of sediment-laden water. It is expressed in grams per liter or parts per million (by weight). During low-flow seasons, the SSC in the rivers is generally low, and water is almost free from sediment, whereas, during snowmelt and precipitation, the SSC becomes high, and water becomes turbid. Studies have shown that SSC poses high temporal variation during precipitation, particularly during the monsoon season (Boes 2009; Rai et al. 2016; Rai and Kumar 2017).

Various factors, including river slope, width, vegetation, and meandering bends, influence the SSC in the river. The major influencing factor in the studied river basin is the change in river gradient. In the ARB, the SSC follows a decreasing pattern from the upstream (u/s) site (A5), where the river gradient is 1 in 200, to the downstream (d/s) site (A7), where the river gradient is 1 in 500. At A5, the SSC is 439 mg/l, whereas, at A7, it is 280 mg/l during the lean season in November (Arora et al. 2024a). The pattern of SSC is quite different for the monsoon season. In a similar study, Chakrapani and Saini (2009) reported an increasing pattern of SSC from u/s to d/s sites. At A1, the SSC is 571 mg/l, whereas, at A7, it is 2356 mg/l during the monsoon season in July 2004. The landslides and bank erosions during the monsoon season are the main factors that increase the SSC in the downstream reaches. In the SRB, the SSC follows a decreasing pattern from the u/s site (S1), where the river gradient is 1 in 30, to the d/s site (S11), where the river gradient is 1 in 200. At S1, the SSC is 1749 mg/l, whereas, at S11, it is 782 mg/l during the monsoon season in July 2023. In the upstream regions, the SRB is distinguished by narrow gorges and high SSC. As the river progresses downstream, it traverses numerous bends where observable sediment deposition occurs along the inner edges, resulting in diminished sediment transport downstream. A similar study by Pandey et al. (1999) reported that in the Bhagirathi river basin, there was a slight increase in SSC in the upper reaches compared to the lower reaches. This variation can be attributed to a change in river slope, transitioning from 1 in 33 to 1 in 100.

4.7.2 PSD

The size of a spherical particle is determined by its diameter, whereas the size of a non-spherical particle is determined by the volume-equivalent sphere diameter, which is a straightforward and comparable parameter (Felix et al. 2016b). In sediment-laden water, numerous particles exist with varying sizes and specific gravity. Therefore, PSD is referred to as the bulk property of suspended sediment. It is measured by d_x, representing the particle diameter equaled or exceeded by x% of the particles in the sample arranged in ascending order of size.

The PSD is influenced by a variety of factors, including geological composition, degree of weathering, climatic conditions, the physical characteristics of the basin, and the river's energy (Xu 2000; Zhang et al. 2006). The Folk and Ward (1957) graphical method is widely used to describe the PSD in sedimentary environments. This method involves the estimation of several statistical terms, namely mean, sorting, skewness, and kurtosis, to describe the variation of PSD in the river. Clay, very fine sand particles, and fine and coarse silt particles are mostly present in the suspended sediment of ARB and SRB, respectively. In the ARB, the mean size varies from 4.60 to 6.17 Φ, whereas, in the SRB, it varies from 5.53 to 7.53 Φ. This indicates that both these river basins carry mostly fine particles. The presence of finer particle sizes can be attributed to the river's incapacity to transport larger particles. This phenomenon may be influenced by extensive hydropower development along the Alaknanda and Sutlej rivers and their tributaries, leading to reduced coarse sediment through accumulation owing to the diminished velocity of flow (Panwar et al. 2016).

Sorting measures the degree to which sediment particles are distributed based on their size. The skewness signifies the size distribution characteristics and measures the degree to which size distribution symmetry. Kurtosis measures the concentration of particles across different sizes in relation to the average particle size. In the ARB, the sediments are characterized by moderate sorting, coarse skewness, and leptokurtic distribution, whereas, in the SRB, the sediments are characterized by very poor sorting, coarse to nearly symmetrical skewness, and platykurtic to leptokurtic distribution.

4.7.3 Shape

The shape of the sediment can be quantified by measuring dimensions such as length, width, diameter, area, volume, and perimeter. The shape can also be classified as one-dimensional, two-dimensional, and three-dimensional. One-dimensional shapes measured through particle lengths using calipers, rulers, and images are referred to as form factors (aspect ratio). Two-dimensional shapes measured through projections using image analysis are called particle projections (aspect ratio). Three-dimensional shapes measured from all three dimensions using image analysis are called sphericity. Recent studies quantified the sediment shape as sphericity and aspect ratio.

The shape of sediment depends upon several factors, including river slope, flow velocity, lithological composition, vegetative cover, and anthropogenic interventions. Sediment shape provides insights into the characteristics of the flow medium, the distance over which sediment travels, and the intensity of its movement (Li et al. 2016). In both the ARB and SRB, the sediments are characterized by high sphericity and slight elongation.

4.7.4 Mineralogy

Erosion and weathering of rocks are the main factors influencing sediment mineralogy in the river. The transport of these minerals in rivers is through both solid and dissolved states (Berner and Kothavala 2001). Research pertaining to sediment mineral characteristics of rivers has become important considering the design of hydraulic turbines concerning sediment erosion. The selection of erosion-resistant materials and hard coatings depends on the prevailing sediment minerals in the river catchment. Studies are being undertaken to determine the impact of sediment minerals on different turbine materials and to identify the most suitable turbine material and coatings to mitigate sediment erosion.

In the ARB, the minerals present in sediments are quartz (69–80%), feldspar (1–10%), biotite (1–6%), muscovite (2–11%), garnet (1–5%), laumontite (3–25%), silicates (1–11%), oxides (1–13%) and clays (1–6%). In the SRB, the minerals present in sediments are quartz (67–82%), feldspar (1–9%), muscovite (2–11%), garnet (1–5%), dolomite (1–6%), chamosite (1–5%), diopside (1–9%), illite (1–20%), and silicates (1–11%). In both river basins, quartz and feldspar are abundant minerals, which indicates physical and chemical weathering of rocks and immature mineral composition. The feldspar content in the Bhagirathi and Brahmaputra river basins varies from 1 to 7% and 5 to 8%, respectively (Rahman et al. 2016; Arora et al. 2023). In the river basins, the presence of muscovite indicates the significant alteration of feldspar, whereas the presence of biotite and garnet indicates the weathering of gneiss-type metamorphic rocks (Nesse 2013). The assemblage of clay minerals in river sediments suggests the intensity of weathering and the source bedrock (Panwar et al. 2017).

The presence of laumontite mineral in ARB indicates the diagenesis process in the sedimentary rocks. Low temperatures in the higher reaches of the ARB give rise to diagenetic processes. In the SRB, there exist chamosite and dolomite minerals. The presence of chamosite in the sediments is due to the hydrothermal alteration of igneous rocks, and dolomite is due to the dolomitization of limestock rocks (Mehmood 2018). Diopside and several silicate minerals may be formed due to glacier melting, weathering, metamorphism, and diagenesis (Bonewitz 2012).

4.7.5 International Standard IEC 62364 (2019)

Sediment erosion in hydraulic turbine parts is influenced by the operating parameters, sediment parameters, and turbine material properties. It is evident that sediment particles with angular shapes and harder minerals cause a higher impact on the turbine part. The International Standard IEC 62364 (2019) from the International Electrotechnical Commission prescribed the guidelines for dealing with sediment erosion in the Pelton, Francis, and Kaplan hydraulic turbines. The standard is widely used in estimating the erosion depth in the underwater components of hydraulic turbines. Under IEC 62364 (2019), a parameter "Particle Load" has been introduced, which is a combination of SSC, PSD, shape, and mineralogy. The PL for a particular site can be determined by

$$PL\left(\mathrm{kg}\cdot\mathrm{h/m^3}\right)=\sum_{i=1}^{n}C_i\cdot K_{\mathrm{size},i}\cdot K_{\mathrm{shape},i}\cdot K_{\mathrm{hardness},i}\cdot t_i \tag{4.1}$$

where

PL represents the particle load derived as the sum of the products of sediment parameters over t time period in h,

C represents the mass concentration of suspended sediment in kg/m^3,

K_{size} represents the sediment particle size coefficient taken as d_{50} in mm,

K_{shape} represents the shape factor varying from 1 to 2, depending on the particle angularity,

$K_{hardness}$ represents the hardness factor considering particle hardness against turbine material hardness taken as the particle fraction harder than the Mohs hardness of 4.5,

i represents the typical time step out of the total *n* number of time steps,

t_i represents the time step duration in h.

For ARB and SRB, the K_{shape} and $K_{hardness}$ and maximum PL for the studied sites are given in Tables 4.2 and 4.3, respectively.

Sediment erosion in hydraulic turbines depends on PL_{max}. The average PL_{max} for ARB and SRB are 0.022 and 0.014 kg/m^3, respectively, which are recommended as the permissible limits for sediment erosion. PL_{max} values exhibit significant variation in both the ARB and SRB, with lower values typically found in the upper reaches of the basin and relatively higher values in the lower reaches. The proposed HPPs in both river basins are situated at both the upper and lower reaches. For instance, Alaknanda (300 MW), Tapovan Vishnugad (520 MW), and Vishnugad Pipalkoti (444 MW) are some of the major HPPs proposed in the ARB. Similarly, in the SRB, some of the proposed HPPs are Khab (636 MW), Shongtong Karcham (450 MW), and Jangi Thopan Powari (804 MW). The sites with PL_{max} values greater than the average are susceptible to sediment erosion.

In HPPs susceptible to significant erosion, it is crucial to take extreme caution to minimize the impact of sediment on hydraulic turbines. Some practices that may be adopted are as follows:

- Selecting innovative techniques for sediment removal or enhancing the capacity of existing desilting structures to eliminate specific particle sizes efficiently. Studies demonstrated that during the peak monsoon and sediment events, the efficiency of sediment

Table 4.2 IEC 62364 based sediment factors for ARB.

Site	K_{shape}	$K_{hardness}$	PL_{max}
KG1	1.42	0.70	0.003
A1	1.42	0.88	0.019
A2	1.47	0.78	0.022
DG1	1.39	0.83	0.025
BG1	1.40	0.84	0.004
N1	1.17	0.85	0.003
A3	1.44	0.88	0.019
A4	1.42	0.83	0.036
P1	1.43	0.79	0.020
A5	1.41	0.81	0.026
A6	1.45	0.84	0.035
M1	1.16	0.87	0.021
M2	1.17	0.89	0.020
A7	1.46	0.88	0.046
G1	1.41	0.84	0.021
G2	1.45	0.87	0.022
G3	1.42	0.86	0.021

Table 4.3 IEC 62364 based sediment factors for SRB.

Site	K_{shape}	$K_{hardness}$	PL_{max}
S1	1.38	0.84	0.019
SP1	1.35	0.89	0.012
T1	1.28	0.74	0.002
S2	1.31	0.86	0.014
R1	1.37	0.83	0.004
S3	1.40	0.81	0.009
K1	1.36	0.88	0.021
S4	1.39	0.85	0.014
S5	1.39	0.91	0.015
KS1	1.40	0.90	0.019
P1	1.32	0.86	0.005
S6	1.34	0.85	0.019
S7	1.39	0.80	0.010
BS1	1.35	0.79	0.004
BB1	1.40	0.85	0.004
S8	1.38	0.86	0.035
S9	1.35	0.87	0.034
S10	1.33	0.89	0.022
N1	1.36	0.82	0.004
S11	1.33	0.87	0.018

removal structures decreases, and high sediment passes through the turbine parts. The advancement in sediment removal structures is a need of hour.

- Measuring the effectiveness of sediment removal structures to efficiently eliminate sediment particles, especially during the monsoon season. This can be achieved by measuring the sediment-laden water on the upstream and downstream of the sediment removal structure.
- Continuous sediment measurement at several sites of HPPs to quantify the sediment load exposed to hydraulic turbines. Recently developed real-time sediment monitoring techniques are one of the key solutions in high erosion prone HPPs.
- Selecting turbine materials resistant to erosion and applying hard coatings can minimize the effects of sediment erosion. Utilizing hard coatings is a cost-efficient strategy for reducing erosion over time (Rai et al. 2019b). Currently, the most suitable turbine material is 13Cr4Ni with a WC-Co-Cr coating applied through the high-velocity oxyfuel (HVOF) process.
- Establishing maintenance and coating facilities tailored to specific sites or regions could mitigate downtime and the expensive logistics of transporting turbines to factories for maintenance. The turbine maintenance workshops are normally not available in remote

areas where the problem of sediment erosion exists. Capacity building of regional turbine maintenance workshops by turbine manufacturers can serve as a cost-effective solution in remote areas.

- Turning off turbines during peak sediment periods, such as when sediment levels at the headworks exceed allowable values, has been shown to be a cost-effective measure for reducing erosion during these events. Research indicates that utilizing real-time sediment monitoring devices in these situations can be advantageous, providing an alert system.

4.8 Conclusion

Significant untapped hydropower potential is available in the Indian Himalayan river. In the present study, the assessment of suspended sediment properties and IEC 62364 based maximum particle load for ARB and SRB has been reported in the present chapter. In both ARB and SRB, the SSC has a wide variation, mainly attributed to changes in river gradient, river width, and bends. In the ARB, the sediments are characterized by moderate sorting, coarse skewness, and leptokurtic distribution, whereas, in the SRB, the sediments are characterized by very poor sorting, coarse to nearly symmetrical skewness, and platykurtic to leptokurtic distribution. Geological composition, degree of weathering, climatic conditions, the physical characteristics of the basin, and the river's energy influence the PSD.

The sediments exhibited a high degree of sphericity and slight elongation for both ARB and SRB. A wide variety of minerals in the river basins indicates a high degree of sedimentary and metamorphic rocks, as well as the processes of diagenesis, metamorphism, and glacier melting. The average PL_{max} for ARB and SRB are 0.022 and 0.014 kg/m^3, respectively, which are recommended as the permissible limits for sediment erosion. The PL_{max} varies significantly, with lower values typically found in the upper reaches of the basin and relatively higher values in the lower reaches. The PL_{max} values indicate the sites with high erosion potential where extreme caution must be taken, including cost-effective design of new sediment removal structures, efficiency measurement of sediment removal structures, real-time sediment monitoring at several sites of HPP, application of hard coatings to underwater turbine parts, and switching off turbines during periods of peak sediment.

References

Amir Khan, A., Pant, N.C., Ravindra, R. et al. (2018). A precipitation perspective of the hydrosphere-cryosphere interaction in the Himalaya. *Geological Society, London, Special Publications* 462 (1): 73–87. https://www.lyellcollection.org/doi/10.1144/SP462.2 (accessed 3 October 2024).

Arora, N., Kumar, A., and Rai, A.K. (2021). Online sediment measurement strategy for hydro power plants. In: *6th IAHR Europe Congress 2020*, 137–138. Warsaw: IAHR.

Arora, N., Kumar, A., and Singal, S.K. (2022a). Measurement of sediment particle size distribution and shape using dynamic imaging analysis. *IOP Conference Series: Earth and Environmental Science* 1079 (1): 012063. https://doi.org/10.1088/1755-1315/1079/1/012063.

Arora, N., Kumar, A., and Singal, S.K. (2022b). Technological advancement in measurements of suspended sediment and hydraulic turbine erosion. *Measurement: Journal of the International Measurement Confederation* 190 (January): 110700. https://doi.org/10.1016/j.measurement.2022.110700.

Arora, N., Kumar, A., and Singal, S.K. (2023). Matrix flushing approach-based sediment mineralogical composition of Bhagirathi river basin in Indian Himalayas. *Environment, Development and Sustainability* https://link.springer.com/10.1007/s10668-023-04005-2.

Arora, N., Kumar, A., and Singal, S.K. (2024a). Spatial variation in hydrosedimentary characteristics of the Alaknanda River Basin in the Indian Himalayas: a field study. *Journal of Irrigation and Drainage Engineering* 150 (3): 06024001. https://ascelibrary.org/doi/10.1061/JIDEDH.IRENG-10237 (accessed 3 October 2024).

Arora, N., Kumar, A. and Singal, S.K. 2024b. Spatial variation of suspended sediment properties in the Sutlej river basin of Indian Himalayas. *Acta Geophysica* doi:https://doi.org/10.1007/s11600-024-01322-6.

Balogh, S.J., Meyer, M.L., and Johnson, D.K. (1997). Mercury and suspended sediment loadings in the lower Minnesota river. *Environmental Science & Technology* 31 (1): 198–202. https://pubs.acs.org/doi/10.1021/es960327t.

Berner, R.A. and Kothavala, Z. (2001). Geocarb III: a revised model of atmospheric CO_2 over phanerozoic time. *American Journal of Science* 301 (2): 182–204. https://doi.org/10.2475/ajs.301.2.182.

Bisht, H., Kotlia, B.S., Kumar, K. et al. (2020). Estimation of suspended sediment concentration and meltwater discharge draining from the Chaturangi glacier, Garhwal Himalaya. *Arabian Journal of Geosciences* 13 (6): 248. https://doi.org/10.1007/s12517-020-5204-4.

Boes, R.M. (2009). Real-time monitoring of suspended sediment concentration and particle size distribution in the headwater way of a high-head hydropower plant. In: *Proceedings of 3rd IAHR Congress* (ed. D. Ferras, A. Varrani, E. Quaranta, et al.), 4037–4044. Beijing, China: IAHR.

Bonewitz, R. (2012). *Rocks and Minerals*, 2e. London: DK Publishing.

Bookhagen, B. and Burbank, D.W. (2010). Toward a complete Himalayan hydrological budget: spatiotemporal distribution of snowmelt and rainfall and their impact on river discharge. *Journal of Geophysical Research* 115 (F3): F03019. http://doi.wiley.com/10.1029/2009JF001426.

Central Electricity Authority (2023). All india installed capacity of power stations installed capacity (in MW). https://cea.nic.in/installed-capacity-report/?lang=en.

Chakrapani, G.J. and Saini, R.K. (2009). Temporal and spatial variations in water discharge and sediment load in the Alaknanda and Bhagirathi Rivers in Himalaya, India. *Journal of Asian Earth Sciences* 35 (6): 545–553. https://doi.org/10.1016/j.jseaes.2009.04.002.

Chakrapani, G.J., Saini, R.K., and Yadav, S.K. (2009). Chemical weathering rates in the Alaknanda-Bhagirathi river basins in Himalayas, India. *Journal of Asian Earth Sciences* 34 (3): 347–362. https://doi.org/10.1016/j.jseaes.2008.06.002.

Chakrapani, G.J., Subramanian, V., Gibbs, R.J., and Jha, P.K. (1995). Size characteristics and mineralogy of suspended sediments of the Ganges river. *India. Environmental Geology* 25 (3): 192–196. https://doi.org/10.1007/BF00768548.

Collins, D.N. (1999). Solute flux in meltwaters draining from a glacierized basin in the Karakoram mountains. *Hydrological Processes* 13 (18): 3001–3015. https://doi.org/10.1002/(SICI)1099-1085(19991230)13:18%3C3001::AID-HYP15%3E3.0.CO;2-N.

Dasgupta, S.P. (1975). *The Flood Plain of the Ganga*. Calcutta: National Atlas Organisation, Department of Science and Technology.

Felix, D., Albayrak, I., Abgottspon, A., and Boes, R.M. (2016a). Hydro-abrasive erosion of hydraulic turbines caused by sediment – a century of research and development. *IOP Conference Series: Earth and Environmental Science* 49 (12): 122001. https://doi.org/10.1088/1755-1315/49/12/122001.

Felix, D., Albayrak, I., and Boes, R.M. (2016b). Continuous measurement of suspended sediment concentration: discussion of four techniques. *Measurement: Journal of the International Measurement Confederation* 89: 44–47. https://doi.org/10.1016/j.measurement.2016.03.066.

Gardner, J.S. (1986). Recent fluctuations of Rakhiot Glacier, Nanga Parbat, Punjab Himalaya, Pakistan. *Journal of Glaciology* 32 (112): 527–529. https://doi.org/10.1017/s0022143000012247.

Garrels, R.M., Mackenzie, F.T., and Hunt, C. (1975). *Chemical Cycle and the Global Environment*. New York: William Kaufman.

Gogoi, A., Taki, K. and Kumar, M. 2020. Seasonal dynamics of metal phase distributions in the perennial tropical (Brahmaputra) river: environmental fate and transport perspective. *Environmental Research* 183, 109265. https://linkinghub.elsevier.com/retrieve/pii/S0013935120301572 (accessed 3 October 2024).

Grove, M.K., Bilotta, G.S., Woockman, R.R. and Schwartz, J.S. 2015. Suspended sediment regimes in contrasting reference-condition freshwater ecosystems: Implications for water quality guidelines and management. *Science of The Total Environment* 502, 481–492. https://linkinghub.elsevier.com/retrieve/pii/S004896971401376X (accessed 3 October 2024).

Gummer, B.J.H. (2018). Combating silt erosion in hydraulic turbines. *Hydropower* 1: 1–14.

Gupta, L.K., Pandey, M., Raj, P.A. and Shukla, A.K. 2023. Fine sediment intrusion and its consequences for river ecosystems: a review. *Journal of Hazardous, Toxic, and Radioactive Waste* 27(1). https://ascelibrary.org/doi/10.1061/%28ASCE%29HZ.2153-5515.0000729 (accessed 3 October 2024).

Gupta, V. and Sah, M.P. (2008). Spatial variability of mass movements in the Satluj Valley, Himachal Pradesh during 1990–2006. *Journal of Mountain Science* 5 (1): 38–51. https://doi.org/10.1007/s11629-008-0038-7.

Gupta, V., Sah, M.P., Virdi, N.S., and Bartarya, S.K. (1993). Landslide hazard zonation in the upper Satluj valley, district Kinnaur, Himachal Pradesh. *Journal of Himalayan Geology* 4 (1): 81–93.

Haritashya, U.K., Singh, P., Kumar, N., and Gupta, R.P. (2006). Suspended sediment from the Gangotri Glacier: quantification, variability and associations with discharge and air temperature. *Journal of Hydrology* 321 (1–4): 116–130. https://doi.org/10.1016/j.jhydrol.2005.07.037.

Hasnain, S.I. and Chauhan, D.S. (1993). Sediment transfer in the glaciofluvial environment - a Himalayan perspective. *Environmental Geology* 22 (3): 205–211. https://doi.org/10.1007/BF00767405.

Hasnain, S.I. and Thayyen, R.J. (1999). Controls on the major-ion chemistry of the Dokriani glacier meltwaters, Ganga basin, Garhwal Himalaya, India. *Journal of Glaciology* 45 (149): 87–92.

Himachal Pradesh Directorate of Energy HPDOE (2023). Status of hydro potential in Himachal Pradesh.

IEC 62364 (2019). *Hydraulic Machines – Guidelines for Dealing with Hydro-abrasive Erosion in Kaplan, Francis, and Pelton Turbines*, 2e. Geneva, Switzerland: IEC.

Jahan, A., Khan, M.U., Rai, N. et al. (2023). Geochemical characterization, its controlling factors, and comparison between the upstream and downstream segments of the Himalayan Satluj River basin, India. *Geochemistry* 83 (2): 125974. https://doi.org/10.1016/j.chemer.2023.125974.

John, C.K., Pu, J.H., Pandey, M., and Moruzzi, R. (2021). Impacts of sedimentation on rainwater quality: case study at Ikorodu of Lagos, Nigeria. *Water Supply* 21 (7): 3356–3369. https://doi.org/10.2166/ws.2021.093.

Joshi, R., Khan, A.A., Pant, N.C. et al. (2023). Geochemical characteristics and suspended sediments dynamics in the meltwater from the Gangotri Glacier, Garhwal Himalaya, India. *Environmental Earth Sciences* 82 (6): 1–16. https://doi.org/10.1007/s12665-023-10802-9.

Joshi, R., Kumar, K., and Adhikari, V.P.S. (2016). Modelling suspended sediment concentration using artificial neural networks for Gangotri glacier. *Hydrological Processes* 30 (9): 1354–1366. https://doi.org/10.1002/hyp.10723.

Khurana, S. and Varun and Kumar, A. (2014). Effect of silt particles on erosion of Turgo impulse turbine blades. *International Journal of Ambient Energy* 35 (3): 155–162. https://doi.org/10.1080/01430750.2013.789985.

Khurana, S., Varun, and Kumar, A. (2016). Silt erosion study on the performance of an impulse turbine in small hydropower. *International Journal of Ambient Energy* 37 (5): 520–527. https://doi.org/10.1080/01430750.2015.1023831.

Kociuba, W. and Janicki, G. (2018). Effect of meteorological patterns on the intensity of streambank erosion in a proglacial gravel-bed river (spitsbergen). *Water* 10 (3): 320. https://www.mdpi.com/2073-4441/10/3/320.

Kumar, A., Verma, A., Gokhale, A.A. et al. (2018). Hydrometeorological assessments and suspended sediment delivery from a central Himalayan glacier in the upper Ganga basin. *International Journal of Sediment Research* 33 (4): 493–509. https://doi.org/10.1016/j.ijsrc.2018.03.004.

Kumar, A., Gokhale, A.A., Shukla, T., and Dobhal, D.P. (2016). Hydroclimatic influence on particle size distribution of suspended sediments evacuated from debris-covered Chorabari Glacier, upper Mandakini catchment, central Himalaya. *Geomorphology* 265: 45–67. https://doi.org/10.1016/j.geomorph.2016.04.019.

Kumar, K., Miral, M.S., Joshi, V., and Panda, Y.S. (2002). Discharge and suspended sediment in the meltwater of Gangotri Glacier, Garhwal Himalaya India. *Hydrological Sciences Journal* 47 (4): 611–619. http://www.tandfonline.com/doi/abs/10.1080/02626660209492963.

Li, D., Li, Y., Wang, Z. et al. (2016). Quantitative, SEM-based shape analysis of sediment particles in the Yellow River. *International Journal of Sediment Research* 31 (4): 341–350. https://doi.org/10.1016/j.ijsrc.2016.05.006.

Maharana, C., Srivastava, D., and Tripathi, J.K. (2018). Geochemistry of sediments of the Peninsular rivers of the Ganga basin and its implication to weathering, sedimentary processes and provenance. *Chemical Geology* 483: 1–20. https://linkinghub.elsevier.com/retrieve/pii/S0009254118300846 (accessed 3 October 2024).

Mehmood, M. (2018). Dolomite and dolomitization model – a short review. *International Journal of Hydrology* 2 (5): 549–553. https://doi.org/10.15406/ijh.2018.02.00124.

Milliman, J.D. and Syvitski, J.P.M. (1992). Geomorphic/tectonic control of sediment discharge to the ocean: the importance of small mountainous rivers. *The Journal of Geology* 100 (5): 525–544. https://www.journals.uchicago.edu/doi/10.1086/629606 (accessed 3 October 2024).

Neopane, H.P., Gunnar Dahlhaug, O., and Cervantes, M. (2011). Sediment erosion in hydraulic turbines. *Global Journal of Researches in Engineering Mechanical and Mechanics Engineering* 11 (6): 17–26. https://pdfs.semanticscholar.org/f219/b5289d91874f8d13090063c34688ebebe978.pdf?_ga=2.265510923.564185937.1556240431-87428635.1550801047 (accessed 3 October 2024).

Nesse, W.D. (2013). *Introduction to Optical Mineralogy*, 4e. New York: Oxford University Press.

Padhy, M.K. and Saini, R.P. (2009). Effect of size and concentration of silt particles on erosion of Pelton turbine buckets. *Energy* 34 (10): 1477–1483. https://doi.org/10.1016/j.energy.2009.06.015.

Panagopoulos, A. and Haralambous, K.-J. (2020). Minimal Liquid Discharge (MLD) and Zero Liquid Discharge (ZLD) strategies for wastewater management and resource recovery – Analysis, challenges and prospects. *Journal of Environmental Chemical Engineering* 8 (5): 104418. https://linkinghub.elsevier.com/retrieve/pii/S2213343720307673.

Pandey, S.K., Singh, A.K., and Hasnain, S.I. (1999). Weathering and geochemical processes controlling solute acquisition in Ganga Headwater-Bhagirathi River, Garhwal Himalaya India. *Aquatic Geochemistry* 5 (4): 357–379. https://doi.org/10.1023/A:1009698016548.

Panwar, S., Agarwal, V., and Chakrapani, G.J. (2017). Morphometric and sediment source characterization of the Alaknanda river basin, headwaters of river Ganga India. *Natural Hazards* 87 (3): 1649–1671. https://doi.org/10.1007/s11069-017-2838-y.

Panwar, S., Khan, M.Y.A., and Chakrapani, G.J. (2016). Grain size characteristics and provenance determination of sediment and dissolved load of Alaknanda River, Garhwal Himalaya India. *Environmental Earth Sciences* 75 (2): 1–15. https://doi.org/10.1007/s12665-015-4785-9.

Poudel, L.P., Thapa, B., Shrestha, B.P., and Shrestha, N.K. (2012). Impact of sand on hydraulic turbine material: a case study of Roshi Khola, Nepal. *Hydro Nepal: Journal of Water Energy and Environment* 10 (Ii): 60–65. https://doi.org/10.3126/hn.v10i0.7116.

Pradhan, P.M.S. (2004). Improving sediment handling in the himalayas. *OSH Research* (October): 1–6. https://www.un.org/esa/sustdev/sdissues/energy/op/hydro_pratik_pradhan.pdf (accessed 3 October 2024).

Rahman, M.A., Pownceby, M.I., Haque, N. et al. (2016). Valuable heavy minerals from the Brahmaputra River sands of Northern Bangladesh. *Applied Earth Science* 125 (3): 174–188. https://www.tandfonline.com/doi/full/10.1080/03717453.2015.1115159 (accessed 3 October 2024).

Rai, A.K. and Kumar, A. (2017). Sediment monitoring for hydro-abrasive erosion: a field study from Himalayas, India. *International Journal of Fluid Machinery and Systems* 10 (2): 146–153. https://doi.org/10.5293/IJFMS.2017.10.2.146.

Rai, A.K., Kumar, A., Hies, T., and Nguyen, H.H. (2016). Field application of a multi-frequency acoustic instrument to monitor sediment for silt erosion study in Pelton turbine in Himalayan region, India. *IOP Conference Series: Earth and Environmental Science* 49 (12): 122004. https://doi.org/10.1088/1755-1315/49/12/122004.

Rai, A.K., Kumar, A., and Staubli, T. (2017). Hydro-abrasive erosion in Pelton buckets: classification and field study. *Wear* 392–393 (April): 8–20. https://doi.org/10.1016/j.wear.2017.08.016.

Rai, A.K., Kumar, A., and Staubli, T. (2019a). Analytical modelling and mechanism of hydro-abrasive erosion in pelton buckets. *Wear* 436–437 (March): 203003. https://doi.org/10.1016/j.wear.2019.203003.

Rai, A.K., Kumar, A., and Staubli, T. (2019b). Financial analysis for optimization of hydropower plants regarding hydro-abrasive erosion: a study from Indian Himalayas. *IOP Conference Series: Earth and Environmental Science* 240 (2): 022025. https://doi.org/10.1088/1755-1315/240/2/022025.

Rai, A.K., Kumar, A., Yexiang, X. et al. (2021). Analysis and visualization of progressive erosion in Pelton buckets. *IOP Conference Series: Earth and Environmental Science* 774 (1): 012105. https://doi.org/10.1088/1755-1315/774/1/012105.

Ren, S., Zhang, B., Wang, W.J. et al. (2021). Sedimentation and its response to management strategies of the Three Gorges Reservoir, Yangtze River, China. *Catena* 199: 105096. https://doi.org/10.1016/j.catena.2020.105096.

Richter, W., Vereide, K., Mauko, G. et al. (2021). Retrofitting of pressurized sand traps in hydropower plants. *Water* 13 (18): 2515. https://www.mdpi.com/2073-4441/13/18/2515 (accessed 3 October 2024).

Richter, W. and Zenz, G. (2020). *Calcite Sediment Flushing in Tunnel Drainage Systems: Physical Model Test and Hydraulic Investigations*. Austria: Graz.

Sangal, S., Singhal, M.K., and Saini, R.P. (2018). Hydro-abrasive erosion in hydro turbines: a review. *International Journal of Green Energy* 15 (4): 232–253. https://doi.org/10.1080/15435075.2018.1431546.

Sarin, M.M., Krishnaswami, S., Sharma, K.K., and Trivedi, J.R. (1992). Uranium isotopes and radium in the Bhagirathi alaknanda river system - evidence for high uranium mobilization in the Himalaya. *Current Science* 62 (12): 801–805.

Schleiss, A.J., Franca, M.J., Juez, C., and De Cesare, G. (2016). Reservoir sedimentation. *Journal of Hydraulic Research* 54 (6): 595–614. https://doi.org/10.1080/00221686.2016.1225320.

Sharma, I. and Dhanze, R. (2013). Ecology of River Spiti, Lahaul-Spiti (Himachal Pradesh), India. *International Journal of Fisheries and Aquaculture Sciences* 3 (2): 131–141. http://www.irphouse.com (accessed 3 October 2024).

Sharma, P.D., Goel, A.K., and Minhas, R.S. (1991). Water and sediment yields into the Sutlej river from the high Himalaya. *Mountain Research and Development* 11 (2): 87–100. https://www.jstor.org/stable/3673569?seq=1&cid=pdf-reference#references_tab_contents (accessed 3 October 2024).

Singh, A.T., Sharma, P., Sharma, C. et al. (2020). Water discharge and suspended sediment dynamics in the Chandra River, Western Himalaya. *Journal of Earth System Science* 129 (1): 206. https://doi.org/10.1007/s12040-020-01455-4.

Singh, D.S., Tangri, A.K., Kumar, D. et al. (2017). Pattern of retreat and related morphological zones of Gangotri Glacier, Garhwal Himalaya, India. *Quaternary International* 444: 172–181. https://linkinghub.elsevier.com/retrieve/pii/S104061821630194X (accessed 3 October 2024).

Singh, P. and Kumar, N. (1997). Effect of orography on precipitation in the western Himalayan region. *Journal of Hydrology* 199 (1–2): 183–206. https://doi.org/10.1016/S0022-1694(96)03222-2.

Singh, P., Ramasatri, K.S., Kumar, N., and Bhatnagar, N.K. (2003). Suspended sediment transport from the Dokriani Glacier in the Garhwal Himalayas. *Nordic Hydrology* 34 (3): 221–244. https://doi.org/10.2166/nh.2003.0005.

Singh, S., Kumar, R., Bhardwaj, A. et al. (2016a). Changing climate and glacio-hydrology in Indian Himalayan region: a review. *WIREs Climate Change* 7 (3): 393–410. https://wires.onlinelibrary.wiley.com/doi/10.1002/wcc.393 (accessed 3 October 2024).

Singh, V.B., Ramanathan, A.L., Mandal, A., and Angchuk, T. (2015). Transportation of suspended sediment from meltwater of the patsio glacier, Western Himalaya, India. *Proceedings of the National Academy of Sciences India Section A – Physical Sciences* 85 (1): 169–175. https://doi.org/10.1007/s40010-015-0198-0.

Singh, V.B., Ramanathan, A.L., and Pottakkal, J.G. (2016b). Glacial runoff and transport of suspended sediment from the Chhota Shigri glacier, Western Himalaya India. *Environmental Earth Sciences* 75 (8): 1–13. https://doi.org/10.1007/s12665-016-5271-8.

Srivastava, D., Kumar, A., Verma, A., and Swaroop, S. (2014). Characterization of suspended sediment in meltwater from glaciers of Garhwal Himalaya. *Hydrological Processes* 28 (3): 969–979. https://doi.org/10.1002/hyp.9631.

Thapa, B., Chaudhary, P., Dahlhaug, O.G. and Upadhyay, P. (2007). Study of combined effect of sand erosion and cavitation in hydraulic turbines. *International Conference on Small Hydropower, Sri Lanka* (22–24 October). https://www.researchgate.net/publication/240622577_Study_of_Combined_Effect_of_Sand_Erosion_and_Cavitation_in_Hydraulic_Turbines/citations (accessed 3 October 2024).

Thapa, B.S., Dahlhaug, O.G., and Thapa, B. (2015). Sediment erosion in hydro turbines and its effect on the flow around guide vanes of Francis turbine. *Renewable and Sustainable Energy Reviews* 49 (September): 1100–1113. https://doi.org/10.1016/j.rser.2015.04.178.

Thapa, B.S., Gjosater, K., Eltvik, M. Dahlhaug, O.G. and Thapa, B. (2012a). Effects of turbine design parameters on sediment erosion of Francis runner. *Proceedings of 2nd International Conference on the Developments in Renewable Energy Technology, ICDRET 2012*, Dhaka, Bangladesh (5–7 January), 167–171. IEEE.

Thapa, B.S., Thapa, B., Eltvik, M. et al. (2012b). Optimizing runner blade profile of Francis turbine to minimize sediment erosion. *IOP Conference Series: Earth and Environmental Science* 15 (PART 3): 032052. https://doi.org/10.1088/1755-1315/15/3/032052.

Truscott, G.F. (1972). A literature survey on abrasive wear in hydraulic machinery. *Wear* 20 (1): 29–50. https://doi.org/10.1016/0043-1648(72)90285-2.

UJVNL (2023). Hydro power projects in Uttarakhand. https://www.ujvnl.com/hydro-power-projects-in-uttarakhand (accessed 3 October 2024).

Wang, G., Wu, B., and Wang, Z. (2005). Sedimentation problems and management strategies of Sanmenxia Reservoir, Yellow River, China. *Water Resources Research* 41 (9): W09417. https://doi.org/10.1029/2004WR003919.

Ward, F. (1957). Brazos river bar: a study in the significance of grain size parameters. *Journal of Sedimentary Research* 27 (1): 3–26.

White, R. 2001. *Evacuation of Sediments from Reservoirs*. London: Thomas Telford Publishing. doi:https://doi.org/10.1680/eosfr.29538.

Whitehead, P.G., Barbour, E., Futter, M.N. et al. (2015). Impacts of climate change and socio-economic scenarios on flow and water quality of the Ganges, Brahmaputra and Meghna (GBM) river systems: low flow and flood statistics. *Environmental Science: Processes & Impacts* 17 (6): 1057–1069. http://xlink.rsc.org/?DOI=C4EM00619D (accessed 3 October 2024).

Winkler, K. (2014). Hydro-abrasive erosion: problems and solutions. *IOP Conference Series: Earth and Environmental Science* 22: 052022. https://doi.org/10.1088/1755-1315/22/5/052022.

Wulf, H., Bookhagen, B., and Scherler, D. (2010). Seasonal precipitation gradients and their impact on fluvial sediment flux in the Northwest Himalaya. *Geomorphology* 118 (1–2): 13–21. https://doi.org/10.1016/j.geomorph.2009.12.003.

Wulf, H., Bookhagen, B., and Scherler, D. (2012). Climatic and geologic controls on suspended sediment flux in the Sutlej River Valley, western Himalaya. *Hydrology and Earth System Sciences* 16 (7): 2193–2217. https://doi.org/10.5194/hess-16-2193-2012.

Xu, J. (2000). Grain-size characteristics of suspended sediment in the Yellow River, China. *Catena* 38 (3): 243–263. https://doi.org/10.1016/S0341-8162(99)00070-3.

Zhang, Q., Xu, C., Becker, S., and Jiang, T. (2006). Sediment and runoff changes in the Yangtze River basin during past 50 years. *Journal of Hydrology* 331 (3–4): 511–523. https://doi.org/10.1016/j.jhydrol.2006.05.036.

5

Hydrochemistry and Quality Assessment of High-Altitudinal Kareri Lake, Northwest Himalaya, Himachal Pradesh, India

Pawan Kumar[1], *Ambrish K. Mahajan*[2], *Deepak Kumar*[3], *and Anil Kumar*[4]

[1] *Department of Natural Resources Management, Maharana Pratap Horticultural University, Karnal, Haryana, India*
[2] *Department of Environmental Sciences, Central University of Himachal Pradesh, Dharamshala, Himachal Pradesh, India*
[3] *Department of Chemistry, Guru Jambheshwar University of Science and Technology, Hisar, Haryana, India*
[4] *Forest Ecology and Climate Change Division, ICFRE-Himalayan Forest Research Institute, Shimla, Himachal Pradesh, India*

5.1 Introduction

The lakes' water chemistry is maintained by several physiochemical parameters including heavy metals and major ions, etc. The concentration of these chemicals depends on the surrounding lithology and anthropogenic activities (Meena et al. 2017). However, the natural factors maintain the equilibrium, but the later deteriorate the water quality by increasing the concentration of different parameters beyond the permissible limit (Sheikh et al. 2014; Kumar et al. 2019a). The lakes of Himalayan regions attract the tourists from all over the world, owing to locations and scenic beauty, and therefore most of lakes are turned polluted because the people visiting there add lot of waste materials (Gaury et al., 2018; Kumar et al. 2019b). However, the lakes situated at the higher altitudes of the Greater Himalayan regions are having extreme climatic conditions and somehow far from reach of all the tourists and local people. The Greater Himalaya is consisting of glacial lakes receive water from glacier melting (Jain et al. 2015), whereas the lakes situated at the Lower and Outer Himalayan regions are the freshwater lakes having rainfall as the main water source (Kumar et al. 2019a).

The western part of the Himalaya is mainly composed of several lakes and their importance lies in the fact that these are a huge source of freshwater for the civilians of the Himalayan regions and instead, the survival of wildlife in surrounding of these lakes should not be ignored. Owing to the changing natural scenario of these lakes due to nutrients and metals enrichment, bacterial contamination, sedimentation, etc., there is a strong need for regular monitoring of lakes form different perspectives (Sarkar et al. 2016; Kumar et al. 2019a, 2019b). The chemical composition of various parameters can be helpful to determine the contribution of erosion and weathering via different means in the surrounding, pollution level, nutrient enrichment, pattern of evaporation, and source of different chemical parameters (Xu et al. 2011).

Water Resources Management in Mountain Regions, First Edition. Edited by Virendra Bahadur Singh, Sughosh Madhav, Rakesh Kumar Gupta, Prerna Diwan, and Amit Kumar.

The weathering of the rocks depends on their types, vegetation cover, atmospheric temperature, and burden of anthropogenic activities. The rocks chemical weathering reduces the carbon dioxide level from the atmosphere and therefore imparts a long-term effect on air temperature (Caldeira 1995; Sun et al. 2010). It is the rock weathering, which also lays impact on the major ions chemistry of the water bodies and has determined by several researchers in past (Caspari et al. 2006; Saleem et al. 2015; Prasanna et al. 2017; Mallick 2017; Gaury et al. 2018; Kumar and Mahajan 2020). The concentration of major ions, which are the chemical elements having concentration greater than 1 ppm and are susceptible to be consumed by the human being through drinking water, food chain, and other direct and indirect contacts (Kumar et al. 2019a). The higher concentration of the heavy metals is also considered to be harmful for human being as these may cause various disorders (Kumar and Kumar 2019; Kumar et al. 2022).

The hydro-geochemical processes also define the source of physiochemical parameters and their dominant sources, catchment weathering, and pollution level. The Ca^{2+} and Mg^{2+} in the lakes are derived from the weathering of carbonate rocks and anthropogenic sources, whereas Na^{+} and K^{+} reflect silicate weathering, halite dissolution, etc. sources of their origin (Gaury et al. 2018; Kumar et al. 2019b). The SO_4^{2-} in the lakes comes from gypsum and dolomite minerals as well anthropogenic activities (Zhijun et al. 2015), and on the other hand, the presence of bicarbonate (HCO_3^{-}) indicates carbonate minerals dissolution (Kumar et al. 2020b). The pH, salinity, and temperature influence various lakes processes and concentration of different physicochemical parameters (Kumar et al. 2019b). The concentration of ionic compounds within the permissible limit of the Bureau of Indian Standards (BIS 2012) is essential, but their concentration more than this limit causes various metabolic disorders in human being (Kumar et al. 2019b).

The changing scenario of lakes has been a matter of serious concern and attracted various researchers for their monitoring to understand the current status. The Wular Lake, which is a Ramsar site, in a study revealed water quality in good condition, but the increased concentration of observed parameters was a matter of serious concern for future (Bhat and Pandit 2014). The Rara Lake of western Nepal, in a study revealed greater influence of carbonate rocks weathering on the water quality and in addition, some signs of dissolved oxygen deficiency were also observed (Kaphle et al. 2021). The hydrochemistry of Ghodaghodi Lake of west Nepal in a study comprising different seasons marked a major influence of the regional carbonate weathering on the water quality of the lake. The marine transport was observed as a main cause of increased concentration of Na^{+} and Cl^{-} in the lake (Bhatta et al. 2022). The Phewa Lake of Pokhara, Nepal, in a study showed higher contribution of carbonate rocks weathering, rather than the silicates rocks on the hydrochemistry. The water quality of the lake was found to be suitable for the irrigation purposes (Kafle et al. 2023).

The Himalayan lakes are facing major challenges due to natural processes, climate change, and anthropogenic activities. Therefore, the lakes require continuously attention for their conservation and management (Najar and Khan 2012; Kumari et al. 2023; Kumar and Mahajan 2023). The water of Beeshazar and associated lakes (Central Nepal) in a study was found to be slightly alkaline and water quality suitable for drinking purposes. However, the concentration of PO_4^{3-} and NO_3^{-} was found to be increased and may create problems at the lakes in future (Pant et al. 2021). Similarly, the hydrochemistry and water quality characterization have also been determined from the Rupa and Begnas Lake, Nepal, by Pant et al. (2022). The study showed the carbonate weathering as a controlling factor of the water quality, which was also found to be suitable both for the drinking and irrigation purposes. Singh et al. (2016) carried out a study at a high-altitudinal Chandra Tal Lake and observed hydrochemistry and water quality. They found major influence of weathering process as compared to anthropogenic activities and atmospheric precipitation on the lake water. Deka et al. (2015) in their study at the high-altitudinal Tsokyo Tso and Sella Lake found the water quality pure and impact of the geochemical processes.

The Himalayan lakes have also been monitored from other different perspectives. Diwate et al. (2020) in a study at the Renuka Lake observed sedimentation at a rate of 0.64 cm/year in the duration of last 61 years. Likewise, Sarkar et al. (2016) in their study at the Rewalsar Lake observed sedimentation having a rate of 3.35 cm/year in the past 50 years. This sedimentation rate was more as observed in other studies at the Dal, Mansar, Nainital, Naukuchiatal, Bhimtal, and Sattal Lake, India. The Himalayan lakes are also found to be eutrophic and hyper-eutrophic as a result of increased nutrients due to human activities in the surrounding and at the lakes (Kumar et al. 2019a; Kumar and Mahajan 2020). The heavy metals and their contamination history have also been monitored from the Himalayan lakes in past (Meena et al. 2017; Kumar et al. 2019c). The dissolved oxygen and hyper-eutrophic status of the lake may also be responsible for fish kills and has been observed from the Rewalsar Lake (Kumar et al. 2020b). The ecological study at Parashar, Rewalsar, and Kuntbhyog Lake has also confirmed the presence of 79 zooplankton species belonging to five groups and 148 phytoplankton species from the nine different groups (Thakur et al. 2013).

The high-altitudinal Kareri Lake is a tourist attraction from all over the country and abroad, owing to its long difficult tract through higher altitudes of Lesser and Higher Himalaya having natural scenic beauty of various lithological features. The lake and its long track generally turn white in color due to heavy snowfall during winter season and that unable tourist's access at this lake for some months. Keeping in view the changing scenario, this lake also needs to be monitored for its preservation. Therefore, the objective of this study at Kareri Lake is to explore the hydrochemistry covering influence of the weathering processes in the lake, in addition to the water quality characteristics based on physicochemical parameters and microbial contamination. This lake is having lot of significances from social and ecological points of view and needs this study to apprehend first base line information for its preservation and maintenance of the qualitative characteristics.

5.2 Information of the Study Site

The Kareri Lake is situated within the Lesser Himalayan rocks at 76°27′31″ E Latitude and 32°32′68″ N Longitude, and an altitude of 2934 m to the south of the Dhauladhar range, which is approximately 9 km northwest of Dharamshala in Kangra district, Himachal Pradesh, India. The Kareri Lake is a shallow freshwater lake and one of the famous tourist destinations for mountaineering people. The regional geology comprises rocks of Subathu formation occurred during Paleocene to Eocene age. It is formed of grey and purple red shales, sandstone, limestone, clay, etc. (Gupta and Thakur 1974). The region contains active tectonics, fragile lithology, and high slopes angles (Mahajan et al. 2021). The climatic condition of the lake includes wet temperate with a mean annual temperature and precipitation around 19 °C and 2900 mm, respectively (Jaswal et al. 2014). This lake during winter season gets covered with fresh snow, which melts up during summer and the extra water thus flows to downward regions. Approaching at the Kareri Lake during winter season is not possible and allowed for tourists by administration due to safety reasons. It provides freshwater to the tourists including residents of Outer Himalaya. Water sources of the lake are the snow melting and monsoon rain. The most common wildlife includes leopard, Himalayan brown and black bear, ibex, wild goat, etc. Hence, a complete food chain can be seen there in the surrounding region of the lake.

5.3 Methodology

The hydrochemistry of the lake has been exposed using water samples, which were randomly collected from surface of the Kareri Lake and used for analysis of different essential physiochemical parameters (Table 5.1). In addition to it, the quality assessment is known using concentration of different parameters including level of the microbial contamination. The narrow mouthed polyethylene sampling bottles were used for the water sample collection and the pH, electrical conductivity (EC), total dissolved solid (TDS), and salinity were determined in situ using water analysis kit. The bicarbonate from the samples was determined using sulfuric acid (H_2SO_4) and titration method. The composition of ionic species was known with the instrument ion chromatography (IC; Gaury et al. 2018; Kumar

Table 5.1 Showing concentration of various parameters and their comparison with the BIS (2012) permissible limit (Kumar et al. 2019b). The pH comprises numerical scale, while the unit of EC is μS/cm, MPN is the number and that of rest all parameters is mg/L.

Sr. no.	Parameters	Observed concentration	BIS (2012)	Permissible limit
1	pH	6.98	6.5–8.5	No-relaxation
2	EC	35.70	2000	2250
3	TDS	150	1000	2000
4	Ca^{2+}	16.29	75	200
5	Mg^{2+}	0.80	30	100
6	Na^+	2.65	200	200
7	TH	43.98	300	600
8	K^+	0.70	12	12
9	NO_3^-	3.52	45	45
10	HCO_3^-	58.60	200	600
11	SO_4^{2-}	1.70	200	400
12	MPN	33	Nil	Nil

and Mahajan 2023) and the microbial contamination was known by most probable number (MPN) test method. The standards permissible limit of physicochemical parameters and microbial contamination prescribed by BIS (2012) was considered for quality assessment of the lake water.

5.4 Results and Discussion

The in situ and ex situ analysis of water sample confirmed concentration of various essential parameters to fulfill objective of the present study. The pH level in the lake was observed in the range of 6.8–7.2 and marked a slight acidic to alkaline scenario of the lake. Similarly, the EC [31–42 μS/cm (35.7 ± 3.73 μS/cm)], TDS [142–156 mg/l (150 ± 4.11 mg/l)], and salinity [14–24 mg/l (17.6 ± 2.9 mg/l)] were also attributed to the normal level. The concentration of major ions such as Ca^{2+} and Mg^{2+} was found to be 14.9–17.4 mg/l (16.29 ± 0.69 mg/l) and 0.58–0.95 mg/l (0.80 ± 0.11 mg/l), respectively. The concentration of these parameter are lower than observed from the Renuka Lake by Kumar et al. (2019b), which may be due to less anthropogenic activities at this lake. The Na^+ and K^+ in the lake marked low content in the range of 2.40–2.90 mg/l (2.65 ± 0.16 mg/l) and 0.43–0.87 (0.70 ± 0.16 mg/l), respectively. The Rewalsar Lake in a study had indicated higher content of alkali metals than the Kareri Lake (Gaury et al. 2018; Kumar et al. 2020b). The content of Cl^- in the lake was observed 0.33–0.67 mg/l (0.44 ± 0.11 mg/l) and in addition the SO_4^{2-} was monitored 1.30–2.36 mg/l (1.70 ± 0.35 mg/l). The nitrogenous compounds such as NH_4^+ and NO_3^- in lake were also observed in the content range of 0.15–0.93 mg/l (0.51 ± 0.21 mg/l) and 2.52–4.79 mg/l (3.52 ± 0.81 mg/l), respectively. The total hardness (TH) in the lake was confirmed 40.02–46.84 mg/l (43.98 ± 1.86 mg/l), while the HCO_3^- was found in the range of 54–64 mg/l (58.6 ± 3.53 mg/l). The charge balance error (CBE) from the observed parameters was found to be 0.04–4.82 (2.52 ± 1.63), which is a direct indication of reliability of the result obtained and analytical accuracy (Kumar et al. 2020a). The MPN number from the water samples (100 ml each) marked the total coliform ranging 26–37 (33 ± 3.09).

5.5 Hydrochemistry of the Lake

There are certain factors, which maintain the hydrochemistry of the lake water during the year round under natural environmental conditions. Among several factors, the surrounding rocks, precipitation, and evaporation are generally found to play a prominent role for maintaining the concentration of most of the parameters including major ions (Ca^{2+}, Mg^{2+}, Na^+, K^+, HCO_3^-, SO_4^{2-}, Cl^-, and NO_3^-) (Kumar et al. 2019a, 2019b). Instead, the parameters such as NH_4^+ and NO_3^- in lakes may also mark the influence of anthropogenic activities. The rocks dominance as inferred by Gibbs (1970) has been specified as the dominant source of ionic species in the Kareri Lake (Figure 5.1), which is also similar to that observed for the other lakes such as Dal Lake Mcleodganj, Renuka, and Rewalsar Lake, Himachal Pradesh, India (Kumar et al. 2019b, 2020b). Similarly, Rupa and Begnas Lake of Nepal in an earlier study also indicated rocks dominance as a major ionic source in the water (Pant et al. 2022).

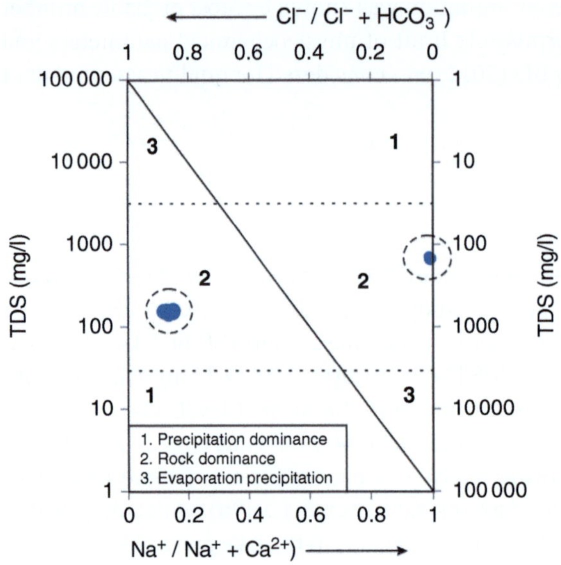

Figure 5.1 Dominant source of ionic species in the Kareri Lake. *Source:* Modified after Gibbs (1970).

The Kareri Lake comprises carbonate and silicate rocks in the surrounding, but which among these having overweight can be understood by using milliequivalent (meq/l) concentration and drawing the scatter plots from different parameters (Figure 5.2a–f). The major contribution of carbonate rocks weathering is proved rather than the silicate rocks, which are playing lesser role in the hydrochemistry of the Kareri Lake (see Figure 5.2a–c). These figures suggest higher participation of alkaline earth metals (Ca^{2+} and Mg^{2+}), which are produced from carbonate rocks as compared to the alkali metals (Na^{+} and K^{+}) mainly generated form silicate rocks. These findings are similar to observed from a high-altitudinal Himalayan Lake, Chandra Tal, by Singh et al. (2016). Instead, the faster dissolution of carbonate minerals than that of the silicate minerals also cannot be neglected for this study at the Kareri Lake (Kumar et al. 2020a). Figure 5.2d infers that the bicarbonate alkalinity is controlled by the carbonate rocks weathering in the catchment and dissolution in this lake. This figure indicates a low content of bicarbonate and to make its ratio equal to that of alkaline earth metals, SO_4^{2-} and Cl^{-} may participate very well (Gaury et al. 2018).

Similarly, Figure 5.2e is indicating the higher participation of the $HCO_3^{-}+SO_4^{2-}$ rather than the alkaline earth metals and that may be due to ion exchange process (Gaury et al. 2018; Kumar et al. 2019b, 2020a). Although the points also seem to attract toward the divider line that may be due to impact of dolomite, calcite and gypsum minerals therein the catchment region of the lake. An earlier study at the Ghodaghodi and Rara Lake, Nepal, having comparison between the similar parameters also showed the points close as well as below the equiline and inferred minor contribution of silicate rocks weathering (Bhatta et al. 2022). The impact of precipitation to major extent is also denied for this lake and is proved by Figure 5.2f, where all the points are lying in upper part of the scatter plot and nearby the vertical axis (Kumar et al. 2020a). The points above the line are also indicating the release of Na^{+} due to the halite dissolution as well as the silicate weathering (Saleem et al. 2015).

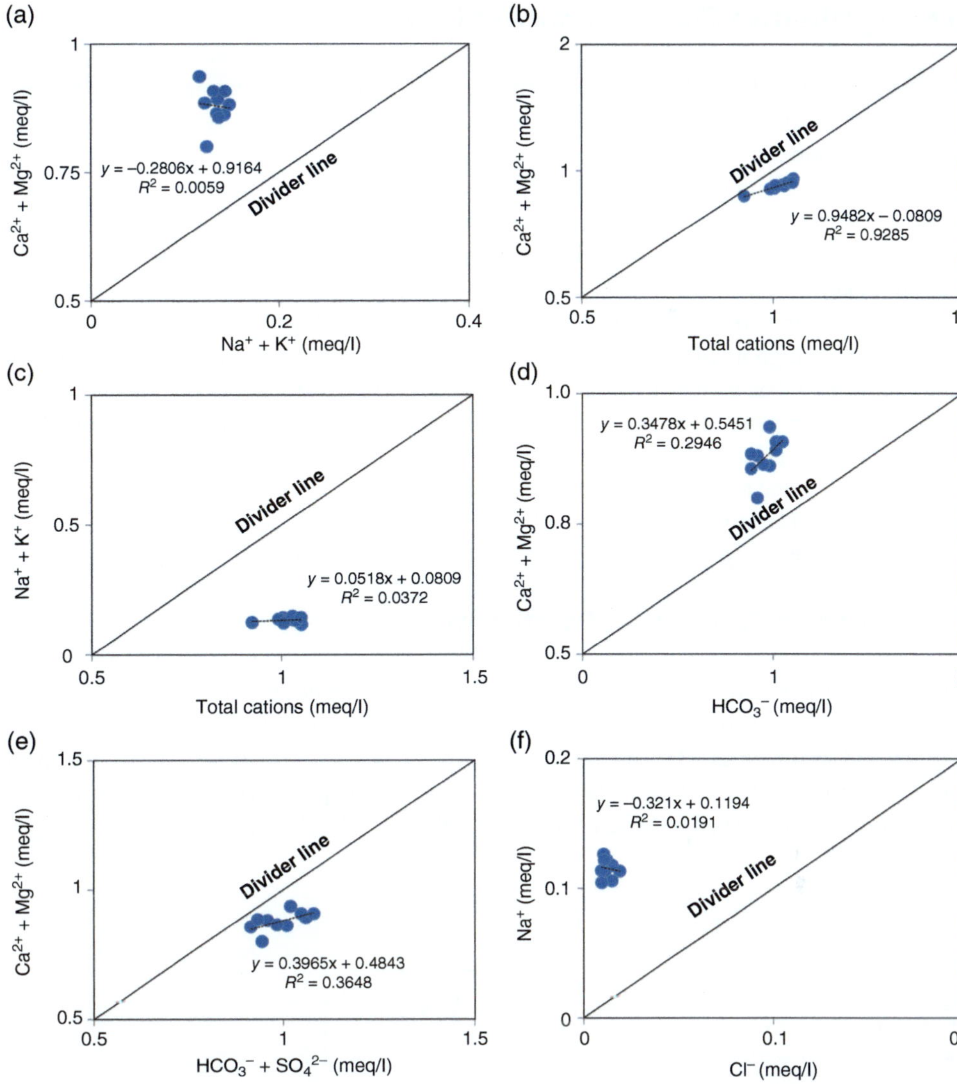

Figure 5.2 Comparison among different parameters of Kareri Lake: (a). $Ca^{2+} + Mg^{2+}$ versus $Na^+ + K^+$; (b). $Ca^{2+} + Mg^{2+}$ versus total cations; (c). $Na^+ + K^+$ versus total cations; (d). $Ca^{2+} + Mg^{2+}$ versus HCO_3^-; (e). $Ca^{2+} + Mg^{2+}$ versus $HCO_3^- + SO_4^{2-}$; (f). Na^+ versus Cl^-. *Source:* Gaury et al. (2018), Kumar et al. (2019b, 2020a).

The decreasing order of arrangement of ionic components in the lake are found to be $HCO_3^- > Ca^{2+} > NO_3^- > Na^+ > SO_4^{2-} > Mg^{2+} > K^+ > Cl^-$. Deka et al. (2015) had also observed HCO_3^- and Ca^{2+} as the dominant ionic species in the water of the Sella and Tsokyo Tso Lake during different seasons. The alkaline earth metals and weak acidic anion are observed to surpass the alkali metals and strong acidic anion, respectively, and that may be due to greater impact of calcite and dolomite rather than the silicate and gypsum minerals.

5.6 Qualitative Characteristics of Lake

The water quality depends on the concentration of different parameters and for the present study, the limit prescribed by the Bureau of Indian standards (BIS 2012) is considered. The concentration of the physicochemical parameters is observed within the permissible limit and thus on that basis, the water in the lake is declared safe (Table 5.1). However, some signature of microbial contamination is also observed in the lake that recommends some treatment before domestic use; otherwise may be harmful for human health.

The diagram inferred by Chadha (1999) delineated Ca^{2+} - Mg^{2+} - HCO_3^- type of water in the lake due to higher contribution of carbonate rocks weathering in the catchment region (see Figure 5.3). The water of the lake is observed soft in nature due to the presence of TH as 43.98 and can generate foam with soap and detergent. The Rewalsar and Renuka Lake in other studies indicated the TH more than 75 and the nature of water as hard (Gaury et al. 2018; Kumar et al. 2019b). The concentration of NH_4^+ and NO_3^- is observed low, which thus infers their natural source of origin.

5.7 Pearson Correlation Coefficient

The Pearson correlation coefficient (r) defines the dependency of one variable over another and degree of the interrelationship exists between them (Singh et al. 2016; Kumar and Mahajan 2023). The correlation may be positive ($r > 0.5$–0.7), strongly positive ($r > 0.8$), negative ($r > -0.5$ to -0.7), and strongly negative ($r > -0.8$). The positive correlations are observed between Salinity-EC, K^+, Cl^-; NH_4^+ $-Ca^{2+}$, TH; HCO_3^- $-Mg^{2+}$, TH, and Na^+ $-K^+$.

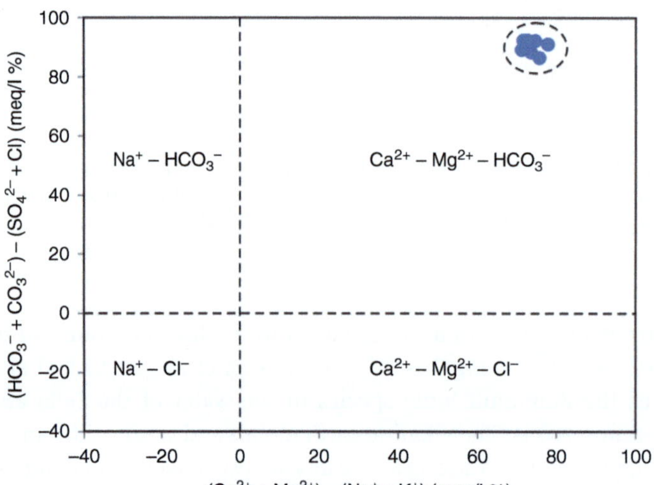

Figure 5.3 Types of water in the Kareri Lake. *Source:* After Chadha (1999), Kumar et al. (2020a).

The strong positive correlations are among Ca^{2+} and TH, which is indicating that hardness is due to Ca^{2+} in the lake water. Likewise, the negative correlation is occurring in the EC $-Ca^{2+}$, NH_4^+; Mg^{2+} $-NO_3^-$, and Na^+ $-NH_4^+$; The strong negative correlation is shown by HCO_3^- and NO_3^- (Table 5.2). The positive value shows similar sources of different parameters, while negative value infers their different sources of origin. However, the parameters are having poor correlation may indicate the anthropogenic sources of different parameters (Kumar and Mahajan 2020).

5.8 Comparison of Different Parameters of Kareri with Other Himalayan Lakes

It is clear from Table 5.3 that most of the parameters observed in the present study are having some variations with lakes of other Himalayan regions. The difference may be due to impact of surrounding lithology and anthropogenic influences at the lakes. The Nainital, Bhimtal, Rewalsar, Renuka, Dal (Mcleodganj), and Dal (Srinagar) are having higher level of anthropogenic activities and hence the concentrations of various parameters are found to be high as compared to the other lakes. The Kareri Lake is a high-altitudinal lake like the Chandra Tal Lake, and hence having low level of anthropogenic activities and concentration of different observed parameters than the other lakes of Himalayan regions (Table 5.3).

5.9 Conclusion

The present study has showed hydrochemistry and water qualitative characteristics of the Kareri Lake situated at the higher altitudes of Lesser Himalaya in India. The hydrochemistry of the lake is mainly found to be controlled by the surrounding rocks and among the carbonate rocks weathering is participating more than the silicates rocks. This lake is a well-known tourist place for the mountaineering people and thus lot of anthropogenic activities can be observed on the lake, but the concentration of the physical and chemical parameters is found to be within permissible limit prescribed by the Bureau of Indian Standards. However, the microbial contamination is observed from the lake that suggests treatment of water before domestic use. The concentration of nitrogenous compounds is observed very low and points out their natural source of origin and their low concentration may be due to lack of catchment related agricultural practice, which is generally considered a main causal factor of increasing concentration of such parameters. The water in the lake is of soft nature and can be well used to create foam with soap and other detergents. The Pearson correlation coefficient has indicated that most of the parameters are having their separate source of origin. The concentrations of most of parameters in the lake are having low concentration; may be due to less influence of anthropogenic activities than other lakes of Himalayan regions. The reason may be its high-altitudinal location and a difficult track for the people to reach at the lake and that is only possible during summer season. The Himalayan lakes are a huge water source and support floral and faunal biodiversity, and thus require regular monitoring for preservation through different

Table 5.2 Pearson correlation coefficient among different parameters observed from the Kareri Lake.

	pH	EC	TDS	Salinity	Ca^{2+}	Mg^{2+}	Na^{+}	K^{+}	NH_4^{+}	HCO_3^{-}	Cl^{-}	NO_3^{-}	SO_4^{2-}	TH
pH	1													
EC	−.456	1												
TDS	.341	−.160	1											
Salinity	−.393	**.669***	−.371	1										
Ca^{2+}	.382	**−.559**	−.070	.042	1									
Mg^{2+}	.207	.429	.267	.347	.164	1								
Na^{+}	−.116	.231	.014	.357	−.249	−.043	1							
K^{+}	.034	.205	−.166	**.708***	.177	.271	**.685***	1						
NH_4^{+}	.363	**−.566**	.070	−.343	**.713***	−.206	**−.520**	−.417	1					
HCO_3^{-}	−.040	.062	−.214	.263	.447	**.528**	.117	.321	.068	1				
Cl^{-}	−.307	.479	−.320	**.648***	−.183	.087	−.138	.376	−.332	−.163	1			
NO_3^{-}	.291	−.212	.317	−.383	−.254	**−.513**	−.275	−.399	.238	**−.910****	.114	1		
SO_4^{2-}	.327	.047	−.196	.258	.333	.246	−.355	.154	.162	−.248	.466	.263	1	
TH	.406	−.419	−.002	.122	**.972****	.391	−.243	.229	**.616**	**.543**	−.150	−.359	.369	1

* Correlation is indicated significant at the 0.05 level (2-tailed).
** Correlation is indicated significant at the 0.01 level (2-tailed).

Table 5.3 Comparison of different parameters of Kareri and other lakes of Himalayan regions. The pH comprises numerical scale, the unit of EC is µS/cm and that of rest all parameters is mg/L.

Name of lake	EC	pH	TDS	Ca^{2+}	Mg^{2+}	Na^{+}	K^{+}	HCO_3^-	Cl^-	NO_3^-	SO_4^{2-}	Reference
Nainital	312.6	8.3	—	492.6	2235	548.8	9.9	1803	—	71	1024	Chakrapani (2002)
Bhimtal	101	7.9	—	190.3	490.2	4.6	4.6	482	—	43.1	62.6	Chakrapani (2002)
Mansar	416.5	8.4	—	21.6	6.45	11.02	2.68	96.83	—	0.63	6.88	Kumar et al. (2006)
Surinsar	545	8.4	350	38	13	10	4	127	—	5	11	Singh et al. (2008)
Begnas	91	7.3	58	7	2	3.9	1.4	25.30	2.6	5.30	7.30	Khadka and Ramanathan (2013)
Sella	5.2	6.1	—	0.90	0.3	2.2	0.3	11.3	0.50	0.70	3.70	Deka et al. (2015)
Chandra Tal	212	8.1	—	22.4	13.2	0.9	1.4	118	0.5	0.5	12.5	Singh et al. (2016)
Rewalsar	216.2	7.36	143.8	40.59	5.35	26.0	12.77	164	13.64	6.58	2.45	Gaury et al. (2018)
Renuka	451	8.1	300.4	71.8	43.82	23.32	2.95	275.68	—	4.1	129.7	Kumar et al. (2019b)
Rupa	110	9.06	64	9.83	3.58	4.12	1	37.60	2.13	0.13	2.40	Pant et al. (2022)
Rara	194.6	8.47	124.2	21.72	10.38	0.68	1.51	122.76	0.26	0.07	0.35	Bhatta et al. (2022)
Ghodaghodi	192.3	8.68	95.68	23.83	3.05	4.84	2.52	104	0.38	0.28	0.06	Bhatta et al. (2022)
Dal (Macleodganj)	240	6.64	160.7	23.93	2.29	4.73	1.34	66	7.47	1.34	21.99	Kumar and Mahajan (2023)
Dal (Srinagar)	278	7.43	185.4	30.08	4.75	5.75	2.06	63.2	6.51	0.83	55.23	Kumar and Mahajan (2023)
Kareri	35.7	6.98	150	15.29	0.80	2.65	0.70	58.60	0.44	3.52	1.70	Present study

perspectives. The physicochemical parameters in the Kareri Lake are not so increased, but the administration should focus on the anthropogenic activities and microbial contamination to maintain qualitative characteristic of the lake in future.

References

Bhat, S.A. and Pandit, A.K. (2014). Surface water quality assessment of Wular Lake, a Ramsar site in Kashmir Himalaya, using discriminant analysis and WQI. *Journal of Ecosystems* 2014 (1): 724728.

Bhatta, R., Gurung, S., Joshi, R. et al. (2022). Spatio-temporal hydrochemistry of two selected Ramsar sites (Rara and Ghodaghodi) of west Nepal. *Heliyon* 8 (11): e11243.

Bureau of Indian Standards (BIS) (2012). *Drinking Water Specification, Second Revision IS:10500*. New Delhi, India: Bureau of Indian Standards.

Caldeira, K. (1995). Long-term control of atmospheric carbon dioxide; low-temperature seafloor alteration or terrestrial silicate-rock weathering? *American Journal of Science* 295 (9): 1077–1114.

Caspari, T., Baumler, R., Norbu, C. et al. (2006). Geochemical investigation of soils developed in different lithologies in Bhutan, Eastern Himalayas. *Geoderma* 136 (1–2): 436–458.

Chadha, D.K. (1999). A proposed new diagram for geochemical classification of natural waters and interpretation of chemical data. *Hydrogeology Journal* 7 (5): 431–439.

Chakrapani, G. (2002). Water and sediment geochemistry of major Kumaun Himalayan lakes, India. *Environmental Geology* 43: 99–107.

Deka, J.P., Tayeng, G., Singh, S. et al. (2015). Source and seasonal variation in the major ion chemistry of two eastern Himalayan high altitude lakes, India. *Arabian Journal of Geosciences* 8: 10597–10610.

Diwate, P., Meena, N.K., Bhushan, R. et al. (2020). Sedimentation rate (Pb-210 and Cs-137), grain size, organic matter and bathymetric studies in Renuka Lake, Himachal Pradesh, India. *Himalayan Geology* 41 (1): 51–62.

Gaury, P.K., Meena, N.K., and Mahajan, A.K. (2018). Hydrochemistry and water quality of Rewalsar Lake of Lesser Himalaya, Himachal Pradesh, India. *Environmental Monitoring and Assessment* 190: 1–22.

Gibbs, R.J. (1970). Mechanisms controlling world water chemistry. *Science* 170 (3962): 1088–1090.

Gupta, V.J. and Thakur, V.C. (1974). Geology of the area around Dharmsala, Kangra District, HP, India. *Geologische Rundschau* 63: 548–558.

Jain, S.K., Rishitosh, K.S., Chaudhary, A., and Shukla, S. (2015). Expansion of glacial lake, Tsho Chubda, Chamkhar Chu Basin, Hindukush Himalaya, Bhutan. *Natural Hazards* 75: 1451–1464.

Jaswal, A.K., Kumar, N., and Khare, P. (2014). Climate variability in Dharamsala-a hill station in western Himalayas. *Journal of Indian Geophysical Union* 18 (3): 336–355.

Kafle, B.K., Sharma, C.M., Gurung, S. et al. (2023). Hydrogeochemistry of two major mid-hill lentic water bodies for irrigation of the Central Himalaya, Nepal. *Environment and Natural Resources Journal* 21 (2): 171–185.

Kaphle, B., Wang, J.B., Kai, J.L. et al. (2021). Hydrochemistry of Rara Lake: a Ramsar lake from the southern slope of the central Himalayas, Nepal. *Journal of Mountain Science* 18 (1): 141–158.

Khadka, U.R. and Ramanathan, A.L. (2013). Major ion composition and seasonal variation in the Lesser Himalayan lake: case of Begnas Lake of the Pokhara Valley, Nepal. *Arabian Journal of Geosciences* 6: 4191–4206.

Kumar, P. and Mahajan, A.K. (2023). Hydrogeochemical facie and solute acquisition at Dal Lake of Kashmir and Dal Lake of Mcleodganj, northwest Himalaya, India. *Journal of Earth System Science* 132 (1): 38.

Kumar, P. and Kumar, P. (2019). Removal of cadmium (Cd-II) from aqueous solution using gas industry-based adsorbent. *SN Applied Sciences* 1 (4): 365.

Kumar, P. and Mahajan, A.K. (2020). Trophic status and its regulating factor determination at the Rewalsar Lake, northwest Himalaya (HP), India, based on selected parameters and multivariate statistical analysis. *SN Applied Sciences* 2 (7): 1–11.

Kumar, P., Mahajan, A.K., and Kumar, A. (2020a). Groundwater geochemical facie: implications of rock-water interaction at the Chamba city (HP), northwest Himalaya, India. *Environmental Science and Pollution Research* 27 (9): 9012–9026.

Kumar, P., Mahajan, A.K., and Kumar, P. (2020b). Determining limiting factors influencing fish kills at Rewalsar Lake: a case study with reference to Dal Lake (Mcleodganj), western Himalaya, India. *Arabian Journal of Geosciences* 13 (17): 872.

Kumar, P., Mahajan, A.K., and Meena, N.K. (2019a). Evaluation of trophic status and its limiting factors in the Renuka Lake of Lesser Himalaya, India. *Environmental Monitoring and Assessment* 191 (2): 105.

Kumar, P., Meena, N.K., and Mahajan, A.K. (2019b). Major ion chemistry, catchment weathering and water quality of Renuka Lake, north-west Himalaya, India. *Environmental Earth Sciences* 78 (10): 1–16.

Kumar, P., Meena, N.K., Diwate, P. et al. (2019c). The heavy metal contamination history during ca 1839–2003 AD from Renuka Lake of Lesser Himalaya, Himachal Pradesh, India. *Environmental Earth Sciences* 78: 1–14.

Kumar, P., Prajapati, K.B., Mahajan, A.K. et al. (2022). Evaluating feasibility of biosorption technique for heavy metals removal: limitations and future perspective. *International Journal of Environmental Analytical Chemistry* 102: 1–25.

Kumar, V., Rai, S.P., and Singh, O. (2006). Water quantity and quality of Mansar Lake located in the Himalayan Foothills, India. *Lake and Reservoir Management* 22 (3): 191–198.

Kumari, N., Thakur, K., Kumar, R. et al. (2023). Freshwater lakes in the Western Himalayan Region: an analysis of the present situation. *Water-Energy Nexus* 06: 18–31.

Mahajan, A.K., Kumar, P., and Kumar, P. (2021). Near-surface seismic site characterization using Nakamura-based HVSR technique in the geological complex region of Kangra Valley, northwest Himalaya, India. *Arabian Journal of Geosciences* 14 (10): 826.

Mallick, J. (2017). Hydrogeochemical characteristics and assessment of water quality in the Al-Saad Lake, Abha Saudi Arabia. *Applied Water Science* 7 (6): 2869–2882.

Meena, N.K., Prakasam, M., Bhushan, R. et al. (2017). Last-five-decade heavy metal pollution records from the Rewalsar Lake, Himachal Pradesh, India. *Environment and Earth Science* 76: 39.

Najar, I.A. and Khan, A.B. (2012). Assessment of water quality and identification of pollution sources of three lakes in Kashmir, India, using multivariate analysis. *Environmental Earth Sciences* 66 (8): 2367–2378.

Pant, R.R., Bishwakarma, K., Pal, K.B. et al. (2022). Comparative analysis of hydrochemical variables of two Ramsar-listed lakes in Pokhara Valley, Nepal. *International Journal of Energy and Water Resources* 7 (1): 1–13.

Pant, R.R., Chalaune, T.B., Dangol, A. et al. (2021). Hydrochemical assessment of the Beeshazar and associated lakes in Central Nepal. *SN Applied Sciences* 3: 1–13.

Prasanna, M.V., Nagarajan, R., Chidambaram, S. et al. (2017). Evaluation of hydrogeochemical characteristics and the impact of weathering in seepage water collected within the sedimentary formation. *Acta Geochimica* 36 (1): 44–51.

Saleem, M., Jeelani, G., and Shah, R.A. (2015). Hydrogeochemistry of Dal Lake and the potential for present, future management by using facies, ionic ratios, and statistical analysis. *Environmental Earth Sciences* 74 (4): 3301–3313.

Sarkar, S., Prakasam, M., Upasana, S. et al. (2016). Rapid sedimentation history of Rewalsar Lake, Lesser Himalaya, India during the last fifty years – estimated using Cs and Pb dating techniques: a comparative study with other north-western Himalayan lakes. *Himalayan Geology* 37 (1): 1–7.

Sheikh, J.A., Jeelani, G., Gavali, R.S., and Shah, R.A. (2014). Weathering and anthropogenic influences on the water and sediment chemistry of Wular Lake, Kashmir Himalaya. *Environment and Earth Science* 71 (6): 2837–2846.

Singh, O., Rai, S.P., Kumar, V., Sharma, M.K. et al. (2008). Water quality and Eutrophication status of some lakes of the western Himalayan region (India). In *Proceeding of Taal-2007: The 12th World Lake Conference*, 28 October to 02 November 2007, Jaipur, India, 286–291.

Singh, V.B., Ramanathan, A.L., and Mandal, A. (2016). Hydrogeochemistry of high-altitude lake: a case study of the Chandra Tal, Western Himalaya, India. *Arabian Journal of Geosciences* 9: 1–9.

Sun, H., Han, J., Li, D. et al. (2010). Chemical weathering inferred from riverine water chemistry in the lower Xijiang basin, South China. *Science of the Total Environment* 408 (20): 4749–4760.

Thakur, R.K., Jindal, R., Singh, U.B., and Ahluwalia, A.S. (2013). Plankton diversity and water quality assessment of three freshwater lakes of Mandi (Himachal Pradesh, India) with special reference to planktonic indicators. *Environmental Monitoring and Assessment* 185: 8355–8373.

Xu, Z., Shi, C., Tang, Y., and Han, H. (2011). Chemical and strontium isotopic compositions of the Hanjiang Basin Rivers in China: anthropogenic impacts and chemical weathering. *Aquatic Geochemistry* 17 (3): 243–264.

Zhijun, Y., Rui, W., Zhaofei, L. et al. (2015). Spatial-temporal patterns of major ion chemistry and its controlling factors in the Manasarovar Basin, Tibet. *Journal of Geographical Sciences 25* (6): 687–700.

6

Water Quality Assessment of Lakes in Mountain Region with Spatial Reference to Amarkantak, Madhya Pradesh, India

Piyali Sabui, Shakshee Soni, Ayushii Yadav, and Pallavi Das

Department of Environmental Science, Indira Gandhi National Tribal University, Amarkantak, Madhya Pradesh, India

6.1 Introduction

The quality of surface water is an important global environmental issue that affects agriculture, human health, the development of society, and the sustainability of ecosystems (Akhtar et al. 2021; Elsayed et al. 2021). Surface water bodies, such as lakes, rivers, and springs, are important sources of fresh water because of their diverse physico-chemical characteristics that vary over temporal and spatial scales (Gupta and Chakrapani 2005; Ganaie et al. 2022). Among them, lakes are the most significant water resources, comprising 0.3% of all surface water bodies and are essential water resources that supply freshwater to life on Earth (Mishra and Prajapati 2011; Vasistha and Ganguly 2020). The water quality of lake is affected by natural and human activities like encroachment, unchecked wastewater discharges, and fertilizer waste which decrease water volume and deteriorate its quality (Bhateria and Jain 2016; Fernando et al. 2023).

In the mountain region, high altitude lakes are essential to the sustainability and way of life of trekkers, sages, and wildlife (Kumar and Sharma 2019; Ahamad et al. 2023). According to the United Nations Conference on Sustainable Development (2012), the mountain ecosystem plays an important role in providing the world with the water it needs (Kumar and Sharma 2019). The water quality of lake is influenced by a complex interaction of biological, chemical, and physical elements (Vasistha and Ganguly 2020). More than half of the world's population depends on freshwater from mountain catchments, which are essential for human health but are being harmed by pollution, climate change, and habitat destruction (Singh and Sharma 2020). At low altitudes, the pollutants from agricultural, urban, and industrial activities degrade aquatic environments. As altitude increases, these sources become less, resulting in limited exposure to chemical pollution (Machate et al. 2023). However, Global pollution spreads in the mountain region due to migratory species, marine currents, and atmospheric long-range transport. Because of this, organic chemicals – permanent organic pollutants, or POPs – have accumulated in mountain regions, posing a

Water Resources Management in Mountain Regions, First Edition. Edited by Virendra Bahadur Singh, Sughosh Madhav, Rakesh Kumar Gupta, Prerna Diwan, and Amit Kumar.

threat to human health and wildlife (Ganaie et al. 2022; Vasistha and Ganguly 2022). Local economic activities in the mountains, such as tourism and natural resource exploitation, can also introduce pollutants into mountain lakes (Ahmed et al. 2022). Therefore, maintaining the health of mountain catchments is crucial to ensuring both the availability of clean freshwater and human welfare. As a result, it is necessary to check the water quality before its widespread use routinely.

In recent times, several studies have been reported on lake water quality in mountain regions all over the world. The water quality of high altitude lake has been evaluated using various physicochemical parameters and water quality index (WQI) by various researchers in India (Sharma and Kumar 2017; Kumari and Sharma 2018; Ganaie et al. 2022). The WQI for Glacier-fed Lake Neel Tal in India was measured and found excellent water quality and suitability for human use (Kumar and Sharma 2019). Comparative assessments of WQI in Dal Lake in Jammu and Kashmir were conducted by Ahamad et al. (2023). In 2022, Vasistha and Ganguly used conventional techniques in conjunction with a GIS modelling system to analyze the water quality in two Panchkula lakes, Haryana, utilizing the inverse distance weighting (IDW) approach. The temperature, dissolved oxygen, nitrates, and total phosphorus were all measured using IDW method (Vasistha and Ganguly 2022). The seasonal variation of water quality of Rewalsar Lake of Lesser Himalaya in Himachal Pradesh was studied (Gaury et al. 2018). Kol et al. (2022) conducted an study on seasonal variation of water quality of Johila reservoir located in Amarkantak, Madhya Pradesh M.P focusing on the phytoplankton abundance and composition. Water quality of Alpine Lake Sevan, Armenia's largest reservoir was assessed and focused on eutrophication and developed a system for resource management (Babayan and Adamovich 2023). The seasonal variation of water quality of high altitude lake of South America was studied by measuring different physiochemical parameters (Avila-Diaz et al. 2023). Water quality from 16 lakes and two lakes-fed rivers was assessed in northern Pakistan (Khan et al. 2022).

The WQI approach is considered one of the best ways to evaluate water quality (Vasistha and Ganguly 2022; Su et al. 2022). Thus, considering this, the present study has been conducted in different lakes in Amarkantak, located in Madhya Pradesh, India. Amarkantak is situated on the Maikal Mountain range, links the Vindhyachal and Satpura mountain ranges, and is one of the major tourist destinations. Amarkantak holds immense religious and ecological importance and is called "Teerthraj" or the "King of Pilgrimages." The rivers, Narmada, Sone, and Johila, converge here and are surrounded by lakes and mountains. Moreover, people widely use lake water for agricultural activities, drinking, and settlement purposes. So far, no study has been done on lake water quality in the Amarkantak Mountain region. Therefore, this study aims to evaluate the water quality of different lakes in the Amarkantak region using the WQI tool.

6.1.1 Sources of Pollution in Mountain Lakes

Pollution in mountain lakes is often associated with human activities and natural sources of pollution, which impact the lake ecosystem. Some natural sources of pollution in mountain lake regions include the following:

6.1.1.1 Geological Processes

Certain geological processes can introduce naturally occurring substances into lakes. For example, volcanic activity can release minerals and gases that may affect water quality. The presence of naturally elevated levels of elements like sulfur or iron in the bedrock can also contribute to water contamination (Bhateria and Jain 2016).

6.1.1.2 Vegetation Decay

The breakdown of plant matter, such as leaves and organic waste, can discharge nutrients and organic compounds into the water in hilly regions (Bhateria and Jain 2016).

6.1.1.3 Wildlife Activities

The natural cycling of nutrients in lakes can be aided by the presence of wildlife, including fish, birds, and other aquatic species (Singh and Rai 2014). Water quality can be impacted by the nutrients and organic matter that animals excrete and by the decomposition of deceased creatures.

6.1.1.4 Wildfires

Wildfires are a natural element of certain ecosystems' cycles, yet they can also be caused by humans. Ash and other organic materials can be released by wildfires into lakes, affecting the quality of the water (Lydersen et al. 2014). The ash has the ability to change the pH of the water and add nutrients.

Anthropogenic (human-induced) sources of lake pollution can have a major negative impact on the ecology in mountainous areas. The following are a few typical man-made sources of pollution in mountain lakes:

6.1.1.4.1 Agricultural Runoff In mountainous agricultural regions, the application of pesticides and fertilizers may cause runoff into neighboring lakes (Akhtar et al. 2021). These sources can provide an abundance of nutrients, such as nitrogen and phosphorus, which can lead to algal blooms, which lower water oxygen levels and damage aquatic life.

6.1.1.4.2 Urbanization and Development Mountain lakes may become more turbid due to sedimentation and erosion caused by construction, deforestation, and urban growth (Akhtar et al. 2021). Water quality and aquatic habitats may suffer as a result of sedimentation.

6.1.1.4.3 Industrial Discharges Industries in or close to mountainous areas may release heavy metals, chemicals, and other pollutants into lakes (Xia et al. 2020). These contaminants have the potential to build up in the water and harm aquatic ecosystems.

6.1.1.4.4 Recreational Activities Activities like boating, fishing, and camping can add to pollution in mountainous areas due to tourism and leisure (Xia et al. 2020). The lake's ecosystem may be harmed by improper waste disposal, fuel spills, and other recreational activities.

6.1.1.4.5 Waste Disposal Improper disposal of solid waste, especially plastics and other non-biodegradable materials, can contribute to pollution in mountain lakes (Mihai 2018). Over time, this pollution can deteriorate water quality and endanger aquatic life.

6.1.1.5 Natural Erosion

Erosion is a natural process that can be brought on by soil movement and rock weathering in mountainous areas. Lake clarity and quality can be impacted by the introduction of sediments, minerals, and other particles caused by erosion (Vale et al. 2023).

6.2 Study Area and Site Description

Amarkantak is a stunning hilltop pilgrimage town located in the east of Madhya Pradesh, India, converging by the Vindhya, Maikal, and Satpura mountain ranges. Amarkantak is a part of the Achanakmar-Amarkantak Biosphere Reserve (AABR), recognized by UNESCO as a world heritage site due to its ecological and economic significance (Thakur et al. 2019; Vishwakarma et al. 2023). It is located between longitude 81° 25 N and 20° 5 E and latitude 22° 15 to 20° 58 N and has an average elevation of 1048 m (3438 ft) above mean sea level. The study area is dominated by four rock types: basalts, Lamets, Gondwanas, Archaean. The primary southwest portion of the Anuppur district exposes the Cretaceous to Eocene Deccan Trap basalt. Dolerite sills and dykes are widespread in the region and trend north east, south west to the eastwest. A substantial expanse of dense basaltic lava flows covers a sizable portion of the district known as Pushprajgarh's southern region (Alam et al. 2021). The major minerals found in Amarkantak region are quartz, agate, calcite, and zeolites. The climatic condition of this region is humid tropical with an average annual rainfall of 1178 mm and relative humidity is 50–85% (Gupta and Chakrapani 2005; Ahirvar et al. 2020). The Narmada, Son, and Johila rivers and their tributaries originate from the hilly region of Amarkantak (Ahirvar et al. 2020). In the present study, five lakes located in the Amarkantak region have been selected: Johila Lake, Pamra1 Lake, Pamra2 Lake, Bhamra Lake, and Sambudhra Lake (Figure 6.1). The Johila lake cover spans 86 hectares situated in Miakel Velly and begins at the opposite location of the Narmada River and travels 15 km before joining the Sone River at Dasrathghat in the Umaria District of Madhya Pradesh (Kol et al. 2022). Sambudhara Lake is located in Maikal Hills, Amarkantak and is surrounded by dense forest. Pamra 1 and Pamra 2 Lakesare situated in village area and water is used for agriculture, bathing, and drinking purposes by local people. Bhamra Lake is located near the village area and is used exclusively for fishing purposes.

6.2.1 Materials and Methods

Water samples were collected from five different lakes in the Amarkantak region in 250 ml HDPE bottles. pH, electrical conductivity (EC), and total dissolved solids (TDS) were determined *insitu* using Multiparameter Water Quality Portable Meter (Hanna model-HI 98194). Turbidity was determined using Nephelometer (Systronics, model.135). The major ion analyses were done as per the standard procedure given by American Public Health

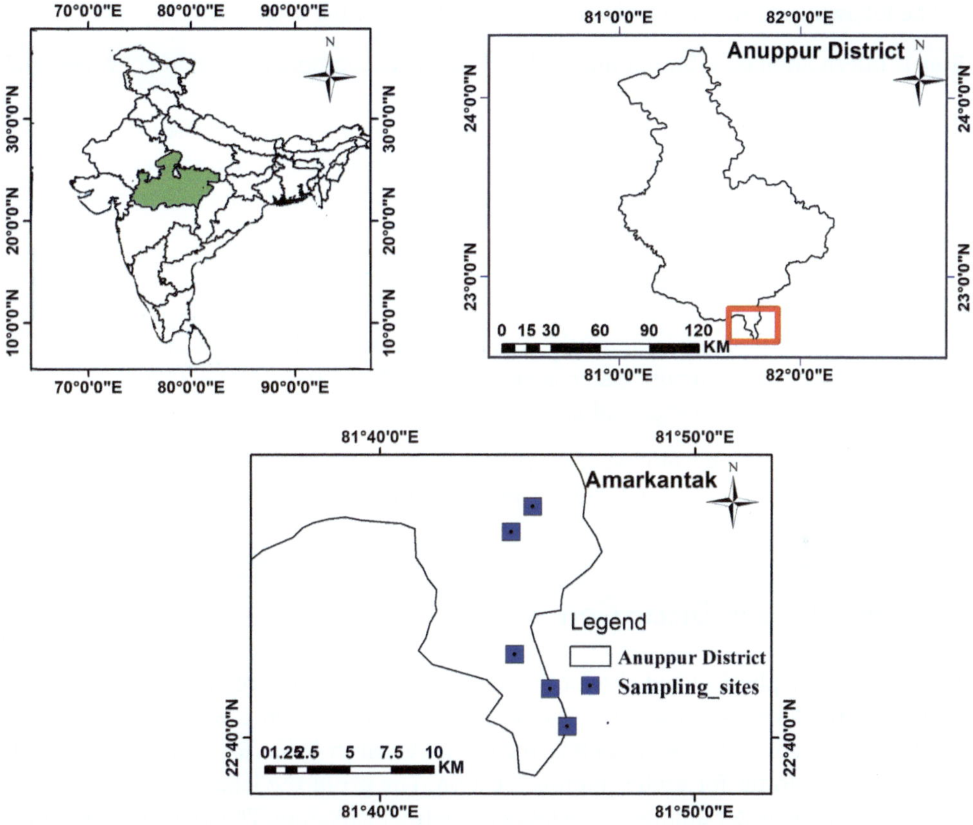

Figure 6.1 Study area showing sampling location in different lakes in Amarkantak, Madhya Pradesh.

Association (APHA 2005). Bicarbonate (HCO_3^-) and chloride (Cl^-) were measured using potentiometric and argentometric titration methods, respectively. The dissolved oxygen (DO) and biological oxygen demand (BOD) were measured using Winkler's titration method. Calcium (Ca^{2+}), magnesium (Mg^{2+}), and total hardness (TH) were measured using the EDTA titrimetric method. Nitrate (NO_3^-) was determined by ultraviolet spectrophotometer screening method. The sulfate (SO_4^{2-}) was determined using turbiditimetric method and phosphate (PO_4^{3-}) using stannous chloride method in UV-Vis spectrophotometer (Model No. 117). The Sodium (Na^+) and Potassium (K^+) were determined in flame photometer (model No. 1385).

6.2.2 Water Quality Index (WQI)

The WQI was calculated using the measured value of the physico-chemical parameter. Based on their significance to water quality, 15 parameters were selected-Ph, EC, TDS, turbidity, HCO_3^-, TH, DO, BOD, Na^+, K^+, Cl^-, Ca^{2+}, Mg^{2+}, SO_4^{2-}, NO_3^-. Each parameter is set to its mean values. The World Health Organization (WHO 2011) and Bureau of Indian (BIS) guidelines were standard for drinking water.

The following steps were involved in the calculation of the WQI:

Step1: Information on different physical and chemical components that affect water quality is compiled

Step2: The formula $k = (1/ (1/\sum S_n))$ is used to determine the value of the proportionality constant "K." Where "S_n" is the standard permissible for the nth parameter.

Step3: Where parameters exist, then the parameter (Q_n) quality rating is determined. This is determined by applying the formula.

$$(V_n - V_0)/(S_n - V_0)\}^* 100 = Q_n$$

Where V_n = Estimated value of the sampling station's nth parameter, V_0 = The nth parameter's ideal value in pure water, ideal value for pH = 7.0 and DO = 14.6, S_n = The nth parameter's standard allowable value.

Step4: The nth parameter unit weight. $W_n = (k/s\ n)$ is determined

Step5: The formula WQI = $\Sigma\ (W_n{}^*Q_n)$ is applied to determine the WQI

6.3 Results and Discussion

Determining the physical and chemical parameters of lake water offers valuable information about the lake's water quality. The results of various physio-chemical parameters of the different lakes in the Amarkantak region are given in Table 6.1 and compared with WHO and BIS norms for each sampling location. The pH of the lakes varies from 7.25 to 7.92, indicate the water quality of the lakes is neutral to alkaline. The increased pH can be caused by a slower rate of photosynthetic activity and the absorption of carbon dioxide and bicarbonates. The higher pH values indicate that carbon dioxide and carbonate–bicarbonate equilibrium are more sensitive to changes in physico-chemical conditions (Bhateria and Jain 2016).

The average temperature of lakes in Amarkantak region is 24.2 °C. The turbidity varies from 2 to 6 NTU. Turbidity in lakes can be caused by slit, mud, algae, plant pieces, wood ashes, etc. (Das Kangabam et al. 2017). A high sediment deposition from the rivers also causes the lake's water quality to deteriorate and become murky. The EC values range from 62 to 283 µs/cm. Highest EC value was found in Pamara Lake 2 which may be due to increased level of anthropogenic activities and because clay soils include elements that ionize as they wash into the water, these places typically have greater conductivity levels (Bhateria and Jain 2016). The TDS value ranges from 31 to 142 mgL^{-1}. TDS in all water samples was found to be less than 500 as per WHO limit infers that water is fit for drinking purposes. The salinity value ranges from 0.03 to 0.13 ppm. Pamara Lake 2 and Bhamra Lake have the highest salinity of the water, which can be due to the presence of common salts, mainly sodium chloride, and weathering of rocks that release ions.

The HCO_3^- concentration varies from 40 to 180 mgL^{-1}. The high HCO_3^- concentration was found in Pamara Lake 2, which could be the weathering carbonate rocks that contain calcite ($CaCO_3$) and dolomite ($CaMg(CO_3{}^{2-}$). DO is the important parameter of water quality in lake indicate pollution status of lake water. DO in the sample varies from 7.47 to

Table 6.1 Results of physico-chemical parameters of Lakes in Amarkantak region.

Parameters	Johilalake	Pamaralake 1	Pamaralake 2	Bhamralake	Sambhu Dhara lake	WHO (2017)	BIS (2012)
pH	7.78	7.92	7.44	7.54	7.25	6.5–8.5	6.5–8.5
Temperature (°C)	24.9	24.7	25.4	22.8	23.8	25–30	25.3
EC (µS/cm)	155	238	283	272	62	1000	—
ORP	178	180	213	100	109	—	—
Turbidity (NTU)	63	16	7	6	2	1	5
Salinity (%)	0.07	0.11	0.13	0.13	0.03	—	—
TDS (mgL^{-1})	78	119	142	136	31	500	500
HCO_3^- (mgL^{-1})	110	100	180	95	40	300	1000
SO_4^{2-} (mgL^{-1})	2.20	5.36	6.45	15.6	2.75	400	—
NO_3^- (mgL^{-1})	8.54	8.74	8.54	8.58	0.88	50	45
Cl^- (mgL^{-1})	35.5	24.8	24.8	67.4	14.2	250	250
PO_4^{3-} (mgL^{-1})	3.62	0	0	0.23	2.82	0.3	—
TH (mgL^{-1})	70	160	160	260	30	—	300
Ca^{2+} (mgL^{-1})	20.0	24.5	32.1	56.1	8.02	75	75
Mg^{2+} (mgL^{-1})	65.1	154	152	186	28.0	50	30
Na^+ (mgL^{-1})	0.76	7.53	7.97	0.37	0.58	200	—
K^+ (mgL^{-1})	2.37	6.46	3.20	9.75	4.59	10	—
DO (mgL^{-1})	7.52	8.50	7.47	8.28	8.52	—	5
BOD (mgL^{-1})	3.58	3.6	3.53	3.58	2.61	6	—

8.52 mgL^{-1}, which infers that the water quality of the Amarkantak region lakes is good, supporting a good rate of photosynthesis by phytoplankton. The highest value of DO was recorded in Sambhu Dhara Lake. The lowest value was recorded in Pamara Lake 2, possibly due to factors like microbial population, temperature, time of sampling, and pressure (Das Kangabam et al. 2017). BOD of the lake water samples varies from 2.61 to 3.6 mgL^{-1}. The highest BOD was found in Pamara Lake 1, which is due to the location of the lake close to the residential waste disposal sites and affected by anthropogenic activities. Low BOD in Sambhu Dhara Lake is due to the presence of forest cover and due to minimum human interference.

The Na^+ concentration ranges from 0.37 to 7.97 mgL^{-1}. Pamara Lake 2 has the highest Na^+ concentration, which may be due to dissolved natural sodium sources originating from Igneous rocks and municipal waste discharges (Mallick 2017). The concentration of K^+ levels in lake water ranges from 2.37 to 9.75 mgL^{-1}. Significant amounts of weathering in the study area could lead to the formation of insoluble secondary minerals (Mallick 2017). The Ca^{2+} concentration values range from 8.02 to 56.1 mgL^{-1}. A high concentration of calcium was found in Bhamra Lake. A high calcium concentration was found in Bhamra Lake due to local people's use of lake water for domestic activity. Moreover, the source of Ca in lake water includes mineral weathering, soil degradation, precipitation, and agricultural runoff (Weyhenmeyer et al. 2019). The Mg^{2+} concentrations ranges from 28.0 to 186 mgL^{-1}. Except in Sambu Dhra Lake, the Mg^{2+} concentration in three lakes is above the permissible limit per WHO drinking quality standards. The presence of various rocks and minerals containing magnesium in the nearby mountain area, which may eventually be washed off rocks and go into the water, is the cause of the high rate of Mg^{2+} in lake water (Mallick 2017). The Cl^- concentration ranges from 14.2 to 67.4 mgL^{-1} in lake water samples. The concentration of Cl^- is found to be below permissible limit in all Lakes. Chloride can enter water from natural sources like precipitation or weathering of rocks and can enter from human-caused sources such as surface runoff, industrial effluent, farming practices, and trash disposal (Thapa et al. 2024). The SO_4^{2-} concentration in the water sample ranges from 2.20 to 15.6 mgL^{-1}. The highest concentration is found in Bhamra Lake, which may be due to precipitating sulfuric acid through airborne pollutants containing sulfur oxides, weathering of sulfide-bearing minerals, decomposition, and combustion of organic matter in natural waters (Zak et al. 2020). PO_4^{3-} concentration values range from 0.0 to 3.62 mgL^{-1}, and NO_3^- concentration ranges from 0.876 to 8.74 mgL^{-1}. High phosphate content is found in Johila Lake, above the permissible limit. Nitrate values in all sampling sites are within permissible limits. The presence of sustainable amounts of nitrate and phosphate is due to the application of inorganic nitrogen fertilizers near agriculture fields. The variation of average cation concentration is in order $Mg^{2+} > Ca^{2+} > K^+ > Na^+$ and concentration of anion is $HCO_3^- > Cl^- > NO_3^- > SO_4^{2-}$. It shows that hardness in lake water is mainly due to presence of Mg^{2+}, Ca^{2+}, HCO_3^- and Cl^-.

The concentration of physico-chemical parameters of several lakes in the mountainous regions is compared with the present study is given in Table 6.2. It was found that the pH value of all lakes, including lake in Amarkantak is alkaline in nature. The DO value of Amarkantak Lake is higher than Loktak Lake and lower than Dodi Tal Lake and Nainital Lake, and the BOD value is lower than that of Nigeen Lake, Loktak Lake, Dal Lake, and Nigeen Lake. The EC concentration in Amarkantak Lake is higher than LoktakLake and lower than Nainital Lake. TDS values in all mountainous lakes are less than the permissible limit. Nitrate concentration is much higher in Amarkantak Lake than in Nigeen Lake and Dal Lake. The total hardness and alkalinity of Amarkantak Lake are higher than Loktak Lake and lower than Nainital Tal Lake. K^+ concentration in Amarkantak region lakes is higher than Dal Lak and lower than other lakes. Cl^-, Mg^{2+} have higher concentration levels in the Amarkantak region lake than in other tabulated lakes. SO_4^{2-} has higher concentration level in the Amarkantak region lake than in Neel Tal and Hemkund Lake. High PO_4^{3-} concentration is found in Amarkantak lake as compared to other lakes (Hemkund Lake, Loktak Lake, and Neel Tal Lake). Overallresult shows that Lakes in the Amarkantak region are polluted with nitrate and have a high concentration of magnesium as compared with other lakes in mountainous regions.

6.3.1 Water Quality Index (WQI)

The calculated WQI values were categorized as 0–25 – excellent water quality, 26–50 – good water quality, 51–75 – poor water quality, 76–100 – very poor water quality, and >100 unsuitable for drinking according to Yadav et al. (2010). The WQI values of different lakes

Table 6.2 Comparisons ofphysio-chemical parameters of several lakes in mountain region in India.

Lake	pH	DO	BOD	EC	TDS	NO_3^-	TH*	TA*	Na^+	K^+	Cl^-	Mg^{2+}	Ca^{2+}	SO_4^{2-}	PO_4^{3-}	References
Nainital	7.55	8.20	3.25	536	400	—	250	253	2.79	12.8	14.5	—	106	—	0.14	Sharma (2014), Negi et al. (2023)
Hemkund	7.9	7.00	1.7	158	95.2	0.45	9.6	35	—	—	2.75	1.23	1.3	0.25	0.07	Deep et al. (2020)
Dal lake	7.97	3.75	9.55	272	148	4.95	170	—	—	2.70	17.5	18.3	43.8	27.7	5.34	Ahamad et al. (2023), Sharma et al. (2015)
Nigeen lake	7.90	5.54	7.86	261	—	5.05	85–305	—	—	—	1–26	3.7–31	64–70	31.5	4.52	Ahamad et al. (2023), Dar et al. (2020)
Neel Tal lake	6.97	6.27	—	138	75.5	0.67	23.0	32.9	—	—	4.0	1.95	6.0	0.32	0.07	Kumar and Sharma (2019)
Dodi Tal lake	7.66	9.15	0.33	2.3	189	2.5	2.3	36.3	2.8	—	2.5	2.3	2.0	3.0	—	Singh and Sharrma (2020),Kumari et al. (2023), Kumar et al. (2019)
Bhimtal lake	7.50	7.9	2.6	313	120	1.13	137	102	13.9	6.06	19.4	13.2	32.8	16.6	9.28	Dalakoti et al. (2018), Lohchab et al. (2017), Kumari et al. (2023)
Beni Tal lake	6.7	6.1	1.8	163	129	1.58	55.0	48.8	—	—	—	5.62	12.8	4.29	—	Singh et al. (2018)
Loktak lake	7.26	7.03	4.99	161	60.0	0.36	43.4	—	8.94	5.98	23.4	8.40	26.9	—	0.54	Roy and Majumder (2019), Das Kangabam et al. (2017)
Amarkantak lake	7.59	8.06	3.38	202	101	7.06	136	105	3.44	5.27	33.4	117	28.1	6.47	1.33	Present Study (Average value)

*TH = Total Hardness, *TA = Total Alkalinity, All parameters are in mgL^{-1}, EC in μSCm^{-1}.

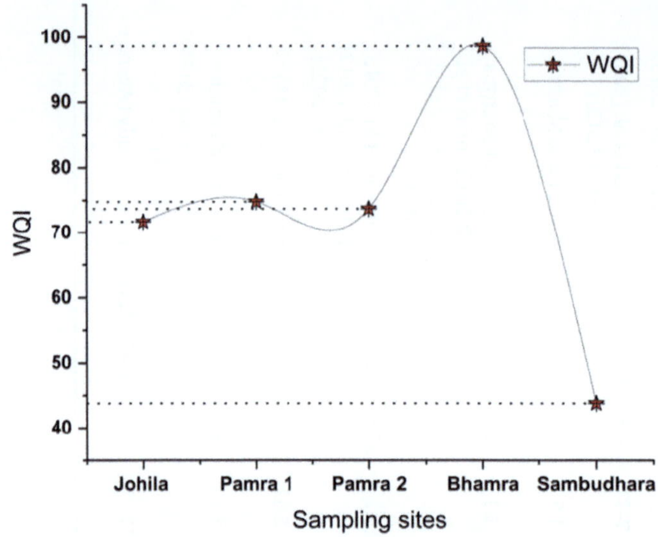

Figure 6.2 Water quality index (WQI) of different Amarkantak region lakes.

in the Amarkantak region are shown in Figure 6.2. It was found that the WQI of lake water ranges from 43.8 to 98.6, which indicates the very poor to good quality of water. It was found that the Bhamra Lake has recorded the highest WQI value of 98.6, indicating very poor quality of water and unfit for drinking. The high WQI value of lake water is due to runoff from agricultural land, soil erosion, household wastewater, and heavy rainfall (Roy and Majumder 2019; Singh and Sharma 2020). The WQI values in Pamra 1 and Pamra 2 were 74.8 and 73.7, respectively, indicating poor water quality. Shambhu Dhara Lake shows the lowest WQI, indicating good quality of water, which is due to the lake's location in a forest area and has minimal anthropogenic activity. Based on this result, Shambhu Dhara has better water quality, followed by Pamra 1, Pamra 2, and Bhamra Lake. The high WQI value is due to the high value of sub-indices of Mg^{2+}, TDS, Cl^-, and PO_4^{3-}.

6.4 Conclusion

The study shows that lake water quality is found to be alkaline. The average cation concentrationisinorder$Mg^{2+} > Ca^{2+} > K^+ > Na^+$andconcentrationofanionis$HCO_3^- > Cl^- > NO_3^- > SO_4^2$ hardness in lake water is mainly due to presence of Mg^{2+}, Ca^{2+}, HCO_3^-, and Cl^-. All other parameters except magnesium, chloride, and phosphate are above permissible limit as per WHO and BIS guidelines. High chloride and phosphate concentration in some lakes is due to the impact of anthropogenic activity by local people. Except in Sambudhara Lake, magnesium concentration in three lakes is found to be above permissible limit which may be due to weathering of rocks and minerals. The hardness in lake water is mainly due to presence of Mg^{2+}, Ca^{2+}, HCO_3^-, and Cl^-. The low DO and high BOD are found in Pamra Lake, which is due to the impact of waste disposal and anthropogenic activities by resident of the area. The WQI values show that the water quality of lakes varies from good to very poor (Sambudhara Lake > Johila Lake > Pamra 2 Lake > Pamra 1 Lake > Bhamra Lake) quality. Water quality in some lakes is not fit for drinking purposes due to the impact of anthropogenic activity. The high WQI values are due to the high valueof sub-indices of Mg^{2+}, TDS, Cl^{-}, and PO_4^{3-}. Among the five lakes, Bhamra Lake is more polluted, and Sambudhara Lake's water quality is better than other lakes. The concentration of physico-chemical parameters lakes in Amarkantak region has been compared with several lakes. It was found that lakes in the Amarkantak region are polluted with nitrate and have a high concentration of magnesium as compared with other lakes in mountainous regions. Moreover, these lakes play a crucial role in the ecosystem of the area and offer essential services, are affected by variety of factors including natural processes, anthropogenic activity, and climate change. The present work provides valuable information about lake water quality in the Amarkantak region. The result also suggests that residents must treat the water before using it. Effective management and continuous monitoring of lake water are needed to stop the lake's deterioration.

Acknowledgement

The authors are thankful to the Department of Science and Technology-Science and Engineering Research Board for funding research grant under EMEQ scheme.

References

Ahamad, F., Sharma, K., and Tyagi, S.K. (2023). A study on comparative assessment of water quality of Dal and Nigeen lakes of Jammu and Kashmir, India. *AgroEnvironmental Sustainability* 1 (1): 48–56. https://doi.org/10.59983/s2023010107.

Ahmed, R., Rawat, M., Wani, G.F. et al. (2022). Glacial Lake outburst flood hazard and risk assessment of Gangabal Lake in the Upper Jhelum basin of Kashmir Himalaya using geospatial technology and hydrodynamic modeling. *Remote Sensing* 14 (23): 5957. https://doi.org/10.3390/rs14235957.

Ahirvar, B.P., Chaudhry, S., Kumar, M., and Das, P. (2020). Climate change impact on forest and agrobiodiversity, a special reference to Amarkantak Area, Madhya Pradesh. In: *Contemporary Environmental Issues and Challenges in Era of Climate Change* (ed. Singh, P., Singh, R.P., Srivastava, V.), 65–76. Singapore: Springer. https://doi.org/10.1007/978-981-32-9595-7_3.

Akhtar, N., Syakir Ishak, M.I., Bhawani, S.A., and Umar, K. (2021). Various natural and anthropogenic factors responsible for water quality degradation. *Water* 13 (19): 2660. https://doi.org/10.3390/w13192660.

Alam, M., Saket, I., Singh, V.P. et al. (2021). Comparative study of Amarkantak, Shahdol, and Umaria basalt lava flows from the Eastern Deccan Volcanic Province (EDVP): field and petrographic characteristics. *Xplore* 12 (2): 88–104.

APHA (2005). *Standard Methods for the Examination of Water and Wastewater*, 21e. Washington DC: American Public Health Association.

Avila-Diaz, A., Torres, R.R., Zuluaga, C.F. et al. (2023). Current and future climate extremes over Latin America and Caribbean: Assessing Earth System Models from High Resolution Model Intercomparison Project (HighResMIP). *Earth Systems and Environment* 7 (1): 99–130. https://doi.org/10.1007/s41748-022-00337-7.

Babayan, G. and Adamovich, B. (2023). Water quality assessment of large alpine Sevan lake. *Environmental Processes* 10: 52. https://doi.org/10.1007/s40710-023-00668-1.

Bhateria, R. and Jain, D. (2016). Water quality assessment of lake water. *Sustainable Water Resources Management* 2 (2): 161–173. https://doi.org/10.1007/s40899-015-0014-7.

Bureau of Indian Standard (2012). 10500, Indian standard drinking water specification, Second revision, 1–24.

Dalakoti, H., Mishra, S., Chaudhary, M., and Singal, S.K. (2018). Appraisal of water quality in the Lakes of Nainital District through numerical indices and multivariate statistics India. *International Journal of River Basin Management* 16 (2): 219–229. https://doi.org/10.1080/15715124.2017.1394316.

Dar, S.A., Bhat, S.U., Aneaus, S., and Rashid, I. (2020). A geospatial approach for limnological characterization of Nigeen Lake, Kashmir Himalaya. *Environ Monit Assess* 192: 121. https://doi.org/10.1007/s10661-020-8091-y.

Das Kangabam, R., Bhoominathan, S.D., Kanagaraj, S., and Govindaraju, M. (2017). Development of a water quality index (WQI) for the Loktak Lake in India. *Applied Water Science* 7: 2907–2918. https://doi.org/10.1007/s13201-017-0579-4.

Deep, A., Gupta, V., Bisht, L., and Kumar, R. (2020). Application of WQI for water quality assessment of high-altitude snow-fed sacred Lake Hemkund, Garhwal Himalaya. *Sustainable Water Resources Management* 6: 89. https://doi.org/10.1007/s40899-020-00449-w.

Elsayed, S., Ibrahim, H., Hussein, H. et al. (2021). Assessment of water quality in lake Qaroun using ground-based remote sensing data and artificial neural networks. *Water* 13 (21): 3094. https://doi.org/10.3390/w13213094.

Fernando, G.A., Pablo, L.S., Christopher, S.J. et al. (2023). Evaluation of the water quality of a high Andean lake using different quantitative approaches. *Ecological Indicators* 154: 110924. https://doi.org/10.1016/j.ecolind.2023.110924.

Ganaie, D.B., Malhotra, A., and Wani, I. (2022). Water quality assessment of Mansbal Lake in Kashmir. *Water Science* 36: 114–124. https://doi.org/10.1080/23570008.2022.2129149.

Gaury, P.K., Meena, N.K., and Mahajan, A.K. (2018). Hydrochemistry and water quality of Rewalsar Lake of Lesser Himalaya, Himachal Pradesh, India. *Environmental Monitoring and Assessment* 190: 84. https://doi.org/10.1007/s10661-017-6451-z.

Gupta, H. and Chakrapani, G.J. (2005). Temporal and spatial variations in water flow and sediment load in Narmada River Basin, India: natural and man-made factors. *Environmental Geology* 48: 579–589. https://doi.org/10.1007/s00254-005-1314-2.

Khan, M., Ellahi, A., Niaz, R. et al. (2022). Water quality assessment of alpine glacial blue water lakes and glacial-fed rivers. *Geomatics, Natural Hazards and Risk* 13 (1): 2597–2617. https://doi.org/10.1080/19475705.2022.2126800.

Kol, S., Payasi, Y., Shukla, J., and Shukla, A. (2022). Seasonal variation in water quality and phytoplankton production in johila reservoir, lalpur, amarkantak (m.p.). *Indian Journal of Applied Research* 23–25. https://doi.org/10.36106/ijar/9612159.

Kumar, R. and Sharma, R.C. (2019). Assessment of the water quality of Glacier-fed Lake Neel Tal of Garhwal Himalaya, India. *Water Science* 33 (1): 22–28. https://doi.org/10.1080/11104929.2019.1631554.

Kumar, R., Singh, S., and Sharma, R.C. (2019). Application of WQI for assessment of water quality of high-altitude lake Dodi TalGarhwal Himalaya, India. *Sustainable Water Resource Management* 5: 1033–1042.

Kumari, N., Thakur, K., Kumar, R. et al. (2023). Freshwater lakes in the Western Himalayan region: an analysis of the present situation. *Water-Energy Nexus* 6: 18–31. https://doi.org/10.1016/j.wen.2023.06.002.

Kumari, R. and Sharma, R. (2018). Assessment of water quality index and multivariate analysis of high-altitude sacred Lake Prashar, Himachal Pradesh, India. *International Journal of Environmental Science and Technology* 16: 6125–6134. https://doi.org/10.1007/s13762-018-2007-1.

Lohchab, R.K., Agrawal, D.K., and Kaushik, C.P. (2017). Tourism impacts on the water quality of Bhimtal lake in central Himalaya. *Pollution Research* 36 (2): 296–305.

Lydersen, E., Høgberget, R., Moreno, C.E., Garmo, O.A., and Hagen, C.P. (2014). The effects of wildfire on the water chemistry of dilute, acidic lakes in southern Norway. *Biogeochemistry* 119: 109–124. https://doi.org/10.1007/s10533-014-9951-8.

Machate, O., Schmeller, D.S., Schulze, T., and Brack, W. (2023). Review: mountain lakes as freshwater resources at risk from chemical pollution. *Environmental Sciences Europe* 35 (1): 1–15. https://doi.org/10.1186/s12302-022-00710-3.

Mallick, J. (2017). Hydrogeochemical characteristics and assessment of water quality in the Al-Saad Lake, Abha Saudi Arabia. *Applied Water Science* 7: 2869–2882. https://doi.org/10.1007/s13201-017-0553-1.

Mihai, F.C. (2018). Rural plastic emissions into the largest mountain lake of the Eastern Carpathians. *Royal Society Open Science* 5: 172396. https://doi.org/10.1098/rsos.172396.

Mishra, R. and Prajapati, R. (2011). Water quality assessment of Rani Lake of Rewa (M.P.) India. *GERF Bulletin of Biosciences* 2: 11–17.

Negi, P.S., Singh, R., and Mishra, A.P. (2023). An assessment of water quality of nainital springs, Uttarakhand, India. *Journal of Indian Association for Environmental Management (JIAEM) 42* (4): 17–23.

Singh, M.M. and Rai, P.K. (2014). A study of depleting water quality and wildlife resources of the Loktak Lake, Manipur, India. *Issues and Trends of Wildlife Conservation in Northeast India* 96–104. ISBN 978-81-924321-7-5.

Roy, R. and Majumder, M. (2019). Assessment of water quality trends in Loktak Lake, Manipur, India. *Environmental Earth Sciences* 78: 383. https://doi.org/10.1007/s12665-019-8383-0.

Sharma, M. (2014). Water quality assessment of the Central Himalayan Lake, Nainital. *Advances in Environmental Chemistry* 2014 (1): 473074. https://doi.org/10.1155/2014/473074.

Sharma, J.N., Kanakiya, R.S., and Singh, S.K. (2015). Limnological study of water quality parameters of Dal lake, India. *International Journal of Innovative Research in Science, Engineering and Technology* 4 (2): 380–386.

Sharma, R.C. and Kumar, R. (2017). Water quality assessment of sacred glacial Lake Satopanth of Garhwal Himalaya, India. *Applied Water Science* 7: 4757–4764. https://doi.org/10.1007/s13201-017-0638-x.

Singh, K., Sharma, R., Tiwari, V., and Kumar, R. (2018). Assessment of surface water quality of the Himalayan Lake Beni Tal, India. *Current Research in Hydrology and Water Resources* 2018: CRHR-102. https://doi.org/10.29011/CRHR-102.

Singh, S. and Sharma, R.C. (2020). Zooplankton diversity and potential indicator species for assessment water quality of high-altitude wetland, Dodi Tal of Garhwal Himalaya, India. *Academia Arena* 12 (5): 1–16.

Su, K., Wang, Q., Li, L. et al. (2022). Water quality assessment of Lugu Lake based on Nemerow pollution index method. *Scientific Reports* 12 (1): 13613. https://doi.org/10.1038/s41598-022-17874-w.

Thakur, T.K., Padwar, G.K., Patel, D.K., and Bijalwan, A. (2019). Monitoring land use, species composition and diversity of moist tropical environ in Achanakmaar Amarkantak Biosphere reserve, India using satellite data. *Biodiversity International Journal* 3 (4): 162–172. https://doi.org/10.15406/bij.2019.03.00141.

Thapa, B., Khanal, L., Pant, R.R. et al. (2024). Hydrochemistry and irrigation quality of high-altitude lakes: a case study of the Ramaroshan lake complex, Nepal Himalayas. *Limnological Review* 24 (1): 30–52. https://doi.org/10.3390/limnolrev24010003.

Vale, S.S., Smith, H.G., Davies-Colley, R.J. et al. (2023). The influence of erosion sources on sediment-related water quality attributes. *Science of the Total Environment* 20 (860): 160452. https://doi.org/10.1016/j.scitotenv.2022.160452.

Vasistha, P. and Ganguly, R. (2020). Water quality assessment of natural lakes and its importance: an overview. *Materials Today: Proceedings* 32: 544–552. https://doi.org/10.1016/j.matpr.2020.02.092.

Vasistha, P. and Ganguly, R. (2022). Water quality assessment in two lakes of Panchkula, Haryana, using GIS: case study on seasonal and depth wise variations. *Environmental Science and Pollution Research International* 29 (28): 43212–43236. https://doi.org/10.1007/s11356-022-18635-y.

Vishwakarma, S.P., Prajapati, R., Dubey, Y. et al. (2023). Documentation of medicinal plants of moist deciduous forest of Amarkantak region, Madhya Pradesh, India. *International Journal of Environment and Climate Change* 13: 2234–2255. https://doi.org/10.9734/ijecc/2023/v13i92457.

Weyhenmeyer, G.A., Hartmann, J., Hessen, D.O. et al. (2019). Widespread diminishing anthropogenic effects on calcium in freshwaters. *Scientific Reports* 9 (1): 1–10. https://doi.org/10.1038/s41598-019-46838-w.

World Health Organization (2011). *Guidelines for Drinking-Water Quality*, 4e. Geneva: WHO.

WHO (2017). *Guidelines for Drinking Water Quality*, 4e, Incorporating the First Addendum. Geneva: World Health Organization.

Xia, W., Wang, R., Zhu, B. et al. (2020). Heavy metal gradients from rural to urban lakes in central China. *Ecological Processes* 9: 47. https://doi.org/10.1186/s13717-020-00251-8.

Yadav, A.K., Khan, P., and Sharma, S.K. (2010). Water quality index assessment of groundwater in Todaraisingh Tehsil of Rajasthan State, India-A greener approach. *Journal of Chemistry* 7: S428–S432.

Zak, D., Hupfer, M., Cabezas, A. et al. (2020). Sulphate in freshwater ecosystems: a review of sources, biogeochemical cycles, ecotoxicological effects and bioremediation. *Earth-Science Reviews* 212: 103446. https://doi.org/10.1016/j.earscirev.2020.103446.

7

Hydrogeochemical Evaluation of Groundwater Quality in Leh Town, Trans-Himalaya, India Using Entropy Water Quality Index

Lobzang Chorol and Sunil K. Gupta

Department of Environmental Science and Engineering, Indian Institute of Technology (Indian School of Mines), Dhanbad, India

7.1 Introduction

The groundwater supply plays a significant role in supporting socioeconomic development and maintaining diversified ecosystems. Human health and well-being are greatly influenced by hydrogeochemistry. The chemistry of groundwater is crucial in understanding how groundwater systems evolve and interact with irrigation water (Chen et al. 2020; Chorol and Gupta 2023b). It is subject to multiple hydrogeological, chemistry, and biological processes that control the transportation, retention, and release of natural or anthropogenic solutes (Giordano 2009). Groundwater is the prime natural resource in semiarid and arid zones due to a shortage of precipitation and temperature regimes which are experiencing a double water crisis due to depletion and deteriorating quality of groundwater (Wu et al. 2018; Gao et al. 2020). Groundwater quality tends to be better than surface water; however, this is not certain since groundwater quality varies owing to natural and artificial processes during recharge and abstraction (Sunkari et al. 2022). The study of hydrogeochemical processes controlling groundwater quality offers insight into groundwater protection and sustainability (Zhang et al. 2019; Wang et al. 2022). The main natural chemical composition of groundwater is derived from weathering of rocks and soils, solutes from the atmosphere, and chemical reactions occurring underground (Sunkari et al. 2021). Several anthropic activities such as agricultural, industrial activities, mining, and unsanitary conditions are the main source of dissolved constituents that can influence the chemistry of groundwater (Panikary and Chander 2022; Sunkari et al. 2022). Groundwater contamination not merely degrades water quality but also threatens human health as well as social and economic growth (Subba Rao et al. 2020). Hydrogeochemical assessment is critical for long-term usage and preservation of groundwater resources. The chemical constituents of groundwater can be identified and utilized for decision-making. It also helps in understanding the mechanism involved and the geochemical alteration in a particular area

Water Resources Management in Mountain Regions, First Edition. Edited by Virendra Bahadur Singh, Sughosh Madhav, Rakesh Kumar Gupta, Prerna Diwan, and Amit Kumar.

which helps understand the quality and its aptness for usage (Qian et al. 2020; Chen et al. 2020; Sunkari et al. 2022).

A study conducted by Subba Rao (2018) reported that in Andhra Pradesh, India, rock weathering is the primary geogenic source influencing groundwater chemistry, while anthropogenic sources (household wastes, agricultural, synthetic fertilizers, etc.) are the secondary source influencing natural groundwater quality. Similarly, Sunkari et al. (2020) investigated the hydrogeochemistry of the region in Ghana and reported that groundwater chemistry is predominantly regulated by silicate mineral weathering, ion exchange processes, and agricultural activities. In another study conducted by Kumar et al. (2014), it was reported that in shoreline zone of South Chennai, India; urbanization and saltwater intrusion had a major influence on groundwater quality. Also, Li et al. (2016) reported that natural activities such as groundwater evaporation, cation exchange, and weathering of parent rocks are the primary variables changing groundwater chemistry in China's Hua County region. Thus, studying the hydrogeochemistry of a region is essential for understanding the chemistry for irrigation and drinking purposes.

Water quality monitoring generates large and complex database which is often difficult to manage and derive conclusions due to large variability of parameters and quality status of water bodies. In India, various studies were carried out to evaluate the quality of potable water and irrigation purposes in Andhra Pradesh (Subba Rao et al. 2020), Telangana (Adimalla and Venkatayogi 2018; Narsimha 2018), and Chandigarh (Ravindra et al. 2019). Although, there is limited information available on the groundwater quality and quantity in the area. The potability evaluation of groundwater using entropy-weighted water quality index (EWQI) model will provide an understanding of major influencing parameters. In addition to being easy and reliable, the EWQI has been shown to offer unbiased results in the determination of the quality of drinking water (Wu et al. 2018; Naik et al. 2022). Researchers around the world use EWQI to study groundwater quality due to its lucid concept and determinable weight (Khanoranga and Khalid 2019; Siddique et al. 2020; Ji et al. 2020; Adimalla 2021). In order to study irrigation potential of groundwater, several indices such as KR, Na%, RSC, magnesium hazard (MH), SAR, and PI (Selvakumar et al. 2014; Amiri et al. 2021; Sunkari et al. 2022). Various researchers have used multivariate statistical analysis together with hydrogeochemical studies to identify pollution sources and examine groundwater dynamics (Loh et al. 2020; Mohana and Velmurugan 2020; Ciner et al. 2021).

Leh is a fast-growing Indian city with a model conventional agricultural society (Muller and Dame 2016). Groundwater act as the primary source for domestic purposes and the quality is prone to degradation because of rapid urbanization, improper waste disposal, and leaching. As of now, it is under severe stress by several factors like limited supply, abstractions, and climatic variations. Due to the growth of the tourism industry, the arable land has been steadily altered into built-up areas such as hotels and guest houses. Li et al. (2019) found that a decrease in irrigated land is affecting groundwater replenishment.

Leh, situated in the trans-Himalayan region, is nestled within a series of mountain ranges that run in a southeast-to-northwest direction, dictating the flow of Ladakh's rivers. Its unique geographical feature and harsh winter conditions, where temperatures drop below freezing, have limited extensive exploration by researchers and scientists. Thus, a large research gap exists on the hydrogeochemical properties of groundwater quality in

Leh town and also in similar geo-climatic regions. Hence, this study aimed to accomplish the following objectives:

i) To evaluate the groundwater suitability for consumption by employing an EWQI
ii) Evaluate the appropriateness of groundwater for irrigation applications.
iii) Conduct a human health risk to estimate the potential risks associated with groundwater consumption.
iv) Identify the sources of contaminants using multivariate statistical analysis.

The study will benefit the government water authority and hydrologists to assess the trend in groundwater quality, detect the origin of contamination, and set preventive rules and regulations to ease the public well-being risk.

7.2 Materials and Methods

7.2.1 Study Area Profile

The study area consists primarily of unconsolidated sediments found within arid high mountains in the rain shadow of mountains. It lies between 32°N and 36°N north latitude and 75°E to 80°E degree longitude, at an elevation of 2300–5000 m above sea level (Figure 7.1). During the Quaternary, the region has been exposed to cycles of glacial, periglacial, and paraglacial processes. This area of the trans-Himalaya is classed as a cold-arid zone (also known as a cold desert) due to low precipitation and hot summers. It is characterized by high land–atmosphere interactions, a thin atmosphere, and intense incoming solar radiation.

Geologically, Leh town is part of the Ladakh Batholith, which is composed of granitic rocks formed during the Late Cretaceous to Early Eocene. The batholith intrudes into older sedimentary and metamorphic rocks of the Ladakh Group and the Indus Suture Zone. The region also features extensive Quaternary deposits, including glacial, fluvial, and lacustrine sediments. These deposits are mainly composed of unconsolidated sand, gravel, clay, and silt with occasional boulders. The movement, occurrence, and physicochemical properties of groundwater are controlled and determined by the permeability, porosity, and geochemistry of prior mentioned lithological units.

7.2.2 Sampling and Analytical Methods

To ensure a comprehensive spatial distribution, groundwater samples were collected in replicate from 28 monitoring sites during the winter season of 2018. The sampling points were selected based on their proximity to the hotel industry, agricultural areas, and the availability of public standpipes (Figure 7.1). Prior to sample collection, each handpump or borehole was purged for 10 minutes to prevent foreign contamination (USGS 2006). Samples were gathered in 1-liter polyethylene containers cleaned with distilled water and rinsed thrice using the sample water during collection. The labelled sample bottles were transported to the lab following protocol, where they were stored at temperatures less than 4 °C until the completion of the analysis. A cellulose acetate filter with a pore size of less than 45 micrometers was used to prepare samples for ion analysis (APHA 2012). Temperature, pH, Electrical

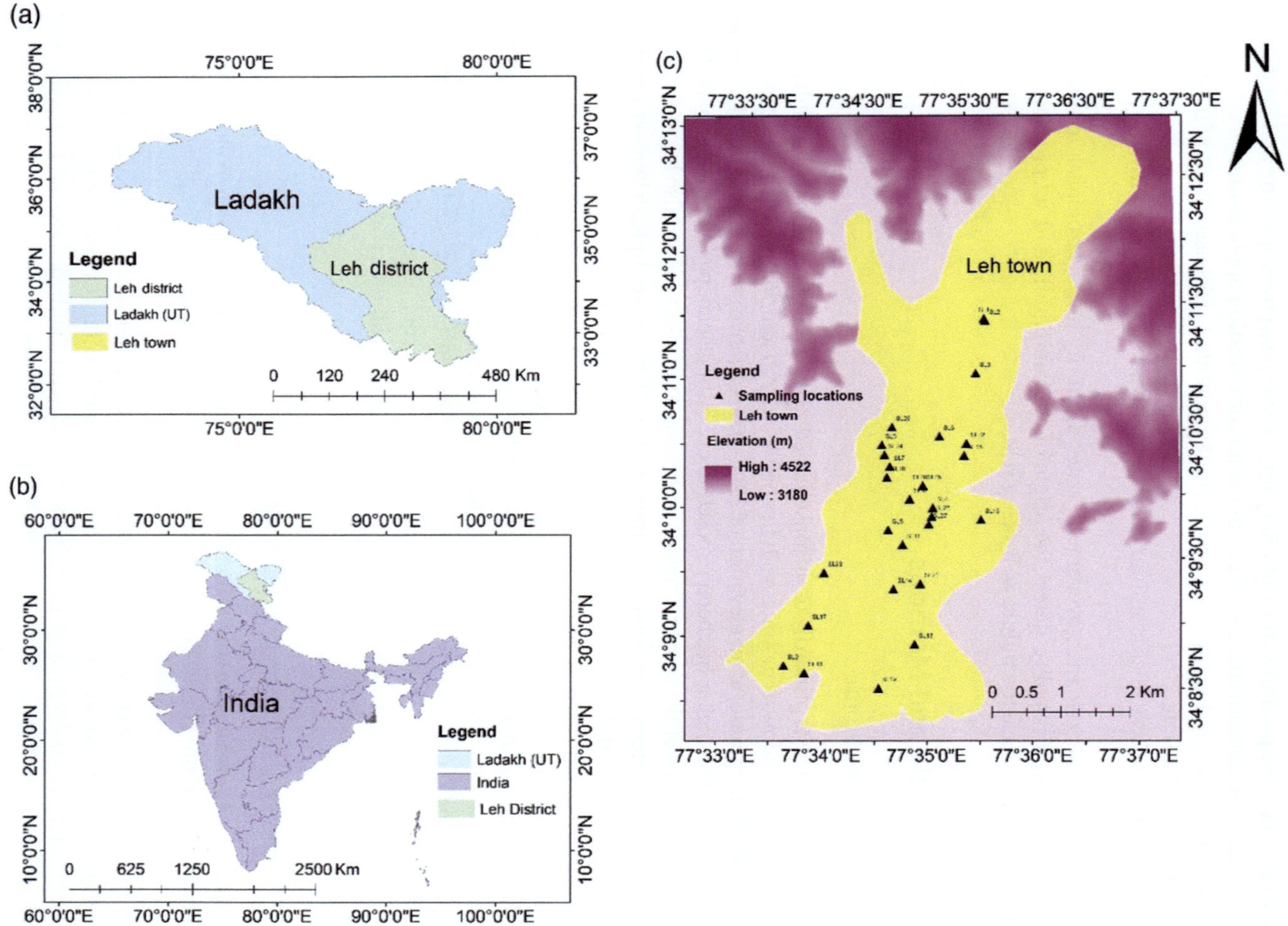

Figure 7.1 Map of Leh town, Ladakh (UT) from India maps with sampling locations. (a) Showing the geographic position of the study area within Leh District. (b) Showing the geographic location of the Leh district in Ladakh UT of India. (c) Showing the sampling locations of study area in Leh town.

conductivity was assessed on-site via calibrated multiparameter display system. Cation analysis (Na^+, K^+, Ca^{2+}) were carried out using flame photometry (Model 1385). While the anions' concentrations were determined using various methods: sulfate by the turbidimetric method, chloride (Cl^-) by the argentometric method, and fluoride (F^-) by the SPADNS method. Total chromium concentrations were measured using atomic absorption spectrophotometry. A Piper diagram was created using Rockware AqQA Scientific software version 1.1.1 to visually depict the hydrochemical properties of the groundwater samples.

7.2.3 Entropy Water Quality Index

The EWQI is a reliable, simple, and unbiased approach for determining water quality (Naik et al. 2022). Because of the lucid concept and determinable weight, the EWQI is commonly used by researchers in assessing groundwater quality throughout the world (Khanoranga and Khalid 2019; Ji et al. 2020; Siddique et al. 2020; Adimalla 2021). The computation of the EWQI involves five stages, each focusing on determining the weight of individual water quality parameters (Egbueri et al. 2020). It classifies water into five distinct categories based on the calculated index values. An EWQI score below 25 indicates excellent water quality, while a score between 25 and 50 suggests good quality. Groundwater with an EWQI value ranging from 50 to 100 is considered to be of average quality. If the index falls between 100 and 150, the water is regarded as poor quality. Lastly, an EWQI score exceeding 150 signifies very poor groundwater quality (Ji et al. 2020; Dashora et al. 2022). A data matrix, X(nm), is constructed assuming evaluation targets (n) and evaluation indicators (m). To assess the weights of each variable, the data undergoes a normalization process to convert it into a dimensionless quantity. The decision matrix is normalized using Eq. (7.1)–(7.4).

$$X = \begin{bmatrix} x_{11}x_{12}\dots x_n \\ x_{21}x_{22}\dots x_{2n} \\ \cdot \\ \cdot \\ \cdot \\ x_{m1}x_{m2}\dots x_{mn} \end{bmatrix} \tag{7.1}$$

A second step in the standardization process is to assess the "y_{ij}" and then use Eqs. (7.2) and (7.3) to derive the standard evaluation matrix "Y."

$$y_{ij} = \frac{x_{ij} - \left(x_{ij}\right)_{\min}}{\left(x_{ij}\right)_{\max} - \left(x_{ij}\right)_{\min}} \tag{7.2}$$

$$Y = \begin{bmatrix} y_{11}y_{12}\dots y_n \\ y_{21}y_{22}\dots y_{2n} \\ \cdot \\ \cdot \\ \cdot \\ y_{m1}y_{m2}\dots y_{mn} \end{bmatrix} \tag{7.3}$$

Normalized decision matrix (p_{ij}) and entropy (e_j) are determined (Dashora et al. 2022) as follows:

$$p_{ij} = \frac{x_{ij}}{\sum_{i-1}^{m} x_{ij}} \tag{7.4}$$

$$e_j = -\frac{1}{\ln(m)} \sum_{i=1}^{m} p_{ij} \ln p_{ij} \tag{7.5}$$

The e_j of each variable (w_j) was calculated:

$$w_j = \frac{1-e_j}{\sum_{j=1}^{n} (1-e_j)} \tag{7.6}$$

In each sample, quality rating scales (q_j) were computed by

$$q_j = \frac{C_j}{S_j} \times 100 \tag{7.7}$$

In this equation, C_j represents the measured concentration of each water quality parameter, while S_j denotes the corresponding acceptable limit as per the guidelines set by the Bureau of Indian Standards (BIS 10500: 2012). Once the values for each parameter have been obtained, the final step involves computing the EWQI:

$$\text{EWQI} = \sum_{j-1}^{n} w_j q_j \tag{7.8}$$

7.2.4 Evaluation Groundwater Quality for Irrigation

The groundwater quality for irrigation depicts its composition of minerals and also shows its impact on plants and soil. So, the determination of groundwater for irrigation is significant to yielding high-quality crops (Riedel et al. 2022). Irrigation water is correlated with plants, types of soils, and their management. It varies with the region, climate, source, and geology. The important parameters for evaluating groundwater quality for irrigation purposes are given in Table 7.1 with formulas and categorization.

7.2.5 Statistical Analysis

Principal component analysis (PCA) is a statistical technique that converts a set of variables into a new, reduced set of components, effectively decreasing the data's dimensionality (Rakotondrabe et al. 2018). IBM's (Statistical Package for Social Sciences (SPSS) version 20 was employed to perform PCA, which helped determine the sources of groundwater contamination. Principal component analysis is a statistical method that

Table 7.1 Parameters and formulae for evaluating groundwater quality for irrigation purposes.

Parameters	Categories			Formula	References
Residual sodium carbonate (RSC)	S. No.	RSC (meq/l)	Class	$RSC = (CO_3^{2-} \mp HCO_3^{-} - Ca^{2+}$	(Richards 1954)
	1.	<1.25	Good		
	2.	1.25–2.50	Doubtful		
	3.	>2.50	Unsuitable		
Sodium percentage (Na%)	S.No.	Na%	Classification	$Na\% = \frac{(Na^{+} + K^{+})}{(Ca^{2+} + Mg^{2+} + Na^{+} + K^{+})} \times 100$	(Eaton 1950)
	1.	0–20	Excellent		
	2.	20–40	Good		
	3.	40–60	Permissible		
	4.	60–80	Doubtful		
	5.	>80	Unsuitable		
Sodium Adsorption Ratio (SAR)	S.No.	SAR	Categorization	$SAR = \frac{Na^{+}}{\sqrt{Ca^{2+} + Mg^{2+} / 2}}$	(USSL 1954)
	1.	<10	Excellent		
	2.	10–18	Good		
	3.	18–26	Fair		
	4.	>26	Poor		
Kelly's ratio (KR)	KR < 1 Bad for irrigation whereas KR > 1 is Good for irrigation.			$KR = \frac{Na^{+}}{(Ca^{2+} + Mg^{2+})}$	(Kelly 1946)
Magnesium hazards (MH)	MH < 50 is Bad for irrigation whereas MH > 50 is Good for irrigation.			$MH = \frac{Mg^{2+} \times 100}{Ca^{2+} + Mg^{2+}}$	(Szabolcs and Darab 1964)
Permeability Index (PI)	S.No.	PI (%)	Class	$PI = \frac{Na^{+} + \sqrt{HCO_3^{-}}}{Ca^{2+} + Mg^{2+} + Na^{+}} \times 100$	(Doneen 1962)
	1.	PI > 75	Excellent		
	2.	25 < PI < 75	Good		
	3.	<25	Unsuitable		

involves transforming a group of variables into a different set, thereby reducing the dimensionality of the data (Rakotondrabe et al. 2018), using IBM's SPSS version 20. PCA was used to identify the origin of groundwater contamination. Pearson correlation was used to assess the linear relationship between variables, has been employed in groundwater studies to identify connections and interactions among different hydrogeological factors. Davis et al. (2002) reported that all eigenvalues of 1.0 or higher are deemed important. And varimax criteria were used to obtain the extraction of the key factors (Kaiser 1958; El Alfy et al. 2017) which contributed to enhancing the sum of the variance of the squared loadings.

7.2.6 Evaluation of Deterministic Potential Risks to Human Health

The assessment of health risks associated with polluted groundwater, influenced by concentration levels and environmental factors, was conducted following the USEPA's four-step approach (Chorol and Gupta 2023a). Noncancer health hazards associated with fluoride to various age groups and genders were evaluated based on USEPA and IARC guidelines, considering ingestion and dermal contact pathways. Chronic daily intakes were calculated using specific formulae:

$$\mathrm{CDI}_{\mathrm{Oral}} = \frac{C_w \times EF \times IR \times ED}{BW \times AT} \tag{7.9}$$

$$\mathrm{CDI}_{\mathrm{Dermal}} = \frac{C_w \times K \times SA \times EF \times ED \times CF \times ET}{BW \times AT} \tag{7.10}$$

where $\mathrm{CDI}_{\mathrm{oral}}$ refers to the chronic daily dose taken orally, while $\mathrm{CDI}_{\mathrm{Dermal}}$ represents the chronic daily dose through skin, expressed in mg/kg/d. The ingestion rate (*IR*) is measured in L/day, while body weight (*BW*) is given in kg. The concentration of fluorides in the groundwater (C_w) is expressed in mg/l. Exposure frequency (*EF*) and exposure duration (*ED*) are measured in days/year and years, respectively. The average lifetime (*AT*) is expressed in days. Additional factors for the dermal pathway involve the skin's permeability coefficient (*K*) measured in cm/hr, skin surface area (*SA*) expressed in cm^2, exposure time (*ET*) calculated in h/event, and a conversion factor (*CF*) stated as $\mathrm{L/Cm}^3$. To assess the noncancer risk caused by fluorides in contaminated groundwater, a hazard quotient (*HQ*) approach was employed. The following equations were used to calculate the *HQ* values for oral and dermal exposure pathways:

$$HQ_{\mathrm{oral}} = \frac{\mathrm{CDI}}{\mathrm{RfD}} \tag{7.11}$$

$$HQ_{\mathrm{Dermal}} = \frac{\mathrm{CDI}}{\mathrm{RfD}} \tag{7.12}$$

$$HI = HQ_{\mathrm{Oral}} + HQ_{\mathrm{Dermal}} \tag{7.13}$$

The reference dose (RfD) for fluoride, established by the USEPA (2012), is 0.06 mg/kg/day, representing the maximum safe daily intake. To assess the total noncancer risk from various exposure pathways, a hazard index (*HI*) was estimated by adding the *HQ*s for each pathway. According to USEPA guidelines (2015), an HI below 1 indicates a safe exposure level, while an HI above 1 suggests potential health risk. This approach allows for a comprehensive evaluation of the risks posed by fluorides in contaminated groundwater, informing decision-making and guiding public health interventions.

7.3 Results and Discussions

7.3.1 Assessing the Suitability of Groundwater for Drinking

To determine the suitability of groundwater for drinking, the measured quality parameters were compared with the standards set by the (BIS 2012) (Table 7.2).

Groundwater in this study region had a pH of approximately 8.18 (Table 7.2), within acceptable limits (BIS 2012), which is slightly alkaline, ranging from 6.5 to 8.6. The average EC value was calculated to be 690.86 μS/cm. While this average falls below the permissible limit of 1000 μS/cm (BIS 2012), some portion of the samples surpassed this threshold, indicating the presence of elevated levels of dissolved ions in certain locations within the study area. The average TDS concentration in the study area was 495.02 mg/l, with 71.43% of the samples falling within the acceptable limits set by BIS (2012). While, 25% of the samples surpassed the permissible limit and rest of the samples were unfit for the consumption. The TDS levels observed in the groundwater samples from the study region can be attributed to

Table 7.2 Statistical summary of the physicochemical properties of groundwater samples collected from the study region.

Parameters	Minimum	Maximum	Mean	Std. deviation	Desirable limits as per BIS (2012)
pH	6.5	8.60	8.18	0.38	6.5–8.5
EC	469.14	1725.43	690.86	253.03	1000
TDS	340	1223.8	495.02	178.44	500
HCO_3^-	147	756.4	228.81	119.00	200
TH	185	794	269.08	118.25	200
Ca	23	120	46.26	22.55	75
Mg	13	105.3	25.08	17.77	30
Na	16.8	59.2	29.36	12.75	—
K	4.8	12.2	7.05	1.78	12
SO_4	54.9	112	83.01	16.57	200
Cl	48	115	64.46	14.89	250
NO_3^-	4.9	18.9	9.28	3.00	45
F	0.02	1.95	0.54	0.56	1
Cr	BDL	0.18	0.06	0.05	0.05

All the cations and anions are in mg/l; except EC in μS/cm.

a combination of anthropogenic activities and natural processes. Factors such as the weathering of rocks, variations in aquifer permeability and porosity, as well as the infiltration of fertilizer residues and sewage effluents from human settlements, can contribute to the dissolution and subsequent accumulation of dissolved solids in the groundwater system (Karunanidhi et al. 2021; Chorol and Gupta 2023a). It was found that bicarbonate (HCO_3) concentrations ranged from 147 to 756.4 mg/l, with the mean concentration reaching 228.81 mg/l. This value is slightly higher than the desirable limit (BIS 2012), suggesting that the groundwater in the study area consists of higher levels of bicarbonate ions, which could be influenced by the local Aquifer's geological and hydrogeochemical characteristics. Based on the total hardness, a significant majority (78.57%) were found to be in the extremely hard category, while the remaining 21.43% were classified as hard (Table 7.3). The high levels of hardness in the groundwater are likely due to the presence of dissolved calcium and magnesium ions, which typically originate from the dissolution of minerals within the aquifer matrix (Chorol and Gupta 2023a).

According to the current study, the most prevalent cations in the Leh groundwater are $Ca^{2+} > Na^+ > Mg^+ > K^+$. Ca^{2+} was found to be in the range of 23–120 mg/l with an average value of 46.26 mg/l (Table 7.1). 14.28 and 17.85% of groundwater samples consist of Ca^{2+} and Mg^{2+}, respectively, which was more than the acceptable limit (BIS 2012). The deficiency of Ca^{2+} in potable water causes hypertension, kidney stones, osteoporosis, and colorectal cancer (WHO 2011; Adimalla et al. 2020). Whereas, Mg^{2+} is an important ion for cell function and enzyme inactivation, and a greater concentration of Mg^{2+} in groundwater acts as a laxative agent (WHO 2011). The mean concentration of K^+ in groundwater samples was 12.2 mg/l and 3.57% of samples exceeded the acceptable limit (BIS 2012). Fertilizers and minerals in rocks, such as orthoclase, biotite, and microcline are the primary source of K^+ in groundwater (Samantaray and Sanyal 2021). Bicarbonate was the most prevalent anion in the groundwater, followed by sulfate (SO_4^{2-}), chloride (Cl^-), nitrate (NO_3^-), and fluoride(F^-). The average value of bicarbonate in groundwater was 228.81 mg/l and 42.86% of samples exceeded the desirable limit (BIS 2012), whereas 3.57% of groundwater samples exceeded the permissible limit and is nonpotable. However, the average value of Cl^- (64.46 mg/l), SO_4^{2-} (83.01 mg/l), F^- (0.54 mg/l), and NO_3^- (9.28%) were observed to be comparable to desirable limits of BIS (2012). This increase is because of the contamination of groundwater by discharge of municipal wastewater into the open land and agricultural practices containing fertilizers (Wang and Li 2021). The average concentration of Cr was

Table 7.3 Classes of groundwater based on hardness (Sawyer and McCarthy 1967).

Total hardness as ($CaCO_3$ mg/l)	Class	Total number of samples	Percentage
<75	Soft water	Nil	Nil
75–150	Moderately hard water	Nil	Nil
150–300	Hard water	6	21.43%
>300	Extremely hard water	22	78.57%

0.06 mg/l and exceeded the permissible limit (BIS 2012). The presence of Cr (VI) is highly influenced by the hydrogeological scenarios of the water level (Coyte et al. 2020). Under an alkaline, highly oxidizing environment, hexavalent chromium predominates, whereas Cr (III) dominates in acidic, slightly oxidizing to reducing conditions. The pH of the groundwater was found to be alkaline hence, there is more probability of the presence of Cr^{6+}. $Cr^{6+}$is harmful for human, and so, the groundwater is unfit for consumption. This increase in Cr^{6+}might be due to the rapidly growing electroplating usages in this region.

7.3.2 Entropy Water Quality Index modelling

In this study, nine groundwater quality parameters, i.e., pH, Ca^{2+}, Mg^{2+}, K^{+}, SO_4^{2-}, Cl^-, F^-, NO_3^-, and Cr, were chosen to implement the water quality evaluation using the EWQI. Based on the study, from 28 groundwater samples, about 7.14% are ranked as excellent water quality that can be consumed without treatment, 71.43% of the studied site ranked as good, 17.86% average that can be utilized for consumption after some preliminary treatment, 3.57% poor water quality which is not recommended for drinking. As a result, indices that have a significant impact on EWQI, such as Cr, F^-, SO_4^-, Cl^-, and Ca^{2+}. Similarly, Li et al. (2021) reported that groundwater quality was influenced by the TDS, NO_3^-, Ca^{2+}, and HCO_3^- concentrations as per EWQI. Wang et al. (2022) investigated the influence of seasonal changes on the EWQI and found that the entropy weights of chemical oxygen demand (COD), chloride (Cl^-), and nitrate-nitrogen (NO_3^-N) varied between the wet and dry seasons. Evaporation is more pronounced in dry season and there is fewer rainfall compared to the wet season.

7.3.3 Potability of Groundwater for Irrigation

It is critical to ensure that the irrigation water quality is not deteriorated since this can have a substantial impact on crop productivity and soil fertility. Some indicators were generated in this research to measure the water quality for irrigation and the statistical parameters of irrigation are given in Table 7.4.

Table 7.4 Descriptive statistic of parameters of irrigation water.

	Descriptive statistics			
	Minimum	**Maximum**	**Mean**	**Std. deviation**
SAR	0.16	0.58	0.32	0.12
RSC	−3.40	1.95	−0.61	1.19
Na%	8.68	41.26	26.41	7.48
PI	23.41	54.07	38.14	8.82
MH	33.90	59.13	46.44	6.12
KR	0.08	0.64	0.32	0.13

In the current study, the SAR values varies from 0.16 to 0.58 meq/l, with an average 0.32 meq/l. According to the water classification for SAR proposed by USSL (1954), entirely samples fell within the good category, representing their suitability for salt-tolerant plants (Selvakumar et al. 2014). The RSC values in the study area ranged from −3.4 to 1.95 meq/l, with the mean of −0.61 meq/l. Negative RSC values suggest an excess of sodium ions, offset by Ca^{2+} and Mg^{2+}, leading to the precipitation of calcium as CO_2 and leaving Na^+ as the dominant cation. Conversely, positive RSC values indicate increased concentrations of Ca^{2+} and Mg^{2+} due to the reaction of HCO_3^- to form Mg–HCO_3^- and Ca–HCO_3^- (Khanoranga and Khalid 2019). The percentage of sodium (Na%) in the groundwater samples ranged from 8.68% to 41.26%, with a mean value of 26.41%. Based on Eaton's (1950) classification, 3.57, 10.71, and 85.71% of the samples were categorized as permissible, excellent, and good, respectively. When irrigation water contains an excess of Na^{2+}, it can exchange Ca^{2+} in the soil through a base exchange reaction (Bhunia et al. 2018), potentially reducing soil permeability and causing internal drainage problems (Al-Tabbal and Al-Zboon 2012). The MH values ranged from 33.90 to 59.13 meq/l, with a mean of 46.44 meq/l. According to the criteria proposed by Szabolcs and Darab (1964), 32.14% of the water samples were found to be unsuitable for irrigation, suggesting potential negative impacts on crop production. The permeability index (PI) values varied from 23.42% to 54.07%, with only 3.57% of the samples being unsuitable for irrigation based on Doneen's (1962) classification. Kelly's ratios for the analyzed samples ranged from 0.08 to 0.64, with all samples having a ratio less than one, indicating their suitability for irrigation as per Kelly's (1946) criteria.

7.3.4 Classification of Water

7.3.4.1 Hydrochemical Facies

Piper's trilinear diagram (1944) shows the allocation of primary cations and anions is valuable for understanding groundwater geochemistry. The Piper diagram (Figure 7.2) shown that the majority of the samples (60.7%) belonged to the Ca-HCO_3^- water type, while 28.57% and 10.71% of the samples were classified as Mg-HCO_3^- and Na-HCO_3^- water types, respectively. The presence of light metals (Na^+, Ca^{2+}, Mg^{2+}, K^+) in groundwater primarily attributed to the erosion of parent rocks. These elements are typically derived from lithogenic rock sources, reflecting the mineralogical composition of the underlying geology (Khanoranga and Khalid 2019). These findings are consistent with the observations made by Selvakumar et al. (2014), who reported that 70% of the samples in their study area were categorized as Ca–Mg–Cl water type, while 25% and 5% of the samples were Ca–HCO_3^- and Ca–Cl facies, respectively.

7.3.4.2 Mechanism Controlling the Hydrochemistry

The processes controlling the hydrogeochemistry of groundwater in the Leh town were investigated using the Gibbs diagram (Gibbs 1970), which illustrates the association between the chemical constituents of water and the lithological characteristics of the aquifer. The results of the present study suggest that rock-water dominance is the primary process of the groundwater chemistry, indicating the influence of mineral dissolution, prolonged irrigation, and poor drainage conditions (Figure 7.3). These findings align with

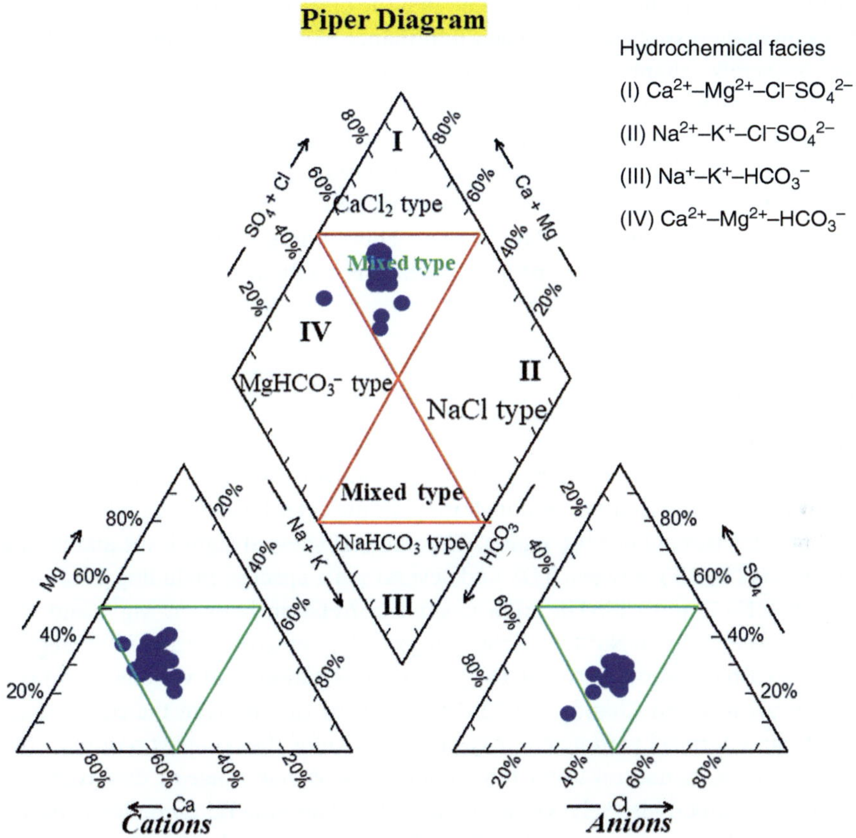

Figure 7.2 Piper diagram shows the categorization of water type in Leh town, Ladakh.

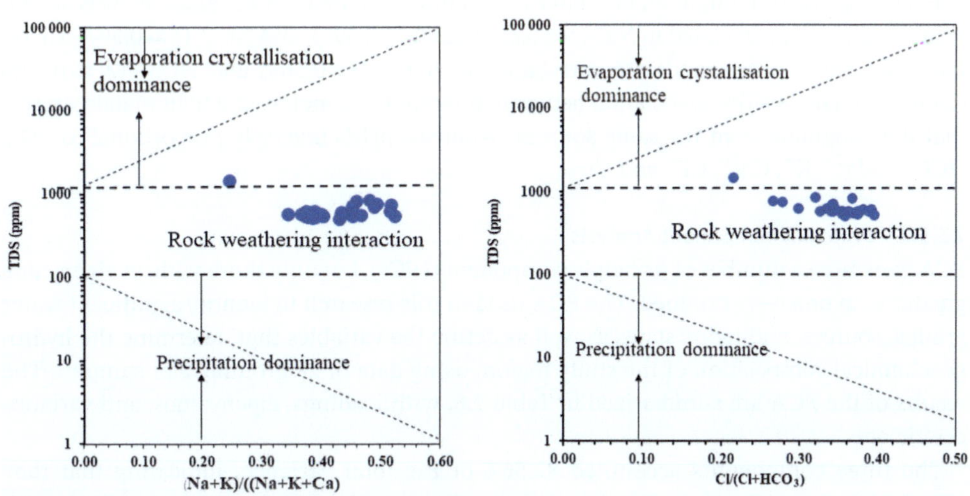

Figure 7.3 Gibbs plot depicting groundwater chemistry.

Amiri et al. (2021) observations, which indicated that the place of groundwater samples on the Gibbs diagram in their study area showed a dominance of rock water and the dissolving of evaporite minerals shaping the hydrochemistry of the groundwater system. Correspondingly, Madhav et al. (2021) observed that the hydrochemical characteristics of groundwater samples from the Trans-Varuna region in Uttar Pradesh were primarily influenced by rock–water interactions. In another study conducted by Paikaray and Chander (2022) in the SW Punjab region of India, the Piper diagram revealed that Na–HCO_3^- was the dominant water type, followed by Na–Cl type. This finding highlights the potential influence of regional geological and anthropogenic factors on the hydrochemical composition of groundwater.

7.3.5 Statistical Analysis

7.3.5.1 Pearson Correlation

Pearson correlation showed an irregular pattern of relationship between most parameters, with significant positive correlations demonstrating strong interdependence, which influenced hydrochemistry throughout the region. The analysis showed significant and strong positive correlations ($r \geq 0.5$) between TDS and several ionic species, including TH, Ca^{2+}, HCO_3^-, Mg^{2+}, K^+, Na^+, and Cl^-. The high positive correlation between Mg^{2+} and Cl^- ($r = 0.853$) and between K^+ and Na^+ ($r = 0.596$) suggests that the groundwater in the region has been influenced by extreme fertilizer usage, overexploitation, and various anthropogenic activities (Paikaray and Chander 2022). The observed variations in the correlations among the hydrogeochemical parameters may be indicative of the complex nature of the groundwater chemistry, as natural water often contains a diverse range of dissolved and suspended mineral compounds (Selvakumar et al. 2014). The mineral exchange between rocks and water was the main source of Mg^{2+} and Na^+ in the groundwater. The NO^-_3 ion had a significant correlation with Na^+ ($r = 0.655$), K^+ ($r = 0.716$), SO_4^{2-} (0.538), and Cl^- ($r = 0.534$), indicating that fertilizers, urban wastewaters, septic systems, and occasionally grassland agriculture might all be sources of pollution (Sunkari et al. 2022). F^- was shown to be significantly correlated to Na^+ ($r = 0.650$), K^+ ($r = 0.693$), and NO_3^- ($r = 0.597$). Cr, on the other hand, has a moderate correlation with K+ (r=0.568) and NO3$^-$ (r = 0.505) (Table 7.5). The positive association between different trace metals and light metals suggest that they originate from the same sources. Whereas, pH is inversely proportional to TDS, HCO_3^-, Mg^{2+}, K^+, Ca^{2+}, Cl^-, and Cr.

7.3.5.2 Principal Component Analysis

PCA generated a number of principle components (PCs), but only those with an eigenvalue greater than one were retained. The PCA used in this research to identify significant water groups, sources, and redox states as well as define the variables that determine the hydrogeochemical composition of the study region, using data of 28 groundwater samples. The results of the PCA are summarized in Table 7.6, with loadings, eigenvalues, and variance coefficients.

The three components accounted 82.56% of the total variance, indicating that they effectively capture the majority of the information contained in the original variables. PC I, which accounts for 36.656% of the total variance, exhibits significant positive loadings

Table 7.5 Pearson correlation matrix for different physicochemical parameters of Leh town's groundwater.

pH	TDS	HCO_3	TH	Ca	Mg	Na	K	SO4	Cl	NO_3^-	F	Cr
pH	1											
TDS	−.66**	1										
HCO_3^-	−.71**	.98**	1									
TH	−.71**	.98**	0.99**	1								
Ca	−.63**	.86**	.81**	.81**	1							
Mg	−.81**	.88**	.88**	.88**	.85**	1						
Na	0.05	.42*	0.30	0.29	0.37	0.15	1					
K	−0.01	.41*	0.31	0.31	0.36	0.16	.60**	1				
SO_4^{2-}	0.20	0.35	0.23	0.22	0.15	0.03	.56**	.49**	1			
Cl^-	−.59**	.89**	.84**	.84**	.77**	.73**	.41*	.43*	.41*	1		
NO_3^-	0.01	.52**	.42*	.42*	.46*	0.36	.65**	.72**	.54**	.53**	1	
F^-	0.08	.44*	0.36	0.36	0.34	0.14	.65**	.69**	.45*	.39*	.59**	1
Cr	−0.13	0.19	0.11	0.12	0.33	0.20	.39*	.57**	0.09	0.17	.51**	0.26

* = 0.05 level (2-tailed); **Correlation is significant at the 0.01 level (2-tailed).

Table 7.6 Principal component analysis.

	Component		
	PC 1	PC 2	PC 3
pH	−.878	.277	−.100
TDS	.895	.406	.002
Ca	.868	.242	.211
Mg	.957	.028	.103
Na	.123	.792	.241
K	.125	.727	.500
SO_4^{2-}	.021	.843	−.225
Cl	.817	.451	−.050
NO_3^-	.259	.754	.373
F	.113	.788	.177
Cr	.111	.212	.921
Eigen value	4.032	3.605	1.444
% Of Variance	36.656	32.776	13.127
Cumulative %	36.656	69.432	82.559

on TDS, Ca^{2+}, Mg^{2+}, and Cl^- as well as a strong negative loading on pH. The positive loadings on TDS, Ca^{2+}, Mg^{2+}, and Cl^- suggest that these parameters are likely influenced by mineral–water interactions, which attributed due to anthropic activities (Chen et al. 2019). The negative loading on pH may be caused by the carbon dioxide gas dissolving in the groundwater. CO_2 is produced by plant respiration and organic matter decomposition in the vadose zone, and as water percolates down to the groundwater, it is constantly replenished with CO_2 from these sources (Gulgundi and Shetty 2016). The varying concentrations of CO_2 in the infiltrating water led to significant variations in the absolute pH value of the groundwater. PC II, which explains 32.776% of the total variance, shows strong positive loadings on Na^+, K^+, NO_3^-, SO_4^{2-}, and F^-. The presence of these ions in the groundwater might be due to the use of chemical fertilizers in agriculture, the release of domestic effluents, and the weathered sedimentary rocks in the study area (Hossain et al. 2016; Chorol and Gupta 2023a). The association of these parameters in PC II suggests that anthropogenic activities and geological processes play a significant role in shaping the groundwater quality in the region. PC III, which accounts for 13.127% of the total variance, is dominated by a strong positive loading on Cr. The appearance of Cr as a separate component in the PCA indicates that it has a distinct source and behavior compared to the other groundwater quality parameters. The elevated Cr concentrations in the groundwater of the region can be attributed to chromite mineralization, which refers to the natural occurrence of chromite ($FeCr_2O_4$), can contribute to increased Cr levels through weathering and geochemical interactions (Naik et al. 2022). Additionally, rapid urbanization, phosphate fertilizers, and municipal sewage sludge, which often contain trace amounts of Cr as impurities, can accumulate in soils and leach into groundwater over time (Huang et al. 2014; Kubier et al. 2019). The combined influence of these factors leads to the observed elevated concentrations of Cr in the groundwater.

7.3.6 Deterministic Health Risk

The HQ analysis revealed that the presence of excess fluoride in groundwater posed significant noncancer risks to the population, particularly via oral pathway. The study found that among the various exposure routes, ingestion of fluoride-contaminated groundwater was the primary contributor to health risks for all age groups, while the dermal pathway posed minimal risk and remained within acceptable limits. The outcomes indicated that infants and children were the most susceptible age groups, with their average HQs exceeding the noncancer risk threshold of 1. This finding confirms that the level of fluoride in the groundwater induced unacceptable health risks to these sensitive populations in the region. The HI values, which represent the cumulative risk from all exposure pathways, followed a distinct pattern: children exhibited the highest risk, followed by infants, women, and men (Table 7.7). The comparative analysis highlighted the heightened susceptibility of infants and children to noncarcinogenic risks associated with fluoride exposure. This vulnerability can be attributed to their BW and immature enzymatic metabolism, which limit their ability to effectively process and eliminate fluoride from their systems (He and Wu 2019). These findings are aligning with the Sheng et al. (2023), who also stated that the oral pathway was the dominant route of fluoride exposure for all age groups, surpassing the contributions of other pathways. The results of this study

Table 7.7 Deterministic noncancer risk of fluoride.

	HQ F⁻		HI
Populations	**Oral**	**Dermal**	**F⁻**
Infant	1.56E+00	4.20E–02	1.60E+00
Children	1.99E+00	6.00E–04	1.99E+00
Women	5.44E–01	3.00E–04	5.44E–01
Men	4.27E–01	2.00E–04	4.28E–01

emphasize the significance of considering age-specific vulnerabilities when assessing the health risks caused by fluoride in groundwater.

7.4 Conclusions

The present study employed entropy techniques to assess the groundwater quality in the trans-Himalayan region, demonstrating their effectiveness in characterizing the hydrochemical properties of the groundwater. The results indicated that areas with high population density, hotel industry presence, and agricultural activities exhibited elevated EWQI values, suggesting increased anthropic contamination in the Himalayan groundwater. The dominant cations and anions in the groundwater followed the order of $Ca^{2+} > Na^{+} > Mg^{2+} > K^{+}$ and $HCO_3^{-} > SO_4^{2-}$, Cl^{-}, NO_3^{-}, and F^{-}, respectively, emphasizing the role of geological setting, human activities, and mineral weathering in shaping the groundwater chemistry. The application of principal component analysis (PCA) in combination with multivariate analysis revealed strong correlations among various ions, such as NO_3^{-}, K^{+}, F, Cl^{-}, SO_4^{2-}, PO_4^{3-}, and Na^{+}, with the lithology of the area. This approach identified the sources, including excessive fertilizer use, wastewater discharges, and agrochemicals. The noncarcinogenic health risk assessment of fluoride showed that considerable potential health risk primarily via the oral pathway, while dermal exposure posed negligible risk. The HQ analysis further revealed that infants and children are at higher noncarcinogenic risk compared to adults. The findings and approaches presented herein can serve as a valuable baseline for the development and implementation of sustainable groundwater management strategies, aiding in the protection of this vital resource for future generations. Furthermore, the study highlights the importance of regular monitoring and evaluation of groundwater quality, particularly in areas experiencing rapid population growth and intensified anthropogenic activities, to ensure the long-term sustainability of groundwater resources and the well-being of the dependent populations.

Acknowledgment

The authors acknowledge the Department of Environmental Science anEngineering, Indian Institute of Technology (ISM), Dhanbad, India, to support the research work.

References

Adimalla, N. (2021). Application of the entropy weighted water quality index (EWQI) and the pollution index of groundwater (PIG) to assess groundwater quality for drinking purposes: a case study in a rural area of Telangana State, India. *Archives of Environmental Contamination and Toxicology* 80 (1): 31–40.

Adimalla, N. and Venkatayogi, S. (2018). Geochemical characterization and evaluation of groundwater suitability for domestic and agricultural utility in the semi-arid region of Basara, Telangana State, South India. *Applied Water Science* 8: 44. https://doi.org/10./s13201-018-0682-1.

Adimalla, N., Dhakate, R., Kasarla, A., and Taloor, A.K. (2020). Appraisal of groundwater quality for drinking and irrigation purposes in Central Telangana, India. *Groundwater for Sustainable Development* 10: 100334. https://doi.org/10.1016/j.gsd.2020.100334.

Al-Tabbal, J.A. and Al-Zboon, K.K. (2012). Suitability assessment of groundwater for irrigation and drinking purpose in the northern region of Jordan. *Journal of Environmental Science and Technology* 5: 274–290.

Amiri, V., Bhattacharya, P., and Nakhaei, M. (2021). The hydrogeochemical evaluation of groundwater resources and their suitability for agricultural and industrial uses in an arid area of Iran. *Groundwater for Sustainable Development* 12: 100527.

APHA (2012). *Standard Methods for the Examination of Water and Wastewater*, 22e. Washington, D.C.: American Public Health Association.

Bhunia, G.S., Keshavarzi, A., Shit, P.K. et al. (2018). Evaluation of groundwater quality and its suitability for drinking and irrigation using GIS and geostatistics techniques in semiarid region of Neyshabur, Iran. *Applied Water Science* 8: 1–16.

BIS (Bureau of Indian Standards) (2012). *Indian Standard Specification for Drinking Water IS 10500: 2012*, 1–4. Indian Standard Institute.

Chen, J., Huang, Q., Lin, Y. et al. (2019). Hydrogeochemical Characteristics and quality assessment of groundwater in an irrigated region, Northwest China. *Water* 1 (1): 96. https://doi.org/10.3390/w11010096.

Chen, J., Qian, H., Gao, Y. et al. (2020). Insights into hydrological and hydrochemical processes in response to water replenishment for lakes in arid regions. *Journal of Hydrology* 581: 124386. https://doi.org/10.1016/j.jhydrol.2019.124386.

Chorol, L. and Gupta, S.K. (2023a). Hybrid analytic network process (ANP)-Entropy model, time series analysis for predicting nitrate and fluoride in groundwater and cumulative health risk assessment. *Journal of Cleaner Production* 428: 139316. https://doi.org/10.1016/j.jclepro.2023.139316.

Chorol, L., & Gupta, S. K. (2023b). Hydrochemical investigation of groundwater in a trans-Himalayan region of Ladakh, India, using geochemical modelling and entropy technique. *Environmental Geochemistry and Health*, 45(8), 6567–6583. https://doi.org/10.1007/s10653-023-01650-4

Çiner, F., Sunkari, E.D., and Şenbaş, B.A. (2021). Geochemical and multivariate statistical evaluation of trace elements in groundwater of Niğde municipality, south-central Turkey: implications for arsenic contamination and human health risks assessment. *Archives of Environmental Contamination and Toxicology* 80: 164–118. https://doi.org/10.1007/s00244-020-00759-2.

Coyte, R.M., McKinley, K.L., Jiang, S. et al. (2020). Occurrence and distribution of hexavalent chromium in groundwater from North Carolina, USA. *Science of the Total Environment* 711: 135135.

Dashora, M., Kumar, A., Kumar, S. et al. (2022). Geochemical assessment of groundwater in a desertic region of India using chemometric analysis and entropy water quality index (EWQI). *Natural Hazards* 112 (1): 747–782. https://doi.org/10.1007/s11069-021-05204-8.

Davis, J. et al. (2002). *Statistics and Data Analysis in Geology*. Wiley.

Doneen, L.D. (1962). The influence of crop and soil on percolating water. *Proceedings of the 1961 biennial conference on groundwater recharge*, pp. 156–163.

Eaton, F.M. (1950). Significance of carbonates in irrigation waters. *Soil Science* 69 (2): 123–134.

Egbueri, J.C., Ukah, B.U., Ubido, O.E., and Unigwe, C.O. (2020). A chemometric approach to source apportionment, ecological and health risk assessment of heavy metals in industrial soils from southwestern Nigeria. *International Journal of Environmental and Analytical Chemistry* 102: 3399–3417. https://doi.org/10.1080/03067319.2020.1769615.

El Alfy, M., Lashin, A., Abdalla, F., and Al-Bassam, A. (2017). Assessing the hydrogeochemical processes affecting groundwater pollution in arid areas using an integration of geochemical equilibrium and multivariate statistical techniques. *Environmental Pollution* 229: 760–770.

Gao, Y., Qian, H., Huo, C. et al. (2020). Assessing natural background levels in shallow groundwater in a large semiarid drainage Basin. *Journal of Hydrology* 584: 124638. https://doi.org/10.1016/j.jhydrol.2020.124638.

Gibbs, R.J. (1970). Mechanisms controlling world water chemistry. *Science* 170 (3962): 1088–1090. https://doi.org/10.1126/science.170.3962.1088.

Giordan, M. (2009). Global groundwater? Issues and solutions. *Annual Review of Environment and Resources* 34: 153–178.

Gulgundi, M.S. and Shetty, A. (2016). Identification and apportionment of pollution sources to groundwater quality. *Environmental Processes* 3: 451–461. https://doi.org/10.1007/s40710-016-0160-4.

He, S. and Wu, J. (2019). Hydrogeochemical characteristics, groundwater quality, and health risks from hexavalent chromium and nitrate in groundwater of Huanhe Formation in Wuqi county, northwest China. *Exposure and Health* 11: 125–137.

Hossain, S., Hosono, T., Yang, H., and Shimada, J. (2016). Geochemical processes controlling fluoride enrichment in groundwater at the western part of Kumamoto area, Japan. *Water, Air, and Soil Pollution* 227: 385. https://doi.org/10.1007/s11270-016-3089-3.

Huang, M., Wang, W., Chan, C.Y. et al. (2014). Contamination and risk assessment (based on bioaccessibility via ingestion and inhalation) of metal (loid) s in outdoor and indoor particles from urban centers of Guangzhou, China. *Science of the Total Environment* 479: 117–124. https://doi.org/10.1016/j.scitotenv.2014.01.115.

Ji, Y., Wu, J., Wang, Y. et al. (2020). Seasonal variation of drinking water quality and human health risk assessment in Hancheng City of Guanzhong Plain, China. *Exposure and Health* 12: 469–485.

Kaiser, H.F. (1958). The varimax criterion for analytic rotation in factor analysis. *Psychometrika* 23: 187–200.

Karunanidhi, D., Aravinthasamy, P., Deepali, M. et al. (2021). Appraisal of subsurface hydrogeochemical processes in a geologically heterogeneous semi-arid region of south India based on mass transfer and fuzzy comprehensive modeling. *Environmental Geochemistry and Health* 43: 1009–1028. https://doi.org/10.1007/s10653-020-00676-2.

Kelly, W.P. (1946). Permissible composition and concentration for irrigation waters. *Proceedings of ASC*, pp. 607

Khanoranga and Khalid, S. (2019). An assessment of groundwater quality for irrigation and drinking purposes around brick kilns in three districts of Balochistan province, Pakistan, through water quality index and multivariate statistical approaches. *Journal of Geochemical Exploration* 197: 14–26. https://doi.org/10.1016/j.gexplo.2018.11.007.

Kubier, A., Wilkin, R.T., and Pichler, T. (2019). Cadmium in soils and groundwater: A review. *Applied Geochemistry* 108: 104388. https://doi.org/10.1016/j.apgeochem.2019.104388.

Kumar, P.J.S., Elango, L., and James, E.J. (2014). Assessment of hydrochemistry and groundwater quality in the coastal area of South Chennai, India. *Arabian Journal of Geosciences* 7: 2641–2653. https://doi.org/10.1007/s12517-013-0940-3.

Li, P., Wu, J., and Qian, H. (2016). Hydrochemical appraisal of groundwater quality for drinking and irrigation purposes and the major influencing factors: a case study in and around Hua County, China. *Arabian Journal of Geosciences* 9: 15. https://doi.org/10.1007/s12517-015-2059-1.

Li, Z., Yang, Q., Yang, Y. et al. (2019). Isotopic and geochemical interpretation of groundwater under the influences of anthropogenic activities. *Journal of Hydrology* 576: 685–697.

Li, X., Huang, X., and Zhang, Y. (2021). Spatio-temporal analysis of groundwater chemistry, quality and potential human health risks in the Pinggu basin of North China Plain: Evidence from high-resolution monitoring dataset of 2015–2017. *Science of the Total Environment* 800: 149568. https://doi.org/10.1016/j.scitotenv.2021.149568.

Loh, Y.S.A., Akurugu, B.A., Manu, E., and Aliou, A.S. (2020). Assessment of groundwater quality and the main controls on its hydrochemistry in some Voltaian and basement aquifers, northern Ghana. *Groundwater for Sustainable Development* 10: 100296. https://doi.org/10.1016/j.gsd.2019.100296.

Madhav, S., Raju, N.J., and Ahamad, A. (2021). A study of hydrogeochemical processes using integrated geochemical and multivariate statistical methods and health risk assessment of groundwater in Trans-Varuna region, Uttar Pradesh. *Environment, Development and Sustainability* 23: 7480–7508.

Mohana, P. and Velmurugan, P.M. (2020). Geochemical characterization and water quality assessment of groundwater in Arani taluk of Tamil Nadu, South India. *Geochemistry International* 58: 598–612. https://doi.org/10.1134/S0016702920050079.

Muller, J. and Dame, J. (2016). Small town, great expectations: urbanization and beautification in Leh. *South Asia Multidisciplinary Academic Journal* 14: 4. https://doi.org/10.4000/samaj.4208.

Naik, M.R., Mahanty, B., Sahoo, S.K. et al. (2022). Assessment of groundwater geochemistry using multivariate water quality index and potential health risk in industrial belt of central Odisha, India. *Environmental Pollution* 303: 119161. 10.1016/j.envpol.2022.119161.

Narsimha, A. (2018). Elevated fluoride concentration levels in rural villages of Siddipet, Telangana State, South India. *Data in Brief* 16: 693–699. https://doi.org/10.1016/j.dib.2017.11.088.

Paikaray, S. and Chander, S. (2022). Geochemical variations in uranium and fluoride enriched saline groundwater around a semi-arid region of SW Punjab, India. *Applied Geochemistry* 136: 105167. https://doi.org/10.1016/j.apgeochem.2021.105167.

Piper, A.M. (1944). A graphic procedure in the geochemical interpretation of water-analyses. *Eos, Transactions American Geophysical Union* 25 (6): 914–928.

Qian, H., Chen, J., and Howard, K.W.F. (2020). Assessing groundwater pollution and potential remediation processes in a multi-layer aquifer system. *Environmental Pollution* 263 (Part A): 114669. https://doi.org/10.1016/j.envpol.2020.114669.

Rakotondrabe, F., Ngoupayou, J.R.N., Mfonka, Z. et al. (2018). Water quality assessment in the Bétaré-Oya gold mining area (East-Cameroon): multivariate statistical analysis approach. *Science of the Total Environment* 610-611: 831–844. https://doi.org/10.1016/j.scitotenv.2017.08.080.

Ravindra, K., Thind, P.S., Mor, S. et al. (2019). Evaluation of groundwater contamination in Chandigarh: source identification and health risk assessment. *Environmental Pollution* 255 (Pt 1): 113062. https://doi.org/10.1016/j.envpol.2019.113062.

Richards, L.A. (ed.) (1954). *Diagnosis and Improvement of Saline and Alkali Soils (No. 60).* US Government Printing Office.

Riedel, T., Kübeck, C., and Quirin, M. (2022). Legacy nitrate and trace metal (Mn, Ni, As, Cd, U) pollution in anaerobic groundwater: quantifying potential health risk from "the other nitrate problem". *Applied Geochemistry* 139: 105254. https://doi.org/10.1016/j.apgeochem.2022.105254.

Samantaray, S. and Sanyal, P. (2021, December). Response of organic matter to short-term failure of monsoon: An insight into the carbon and nitrogen cycle in a hypersaline lagoon. *AGU Fall Meeting Abstracts*, vol. 2021, B25C–B1471C.

Sawyer, C.N. and McCarty, P.L. (1967). *Chemistry for Sanitary Engineers.* McGraw Hill.

Selvakumar, S., Ramkumar, K., Chandrasekar, N. et al. (2014). Groundwater quality and its suitability for drinking and irrigational use in the Southern Tiruchirappalli district, Tamil Nadu, India. *Applied Water Science* 7: 411–420. https://doi.org/10.1007/s13201-014-0256-9.

Sheng, D., Meng, X., Wen, X. et al. (2023). Hydrochemical characteristics, quality and health risk assessment of nitrate enriched coastal groundwater in northern China. *Journal of Cleaner Production* 403: 136872.

Siddique, J., Menggui, J., Shah, M.H. et al. (2020). Integrated approach to hydrogeochemical appraisal and quality assessment of groundwater from Sargodha District, Pakistan. *Geofluids* 2020 (1): 6621038.

Subba Rao, N. (2018). Groundwater quality from a part of Prakasam district, Andhra Pradesh, India. *Applied Water Science* 8: 1–18. https://doi.org/10.1007/s13201-018-0665-2.

Subba Rao, N., Ravindra, B., Wu, J., (2020) Geochemical and health risk evaluation of fluoride rich groundwater in Sattenapalle Region, Guntur district, Andhra Pradesh, India. *Human and Ecological Risk Assessment: An International Journal*, 26: 9, 2316–2348, DOI:https://doi.org/10.1080/10807039.2020.1741338

Sunkari, E.D., Abu, M., Zango, M.S., and Wani, A.M.L. (2020). Hydrogeochemical characterization and assessment of groundwater quality in the Kwahu-Bombouaka Group of the Voltaian Supergroup, Ghana. *Journal of African Earth Sciences* 169: 103899. https://doi.org/10.1016/j.jafrearsci.2020.103899.

Sunkari, E.D., Abu, M., and Zango, M.S. (2021). Geochemical evolution and tracing of groundwater salinization using different ionic ratios, multivariate statistical and geochemical modeling approaches in a typical semi-arid basin. *Journal of Contaminant Hydrology* 236: 103742. https://doi.org/10.1016/j.jconhyd.2020.103742.

Sunkari, E.D., Seidu, J., and Ewisi, A. (2022). Hydrogeochemical evolution and assessment of groundwater quality in the Togo and Dahomeyan aquifers, Greater Accra Region, Ghana. *Environmental Research* 208: 112679. https://doi.org/10.1016/j.envres.2022.112679.

Szabolcs, I. and Darab, C. (1964). The influence of irrigation water of high sodium carbonate content of soils. *Proceedings of 8th International Congress of ISSS, Transaction II*, pp. 803–881.

U.S. Geological Survey (2006). 2006 U.S. Geological Survey Collection of water samples (ver. 2.0) U.S. Geological Survey Techniques and Methods (2006) book 9, chap. A4. Variously paged http://pubs.water.usgs.gov/twri9A4/

US Environmental Protection Agency, USEPA (2012). 2012 Edition of the Drinking Water Standards and Health Advisories. EPA 822-S-12-001 (accessed November 2013).

USEPA, U. (2015). Environmental protection agency. *2003. Final NPDES General Permit for New Existing Sources and New Discharges in the Offshore Subcategory of the Oil and Gas Extraction Category for the Western Portion of the Outer Continental Shelf of the Gulf of Mexico*, pp. 1–117.

USSL (1954), Diagnosis and improvement of saline and alkali soils, USDA, Handbook, 60, 147.

Wang, Y. and Li, P. (2021). Appraisal of shallow groundwater quality with human health risk assessment in different seasons in rural areas of the Guanzhong Plain (China). *Environmental Research* 207: 112210. https://doi.org/10.1016/j.envres.2021.112210.

Wang, Z., Guo, X., Kuang, Y. et al. (2022). Recharge sources and hydrogeochemical evolution of groundwater in a heterogeneous karst water system in Hubei Province, Central China. *Applied Geochemistry* 136: 105165. https://doi.org/10.1016/j.apgeochem.2021.105165.

WHO (2011). *Guidelines for Drinking Water Quality*, 4e. Geneva: World Health Organization.

Wu, C., Wu, X., Qian, C., and Zhu, G. (2018). Hydrogeochemistry and groundwater quality assessment of high fluoride levels in the Yanchi endorheic region, northwest China. *Applied Geochemistry* 98: 404–417. https://doi.org/10.1016/j.apgeochem.2018.10.016.

Zhang, X., He, J., He, B. et al. (2019). Assessment, formation mechanism, and different source contributions of dissolved salt pollution in the shallow groundwater of Hutuo River alluvial-pluvial fan in the North China Plain. *Environmental Science and Pollution Research* 26: 35742–35756. https://doi.org/10.1007/s11356-019-06502-2.

8

Revival of Drying Springs in Mountainous Regions of the Himalayas – Inferences from an Isotope Hydrochemical Study

Tirumalesh Keesari[1,2], Diksha Pant[1,2], Hemant Mohokar[1], and Vinod Khatti[3]

[1] *Isotope and Radiation Application Division, Bhabha Atomic Research Centre, Mumbai, India*
[2] *Department of Atomic Energy, Homi Bhabha National Institute, Mumbai, India*
[3] *Himalayan Environmental Studies and Conservation Organization (HESCO), Dehradun, India*

8.1 Introduction

The Himalayan range covers an area of 43 lakh km^2 housing nine major perennial rivers widely spread in eight countries, India being one of them and its home to nearly 1.5 billion people. The Indian Himalayan Region (IHR) spans an approximate area of 5.3 lakh km^2 covering six states and Jammu and Kashmir and Ladakh union territories, with a population of about 50 million people. The IHR is one of the major sources of water for rivers (Ganga, Brahmaputra, and Indus) and has a many glaciers, lakes, and springs. Spring is the point from which the aquifer water flows out on to the surface through some fissure or openings. The main source of water for domestic and livelihood needs is spring in the mountains. The importance of spring to IHR is clear from the fact that out of the five million springs across India; nearly 75% of the springs are homed in the IHR alone. With increased water demand, land use change and ecological degradation, the stress on mountainous springs is increasing. The other major factor that impacts spring discharge rate is climate change, i.e. rising temperatures, erratic and extreme rainfalls, etc. Springs play a major role in river hydrology as well, at river head, it manifests as sources of water to the rivers, while at lower plains, they contribute to the base flows so spring hydrology has clear implications on rivers as well. Springs are classified based on type of opening, discharge rate characteristics (i.e. volume of water), temperature, type of flow, etc. (Bryan 1919, Meinzer 1923). Himalayan springs are mainly nonartesian springs with 6th and 7th magnitude and have seepage-type outlets.

Studies on spring discharges, spring catchment system and effect of changing climatic conditions, were taken up by many researchers (Bartarya 1991; Bonacci 1993). These studies indicated that the spring discharge are affected by rainfall pattern and recharge area characteristics, surface cover characteristics, geology, and permeability of topsoil. A research by Agarwal et al. (2012) suggests that natural stressors, i.e. changing climate, decreasing rain, changed rainfall patterns, etc. and improper watershed management have

Water Resources Management in Mountain Regions, First Edition. Edited by Virendra Bahadur Singh, Sughosh Madhav, Rakesh Kumar Gupta, Prerna Diwan, and Amit Kumar.

led to a decrease in spring discharges with about 40% reduction in discharge rate over a period of three decades (Valdiya and Bartarya 1989) in the Kumaun Himalaya region. Mahamuni and Kulkarni (2012) reported that 8,000 villages face water shortage due to the drying up of springs. The NITI ayog reported that apart from climatic change, change in land use pattern has increased water demand and enhanced surface run offs leading to ecological degradation as well as impacting the spring. A study by Vashisht (2008) in the foothill's region of Siwaliks suggested that site specific watershed interventions like management of small cannels and springs can aid in augmenting spring discharge rate.

Numerous watershed management intervention was carried out to improve the springs but these interventions did not consider the geology of the watershed nor considered preferential pathways of the present aquifer system (Agarwal et al. 2012). The management strategies based on inadequate information did not solve water scarcity problems (Bruijnzeel and Bremmer 1989; Alford 1992). Therefore, drying up of springs and water scarcity issues underscore the need to increase the understanding of spring hydrology, especially in the Himalayan region. One important aspect is to understand the recharge dynamics of spring so that rejuvenation or conservation strategies can be planned.

For understanding recharge dynamics, environmental isotopes of water both stable, i.e. deuterium (^{2}H), oxygen-18 (^{18}O), and radioactive, i.e. tritium (^{3}H), play a vital role. All waters carry fingerprints of naturally occurring isotopes that provide information about their source and mechanism of recharge, circulation and renewability, recharge areas, and transit times of aquifer and hydraulic inter-relationships, etc. These fingerprints are derived due to isotopic fractionation, defined as the phenomenon in which the isotopic composition of an element in a certain compound changes by the transition of the compound from one physical state or chemical composition to another, e.g. water transition from vapor (cloud) to liquid (rain). The isotope fractionation is controlled by thermodynamic (equilibrium, i.e. slow process) or kinetic (nonequilibrium, i.e. fast transitions) processes (Fritz and Fontes 1980; IAEA Guide Book 1983; Kendall and McDonnell 1998; Keesari et al. 2021). Many climatic and meteorological factors control the isotopic composition of rain water (IAEA Guide Book 1983; Mook 2000). In this article, the isotope hydrological investigation carried out in two sites in Rudraprayag is briefed, demonstrating the applicability of isotope technology in identification of the potential recharge altitudes of the drying springs and providing insights into recharge sources and mechanism of springs. This article also highlights the impact of artificial recharge structures on spring discharge rates and duration during summer season.

8.2 Study Area

8.2.1 General Geology

Geologically, Rudraprayag district comprises diverse rock types ranging in age from Palaeoproterozoic to Mesoproterozoic. The rock succession exposed in the district mainly falls into two groups that are the rock sequences exposed between main boundary fault (MBF) and main central thrust (MCT) – constituting the lesser Himalaya and the rock sequences exposed to the north of MCT – Constituting the Higher Himalaya. The lithological sequence of this area belongs to Late Precambrian to Cambrian ages under

Rudraprayag formation of Garhwal Group. The lithostratigraphy of this formation suggests that it belongs to argillo-arenaceous facies (Srivastava and Ahamad 1977). The various lithological units of the Rudraprayag formation are sandwiched between the Alaknanda Fault in the north and north Almora thrust in the south. This formation is divided into five members, Uttyasu, Quartzite, Lameri Member, Haryali Quartzite, Thallassu Schistose Grit, and the Karanprayag Metavolcanics. Haryali quartzite of Rudraprayag formation occupies a greater part of the area. The Haryali quartzite is characterized by white, fine to medium-grained quartzite with epidiorite/chlorite schist. Thalassu schistose grit is found all along the Alakananda Valley. Thalassu schistone grit is composed of schistose grit with basic dykes, purple and greenish grey and green phyllites, gritty to conglomerates, banded white and purple, maroon quartzite with beds of purple slate. The area is tectonically highly disturbed with faults and folds.

8.3 Rainfall and Climate

Rudraprayag district has subtropical to tropical upland type climate. Northern and western parts of the district are under snow cover perennially with subarctic type climatic conditions. Severe winters with high rainfall are the characteristic features of the northern part. The district has a larger spread on the southern slopes that receive highly variable rainfall depending upon the altitude. 83% of the rain occurs during monsoon and 17% occurs during the winter season. The southern part of the district receives an average rainfall of ~ 1220 mm, while in the central and northern parts receive ~ 1751 and ~ 1995 mm, respectively, for the rainy months (July–August). The study area lies in the Lesser Himalayas with an average rainfall of 1700 mm (http://planning.up.nic.in). The distribution of rainfall for the years 2004–2009 of the study area is given in Table 8.1 (www.imd.gov.in).

8.3.1 Isala

The spring locations of Isala site are shown in Figure 8.1a. There are four springs located in one valley with altitude ranging from 1250 to 1450 m amsl, which falls between longitudes 79° 7′ and 79° 8′ and latitudes 30° 18′ and 30° 20′ (Figure 8.1b). The digital elevation map of

Table 8.1 Rainfall data of project sites from 2004 to 2009.

Year	Rudraprayag (Isala/Kakodakhal)
2004	901
2005	976
2006	651
2007	678
2008	1400
2009	970

Source: Adapted from www.imd.gov.in.

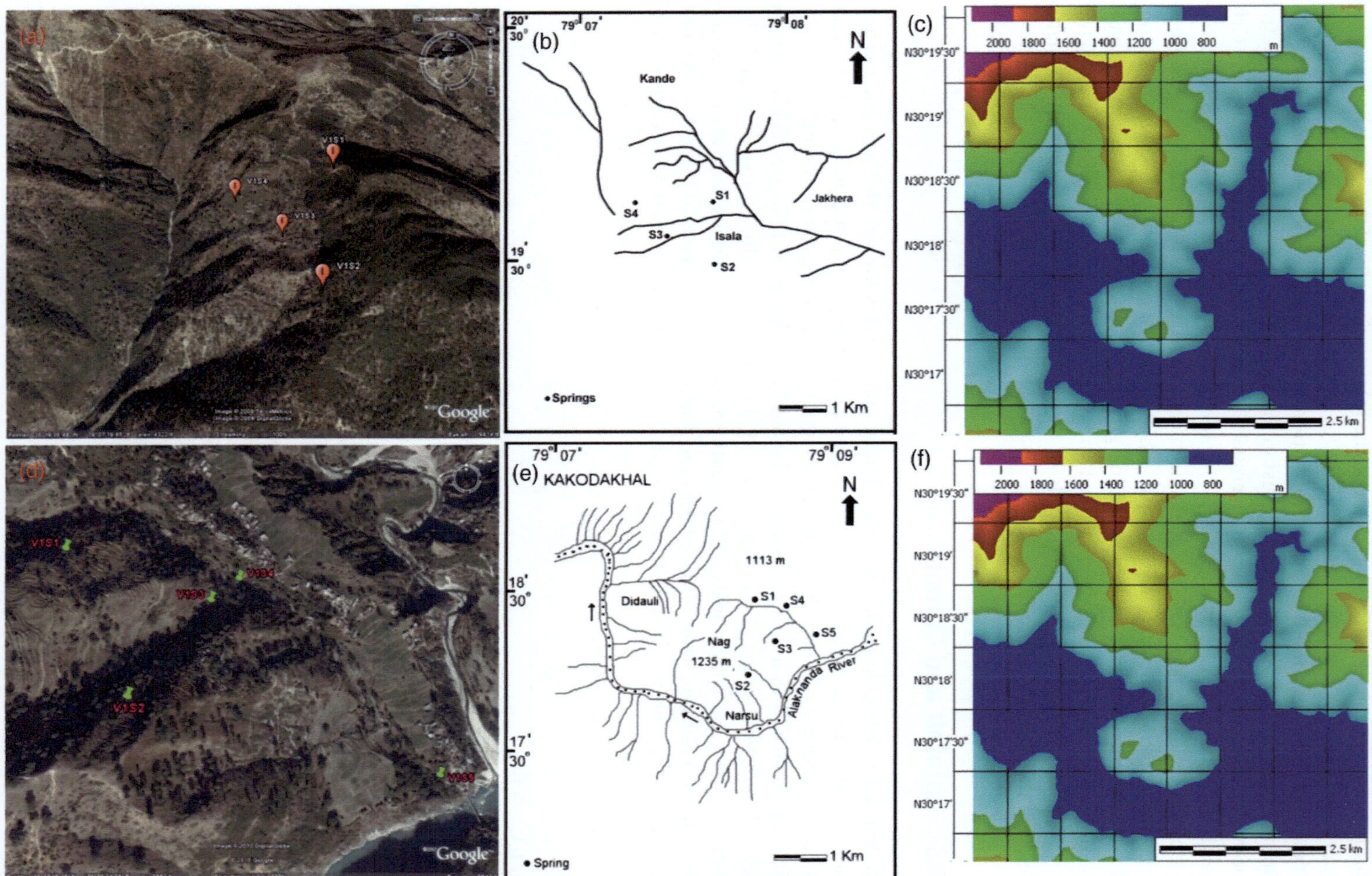

Figure 8.1 (a) Google Earth map of Isala site with sampling points, (b) Location of springs and drainage pattern of Isala site, (c) Digital elevation map of Isala site, (d) Google Earth map of Kakodakhal site with sampling points, (e) Location of springs and drainage pattern of Kakodakhal site, and (f) Digital elevation map of Kakodakhal site.

Table 8.2 Geographical coordinates of springs at Isala and Kakodakhal sites.

Location name	Sample ID	Latitude [N]	Longitude [E]	Altitude (m amsl)
		Isala site		
Pipalpani	V1S1	30° 19.198′	79° 7.455′	1444
Panchpani	V1S2	30° 18.871′	79° 7.423′	1383
Jogshir	V1S3	30° 19.049′	79° 7.417′	1280
Dabar	V1S4	30° 19.184′	79° 7.255′	1250
Ganman	RGS1	30° 19.242′	79° 7.505′	1527
Dabar	RGS2	30° 19.182′	79° 7.172′	1262
		Kakodakhal site		
Mangri	V1S1	30° 17.875′	79° 08.337′	925
Danda	V1S2	30° 17.743′	79° 08.426′	895
Sari Roli	V1S3	30° 17.825′	79° 08.530′	820
Sagadhera II	V1S4	30° 17.855′	79° 08.552′	800
Jhalimath Amdala	V1S5	30° 17.628′	79° 08.818′	720
Near Bridge	RGS1	30° 17.615′	79° 08.866′	720
Sari Village	RGS2	30° 17.733′	79° 08.456′	833

the study area is shown in Figure 8.1c. The geographical coordinates of the springs are given in Table 8.2. The lowest and highest discharge rates of the springs are 0.2 and 5 L/min, respectively. The population of this area is about 1000 and springs are the main source of drinking and irrigation water.

8.3.2 Kakodakhal

In Kakodakhal site, springs are located along two valleys. Five springs with altitude ranging from 720 to 920 m amsl (Figure 8.1e) are located in one valley between longitudes 79° 8′ and 79° 9′ and latitudes 30° 17′ and 30° 18′ (Figure 8.1d). The digital elevation map area is shown in Figure 8.1f. The geographical coordinates of these springs are given in Table 8.2. The discharge rates of the springs vary from 0.7 to 27 L/min. More than 2000 population residing in this area uses spring water for drinking and irrigation purposes.

8.4 Sampling and Analytical Techniques

Four sets of samples were collected from the aforementioned springs during February, June, September, and November for physicochemical parameters, chemical parameters, and isotopes. The physical parameters like pH, electrical conductivity (EC), temperature (temp) were measured in situ using Eutech instruments cyberscan CON 11, pH 11 meter. All the probes were calibrated daily with appropriate standard buffers. Water samples were filtered through 0.45-μm pore size membrane filters and stored in 60-mL Tarson bottles for anion measurements, whereas for cation measurements, water samples filtrated and acidified using high pure nitric acid to pH 2. For stable isotopes, spring waters were collected directly into the 20 mL bottle up to the brim with minimal exposure to the atmosphere. For tritium, the water sample was collected in 1 L Tarson bottle. The spring discharge rates were also measured monthly for 17 months, i.e. from July 2008 to November 2009 (Table 8.4). This simple, direct method of discharge rate measurement based on the time to fill a graduated measuring cylinder of known volume was used to estimate discharge rates (Herschy 1995). In addition to spring water, two sets of rainwater were also collected during monsoon (June and September 2009) from different altitudes.

8.5 Chemical Species

The chemistry (anions and cations) of the water samples was measured using the Ion Chromatograph system (Dionex-500). For cation analysis (Na^+, K^+, Mg^{2+}, and Ca^{2+}) 25 μL of the filtered and acidified sample was introduced into a cation exchange column (CS 12A), where the ions were separated based on their charge to size ratio and were quantified based on the peak area obtained from conductivity detector (ED 40). For anionic (F^-, Cl^-, SO_4^{2-}, and NO_3^-) determination, 25 μL of the filtered and un-acidified sample was introduced into an anion exchange column (AS 11) and measured. A known mixed standard is used for the calibration of the peak area. The accuracy of the chemical data was cross-checked by calculating the charge balance error (CBE) given by

$$\mathrm{CBE}(\%) = \frac{\mathrm{meq}(\mathrm{cations}) - \mathrm{meq}(\mathrm{anions})}{\mathrm{meq}(\mathrm{cations}) + \mathrm{meq}(\mathrm{anions})} \times 100 \tag{8.1}$$

The CBE values were found to be falling within the allowed limits of ±5% (Hounslow 1995).

8.6 Stable Isotopes (δ^2H, and δ^{18}O)

Isotopic ratio differences in waters from different sources are very small and hence, δ^2H and δ^{18}O values are reported with respect to a standard. These relative isotope concentrations can be determined easily with better accuracy by a differential isotope ratio measurement using double-collecting mass spectrometers or continuous flow isotope ratio mass spectrometers. The relative difference is called delta "δ" value and is defined as (Craig 1961a, 1961b)

$$\delta = \left(\frac{R - R_{\mathrm{std}}}{R_{\mathrm{std}}}\right) \times 10^3 \tag{8.2}$$

where R represents the isotope ratio of a sample ($^2H/^1H$, $^{18}O/^{16}O$, etc.) and R_{std} represents the corresponding ratio in the standard. The δ value is generally expressed in parts per thousand (permil, ‰). Stable isotopes δ^2H and δ^{18}O have been measured using IRMS GEO 20–20 (PDS, Europa, UK). The GEO 20–20 dual inlet mass spectrometer is a highly precise and sensitive instrument, computer controlled and designed for ease of operation as well as dual inlet mode. For analysis, hydrogen gas is equilibrated with the water sample in a closed bottle with zinc catalyst whereby isotopic exchange takes place. The $^{18}O/^{16}O$ ratio in water samples is measured by equilibrating with carbon dioxide at 30°C. The equilibrated hydrogen and CO_2 gases are isotopically representative of the water sample. These gases are introduced into mass analyzer and measured. Error of measurement (2σ) was found to be ±0.5‰ and 0.1‰ for deuterium and oxygen-18, respectively.

8.7 Radioactive Isotope (^3H)

The steady state natural concentration of tritium is very low, with a ratio $^3H/^2H$ of the order of 10^{-18}. With a conventional mass spectrometer only concentration ratios down to 10^{-8} can be measured. ^3H isotope in natural waters is measured using their radioactive decay. For measuring ^3H several instruments are available, each with their specific advantages and disadvantages. Low-level tritium concentrations are expressed in tritium unit (TU) where 1 TU is equivalent to a specific activity of 0.118 Becquerel (Bq) per litre (L) of water. Tritium measurement involves distillation of sample followed by electrolytic enrichment, neutralization, and distillation of electrolyzed samples. The enriched sample (8 mL) is then mixed with Hisafe 3 scintillator (14 mL) (Perkin Elmer) in 24-mL high density polythene vial and counted in ultra-Low Level Quantulus 1220 Liquid Scintillation Counter for 500 minutes (Nair 1983) with a minimum detection limit is 0.5 TU.

8.8 Results and Discussion

The summary of physicochemical parameters, chemical and isotopic data of both Isala and Kakodakhal sites for different months is provided in Tables 8.3 and 8.4.

8.9 Discharge Rate Analysis

8.9.1 Isala

The discharge rates of the springs at the Isala site from July 2008 to November 2009 are provided in Table 8.5. The discharge rate of a spring V1S4 could not be measured as the outlet of the spring was not accessible. The lowest discharge rate of 0.2 L/min was found in V1S2 in June 2009 and the highest rate discharge rate of 5 L/min was found in V1S1 during August 2009. In general, the discharge rates are of the same order in all the springs throughout the year except V1S1, which shows higher discharge rates than other springs during August–September. The spring discharge rates were higher in August/September 2008 than August/September 2009 but the variations are small. The high discharge rates of the springs in 2008 during August/September compared to the same period in 2009 is due to high rainfall during 2008 (Table 8.1). The discharge rates of the springs were maximum in September after the high precipitation in July/August (Figure 8.2a). This indicates that all the springs are mainly recharged by precipitation and the residence time of the recharged waters is a few months.

Table 8.3 Summary of physicochemical and isotope data for Isala for different months.

	February			June			September			November		
Parameters/Month	**Min.**	**Max.**	**Mean**	**Min.**	**Max.**	**Mean**	**Min.**	**Max.**	**Mean**	**Min.**	**Max.**	**Mean**
EC (μS/cm)	147	187	168	163	220	197	171	291	216	163	223	191.25
pH	7.2	7.7	7.4	8	8.4	8.2	7.1	7.6	7.3	6.9	7.6	7.4
HCO_3^- (mg/L)	95	122	111	95	127	116	104	178	137	73	115	95
F^- (mg/L)	0	0	0	1.8	2.7	2.2	0	0.2	0.05	0	0	0
Cl^- (mg/L)	0.6	5.4	2.8	0.8	7.2	4.1	0.5	5.3	3.3	1	6	4
NO_3^- (mg/L)	0.1	5.3	1.7	0	4.8	2.0	0	1.6	0.7	0.3	2.2	1.1
SO_4^{2-} (mg/L)	0.9	5.7	3.4	1.2	5	3.5	1.1	6.5	3.95	1.1	4.6	3.4
Na^+ (mg/L)	7.5	11.1	9.3	7.7	12.7	10.4	9	13.3	11.4	8.7	14	11.4
K^+ (mg/L)	1	3.5	2.0	1.2	14.8	4.8	0.7	4.8	2.2	1	2	1.5
Mg^{2+} (mg/L)	1.5	8.9	3.5	6.2	8.4	7.4	0.7	14	7.7	5	8	6.5
Ca^{2+} (mg/L)	13.7	23.7	18.7	15.3	26.1	20.4	24	45.9	33.3	12	23	16.5
$\delta^{18}O$ (‰)	−9.42	−9.13	−9.31	−8.44	−8.09	−8.23	−8.1	−7.81	−7.95	−7.56	−7.53	−7.55
$\delta^{2}H$ (‰)	−52.08	−50.71	−51.18	−59.23	−55.76	−56.98	−54.67	−47.25	−51.58	−52.33	−48.84	−50.88
^{3}H (TU)	6	8.3	7.3	—	—	—	—	—	—	—	—	—

Table 8.4 Summary of physicochemical and isotope data for Kakodakhal for different months.

	February			June			September			November		
Parameters/Month	**Min.**	**Max.**	**Mean**	**Min.**	**Max.**	**Mean**	**Min.**	**Max.**	**Mean**	**Min.**	**Max.**	**Mean**
EC (μS/cm)	115	251	168.2	145	305	232.2	155	310	236.8	147	352	239.8
pH	6.8	7.8	7.2	8.1	8.5	8.2	7.1	7.5	7.3	6.9	7.9	7.4
HCO_3^- (mg/L)	67.0	143.0	107.4	87.0	166.0	139.2	99.0	171.0	143.6	85.0	150.0	119.8
F^- (mg/L)	0.0	0.0	0.0	0.2	0.4	0.3	0.0	0.0	0.0	0.0	0.0	0.0
Cl^- (mg/L)	2.0	9.9	7.6	1.2	9.7	6.3	0.9	9.8	6.0	1.0	12.0	6.4
NO_3^- (mg/L)	0.2	17.3	5.0	0.1	14.5	4.2	0.0	18.7	5.6	0.0	18.9	6.9
SO_4^{2-} (mg/L)	2.6	10.7	5.6	0.3	8.9	3.9	0.5	9.2	4.2	0.4	9.2	4.4
Na^+ (mg/L)	3.5	7.0	5.1	5.1	8.9	6.8	6.4	11.6	8.5	6.0	10.0	7.6
K^+ (mg/L)	2.1	5.2	3.1	0.3	2.4	1.3	0.7	3.0	1.9	0.0	3.0	1.8
Mg^{2+} (mg/L)	1.3	2.5	1.8	4.0	14.5	9.5	5.0	20.4	13.1	4.7	16.0	11.1
Ca^{2+} (mg/L)	12.8	30.4	21.4	14.8	33.0	26.2	22.9	48.7	39.7	12.0	32.0	24.3
$\delta^{18}O$ (‰)	−9.21	−8.84	−9.03	−8.18	−7.2	−7.75	−8.3	−6.19	−7.31	−7.94	−6.23	−7.28
$\delta^{2}H$ (‰)	−53.78	−51.02	−52.38	−59.3	−51.75	−55.79	−55.98	−46.92	−52.92	−56.81	−45.36	−51.88
^{3}H (TU)	8.1	11.7	9.3	—	—	—	—	—	—	—	—	—

Table 8.5 Spring discharge rate (in L/min) of Isala and Kakodakhal sites.

Spring ID	Jul-08	Aug-08	Sep-08	Oct-08	Nov-08	Dec-08	Jan-09	Feb-09	Mar-09	Apr-09	May-09	Jun-09	Jul-09	Aug-09	Sep-09	Oct-09	Nov-09
									Isala site								
V1S1	2	5	3.9	3.3	2.9	2	1.9	1.6	1.6	1.7	1.9	1.6	1.5	3.5	4.2	2.1	1
V1S2	1.5	3.8	3.85	3	2.9	2.1	1.8	1.2	1.1	1	1.1	0.5	0.2	1.3	2	1	0.4
V1S3	1.2	3.2	3.15	2.7	2.7	1.3	1.1	0.9	0.85	0.85	0.9	0.54	0.35	1.55	1.95	0.8	0.35
V1S4									Stagnant								
									Kakodakhal site								
V1S1	7.5	17.8	25.38	19.3	12.3	9.8	9.2	8.7	6.6	3.5	2.1	1.2	9	2.5	3	1.9	1.2
V1S2									Stagnant								
V1S3	11	18.9	27	17.5	15.2	15.2	13.5	12	2	4.5	4.4	4.28	3.2	5	5.5	2.1	1.4
V1S4	3.9	7.3	8.9	8.9	7.35	4.5	3.1	2.5	2	1.8	1.7	1.1	0.8	2.1	2.7	1.3	0.7
V1S5	n.d.	11.5	16.6	9.8	9.2	7.9	7	6	5.1	3	2.7	0.7	0.5	3.2	4	3.5	3.2

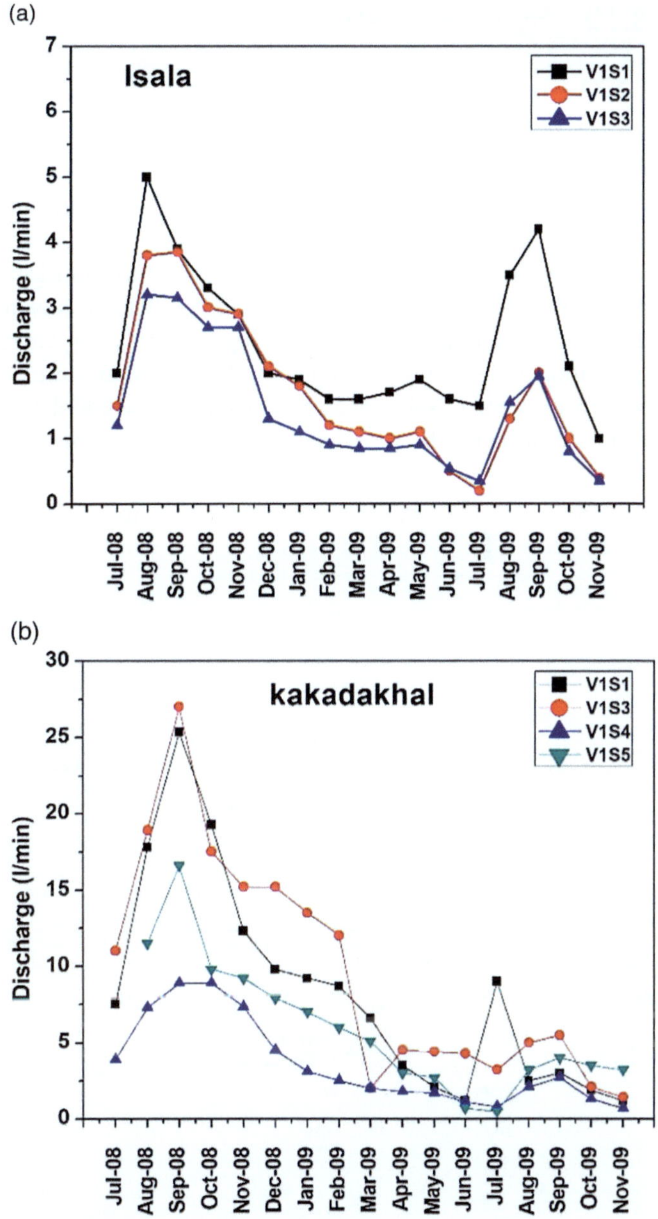

Figure 8.2 Discharge rates of the springs from June 2008 to November 2009 at (a) Isala and (b) Kakodakhal sites.

8.9.2 Kakodakhal

The spring discharge rates are given in Table 8.5. The maximum discharge rate of 27 L/min is found in V1S3 during September 2008 and the minimum of 0.7 L/min is observed in V1S4 in July 2009. The spring discharge rate of V1S2 could not be measured as it was stagnant. There is a wide variation in the maximum discharge rates of the springs during August/September 2008 compared to August/September 2009. It can be seen from Figure 8.2b that spring discharge rates respond to the amount of precipitation in respective years.

8.10 Hydrochemistry

8.10.1 Isala

All the spring water samples show pH between 7 and 8 which is within the permissible limit of WHO (2011) and BIS (2012), while the spring water samples showed increased pH values of more than 8 during June 2009 that may be attributed to local contamination. The EC values in spring waters range from 150 to 190 µs/cm in February 2009 and increase gradually till September (Table 8.3). There is no significant change in EC values of the spring water samples collected during September and November. 2009. Since the rainfall in 2008 was more than in 2009, the spring waters during February 2009 showed low EC due to dilution and faster percolation with less soil water interaction. The Trilinear piper plots of spring waters of Isala site for February, June, September, and November 2009 are shown in Figure 8.3a–d. The piper classification shows that the spring waters are dominantly

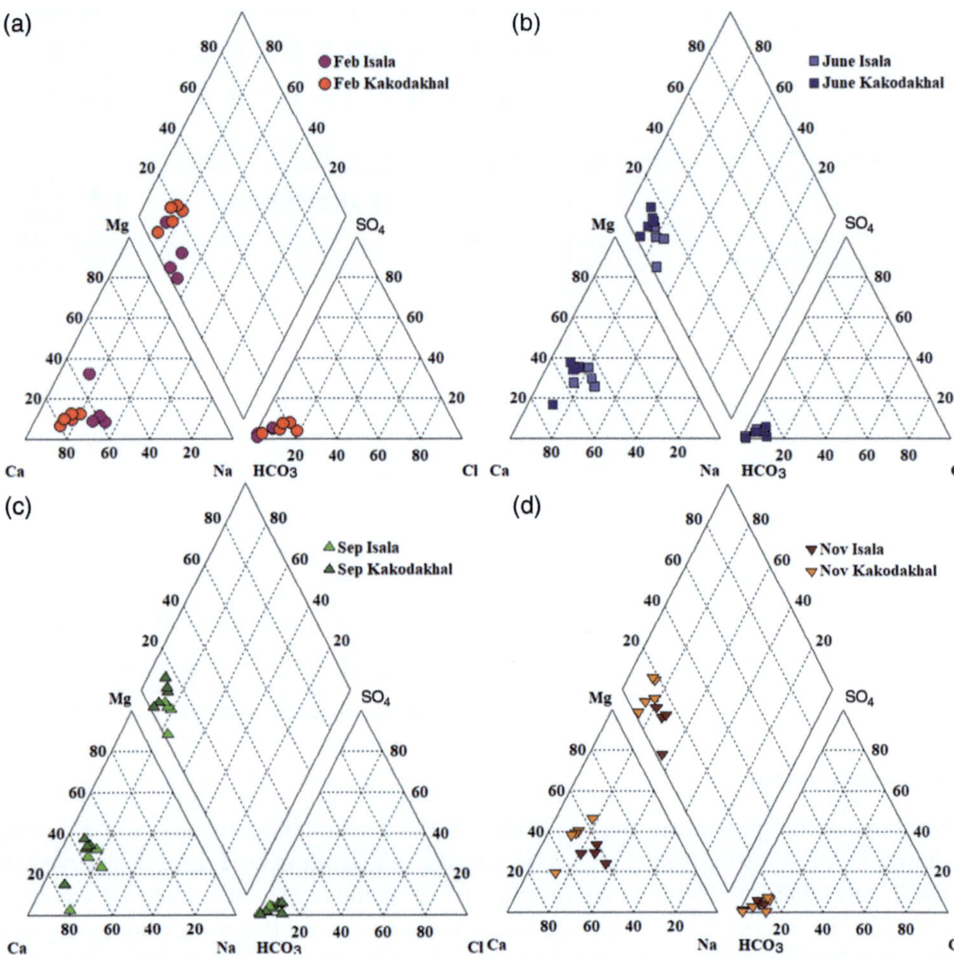

Figure 8.3 Piper plots of the spring waters collected from Isala and Kakodakhal site during (a) February (b) June, (c) September, and (d) November 2009.

Ca–Na–HCO_3 type during February except V1S1 that is Ca–Mg–HCO_3 type (Piper 1953). This changes to Ca–Mg–HCO_3 and Ca–Mg–Na–HCO_3 type during June and September. During November, the dominant water type changes to Ca–Mg–Na–HCO_3. The change in chemical facies can be attributed to the increased contribution of salts from the formation that is corroborated with increased EC value during the postmonsoon season.

8.10.2 Kakodakhal

The pH of the spring waters is neutral to alkaline (pH: 6.9–8.5) in all the samples and the highest pH is observed during June 2009. The electrical conductivity variations are similar to Isala spring waters, i.e. increased EC during the postmonsoon compared to premonsoon (Table 8.4). Major ion data of the spring waters of this site was plotted in Piper's trilinear diagram (Figure 8.3a–d). From the piper classification, these waters were found to be dominantly Ca–HCO_3 type in February that changes to Ca–Mg–HCO_3 type during June, September, and November. This modification in chemical type can be attributed to addition of Ca^{2+} and HCO_3^- ions during postmonsoon.

8.11 Environmental Isotopes

The variations in the stable isotopic ratios of precipitation in the catchment areas play a key role in deducing various hydrological aspects of springs. The isotopic variations with respect to altitude and seasons find potential applications in small-scale studies such as in this case, whereas latitude and continental effects on isotopic signatures of precipitation find applications on a global scale (IAEA Guide Book 1983; Clark and Fritz 1997). The oxygen and hydrogen stable isotope composition in precipitation serves as a benchmark in most isotope atmospheric, (eco-)hydrological, and paleoclimatological applications. However, in the absence of isotope data of precipitation, isotope signatures of shallow groundwaters were used as a proxy for precipitation (Gupta et al. 2005). Similarly, spring water isotope data soon after the monsoon can also be considered a proxy for precipitation. The altitude effect is important, particularly in mountainous areas where altitude varies greatly, to find out the origin and source of springs. For interpreting the precipitation and spring linkages, isotope correlations (plot of $\delta^{18}O$ versus δ^2H), in water samples and rainwaters are plotted. In the present study, the June 2009 isotope values of the springs for Isala and Kakodakhal sites were considered a proxy for rainwater for altitude effect.

8.11.1 Isala

The isotope values of this site range from −9.29‰ to −9.42‰ and −50.71‰ to −52.08‰ for $\delta^{18}O$ and δ^2H, respectively, for February 2009. From the plot $\delta^{18}O$ versus δ^2H, it is seen that all the samples fall above the Global Meteoric Water Line (GMWL Eq. 8.3) and the Northern India Meteoric Water Line (NIMWL, Kumar et al. 2010, Eq. 8.4), shown in Figure 8.4a. This is due to winter rains (western disturbances rain) contribution to the spring water that has high Dexcess value compared to monsoonal rains (Jeelani and Deshpande 2017). In the case of spring water samples collected during June 2009, the values of $\delta^{18}O$ and δ^2H range

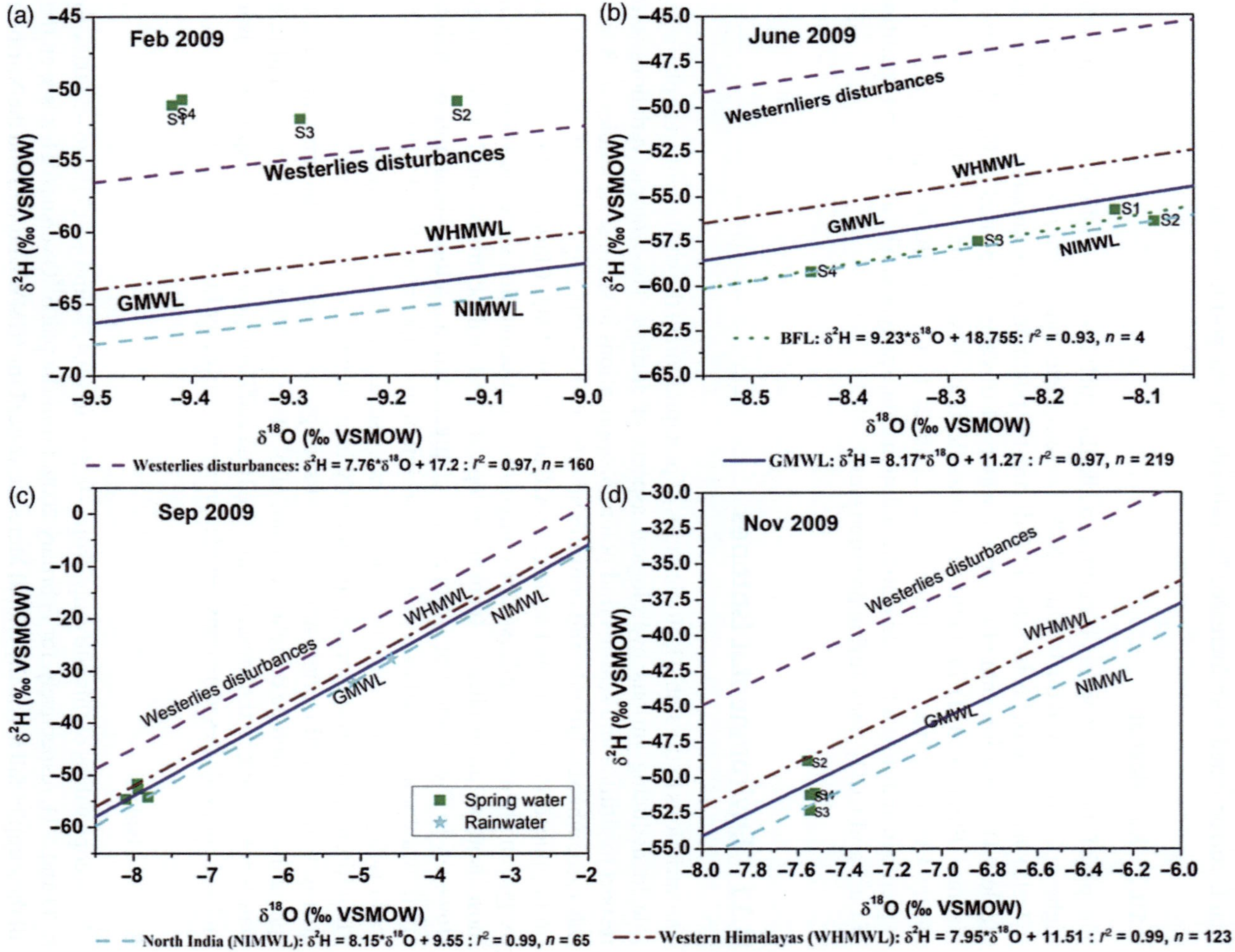

Figure 8.4 Plot of δ^2H versus $\delta^{18}O$ of spring water samples collected from Isala site during (a) February 2009, (b) June 2009, (c) September 2009, and (d) November 2009.

from −7.28 ‰ to −8.52 ‰ and −48.45 ‰ to −55.9, respectively. It has been seen from the plot of $\delta^{18}O$ versus $\delta^{2}H$ that most of the samples fall along the GMWL and NIMWL (Figure 8.4b). The linear fit was constructed for the spring water data and showed a slope of 8.7 and intercept 14.9 (Eq. 8.5).

$$\text{GMWL}: \delta^{2}H = 8.17 \times \delta^{18}O + 11.27 \left(r^{2} = 0.97,\ n = 210\right) \tag{8.3}$$

$$\text{NIMWL}: \delta^{2}H = 8.15 \times \delta^{18}O + 9.55 \left(r^{2} = 0.99,\ n = 65\right) \tag{8.4}$$

$$\text{BFL}: \delta^{2}H = 8.7 \times \delta^{18}O + 14.9 \left(r^{2} = 0.9, n = 4\right) \tag{8.5}$$

$$\text{WHMWL}: \delta^{2}H = 7.95 \times \delta^{18}O + 11.51 \left(r^{2} = 0.99, n = 123\right) \tag{8.6}$$

Since the spring water samples fall along the GMWL and NIMWL, they are taken as proxy for rainwater. The rainwater samples also fall along the NIMWL hence it can be considered the local meteoric water line (LMWL). All four spring water samples fall within a very narrow range of isotope values of $\delta^{18}O$ and $\delta^{2}H$ −7.81 to −8.10‰ and −47.25 to −54.67‰, respectively, for September 2009 sampling. From Figure 8.4c, it was observed that the samples fall very near to the GMWL, NIMWL, and Western Himalayas Meteoric Water Line (WHMWL, Kumar et al. 2010, Eq. 8.6), shown in Figure 8.4c. During November 2009, the isotope values of the spring water samples show $\delta^{18}O$ almost the same value of −7.5‰ and $\delta^{2}H$ varies from −48.84 ‰ to −52.33‰. All the samples fall as a single cluster between WHMWL and NIMWL indicating rainwater as the source of recharge (Figure 8.4d). There is a gradual enrichment in the isotope values of the spring waters from February to November 2009 (Figure 8.5).

From Figure 8.5, it is observed that there is no significant enrichment or depletion in $\delta^{18}O$ values of spring samples with an increase in altitude for all the 4 months of sampling. Therefore, it can be inferred that the springs are being recharged mostly from same rain events. Absence of any dominant altitude variation in isotopic ratios of spring water also

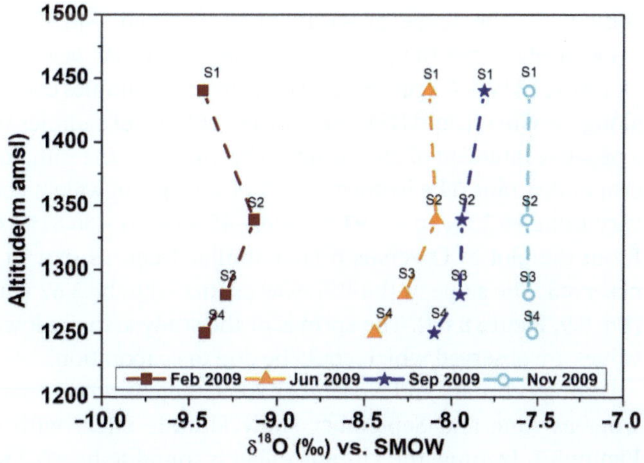

Figure 8.5 Altitude versus $\delta^{18}O$ composition of spring water samples collected in different months from Isala site.

lies in the fact that springs are located within very a small range of altitudes (1250–1350 m amsl). The tritium values of the spring samples vary from 6 to 8.3 TU. It is observed that high-altitude springs show high tritium compared to low-altitude springs, indicating a longer residence of groundwater at lower altitudes. This is further corroborated by high electrical conductivity in the low-altitude springs. Another probable reason for the observed variations in isotope data with respect altitude could be possibility of evaporated surface water contribution. This region being hilly and undulating, multiple small scale depressions exists in and around spring outlets. The stored water in these depressions/pits can contribute to enriched isotopic signatures and the magnitude depends on the number of such pits present within the proximity of spring outlets.

8.11.2 Kakodakhal

The spring water samples collected during February 2009 have isotope values −8.9‰ to −9.2‰ and −46.02‰ to −53.78‰ for $\delta^{18}O$ and δ^2H, respectively. From the plot $\delta^{18}O$ versus δ^2H, it is seen that all the samples fall above GMWL, as shown in Figure 8.6a. This is due to the contribution from winter monsoon, i.e. western disturbances characterized by highly depleted isotopic signature. The isotope values of the spring water samples collected during June 2009 range from -7.2‰ to −8.18‰ and −51.75‰ to −57.92‰ for $\delta^{18}O$ and δ^2H, respectively. All the spring water stable isotope values fall slightly below GMWL and NIMWL (Figure 8.6b). The linear fit was constructed for these data and shown in Eq. 8.7. The samples show enrichment compared to the rainwater samples.

$$\text{BFL}(\text{June}): \delta^2\text{H} = 7.4 \times \delta^{18}\text{O} + 1.5 \left(r^2 = 0.95, n = 5\right) \tag{8.7}$$

$$\text{BFL}(\text{Sep}): \delta^2\text{H} = 3.38 \times \delta^{18}\text{O} - 28.78 \left(r^2 = 0.54, n = 5\right) \tag{8.8}$$

$$\text{BFL}(\text{Nov}): \delta^2\text{H} = 5.67 \times \delta^{18}\text{O} - 10.56 \left(r^2 = 0.92, n = 5\right) \tag{8.9}$$

The water samples collected during September 2009 from all the springs have isotope contents in the range of −6.19 to −8.3‰ and −46.92 to −58.71‰ for $\delta^{18}O$ and δ^2H, respectively. From the plot $\delta^{18}O$ versus δ^2H, it has been seen that all the samples scatter along the LMWL (Figure 8.6c). Two rainwater samples collected from the same location fall along the GMWL and NIMWL. The best fit line of samples from Sep has a slope of 3.38 and a negative intercept of 28 (Eq. 8.8). This indicates the samples are getting contribution from evaporated rain. The isotope values of the spring waters collected during November 2009 vary from −6.23‰ to −7.94 ‰ and −45.36‰ to −56.81‰ for $\delta^{18}O$ and δ^2H, respectively. From the plot $\delta^{18}O$ versus δ^2H, a similar trend as that in the month of September was observed. The slope of the BFL was estimated to be 5.67 with a negative intercept of 10.56 (Eq. 8.9, Figure 8.6d). The springs of the study area are low altitude springs thus enriched values are observed which could be due to evaporation.

During February and November 2009, there is no correlation between altitude and $\delta^{18}O$, whereas June and September show altitude effect with different slopes, as shown in Figure 8.7. In June, the altitude effect is found to be −0.2‰ $\delta^{18}O$ per 100 m, whereas it is

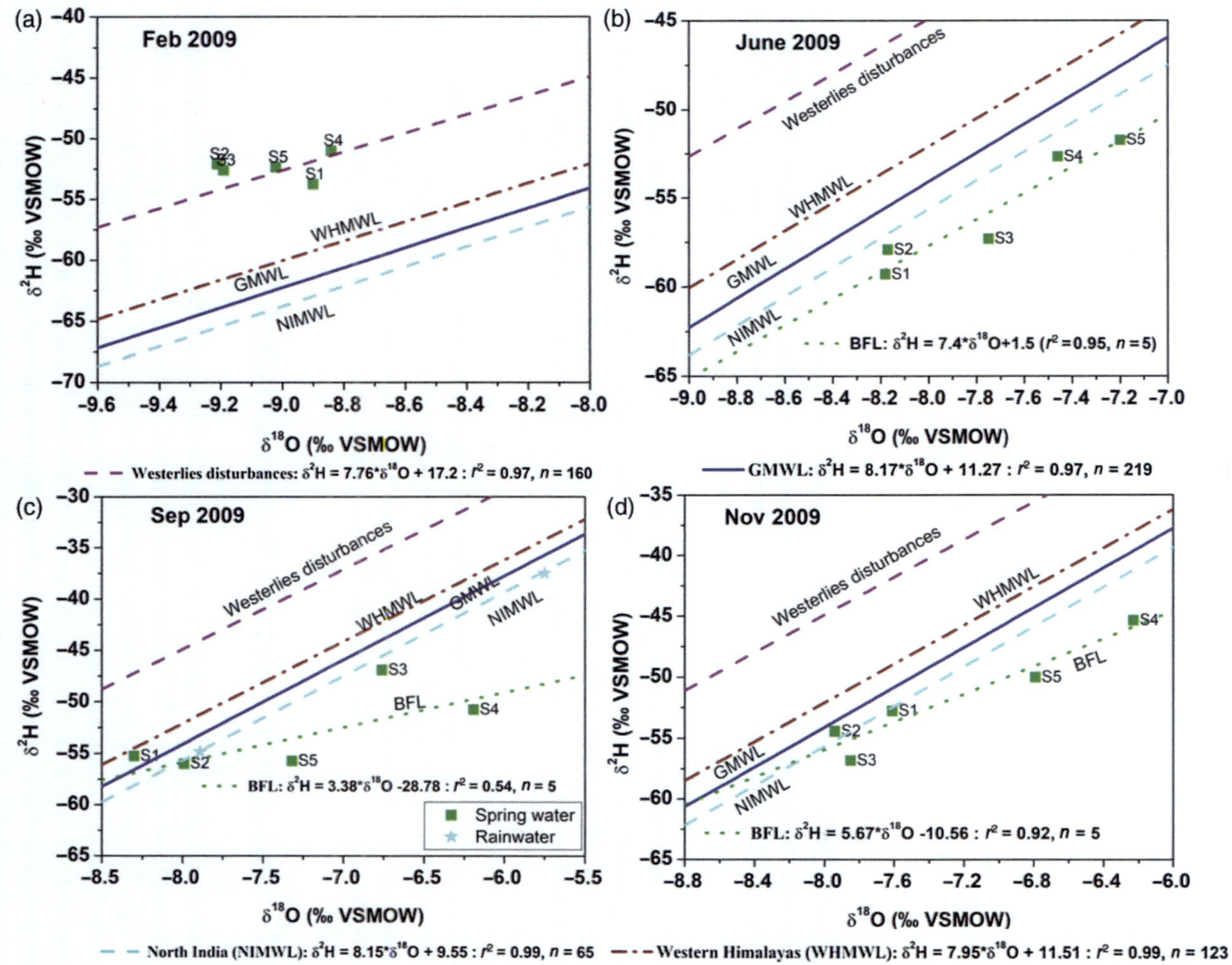

Figure 8.6 Plot of δ2H versus $\delta^{18}O$ of spring water samples collected from Kakodakhal site during (a) February 2009, (b) June 2009, (c) September 2009, and (d) November 2009.

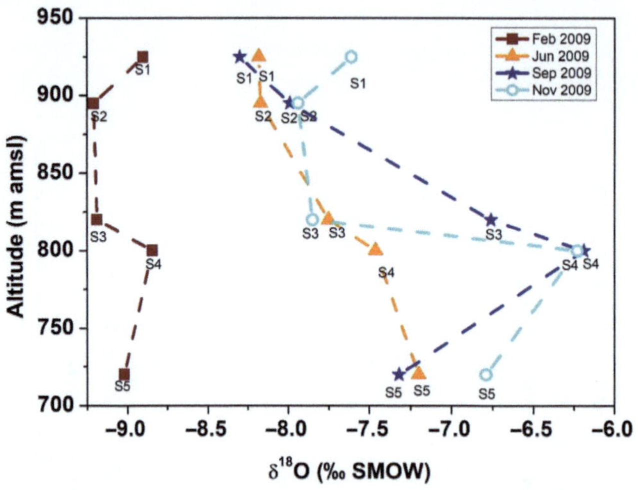

Figure 8.7 Altitude versus $\delta^{18}O$ composition of spring water samples collected in different months from Kakodakhal site.

−0.5‰ $\delta^{18}O$ per 100 m in September 2009. This difference can be attributed to the contribution of evaporated waters to the lower-altitude springs, which is also shown in Figure 8.6b–d. Tritium values of the spring waters of this site fall in the range of 8.1–11.6 TU indicating modern recharge. Shivanna et al. (2008) have reported a tritium value of 12 TU for rainwater in Rudraprayag district, Uttarakhand.

The tritium content of the precipitation varies with the season, latitude and altitude of the region (IAEA Guide Book 1983). In addition, local moisture also contributes to tritium in precipitation. From the plot of tritium versus altitude, an increase in tritium with increasing altitude is observed in accordance with the altitude effect (Figure 8.8a,b). One spring, i.e. S4 from Kakodakhal site shows high concentration at lower altitude that may be due to some local source addition (Figure 8.8b). The spring S1 of Kakodakhal is showing lower tritium which may be due to longer residence of groundwater. An increase in tritium concentration is observed with depletion in $\delta^{18}O$ values for both Isala and Kakodakhal site except from the spring S4 which is enriched and having higher tritium concentration (Figure 8.8c,d). This observation further confirms the contribution of local contamination to the spring.

8.12 Isotopic Inferences and Recommendations

The isotopic signatures of spring water suggest that main source of recharge to springs is precipitation water. The tritium concentration of spring water suggests modern recharge. The contribution of winter monsoon is observed in springs during February month. During September, various subsurface channels contribute to the spring discharge rates leading to high discharge rates and wide variation in isotopic content. However, in the following months, the isotopic signatures of the spring water demonstrate a narrow distribution, suggesting a common source of recharging water. All the spring waters have same water type at Kakodakhal site during pre and postmonsoon season. At Isala site, the hydrochemistry of spring 1 varies from all the other springs, which could be due to variation in local geology.

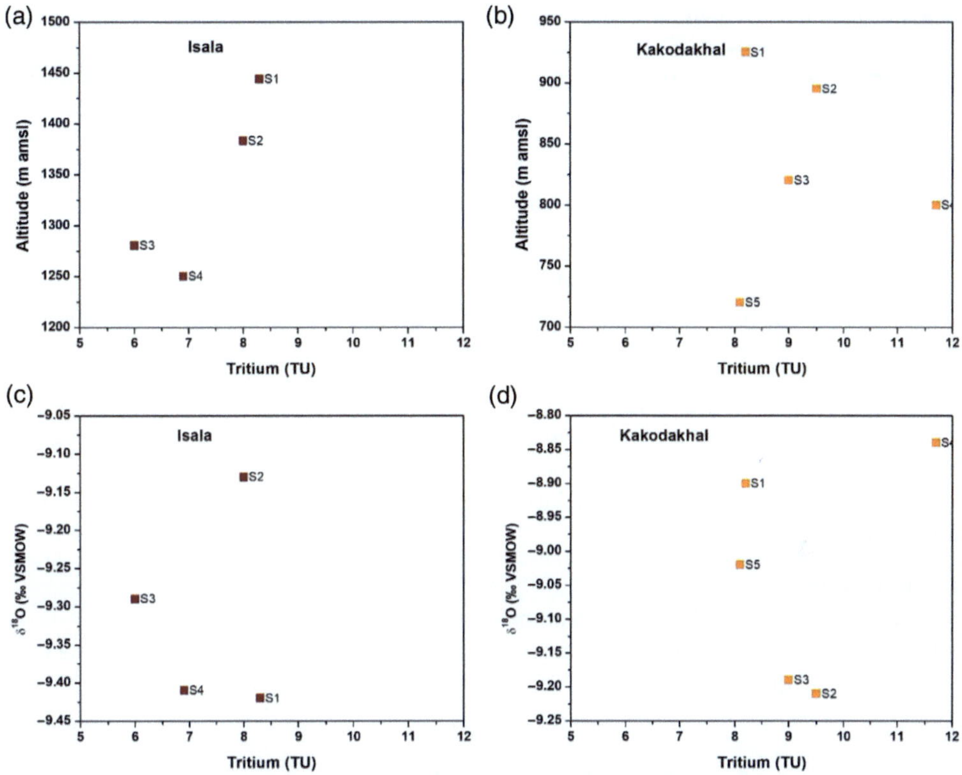

Figure 8.8 Tritium versus altitude for (a) Isala and (b) Kakodakhal and tritium versus oxygen −18 for (c) Isala and (d) Kakodakhal site.

In Kakodakhal, based on altitude effect and geomorphology, the recharge area was found to be 1100 m amsl for the springs S1, S2, S3, and S4 and 800 m amsl for S5. While in the case of Isala since no altitude effect is observed, it is concluded that a common source is responsible for spring recharge for all the seasons. Hence based on geomorphology recharge altitudes were identified and suitable recharge structures were suggested (Tables 8.6 and 8.7). All the springs have their local catchments; however, low-altitude springs also

Table 8.6 Recommendations for spring rejuvenation at Isala site.

Sample ID	Location	Altitude	Suitable recharge structures
S1	Pipalpani	1444	Trenches 150 m above spring and dykes also, may help S3 and S4
S2	Panchpani	1383	Trenches and dykes above 200 m above the spring
S3	Jogshir	1280	Structures for S1 will help, additional trenches 60 m above the spring
S4	Dabar	1250	Structures for S1 will help, additional trenches 30 m above the spring

Table 8.7 Recommendations for spring rejuvenation at Kakodakhal site.

Sample ID	Location	Altitude	Suitable recharge structures
S1	Mangri	925	Checkbund on the visible channel and trenches above the spring, it will help S4
S2	Danda	895	Trenches above the spring, it may help S3 and S4
S3	Sari Roli	820	Checkbund across the visible channel
S4	Sagadhera II	800	Structures at S1, S2, and S3 will help
S5	J. Amdala	720	Checkbund across channel above 125 m above spring and trenches between channels

receive contributions from high-altitude catchments. Therefore, to augment the spring discharge rate, artificial structures are to be constructed at each spring site.

In addition to the above recommendations, the following measures can be beneficial in long-term scenarios.

1) Afforestation
2) Small-scale reservoirs in plateau regions
3) Educating the villagers about the ill effects of grazing their cattle on the hilltop regions on the spring discharge rates.

8.13 Impact of Artificial Recharge Measures

The construction of recharge structures was carried out as per the recommendation and completed by February 2010. The construction sites of Isala and Kakodakhal with check dams and bunds are shown in Figures 8.9 and 8.10.

The major impact of intervention was observed after April 2010. The discharge rate analysis showed a positive result although the rainfall in the present year was relatively low as compared to the year of study. The water discharge rate was found higher as compared to the previous year.

At the Isala site, the increase in discharge rate was estimated to vary from 23% to 48% in springs (Figure 8.11a). At the Kakodakhal site, a visible increase in all the locations was observed and the increase ranged between 6% and 100% (Figure 8.11b). A slight decrease was noticed in S1 and S3, and this could be due to lower rainfall of the preceding year.

8.14 Conclusions

The lithological sequence of both sites Isala and Kakodakhal belongs to Rudraprayag formation of Garhwal Group (Late Precambrian to Cambrian age), which is characterized by argillo-arenaceous facies. The altitude range of the springs at Isala and Kakodakhal is 1250–1450 m amsl and 720–920 m amsl, respectively. The average rainfall is 930 mm for the past 5 years. From the spring discharge rate data during the past few years, it can be

Figure 8.9 Constructed check dam and bund at Isala site.

Figure 8.10 Constructed check dam and bund at Kakodakhal site.

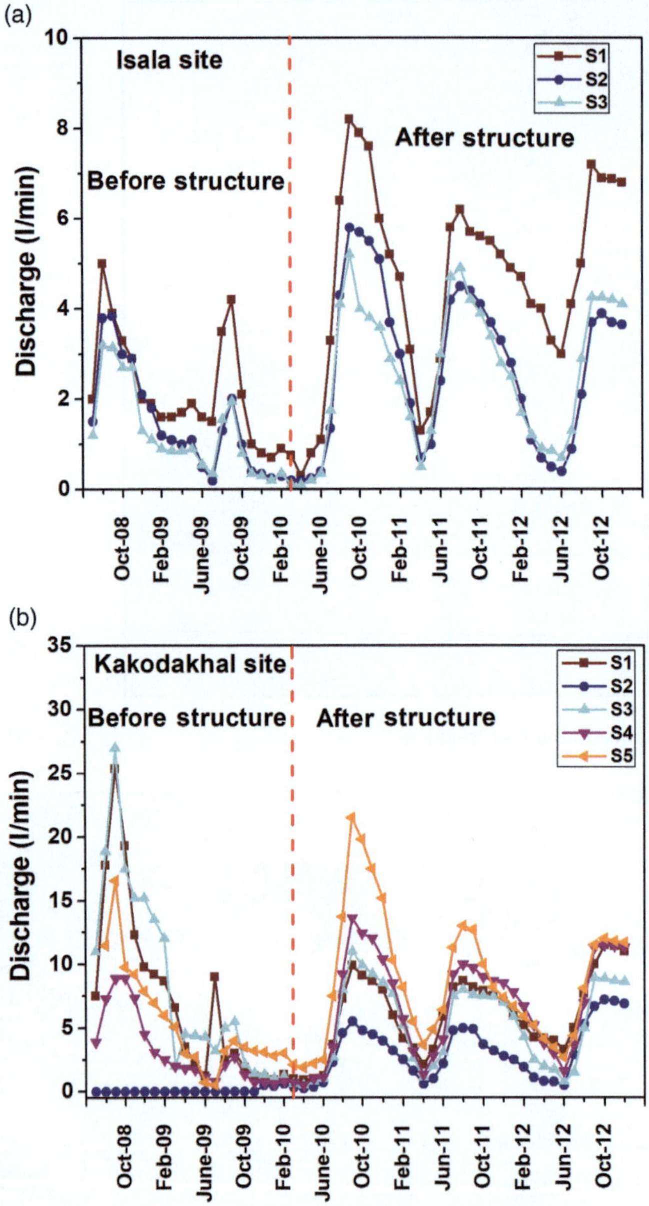

Figure 8.11 The discharge rate variation before and after recharge structures at (a) Isala and (b) Kakodakhal sites.

concluded that maximum discharge rates are observed during August–September (Isala 2–5 L/min, Kakodakhal: 10–25 L/min) and decline to record lowest discharge rates during May–June. However, it is observed that the maximum spring discharge rates during August–September 2008 were much higher than in the same period during 2009. This change in spring discharge rates can be attributed to the amount of rainfall in respective years. This indicates that the discharge rate of the springs is highly dependent on rainfall. It was also found that low-altitude springs were yielding more discharge rates compared to high-altitude springs in the respective valleys after construction of artificial recharge structure at all sites suggesting extent of catchment area is an important parameter contributing to spring discharge rate.

Spring water quality was found to be fresh with electrical conductivity of less than 350 μS/cm and neutral to slightly alkaline (pH: 6.8–8.5). The water samples represent Ca–Na–HCO_3 and Ca–Mg–HCO_3 types and show seasonal variations suggesting impact of rainwater recharge. Chemistry of spring water suggest that lithology of the formation plays a major role in controlling the hydrochemical behavior of spring water.

All spring water samples of Isala and Kakodakhal fall on or along the meteoric water line (GMWL and NIMWL) except in February months. In the plots of δ^2H versus $\delta^{18}O$, spring water samples of all the sites collected during February 2009 fall well above the meteoric water line, this may be due to the contribution of western disturbance that are known to have depleted values and higher Dexcess values. This is also corroborated by the observed isotope composition of spring waters collected during February 2009, which was found to be most depleted among the four sets. This suggests that winter rainfall contributes to February spring discharge rates. There is a gradual enrichment in the isotope values of the spring waters from February to November 2009 this may be attributed to combined effect of local climatic variation and seasonality. The isotope trends indicate that a major component of spring waters is freshly recharged precipitation. Low altitude springs of Kakodakhal show some evaporation effect. In the case of Isala, all the samples fall as a single cluster during September 2009 and November 2009, indicating the source of recharge to be common, probably originating from same altitude.

There is no correlation found between the isotope content of the spring water and the altitude of the spring at Kakodakhal during February 2009 and November 2009, but altitude effect was found during June 2009 and September 2009. The isotope versus altitude trends indicate that spring water isotope signatures do not vary as per the altitude of springs. This suggests that springs are recharged from a common location in all the seasons. The seasonal variation in isotope data of Isala springs is mainly attributed to seasonal variations in rainwater isotopic composition and not to altitude variation. The environmental tritium content of all the samples is generally between 7.5 and 11.6 TU except a few samples which show about 6 TU, which indicates that all the spring waters receive modern recharge.

Based on the isotope hydrogeochemical investigations and considering the geomorphological features of each spring site, artificial recharge structures for rainwater harvesting were suggested. After construction of artificial recharge structures, an overall increase in spring discharge rate is reported. In Isala, the mean spring discharge rate has increased from 2.2 to 4.5 L/min, while in Kakodakhal the increase was from 8 to 8.2 L/min. Some anomalies were observed in the discharge rates of springs after construction, which can be attributed to changes in land use land cover or improper maintenance of spring water collection points.

It is evident from the above study that environmental isotopes play a vital role in getting insight into the recharge area/s and sources of springs in the mountainous region. Integration of isotope interpretation along with the available hydrogeology and geomorphology can help demarcate the recharge zones of the springs. Accordingly, suitable measures can be taken up to build the required artificial recharge structures and to increase the longevity of the springs with increased discharge rate. Information gathered through isotope approach, along with the other complementary data, will help in comprehensive understanding of the hydrogeologic processes associated with Himalayan springs, and can lead to scientifically validated watershed intervention and monitoring plans.

Acknowledgment

Dr. P.K. Mohapatra, Associate Director, Radiochemistry and Isotope Group and Dr. H. J. Pant, Head, Isotope and Radiation Application Division, Bhabha Atomic Research Centre are thanked for their constant support and encouragement. Prof. Anil Joshi, Director, HESCO, Dehradun, for the local support. Dr. Sitangshu Chatterjee, Mr. Annadasankar Roy, Mr. Ajay Jaryal and Mr. Kamble of the Isotope Hydrology Section have offered their help at various stages of this project and are sincerely acknowledged. The staff members of the HESCO, Dehradun, field staff at Isala and Kakodakhal sites have offered their valuable support during the project fieldwork, their contribution is greatly acknowledged.

References

Agarwal, A., Bhatnagar, N., Nema, R.K., and Agrawal, N.K. (2012). Rainfall dependence of springs in the Midwestern Himalayan hills of Uttarakhand. *Mountain Research and Development* 32 (4): 446–455.

Alford, D. (1992). Stream flow and sediment transport from mountain watersheds of the Chao Phraya Basin, northern Thailand: a reconnaissance study. *Mountain Research and Development* 12 (3): 257–268.

Bartarya, S.K. (1991). Watershed management strategies in central Himalaya: the Gaula River Basin, Kumaun, India. *Land Use Policy* 8 (3): 177–184.

BIS (Bureau of Indian Standards) (2012). *Indian Standards Specification for Drinking Water*. IS 10500. New Delhi, India: BIS.

Bonacci, O. (1993). Karst springs hydrographs as indicators of karst aquifers. *Hydrological Sciences Journal* 38 (1): 51–62.

Bruijnzeel, L.A. and Bremmer, C.N. (1989). *Highland-lowland Interactions in the Ganges Brahmaputra River Basin: A Review of Published Literature*. ICIMOD Occasional Paper No. 11, 152p. Kathmandu, Nepal: International Centre for Integrated Mountain Development (ICIMOD).

Bryan, K. (1919). Classification of springs. *USGS Staff – Published Research* 27: 552–561.

Clark, I. and Fritz, P. (1997). *Environmental Isotopes in Hydrogeology*, 328. Boca Raton, FL: Lewis Publishers.

Craig, H. (1961a). Standard for reporting concentration of deuterium and oxygen-18 in natural waters. *Science* 133: 1833–1834.

Craig, H. (1961b). Isotopic variations in meteoric waters. *Science* 133: 1702–1703.

Fritz, P. and Fontes, J.C. (1980). *Handbook of Environmental Isotope Geochemistry, The Terrestrial Environment, A*. Amsterdam: Elsevier https://doi.org/10.1016/B978-0-444-41780-0.50002-X.

Gupta, S.K., Deshpande, R.D., Bhattacharya, S.K., and Jani, R.A. (2005). Groundwater $\delta^{18}O$ and δD from central Indian Peninsula: influence of the Arabian Sea and the Bay of Bengal branches of the summer monsoon. *Journal of Hydrology* 303 (1–4): 38–55.

Herschy, R.W. (1995). *Streamflow Measurement*. London: CRC Press.

Hounslow, W.A. (1995). *Water Quality Data- Analysis and Interpretation*. New York, NY: Lewis Publishers.

IAEA Guide Book (1983). *Guidebook on Nuclear Techniques in Hydrology*. Technical Report Series No. 91, 439. Vienna: International Atomic Energy Agency.

Jeelani, G. and Deshpande, R.D. (2017). Isotope fingerprinting of precipitation associated with western disturbances and Indian summer monsoons across the Himalayas. *Journal of Earth System Science* 126: –108. https://doi.org/10.1007/s12040-017-0894-z.

Keesari, T., Sinha, U.K., Pant, H.J., and Pujari, P.K. (2021). *Societal Benefits of Isotope Techniques in Water Resources*. Mumbai, India: Non-Power Application of Nuclear Technologies ISBN-978-81-954733-2-8.

Kendall, C. and McDonnell, J.J. (1998). *Isotope Tracers in Catchment Hydrology*. Amsterdam: Elsevier https://doi.org/10.1016/B978-0-444-81546-0.50007-0.

Kumar, B., Rai, S.P., Kumar, U.S. et al. (2010). Isotopic characteristics of Indian precipitation. *Water Resources Research* 46 (12): https://doi.org/10.1029/2009wr008532.

Mahamuni, K. and Kulkarni, H. (2012). Groundwater resources and spring hydrogeology in South Sikkim, with special reference to climate change. In: *Climate Change in Sikkim - Patterns, Impacts and Initiatives* (ed. T.M.L. Arrawatia and S. Gangtok), 261–274. Sikkim, India: Information and Public Relations Department, Government of Sikkim.

Meinzer, O.E. (1923). *Outline of Ground-Water Hydrology, with Definitions*. Washington, D.C.: Government Printing Office US Geological Survey Water Supply Paper 494, 69 p.

Mook, W.G. (2000). *Environmental Isotope in Hydrological Cycle: Principles and Applications*. Technical document in Hydrology No. 39, Vol. 1. Paris: UNESCO.

Nair, A.R. (1983). Possibilities of liquid scintillation counting for tritium and radiocarbon measurements in natural water. *Proceedings of the Workshop on Isotope Hydrology*. Mumbai: BARC, 41–56.

Piper, A.M. (1953). *A Graphic Procedure in the Geochemical Interpretation of Water Analyses*. Washington, D.C.: U.S.G.S, Water Resources Division, Groundwater Branch Geochemistry, No. 12, 14 p.

Shivanna, K., Keesari, T., Jacob, N. et al. (2008). Isotope techniques to identify recharge areas of springs for rainwater harvesting in mountainous regions of Gaucher area, Chamoli District, Uttarakhand. *Current Science* 94 (8): 1003–1011.

Srivastava, R.N. and Ahamad, A. (1977). *Geology and Structure of Alknanda Valley, Garhwal Himalaya*. GSI.

Valdiya, K.S. and Bartarya, S.K. (1989). Diminishing discharge rates of mountain springs in a part of Kumaun Himalaya. *Current Science* 58 (8): 417–426.

Vashisht, A.K. (2008). Ingenious techniques for irrigation sustainability in Himalayan and Shiwalik foothill regions. *Current Science* 95 (12): 1688–1693.

WHO (World Health Organization) (2011). *Guidelines for Drinking Water Quality*. Geneva: WHO.

9

Biogeochemical Characterization of Water Resources in the Indian Himalayan Regions

Sayantan Das[1], *Aman Verma*[2], *Saurabh K. Sharma*[1], *and Nidhi Verma*[3]

[1] *School of Computer and Systems Sciences, Jawaharlal Nehru University, New Delhi, India*
[2] *School of Computational and Integrative Sciences, Jawaharlal Nehru University, New Delhi, India*
[3] *Department of Microbiology, Ram Lal Anand College, University of Delhi, New Delhi, India*

9.1 Introduction

Water resources play a significant role in the survival of lives and in maintaining the ecosystem in mountain regions. Mostly, mountains are considered arid or semiarid, thus, clean water is a crucial factor for drinking, hygiene maintenance, agricultural and industrial purposes, and hydropower generation to support millions of lives. Particularly, India consists of the Eastern and Western Himalayan ranges, in which water resources are glaciers, snow, lakes, waterfalls, hot springs, and groundwater. State of Sikkim and Arunachal Pradesh lie in the Eastern Himalayas, whereas, Jammu and Kashmir, Himachal Pradesh, and Uttarakhand are in the Western Himalayan range. The state of Sikkim covers 7096 km^2 area and is situated between 27.5° and 28.10° N latitudes, and, 88.4° and 88.58° E longitudes (Choudhury 2006). The total area of Arunachal Pradesh is 83 850 km^2 and its geographical coordinates are between 26.40° and 29.25° N latitudes and 91.35° and 97.35° E longitudes (Kumar 2013). Jammu and Kashmir is located between latitudes 32° 28′ N and 37° 6′ N, and 72° 53′ E and 80° 32′ E longitudes, with land acquisition of 222 236 km^2 (Romshoo et al. 2020). The state of Himachal Pradesh comprises an area of 55 673 km^2 and is present in the latitudes of 30° 22′–33° 12′ N and longitude 75° 42′–79°04′ E (Singh and Kumar 2014). Lastly, Uttarakhand lies in the latitudes between 28.43° and 31.28° N and longitudes between 77.34° and 81.03° E, and acquires 53 483 km^2 of land (Sati and Bandooni 2018). These states are storehouses for popular hill stations and holiday destinations, which often require a lot of water supply for usage, and water bodies present in these places come in contact with fragile environments. The contamination of water resources presents a significant danger to both individuals and the aquatic ecology, while the rapid growth of the population accelerates climate change (Palmate et al. 2017). For instance, water pollution caused by the addition of man-made hazardous chemicals and by-products can create poisonous and

Water Resources Management in Mountain Regions, First Edition. Edited by Virendra Bahadur Singh, Sughosh Madhav, Rakesh Kumar Gupta, Prerna Diwan, and Amit Kumar.

greenhouse gases, which in turn result in global warming and more severe environmental risks. Mountains function as initial climate predictors (Singh et al. 2010) and any possible change could lead to natural disasters. Many cases of water impurity were reported in different parts of the Indian Himalayan Regions (Chhimwal et al. 2022; Dutta et al. 2022; Sharma et al. 2022), and most of them are recognized as tourist spots.

Due to the overgrowing demand for freshwater, there is an occurrence of water scarcity in the Himalayan regions (Randev 2016; Thakur et al. 2023). Furthermore, the lack of management of water streams leads to pollution and contamination of drinking water. Water is a major cause of nearly 80% of human diseases, according to the WHO (Lin et al. 2022). According to Niti Aayog, nearly 2 lakh Indians pass away from waterborne illnesses every year, and these diseases include typhoid, cholera, shigellosis, hepatitis, and diarrhea (Sharma et al. 2023). There could be many other instances of changing water chemistry in mountain regions, for example, contamination by bacteria or parasites, solid particles, metal ions, geological factors, and other chemical factors. Thus, proper assessment and treatment should be imposed to prevent these circumstances in India. In this paper, you will get an idea about the incidents occurring in the Indian Himalayan Region and are contributing to water chemistry.

9.2 Factors Impacting Water Resources

9.2.1 Biological Factors

The chemistry of water, originating from glaciers and flowing into lakes and rivers, is both influenced and affected by living things, including bacteria, fungi, and other aquatic life forms (phytoplankton and zooplankton). Snow, rivers, and lakes are the potential sources of water supply in the mountain origins. Due to human intervention and anthropogenic activities, the water bodies get contaminated, causing the deterioration of potable water levels. The existence of coliform in water is a sign that human or animal feces have contaminated the water supply. The coliform family of bacteria are markers of harmful organisms. Typhoid, hepatitis, and diarrhea are among the waterborne illnesses brought on by human waste contamination (Theron and Cloete 2002; Elko et al. 2003; Sood et al. 2008). The presence of fecal coliform in Indian mountain streams has been reported in areas of Sikkim (Singh et al. 2019), Uttarakhand (Kothari et al. 2021), Kashmir (Khadse et al. 2016; Qayoom et al. 2023), Himachal Pradesh (Singh et al. 2021), and Arunachal Pradesh (Gaur et al. 2022). These studies suggested that the colony forming unit (CFU) level in these water bodies is beyond Bureau of Indian Standards (BIS) standards. *Escherichia coli* (*E. coli*) and Enterococcus are the most prevalent coliform bacteria found in most water samples. In addition, there is a close relationship between gut health and fecal coliform (Govindarajan et al. 2020). Elevated levels of *E. coli* in the gut promote cognitive decline, colitis, and depression (Kim et al. 2020). Evidence also suggests that dysbiosis also promotes colorectal carcinogenesis (Gagnière et al. 2016). The most prevalent hallmarks of dysbiosis are a decreased diversity of the microbiota, the depletion of beneficial bacteria, and the overgrowth of harmful microbiota (Hrncir 2022). Thus, the consumption of water from mountain resources elevates the bacterial presence in the gut microbiome which leads to several diseases. Other than that, there are also findings of cold-tolerant microorganisms present in high-altitude regions from different states of India (Dhakar and Pandey 2020).

In different studies on Indian water streams from Himalayan-linked states, phytoplanktons such as Bacillariophyceae, Chlorococcales, and Cyanophyceae are common, among which Bacillariophyceae is the most dominant phytoplankton species (Jindal et al. 2014; Ganie et al. 2018; Sharma et al. 2016; Islam et al. 2021). Some species of Bacillariophyceae produce a harmful neurotoxin, Domoic acid, which influences stomach pain, disorientation, confusion, memory loss, coma, and even death (Lefebvre and Robertson 2010). Cyanophyceae is a type of cyanobacteria (Anand et al. 2019) that produces cyanotoxins and is linked to neurodegenerative diseases (Sini et al. 2021). The neurotoxic β-N-methylamino-L-alanine (L-BMAA), synthesized by cyanobacteria and algae species, has been detected in the brain and cerebrospinal fluid of individuals diagnosed with Alzheimer's disease (AD) and amyotrophic lateral sclerosis (ALS). These findings suggest that being exposed to cyanotoxins may have a major impact on the development of neurodegenerative disorders in people (Berntzon et al. 2015; Cox et al. 2018; Ra et al. 2021). There is no reporting of such neurological cases related to these toxin products in Indian regions, however, the agitation of harmful toxins in drinking water could be a potential threat to humans. Results from a case study indicated that during the monsoon season, at the highly disturbed site where maximum disturbances were recorded, an increased concentration in physicochemical parameters [total dissolved solids (TDSs), turbidity, phosphates, and nitrates] harmed phytoplankton density (Sharma et al. 2016). Apart from that, Zooplankton families such as *Protozoa*, *Hydrozoa*, *Rotifers*, *Copepod*, *Cladocera*, *Crustacea*, and *Ostracoda* were identified in the water bodies of Himachal Pradesh (Rana et al. 2013; Thakur et al. 2013; Devi et al. 2017). In the state of Jammu and Kashmir, an abundance of *Protozoa*, *Rotifera*, *Cladocera*, and *Copepoda* were reported in two different study sites (Sharma and Sharma 2019; Gahlaut et al. 2023). *Protozoa*, *Rotifers*, *Cladocera*, and *Copepoda* are common in most studies and two of them carry parasites that cause severe health issues (Bass et al. 2021; Ma et al. 2022). In Bhimtal Lake at Uttarakhand, *Rotifera* (16 sp.), *Cladocera* (8 sp.), and *Cepepoda* (5 sp.) were isolated (Panwar and Malik 2016). Similarly, almost the same kind of organisms were reported from different rivers in the Dehradun district (Negi and Mamgain 2013). In the Eastern Himalayan Range, these phytoplankton species are widespread in Arunachal Pradesh (Sarma et al. 2017; Dai Nimasow et al. 2022).

9.2.2 Chemical Factors

The main contributors to water hardness include bicarbonates, silicates, sulfates, dissolved other metal ions, and salts of calcium and magnesium (Dubey 2022). According to the BIS, the combined measurement of total hardness (TH) and alkalinity, expressed as calcium carbonate ($CaCO_3$), should be 200 mg/l. This measurement is considered within a healthy range for the human body. Elevated levels of water hardness can contribute to the development of cardiovascular disease and kidney stone complications (Napacho and Manyele 2010). There is a direct link between water hardness, alkalinity, and pH. Typically, hard water is considered alkaline, thus increasing pH levels (>6.5). Studies have shown that some regions contain hard water in Himachal Pradesh (Sharma 2004; Sharma 2008; Dev and Bali 2019), Jammu and Kashmir Kouser et al. 2022; (Sudan 2018), Uttarakhand (Chhimwal et al. 2022), Sikkim (Vishwakarma et al. 2018), and Arunachal Pradesh (Goswami et al. 2023).

The sources of nitrogen and phosphorus in the water come from fertilizers and pesticides, sewerage water, and human and animal waste, which runoff through rainfall and mix in water streams. Their high levels in water can lead to eutrophication, where there is an overgrowth of algae, forming algal blooms (Chislock et al. 2013). Eutrophication is the major cause of lowering inorganic carbon and dissolved oxygen (DO), raising pH, and microbial decomposition, resulting in a decline in water quality. The surface of the water is covered with algal biomass, thus, there is a limitation of sunlight entry inside the water, and reduces CO_2 and DO levels (Zagarese et al. 2021). After a while, there is a limitation of nutrients and oxygen to microbes and fishes, which leads to their death and decomposition, causing the water to become turbid and unsuitable for use. The most cases of eutrophication are reported in Jammu and Kashmir (Abubakr and Balkhi 2018; Hassan et al. 2018; Ganaie et al. 2022; Kumar et al. 2022; Mushtaq et al. 2022), and Himachal Pradesh (Kamaldeep et al. 2011; Kumar and Mahajan 2020; Sharma et al. 2021; Choden et al. 2022; Kumar et al. 2023). There is another incidence where no eutrophication happens but still follows the same process to agitate nitrogen and phosphorus in water. Water rich in nitrogen and phosphorus is being consumed by human beings. Elevated levels of nitrates in drinkable water can cause stomach cancer, methemoglobinemia, and negative effects on the neurological and cardiovascular systems (Ward et al. 2011; Isiuku and Enyoh 2020). Whereas, high phosphate concentrations are linked with cardiovascular diseases (CVDs) and kidney disease (Vervloet et al. 2017).

Both Eastern and Western Himalayan regions have many popular holiday destinations, where tourists can enjoy the scenic beauty of the Himalayas. These spots witness a lot of plastic pollution due to the lack of dumping zones in these sites. These plastics are later submerged in water bodies and break into small pieces to produce microplastics. Microplastics are believed to induce several harmful consequences, including DNA damage, metabolic dysfunction, immune system activation, oxidative stress, neurotoxicity, and adverse effects on development and reproduction (Li et al. 2023). Other than that, microplastics are also associated with cancers (Zarus et al. 2023; Chen et al. 2024), Dementia (Gaspar et al. 2023), and cardiovascular diseases (Persiani et al. 2023).

9.2.3 Geological Factors

The underlying forces of the earth's surface are largely explained by the chemistry of lake and river water as well as sediment. Rocks in continuous contact with groundwater, water streams, and glaciers get weakened and are broken down into small pieces by the sedimentation process, and these pieces from one place to another by erosion. Continuous occurrence of this event agitates the rock minerals with water molecules, turning into hard water with an abundance of chemical elements present in rocks. Indian mountain regions contain mostly sedimentary rocks, which are primarily made up of limestone, in the form of $CaCO_3$ and are relevant to water hardening. In addition, soil composition has tremendous effects on water chemistry, both at surface and groundwater levels, because of pesticides, plastics, and cyanobacteria and the decomposition of organic materials. Many hill stations in India are very popular, and the occurrence of such incidents is very common, causing pollution in landfills, and the water near these contaminated sites is affected. Water clarity is measured by turbidity, which is caused by suspended solids including silt, clay, organic matter in colloidal form, plankton, and other creatures. Drinking turbid water can lead to an issue with gastroenteritis (Morris et al. 1996). In a study, water quality analysis of five different rivers in the Kumaun regions of Uttarakhand revealed that the main causes of the decline in the quality of the water in all the rivers may be tourism, anthropogenic processes, eutrophication, and geogenic processes (Seth et al. 2016). Climate change (Pastorino and Prearo 2020), overgrazing (Pastorino et al. 2019), mountain tourism (Rupakheti et al. 2017), and mountain farming (Tiberti et al. 2014) are some of the possible threats to them.

Another perspective is that the Himalayan mountains are rich in medicinal plants which are close to waterfalls, lakes, and hot springs. There could be a fair enough chance that the secondary metabolites of plants mix with water molecules and enhance the properties of water. The extraction process of secondary metabolites is done under hot conditions, however, these kinds of studies have not been done yet. There could be a fair enough chance that secondary metabolites might get combined with water under normal or cold conditions if dipped for a long time.

9.3 Monitoring, Assessment, Management, and Treatment

9.3.1 Monitoring and Assessment

There are four water quality index (WQI): National Sanitation Foundation Water Quality Index (NSFWQI), Oregon Water Quality Index (OWQI), Canadian Council of Ministries of the Environment Water Quality Index (CCMEWQI), and Weighted Arithmetic Water Quality Index Method (WAWQI) (Chidiac et al. 2023). The WQI offers a comprehensive understanding of surface and groundwater quality and whether it is appropriate for human consumption. The main objective of the WQI is to convert intricate data on the health of water into accessible information that can be comprehended by the general public, enabling them to grasp the state of water supplies in a certain region. In this case, the WQI is calculated using eleven water quality parameters, including turbidity, pH, TH, alkalinity, TDSs, chloride, calcium, sulfate, magnesium, nitrate, iron, etc., as listed in Table 9.1. According to Dalakoti et al. (2018), heavy metal contamination is more common in Nainital Lake. Analyzing physical, chemical, and microbiological data that represent the abiotic and biotic circumstances of the ecosystem is part of the process of assessing the quality of water (Bisht et al. 2018). The majority of pollutants found in drinking water are inorganic compounds. Initially, it can be easily identified by looking at water turbidity or transparency. Monitoring groundwater is not only expensive and time-consuming but also challenging. A lot of the time, the contamination was not discovered until it had spread widely. A region's groundwater quality is influenced by several variables, including the type of rocks present, residency time, recharge water composition, soil gas interaction, and soil water interaction.

9.3.2 Management and Treatment

Due to the emerging effects of changing water chemistry due to several interruptions, a close evaluation and strong actions should be taken. Governments should take action against anthropogenic activities, especially in mountain areas, and impose strict rules on

Table 9.1 Drinking water according to BIS standards (IS 10500: 2012).

Serial number	Parameters	Permissible limits
1.	TDS	500 mg/l
2.	Turbidity	1 NTU
3.	pH	6.5–8.5
4.	Hardness (as $CaCO_3$)	200 mg/l
5.	Alkalinity (as $CaCO_3$)	200 mg/l
6.	Fecal Coliform	0 cfu
7.	Calcium	75 mg/l
8.	Magnesium	30 mg/l
9.	Nitrate	45 mg/l
10.	Sulfate	200 mg/l
11.	Fluoride	1 mg/l
12.	Chloride	250 mg/l
13.	Copper	0.05 mg/l
14.	Cadmium	0.003 mg/l
15.	Chromium	0.05 mg/l
16.	Lead	0.01 mg/l
17.	Iron	0.3 mg/l
18.	Zinc	5 mg/l
19.	Arsenic	0.01 mg/l
20.	Sulfide	0.05 mg/l
21.	Selenium	0.01 mg/l
22.	Nickel	0.02 mg/l
23.	Mercury	0.001 mg/l
24.	Boron	0.5 mg/l
25.	Barium	0.7 mg/l
26.	Aluminum	0.03 mg/l
27.	Ammonia	0.5 mg/l

visitors as well as localities. It is also necessary to aware people of these devastating circumstances, which will lead to the reduction of offensive activities near water resources. Other than that, as many water bodies get polluted and contaminated, proper treatment procedures are needed to make water suitable for human consumption. The process of water treatment primarily involves six major steps, which include screening, coagulation, flocculation, sedimentation, filtration, and disinfection as described in Figure 9.1. Screening is necessary to remove large suspended and floating debris from the input. These objects could harm machinery or impede flow through the plant (EPA 2002). The items encompass twigs, leaves, paper, clothing, and other refuse. Chlorine dioxide, an oxidizing agent, is the initial chemical introduced in the process of coagulation. It is used to break down organic

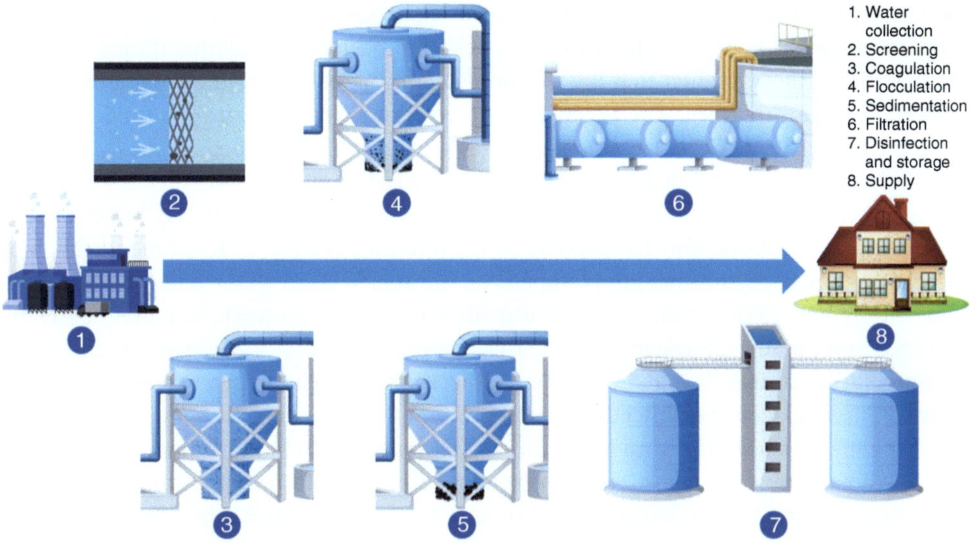

Figure 9.1 The process of water management and treatment. Water treatment involves a comprehensive six-step process after the collection of water which includes screening, coagulation, flocculation, sedimentation, filtration, and disinfection. Screening removes large suspended materials to protect machinery and ensure smooth flow. Coagulation introduces chemicals like ammonium sulfate to clump particles, while flocculation encourages particle aggregation. Sedimentation allows the settling of large flocs, filtration removes remaining sediments, and disinfection, often with chlorine, ensures the water is safe for consumption by eliminating harmful microorganisms before proper storage and supply of water.

materials present in natural sources (Xu et al. 2022), such as decaying leaves and other plant material. The primary coagulant used is aluminum sulfate, which is a chemical coagulant (Brandt et al. 2017). To enhance the coagulation process, a polymer, which is a long chain of synthetic organic compounds, is added to the water. This polymer helps strengthen the bonding chains of the original coagulant (Kurniawan et al. 2020). The rapid mix device, which creates a turbulent mixing force to assist the thorough distribution of artificial coagulants into the untreated water and commence the coagulation process, is the location where coagulants are administered. Coagulants are chemicals that agglomerate minute particles into bigger particles, which can be removed by filtration, draining, settlement, or skimming subsequently in the remediation method. The flocculation is the next fundamental process that the coagulated water travels to. The process of slowly spinning the tiny coagulated particles to form a floc (clump) is referred to as flocculation (Gregory and O'Melia 1989). Interaction between clump and water particles (sediment) is enhanced by the flocculation process. These particle contacts or collisions are often the product of modest churning created by hydraulically or mechanically mixing techniques. The water moves through a pair of flocculation troughs with stirring paddles to gently stir the coagulated liquid (Gregory and O'Melia 1989). The floc that develops creates a surface where the waterborne particulates adhere, or attach, to form larger settleable particulates which are simpler to remove by filtering and sedimentation. Large flocs have to be removed after they develop, and it is done via a method called sedimentation, where the particles sink to the bottom inside a settling chamber (Shammas et al. 2005). After flocculation and coagulation, the water is kept in a chamber for a brief period to allow sedimentation to occur. Sludge is the waste product that has accumulated at the bottom of the compartment and gets removed for disposal. The procedure used to extract particles from a fluid is termed filtering (Gray 2014). Water treatment involves passing water over beds of gravel and sand to eliminate debris that failed to separate inside the sedimentation chamber. Rapid gravity filters are widely utilized due to their rapid rate of discharge of 4–8 m^3/sq. meter of filter area per hour. Back-washing happens while the filters are stuffed with sediments that have become retained. To eliminate the trapped contaminants, clean water, and air are pumped out of the filter via this technique. The dirt-filled water, or backwash, is subsequently discharged into the sewage system if one exists. Alternatively, after a phase of settlement in a sedimentary chamber, it may be discharged back into the parent river to eliminate sediments. The final step is disinfection or chlorination. In this stage, the water gets sterilized to eliminate any hazardous microorganisms (Ghernaout 2017) that may have remained after sedimentation. The most often used chemical for disinfection is chlorine, which can be either a gas or a liquid (Shamrukh and Hassan 2005) (like sodium hypochlorite (NaOCl)). Chlorine interacts with all pollutants in water, including microbes, over a certain length of time termed the contact time (Shamrukh and Hassan 2005). Leftover chlorine indicates the level of chlorine which remains after this. This stays in the water across the distribution system, defending against any microbial entry until the water is drunk. The World Health Organisation (WHO) recommended that water have a maximum residual chlorine concentration of 5 mg/l. Other ways are available for sterilizing water, including the use of UV radiation or ozone gas. These nonetheless do not prevent it from contamination by bacteria once it departs the water purification facility. The purified water is then fed into the supply system after disinfection. The research area's rainfall trend for the previous 3 years indicates that it begins to increase in March and ends in August, then begins to decrease in September and ends in February. August 2010 and August 2011 recorded the highest average rainfall amounts, in addition, the maximum spring discharge was also observed in August.

9.4 Case Studies

9.4.1 Evaluation of the Biotic and Abiotic Condition of the Alaknanda River in Uttarakhand (Biological)

An aquatic ecosystem is driven by the biotic and abiotic elements of a river; these elements are interdependent. A sustainable ecosystem of freshwater indicates a body of water's ability to assimilate diversity. The decline of biological elements in a water stream can be attributed to various environmental variables, including changes in climate, human activity, competition for survival, loss of aquatic habitat, etc. The river Alaknanda is regarded as the Ganges' primary source due to its longer length and discharge (Mittal et al. 2008). It travels 190 km to reach the Bhagirathi River in Devprayag, and the basin size is 10 882 km^3. In the case study done between 2017 and 2018, water samples from a 172 km section of the Alaknanda River were taken from four different zones (Tomar et al. 2022). The standards approach was used to assess the physicochemical parameters, which included water temperature, air temperature, relative humidity, light intensity, rainfall, depth, TDSs, DO, biochemical oxygen demand (BOD), pH, and chemical oxygen demand (COD). The levels of pH, TDS, BOD, and COD were under acceptable standards, however, DO was at the higher site, which is good for biological communities. Using identification keys, biological parameters such as fish, macroinvertebrates, phytoplankton, and zooplankton were identified. In the river's sampling zones, a total of 32 phytoplankton species, 19 zooplankton species, and 33 macrobenthos species, including 21 species of fish were discovered. Canonical correspondence analysis (CCA) is used to analyze all of the biotic and abiotic parameters based on observed data. CCA is a valuable technique for assessing the important relationship between biotic components and the physio-chemical characteristics of an underwater environment. It is appropriate for usage with aquatic biodiversity. The Eigenvector axis denoted the strong relation between phytoplankton and environmental factors of 47% on axis 1 and 29% on axis 2. Several biotic species that had a favorable link with physicochemical parameters were discovered in the river during the investigation. A few fish, macrobenthos, and phytoplankton were shown to be significantly correlated with TDS, BOD, DO, COD, and pH. The river's biotic and abiotic components showed positive and negative correlations with them. It was observed that the river is very diversified, which is a sign of excellent water quality.

9.4.2 Pollution of Heavy Metals at Kameng River in Arunachal Pradesh (Chemical)

The Kameng River is a high-altitude river system that rises in the Indian state of Arunachal Pradesh in the snow-covered peaks of the eastern Himalayas. The river divides the Indian state of Arunachal Pradesh into the East and West Kameng Districts (Arora et al. 2021). The river rises in the snow-covered plains of the Kangto Peak and Gori Chen mountain ranges at an elevation of roughly 5400 m above sea level. It then flows for about 264 km from its starting point in the upper Himalayas to the Brahmaputra River, which it meets close to Tezpur town (Arora et al. 2021). The indigenous populations that live close to the Kameng River rely heavily on it for their water supply. Using seasonal and regional variations, this study presented the first comprehensive assessment of heavy metal concentrations in the river inside the administrative borders of Arunachal Pradesh (Sharma and Sarma 2018). Large- and medium-sized hydropower project construction is anticipated to split the river's course entirely and change the river's natural flow, both of which will inevitably have an impact on the regime for water quality. Water samples from 27 sites, showing four watersheds within the river, the levels of Cd, Pb, Mn, and Zn in the river water were higher above the permissible limits set by regulatory bodies (WHO and BIS). The multivariate statistical analysis, which employed hierarchical cluster analysis (HCA) and principal component analysis (PCA) tools, revealed that geochemical processes, such as base metal mineralization, phosphate rock dissolution, and carbonate rock weathering, were the main factors controlling the concentration of the heavy metals. With the use of far fewer variables, known as factors, PCA can explain a vast and complicated collection of variables. The information included in the huge set of initial observations is represented by the same number of factors (Mishra et al. 2017). Whereas, the internal variable combinations and relationships form the foundation of the HCA approach. HCA's primary goal is to provide a mechanism for rearranging the collection of observations and classifying them using a dendrogram based on predetermined features into groups of variables (cases) (Granato et al. 2018). The graphical representation of the hierarchical clustering is interpreted by the dendrogram. The study also showed the anthropogenic sources of heavy metals, which mostly came from pesticide leaching from agricultural soils.

9.4.3 Ritualistic Practices and Pollution of the Tawi River in Jammu District (Geological)

The Tawi River traverses Jammu, dividing the city into the old and new cities. The Tawi River serves as the city's primary supply of water (Kour et al. 2022) as mentioned in Figure 9.2. There are five bridges across the river: two lead from Gumat to Vikram Chowk, one is on the city bypass at Nagrota, another connects Gujjar Nagar with the Bahu fort region, and one is from Bhagwati Nagar to Balli Nallah. Tawi enters the Chenab River after passing through Jammu City and Pakistan. The Tawi River in Jammu and Kashmir holds significant religious and historical value. The Tawi River follows the Chenab River's left bank through the city of Jammu (Kour et al. 2022). The River Tawi encompasses a catchment area of greater than 2500 km^2 (Ashraf and Jamwal 2022). The river flows through the districts of Jammu, Doda, and Udhampur, covering a total distance of 141 km (Ashraf and Jamwal 2022) highlighted in Figure 9.2. Due to improper use by the locals and certain ritualistic activities, the river became contaminated. The river was well-known for its rainbow trout, but as time went on,

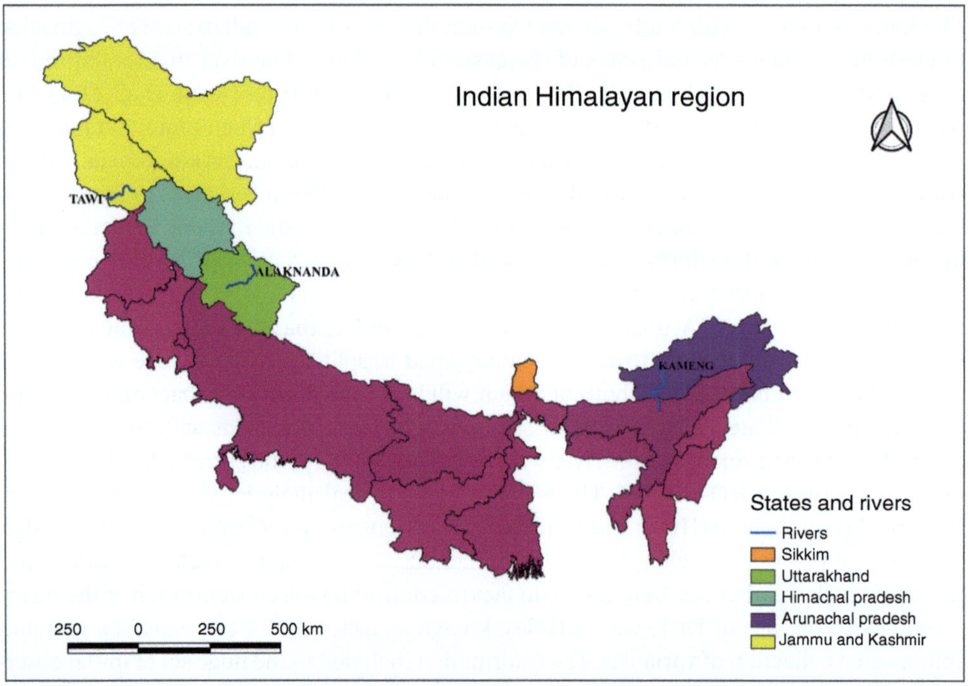

Figure 9.2 The map of the Indian rivers covered in Himalayan region. This map contains Indian Eastern and Western Himalayan regions. The Tawi River crosses through the Jammu district, next to that is the Alaknanda River which flows in the Uttarakhand region, and lastly, the river Kameng which is located in the state of Arunachal Pradesh.

the fish went extinct and the water's quality declined to the point that it was unfit for human use. Previously, the town's only source of drinkable water came from the river. In this case study, their goal was to discuss how religious activities affect the Tawi River's two busiest river ghats, Bikram Chowk Ghat and Tawi Har Ki Poodhi Ghat (WaZza et al. 2018). The Tawi River's water clearly shows that, following religious human activities, the value of the water quality index increased due to increases in turbidity, BOD, TDS, TH, total alkalinity, nitrates, and MPN values, and decreases in DO values. After the religious activities, three of the eight sites under investigation were determined to have high water quality, four (sites 1, 2, 6, and 7) were classified as having medium water quality, and site 8 was classified as having low water quality. The three, four, and five spots that have satisfactory WQI status could be deemed appropriate for swimming. People carry all of the rubbish from their homes that are associated in some way with religious rituals or other holy acts to the holy deity. It is largely made up of remnants of religion, plastic bottles, coconuts, Chris, flowers, leaves, and old glass frames, among other things. These things are all individually wrapped in polythene and thrown into the Tawi River and thus degraded the water quality over years.

9.4.4 Water Conservation and Management Practices in the Duga Region in Sikkim (Management)

Eighty percent of the rural population of Sikkim, Himalaya, depends on natural springs for their water supply (Tambe et al. 2012). Since Duga is one of the state's eight driest regions, springs serve as the main source of water. There is a growing shortage of water in the region. Such scarcity conditions are caused by anthropogenic sources, changing land use and land cover, changing climate conditions that result in unpredictable precipitation patterns, and external pressures. The goal of the study, which involved nine natural springs in the watershed of rural Central Pendam, Duga Block, Eastern Sikkim Himalaya, was to determine how spring discharge and rainfall patterns relate to one another. In addition, it will address the region's water crisis and record the locals' and the water user group's adaptive water management efforts (Rai 2014). Locals and the water user group have started various water management practices and strategies to deal with the situation. Focus groups involving various age groups, farmers, panchayat members, NGOs, government officials, and water user organizations provided details about local water preservation and management techniques. Rooftop water harvesting is important in the region along with other water harvesting practices such as fencing and plantation, scared grooves, and spring-shed development, and there is a need to increase the catchment area used for water conservation. The inhabitants living in the mountains have been using springs as a source of water for both agricultural and home use. Numerous studies from the Eastern and Western Himalayas demonstrate that anthropogenic and climatic factors are likely to be responsible for the marked reduction in spring discharge and the likelihood of natural springs becoming seasonal; anthropogenic factors contribute more to the death of springs, particularly as a result of development activities like building constructions, roads, and bridges (Tambe et al. 2012). As a result, there will be insufficient rainwater to seep into the aquifers, which would hinder spring recharge during the monsoon season and result in a scarcity of water throughout the dry winter. The majority of research has also addressed climate change as a novel threat to Himalayan springs and the notable decrease in their number (Azhoni and Goyal 2018; Bhat et al. 2020; Panwar 2020; Sahu et al. 2020). Local orographic factors typically influence the region's precipitation pattern. Because the lower southern and western regions of Sikkim are situated in the rain shadow of the Darjeeling Hills, they receive relatively less rainfall than other parts of the district. The Dhara Vikas study states that 700 springs are in critical condition, 8 blocks, 53 Gramme Panchayat units, and 293 Gramme Panchayat wards in Sikkim are experiencing dry conditions. In this case study, the reference data related to rainfall, temperature, and humidity were collected from the automatic weather station of ISRO, which was further used to establish the relation with spring discharge. Observation suggested October through February had the lowest mean rainfall, with the maximum mean rainfall recorded in July and August in the Central Pendam watershed.

9.5 Conclusion

This study displays the individual biogeochemical fingerprints of various water bodies, including rivers, lakes, and springs, illustrating the delicate interplay between geological substrates, hydrological cycles, and human interventions. Key findings suggest substantial geographical diversity in water quality measures such as pH, TDS, turbidity, and concentrations of main ions and trace elements. This variety is related to the varied geological formations, varying climatic conditions, and differing levels of human activity across the region. The study also underlines the impact of seasonal fluctuations, notably the monsoon season, in changing the chemical composition of water supplies. Increased runoff during the monsoon leads to elevated amounts of suspended particles and nutrients, compromising water quality downstream. Moreover, the research reveals important locations where water quality is adversely damaged by anthropogenic activity, such as agriculture, urbanization, and industry. High levels of nitrates, phosphates, and heavy metals in certain places emphasize the negative implications of these activities, demanding urgent action to limit pollution sources and implement sustainable water management strategies.

Overall, the work gives useful insights into the biogeochemical dynamics of water resources in the Indian Himalayan Regions, underlining the necessity for integrated management methods that account for both natural processes and human effects. The findings serve as a significant reference for policymakers, environmental managers, and researchers trying to protect ecological integrity and ensure the sustainable use of water resources in this ecologically sensitive and socioeconomically vital region.

9.6 Future Directions

The study on the biogeochemical characterization of water supplies in the Indian Himalayan Regions gives major insights into the processes impacting water quality and recommends various future possibilities for research and policy interventions.

Firstly, the adoption of modern monitoring systems is vital. The development and implementation of high-quality, real-time monitoring systems for continuous evaluation of water quality indicators across the Himalayan areas would be important. These sensors can give detailed temporal and geographical data, assisting in the early identification of pollutants and enabling prompt responses. In addition, using remote sensing technology and Geographic

Information Systems (GIS) to map and monitor water bodies can assist track changes over time, identifying pollution sources, and assessing the impact of anthropogenic activities on water quality. Secondly, significant climate change effect assessments are needed. Conducting long-term research to understand the impact of climate change on water resources, including evaluating patterns of glacier melt, variations in precipitation, and their implications on river flow and water quality, is vital. Developing and verifying simulation models to forecast future scenarios of water availability and quality under different climate change projections would help inform policymakers and stakeholders for improved water resource management.

In terms of biogeochemical interactions, a careful investigation of microbial populations inside the water bodies is important. Investigating the significance of microbial communities in biogeochemical cycles and understanding the relationships between various microbial species and their effect on water chemistry might assist create measures to prevent microbial pollution. Additionally, research on secondary metabolites from medicinal plants and their impact on water quality should be addressed, focusing on the extraction and interaction of these substances under varied environmental settings. Policy and management strategies also need to be enhanced. Reviewing and strengthening current water quality legislation and standards, maintaining stringent enforcement of pollution control measures, and encouraging sustainable behaviors among industry and communities are vital. Encouraging community-based management by teaching and enabling local people to monitor and manage their water resources can lead to more sustainable and effective outcomes.

Innovative therapeutic technologies should be developed and applied. Researching and adopting eco-friendly water treatment methods that employ natural processes and materials, such as bio-filtration, built wetlands, and phytoremediation, can be advantageous. Exploring sophisticated disinfection technologies like ultraviolet (UV) radiation and ozonation, which are effective in eradicating infections without leaving toxic residues, is also crucial.

Finally, educational and awareness activities are crucial. Launching extensive public awareness programs to increase awareness about the necessity of water conservation and pollution control will educate the public on the origins and repercussions of water pollution and promote behavioral changes. Integrating water resource management principles into school curricula at all levels helps generate a higher understanding and commitment to sustainable water practices among future generations. By addressing these future goals, this study might help greatly in the sustainable management and protection of these essential water resources.

References

Abubakr, A. and Balkhi, M.H. (2018). Trophic status assessment of Anchar Lake, Kashmir. *Journal of Pharmacognosy and Phytochemistry* 7 (4): 1154–1156.

Anand, N., Thajuddin, N. and Dadheech, P.K. 2019. Cyanobacterial taxonomy: morphometry to molecular studies. In *Cyanobacteria* (by Mishra A.K., Tiwari D.N. and Rai A.N.); Academic Press: Cambridge, MA, USA. Elsevier (pp. 43–64).

Arora, S., Kulkarni, A.V., Ghosh, P., and Satheesh, S.K. (2021). Estimating contribution of water flow components to Kameng river basin using hydrological modelling. *The International Archives of the Photogrammetry, Remote Sensing and Spatial Information Sciences* 43: 431–436.

Ashraf, S. and Jamwal, R.S. (2022). Conservation plan for Tawi river, India, using geoinformatics techniques. *Arabian Journal of Geosciences* 15 (1): 122.

Azhoni, A. and Goyal, M.K. (2018). Diagnosing climate change impacts and identifying adaptation strategies by involving key stakeholder organisations and farmers in Sikkim, India: challenges and opportunities. *Science of the Total Environment* 626: 468–477.

Bass, D., Rueckert, S., Stern, R. et al. (2021). Parasites, pathogens, and other symbionts of copepods. *Trends in Parasitology* 37 (10): 875–889.

Berntzon, L., Ronnevi, L.O., Bergman, B., and Eriksson, J. (2015). Detection of BMAA in the human central nervous system. *Neuroscience* 292: 137–147.

Bhat, S.U., Mushtaq, S., Qayoom, U., and Sabha, I. (2020). Water quality scenario of Kashmir Himalayan Springs – a case study of Baramulla District, Kashmir Valley. *Water, Air, & Soil Pollution* 231: 1–17.

Bisht, S., Sharma, R.C., Rawat, S., and Kumar, R. (2018). Physico–chemical attributes and bacterial diversity of river water at Rudraprayag, Garhwal Himalaya. *MOJ Ecology & Environmental Science* 3 (4): 277–282.

Brandt, M.J., Johnson, K.M., Elphinston, A.J., and Ratnayaka, D.D. (2017). Storage, clarification and chemical treatment. *Twort's Water Supply* 7: 323–366.

Chen, G., Shan, H., Xiong, S. et al. (2024). Polystyrene nanoparticle exposure accelerates ovarian cancer development in mice by altering the tumor microenvironment. *Science of the Total Environment* 906: 167592.

Chhimwal, M., Kaur, S., Srivastava, R.K. et al. (2022). Water quality of springs and lakes in the Kumaon Lesser Himalayan Region of Uttarakhand, India. *Journal of Water and Health* 20 (4): 737–754.

Chidiac, S., El Najjar, P., Ouaini, N. et al. (2023). A comprehensive review of water quality indices (WQIs): history, models, attempts and perspectives. *Reviews in Environmental Science and Bio/Technology* 22 (2): 349–395.

Chislock, M.F., Doster, E., Zitomer, R.A., and Wilson, A.E. (2013). Eutrophication: causes, consequences, and controls in aquatic ecosystems. *Nature Education Knowledge* 4 (4): 10.

Choden, Y., Sharma, M.P., Pandey, G. et al. (2022). Ecological health assessment of Renuka Lake, Himachal Pradesh, India. *Nature, Environment and Pollution Technology* 21 (1): 167–174.

Choudhury, M. (2006). *Sikkim: Geographical Perspectives*. New Delhi: Mittal Publications.

Cox, P.A., Kostrzewa, R.M., and Guillemin, G.J. (2018). BMAA and neurodegenerative illness. *Neurotoxicity Research* 33 (1): 178–183.

Dai Nimasow, O., Jongkey, K., Abhujam, S.K. et al. (2022). An assessment of phytoplankton distribution in the different altitudinal lakes of Arunachal Himalayan Region, India. *International Journal of Ecology and Environmental Sciences* 48 (4): 449–462.

Dalakoti, H., Mishra, S., Chaudhary, M., and Singal, S.K. (2018). Appraisal of water quality in the Lakes of Nainital District through numerical indices and multivariate statistics, India. *International Journal of River Basin Management* 16 (2): 219–229.

Dev, R. and Bali, M. (2019). Evaluation of groundwater quality and its suitability for drinking and agricultural use in district Kangra of Himachal Pradesh, India. *Journal of the Saudi Society of Agricultural Sciences* 18 (4): 462–468.

Devi, S., Sharma, D., Bhagat, S., and Sharma, A. (2017). Zooplanktonic diversity of Beas river near Indora, Kangra. *Proceeding of National Conference on Role of Biological Science in*

Sustainable Development and Bio Diversity Conservation, Arni University, Kagra, India (pp. 100–107).

Dhakar, K. and Pandey, A. (2020). Microbial ecology from the himalayan cryosphere perspective. *Microorganisms* 8: 1–17.

Dubey, A. (2022). A study on effects of hard water on human health. *Research Ambition: An International Multidisciplinary e-Journal* 6 (IV): 15–16.

Dutta, N., Thakur, B.K., Nurujjaman, M. et al. (2022). An assessment of the water quality index (WQI) of drinking water in the Eastern Himalayas of South Sikkim, India. *Groundwater for Sustainable Development* 17: 100735.

Elko, L., Rosenbach, K., and Sinnott, J. (2003). Cutaneous manifestations of waterborne infections. *Current Infectious Disease Reports* 5 (5): 398–406.

EPA. (2002). Wastewater technology fact sheet. Screening and grit removal. United States Environmental Protection Agency. Municipal Technology Branch. Washington, United States.

Gagnière, J., Raisch, J., Veziant, J. et al. (2016). Gut microbiota imbalance and colorectal cancer. *World Journal of Gastroenterology* 22 (2): 501.

Gahlaut, M., Sharma, A., Singh, M., and Kumar, M. (2023). Study of Zooplankton community in a freshwater pond in Jammu and Kashmir. *International Journal of Emerging Technologies and Innovative Reseach* 10 (1): 197–199.

Ganaie, D.B., Malhotra, A., and Wani, I.A. (2022). Water quality assessment of Mansbal Lake in Kashmir. *Water Science* 36 (1): 114–124.

Ganie, P.A., Baruah, D., Kunal, K. et al. (2018). Phytoplankton diversity and abundance in upland streams of Kameng drainage, Arunachal Pradesh. *Journal of Entomology and Zoology Studies* 6 (6): 1166–1173.

Gaspar, L., Bartman, S., Coppotelli, G., and Ross, J.M. (2023). Acute exposure to microplastics induced changes in behavior and inflammation in young and old mice. *International Journal of Molecular Sciences* 24 (15): 12308.

Gaur, N., Sarkar, A., Dutta, D. et al. (2022). Evaluation of water quality index and geochemical characteristics of surfacewater from Tawang India. *Scientific Reports* 12 (1): 11698.

Ghernaout, D. (2017). Water treatment chlorination: an updated mechanistic insight review. *Chemistry Research Journal* 2: 125–138.

Goswami, R., Bhagat, C., Lollen, I. et al. (2023). Potential arsenic–chromium–lead Co-contamination in the hilly terrain of Arunachal Pradesh, north-eastern India: genesis and health perspective. *Chemosphere* 323: 138067.

Govindarajan, D.K., Viswalingam, N., Meganathan, Y., and Kandaswamy, K. (2020). Adherence patterns of *Escherichia coli* in the intestine and its role in pathogenesis. *Medicine in Microecology* 5: 100025.

Granato, D., Santos, J.S., Escher, G.B. et al. (2018). Use of principal component analysis (PCA) and hierarchical cluster analysis (HCA) for multivariate association between bioactive compounds and functional properties in foods: a critical perspective. *Trends in Food Science & Technology* 72: 83–90.

Gray, N. F. (2014). Filtration methods. In *Microbiology of Waterborne Diseases* (by Percival S.L., Yates M.V., et al.); Academic Press: Cambridge, MA, USA. Elsevier (pp. 631–650).

Gregory, J. and O'Melia, C.R. (1989). Fundamentals of flocculation. *Critical Reviews in Environmental Science and Technology* 19 (3): 185–230.

Hassan, T.B., Arnold, R., and Mishra, R.M. (2018). Comparative physico-chemical limnology of two lakes of Kashmir Himalaya. *International Journal of Multicultural and Multireligious Understanding* 5 (2): 128–135.

Hrncir, T. (2022). Gut microbiota dysbiosis: triggers, consequences, diagnostic and therapeutic options. *Microorganisms* 10 (3): 578.

Isiuku, B.O. and Enyoh, C.E. (2020). Pollution and health risks assessment of nitrate and phosphate concentrations in water bodies in South Eastern, Nigeria. *Environmental Advances* 2: 100018.

Islam, S.T., Dar, S.A., Sofi, M.S. et al. (2021). Limnochemistry and plankton diversity in some high altitude lakes of Kashmir Himalaya. *Frontiers in Environmental Science* 9: 387.

Jindal, R., Thakur, R.K., Singh, U.B., and Ahluwalia, A.S. (2014). Phytoplankton dynamics and water quality of Prashar Lake, Himachal Pradesh, India. *Sustainability of Water Quality and Ecology* 3: 101–113.

Kamaldeep, M.S.R., Kochhar, N., and Ghosh, N. (2011). Impact of industrialization on groundwater quality: a case study of Baddi-Barotiwala industrial belt, Distt. Solan, Himachal Pradesh, India. *Journal of Industrial Pollution Control* 27 (2): 153–159.

Khadse, G.K., Patni, P.M., and Labhasetwar, P.K. (2016). Water quality assessment of Chenab river and its tributaries in Jammu Kashmir (India) based on WQI. *Sustainable Water Resources Management* 2: 121–126.

Kim, J.K., Lee, K.E., Lee, S.A. et al. (2020). Interplay between human gut bacteria *Escherichia coli* and *Lactobacillus mucosae* in the occurrence of neuropsychiatric disorders in mice. *Frontiers in Immunology* 11: 273.

Kothari, V., Vij, S., Sharma, S., and Gupta, N. (2021). Correlation of various water quality parameters and water quality index of districts of Uttarakhand. *Environmental and Sustainability Indicators* 9: 100093.

Kour, G., Kothari, R., Dhar, S. et al. (2022). Impact assessment on water quality in the polluted stretch using a cluster analysis during pre-and COVID-19 lockdown of Tawi river basin, Jammu, North India: an environment resiliency. *Energy, Ecology and Environment* 7 (5): 461–472.

Kouser, B., Bala, A., Verma, O. et al. (2022). Hydrochemistry for the assessment of groundwater quality in the Kathua region, Jammu and Kashmir, India. *Applied Water Science* 12 (7): 143.

Kumar, G. (2013). *Geology of Arunachal Pradesh*, 217. Bangalore: Geological Society of India.

Kumar, R., Jindal, R., Kulharia, M. et al. (2023). Knowledge mining for the differentiation pattern with respect to the productivity and trophic status of some selected lakes of Western Himalayas. *Frontiers in Environmental Science* 11: 125.

Kumar, P. and Mahajan, A.K. (2020). Trophic status and its regulating factor determination at the Rewalsar Lake, northwest Himalaya (HP), India, based on selected parameters and multivariate statistical analysis. *SN Applied Sciences* 2: 1–11.

Kumar, R., Parvaze, S., Huda, M.B., and Allaie, S.P. (2022). The changing water quality of lakes – a case study of Dal Lake, Kashmir Valley. *Environmental Monitoring and Assessment* 194 (3): 228.

Kurniawan, S.B., Abdullah, S.R.S., Imron, M.F. et al. (2020). Challenges and opportunities of biocoagulant/bioflocculant application for drinking water and wastewater treatment and its potential for sludge recovery. *International Journal of Environmental Research and Public Health* 17 (24): 9312.

Lefebvre, K.A. and Robertson, A. (2010). Domoic acid and human exposure risks: a review. *Toxicon* 56 (2): 218–230.

Li, Y., Tao, L., Wang, Q. et al. (2023). Potential health impact of microplastics: a review of environmental distribution, human exposure, and toxic effects. *Environment & Health* 1 (4): 249–257.

Lin, L., Yang, H., and Xu, X. (2022). Effects of water pollution on human health and disease heterogeneity: a review. *Frontiers in Environmental Science* 10: 880246.

Ma, J.Y., Li, M.Y., Qi, Z.Z. et al. (2022). Waterborne protozoan outbreaks: an update on the global, regional, and national prevalence from 2017 to 2020 and sources of contamination. *Science of the Total Environment* 806: 150562.

Mishra, S.P., Sarkar, U., Taraphder, S. et al. (2017). Multivariate statistical data analysis-principal component analysis (PCA). *International Journal of Livestock Research* 7 (5): 60–78.

Mittal, S., Tripathi, G., and Sethi, D. (2008). Development strategy for the hill districts of Uttarakhand. Working Paper (No. 217).

Morris, R.D., Naumova, E.N., Levin, R., and Munasinghe, R.L. (1996). Temporal variation in drinking water turbidity and diagnosed gastroenteritis in Milwaukee. *American Journal of Public Health* 86 (2): 237–239.

Mushtaq, F., Nee Lala, M.G., and Mantoo, A.G. (2022). Trophic state assessment of a freshwater Himalayan Lake using Landsat 8 OLI satellite imagery: a case study of Wular Lake, Jammu and Kashmir (India). *Earth and Space Science* 9 (3): e2021EA001653.

Napacho, Z.A. and Manyele, S.V. (2010). Quality assessment of drinking water in Temeke District (part II): characterization of chemical parameters. *African Journal of Environmental Science and Technology* 4 (11): 775–789.

Negi, R.K. and Mamgain, S.H.E.E.T.A.L. (2013). Zooplankton diversity of tons river of Utarakhand state India. *International Journal of Zoology and Research* 3 (2): 1–8.

Palmate, S.S., Pandey, A., Kumar, D. et al. (2017). Climate change impact on forest cover and vegetation in Betwa Basin, India. *Applied Water Science* 7: 103–114.

Panwar, S. (2020). Vulnerability of Himalayan springs to climate change and anthropogenic impact: a review. *Journal of Mountain Science* 17 (1): 117–132.

Panwar, S. and Malik, D.S. (2016). Zooplankton diversity, species richness and their distribution pattern in Bhimtal Lake of Kumaun region, (Uttarakhand). *Hydrology Current Research* 7 (1): 219.

Pastorino, P., Polazzo, F., Bertoli, M. et al. (2019). Consequences of fish introduction in fishless Alpine lakes: preliminary notes from a sanitary point of view. *Turkish Journal of Fisheries and Aquatic Sciences* 20 (1): 1–8.

Pastorino, P. and Prearo, M. (2020). High-mountain lakes, indicators of global change: ecological characterization and environmental pressures. *Diversity* 12 (6): 260.

Persiani, E., Cecchettini, A., Ceccherini, E. et al. (2023). Microplastics: a matter of the heart (and vascular system). *Biomedicine* 11 (2): 264.

Qayoom, U., Islam, S.T., Sabha, I. et al. (2023). Coliform pollution mapping in major watersheds along Jhelum River Basin of Kashmir Himalaya. *Environmental Science and Pollution Research* 30 (3): 7930–7941.

Ra, D., Sa, B., Sl, B. et al. (2021). Is exposure to BMAA a risk factor for neurodegenerative diseases? A response to a critical review of the BMAA hypothesis. *Neurotoxicity Research* 39: 81–106.

Rai, B., 2014. Conservation and management of water resources: a case study of Duga, Sikkim Himalaya. In: *Management of Natural Resources for Sustainable Development; Challenges and Opportunities*, (by Sati V.P.) Aizwal, India. Excel India Publisher (pp. 108–114).

Rana, P., Dhanze, R., and Sharma, I. (2013). Plankton diversity of four village ponds of Himachal Pradesh. Himalayan Studies Journal. Shimla, India.

Randev, A.K. (2016). Socio-economic issues and national water policy linkages for sustainable development of irrigation in India–a study of Himachal Pradesh. 2nd World Irrigation Forum (6–8 November, 2016). ICID Journal for Irrigation and Drainage. Chiang Mai, Thailand.

Romshoo, S. A., Rashid, I., Altaf, S., & Dar, G. H. (2020). Jammu and Kashmir state: an overview. *Biodiversity of the Himalaya: Jammu and Kashmir State* (by Dar G.H. and Khuroo A.A.). Topics in Biodiversity and Conservation. Springer, Singapore (pp. 129–166).

Rupakheti, D., Tripathee, L., Kang, S. et al. (2017). Assessment of water quality and health risks for toxic trace elements in urban Phewa and remote Gosainkunda lakes, Nepal. *Human and Ecological Risk Assessment: An International Journal* 23 (5): 959–973.

Sahu, N., Sayama, T., Saini, A. et al. (2020). Understanding the hydropower and potential climate change impact on the Himalayan river regimes – a study of local perceptions and responses from himachal Pradesh, India. *Water* 12 (10): 2739.

Sarma, D., Baruah, D., Sharma, P. et al. (2017). Ichthyofaunal diversity and ecological characteristics of upland Mehao and ganga lakes, Arunachal Pradesh. *SKUAST Journal of Research* 19 (2): 192–202.

Sati, V.P. and Bandooni, S.K. (2018). Forests of Uttarakhand: diversity, distribution, use pattern and conservation. *ENVIS Bulletin Himalayan Ecology* 26: 21–27.

Seth, R., Mohan, M., Singh, P. et al. (2016). Water quality evaluation of Himalayan rivers of Kumaun region, Uttarakhand, India. *Applied Water Science* 6: 137–147.

Shammas, N.K., Kumar, I.J., Chang, S.Y., and Hung, Y.T. (2005). *Sedimentation*, 379–429. Totowa, NJ: Humana Press.

Shamrukh, M. and Hassan, Y. (2005). Chlorination and optimal chlorine dosage for Nile water (9th – 11th April, 2005). *Proc. First Ain Shams University Int. Conf. on Environmental Engineering, ASCEE, Egypt*, Vol. 1, 644–654.

Sharma, M.R. (2004). Assessment of ground water quality of Hamirpur area in Himachal Pradesh. *Pollution Research* 23: 131–134.

Sharma, M.R. (2008). Water quality of traditional drinking water sources in outer Himalayas-a case study of Hamirpur district. *HP Nature Environment and Pollution Technology* 7 (4): 677–681.

Sharma, M.D., Mishra, P., Ali, A. et al. (2023). Microbial waterborne diseases in india: status, interventions, and future perspectives. *Current Microbiology* 80 (12): 400.

Sharma, B.B. and Sarma, H.P. (2018). Assessment of heavy metal pollution in a Himalayan river using multivariate statistical tools: a case study of the Kameng River in Arunachal Pradesh, India. *Arabian Journal of Geosciences* 11: 1–12.

Sharma, A. and Sharma, M. (2019). Zooplankton diversity in relation to physico-chemical parameters in subtropical pond of Jammu, Jammu and Kashmir, India. *Biosciences, Biotechnology Research Asia* 16 (2): 425–439.

Sharma, R., Sharma, N., Singh, P., and Samant, S. (2022). Assessment of mountain water quality with special emphasis on nitrate contamination of Kullu Valley, Himachal Pradesh, India. *International Journal of Environmental and Ecology Research* 3: 11–15.

Sharma, R.C., Singh, N., and Chauhan, A. (2016). The influence of physico-chemical parameters on phytoplankton distribution in a head water stream of Garhwal Himalayas: a case study. *The Egyptian Journal of Aquatic Research* 42 (1): 11–21.

Sharma, M.K., Singh, R., Singh, O. and Durbude, D.G., 2021. Contamination in drinking water supply: a case study of Shimla City, Himachal Pradesh, India. *Climate Impacts on Water Resources in India: Environment and Health* (by Pandey A., Mishra S.K., Kansal M.L., Singh R.D. and Singh V.P.). Water Science and Technology Library. Springer, Cham (pp. 11–20).

Singh, A.K., Bhardwaj, S.K., and Devi, S. (2021). Microbiological status of drinking water sources and its relationship with human health in Solan, India. *Environmental Monitoring and Assessment* 193 (1): 32.

Singh, A.K., Das, S., Singh, S. et al. (2019). Physicochemical parameters and alarming coliform count of the potable water of Eastern Himalayan state Sikkim: an indication of severe fecal contamination and immediate health risk. *Frontiers in Public Health* 7: 174.

Singh, R.B. and Kumar, P. (2014). Geographic and socio-economic realities of Himachal Pradesh, Northwestern Himalaya. *Livelihood Security in Northwestern Himalaya: Case Studies from Changing Socio-economic Environments in Himachal Pradesh, India*. (by Singh R.B. and Hietala R.). Advancesin Geographical and Environmentaal Sciences. Springer, Tokyo (pp. 11–26).

Singh, S.P., Singh, V., and Skutsch, M. (2010). Rapid warming in the Himalayas: ecosystem responses and development options. *Climate and Development* 2 (3): 221–232.

Sini, P., Dang, T.B.C., Fais, M. et al. (2021). Cyanobacteria, cyanotoxins, and neurodegenerative diseases: dangerous liaisons. *International Journal of Molecular Sciences* 22 (16): 8726.

Sood, A., Singh, K.D., Pandey, P., and Sharma, S. (2008). Assessment of bacterial indicators and physicochemical parameters to investigate pollution status of Gangetic river system of Uttarakhand (India). *Ecological Indicators* 8 (5): 709–717.

Sudan, S. (2018). Assessment of groundwater quality: a case study of Jammu district. *International Journal of Science and Research* 7 (2): 1151–1162.

Tambe, S., Kharel, G., Arrawatia, M.L. et al. (2012). Reviving dying springs: climate change adaptation experiments from the Sikkim Himalaya. *Mountain Research and Development* 32 (1): 62–72.

Thakur, B.K., Bal, D.P., Nurujjaman, M., and Debnath, K. (2023). Developing a model for residential water demand in the Indian Himalayan Region of Ravangla, South Sikkim, India. *Groundwater for Sustainable Development* 21: 100923.

Thakur, R.K., Jindal, R., Singh, U.B., and Ahluwalia, A.S. (2013). Plankton diversity and water quality assessment of three freshwater lakes of Mandi (Himachal Pradesh, India) with special reference to planktonic indicators. *Environmental Monitoring and Assessment* 185: 8355–8373.

Theron, J. and Cloete, T.E. (2002). Emerging waterborne infections: contributing factors, agents, and detection tools. *Critical Reviews in Microbiology* 28 (1): 1–26.

Tiberti, R., Von Hardenberg, A., and Bogliani, G. (2014). Ecological impact of introduced fish in high altitude lakes: a case of study from the European Alps. *Hydrobiologia* 724: 1–19.

Tomar, G., Malik, D.S., Sharma, A.K., Kamboj, V. and Kumar, V., 2022. Assessment of water quality and biodiversity status of Alaknanda River at Garhwal, Uttarakhand: a case study. In *Environmental Pollution and Natural Resource Management* (by Bahukhandi K.D., Kamboj N., and Kamboj V.). Springer Proceedings in Earth and Environmental Sciences. Springer, Cham (pp. 121–136).

Vervloet, M.G., Sezer, S., Massy, Z.A. et al. (2017). The role of phosphate in kidney disease. *Nature Reviews Nephrology* 13 (1): 27–38.

Vishwakarma, C.A., Sen, R., Singh, N. et al. (2018). Geochemical characterization and controlling factors of chemical composition of spring water in a part of Eastern Himalaya. *Journal of the Geological Society of India* 92: 753–763.

Ward, M.H., Kilfoy, B., Sinha, R. et al. (2011). Ingestion of nitrate and nitrite and risk of stomach cancer in the NIH-AARP diet and health study. *Epidemiology* 22 (1): S107–S108.

WaZza, M.U.D., Sharma, A., and Sharma, S. (2018). Ritualistic practices and river pollution: a case study of river Tawi in Jammu District. *International Journal of Innovative Studies in Sociology and Humanities* 3: 33–38.

Xu, M.Y., Lin, Y.L., Zhang, T.Y. et al. (2022). Chlorine dioxide-based oxidation processes for water purification: a review. *Journal of Hazardous Materials* 436: 129195.

Zagarese, H.E., Sagrario, M.D.L.Á.G., Wolf-Gladrow, D. et al. (2021). Patterns of CO_2 concentration and inorganic carbon limitation of phytoplankton biomass in agriculturally eutrophic lakes. *Water Research* 190: 116715.

Zarus, G.M., Muianga, C., Brenner, S. et al. (2023). Worker studies suggest unique liver carcinogenicity potential of polyvinyl chloride microplastics. *American Journal of Industrial Medicine* 66 (12): 1033–1047.

10

Mountainous Water Resources: Understanding Microbial Dynamics, Ecological Resilience, and Sustainable Management Strategies amidst Climatic Challenges

Deepansha Raina, Devansh Kumar, Ashita S. Khasa, Mohit Marwah, Sunila Hooda, and Shalini Swami

Department of Microbiology, Ram Lal Anand College, University of Delhi, New Delhi, India

10.1 Introduction

10.1.1 Exposition of Mountainous Water Resources

Mountains are often referred to as "water towers" due to their ability to regulate water flow and distribution. In lower-lying areas, the slow melting of snow and ice during warmer seasons releases water downstream, guaranteeing a consistent supply for domestic consumption, industry, and agriculture (Viviroli et al. 2007). Mountains have glaciers that retain a lot of freshwaters as ice. Throughout the warmer months, these glaciers act as natural water reservoirs, progressively releasing water into places downstream.

Water resource management depends on an understanding of glacial water dynamics since climate change is still influencing these areas, as temperatures rise, snowpacks melt, and water flows downstream to refill rivers and streams (Halofsky et al. 2018). This periodic runoff has a significant role in regulating river flow and providing a dependable source of water for different applications (Immerzeel et al. 2020). In addition to contributing to the overall water supply, these bodies of water provide as essential habitat for a wide range of plant and mammal species (Biggs et al. 2017).

The water vapor released from mountainous areas influences precipitation patterns and atmospheric circulation, shaping weather events across vast distances (Beniston 2006). Because of orographic precipitation, mountains produce far more runoff from rain than areas downstream and slow down the flow of water by storing it in snow and glaciers. Because of their potential to act as a buffer, mountains can supply downstream communities with a reasonably consistent supply of water (Immerzeel et al. 2020). Most of the hydropower produced is produced with this water, which is also utilized for industrial processes, agricultural irrigation, and ecosystem maintenance (Immerzeel et al. 2020).

Water Resources Management in Mountain Regions, First Edition. Edited by Virendra Bahadur Singh, Sughosh Madhav, Rakesh Kumar Gupta, Prerna Diwan, and Amit Kumar.

10.1.2 Imperative of Microbial Dynamics in Environmental Ecosystem

Microbial dynamics play a crucial role in shaping environmental ecosystems. The intricate interplay of various microorganisms influences nutrient cycling, soil fertility, and overall ecosystem stability (Fierer 2012). These dynamics are influenced by bacteria, viruses, fungi, and archaea, which influence the nitrogen and carbon cycles, among other activities (Delgado-Baquerizo 2018). Current research emphasizes how important microbial diversity is to preserving the resilience and functionality of ecosystems (Bik 2019). Microbial communities also contribute to environmental detoxification through their participation in pollutant breakdown (Dubey 2019). For forecasting how ecosystems will react to changes in the environment, such as variations in the climate and human activity, an understanding of microbial dynamics is necessary (Allison and Martiny 2008).

10.1.3 A Brief Account of Challenges Climate Change Poses to Mountainous Water Coffers

Mountain glaciers serve as natural reservoirs, but they are melting at an alarming rate as global temperatures rise (Yarzábal et al. 2021). A serious and complex problem, the effects of climate change on highland water reservoirs negatively impact ecosystem's delicate balance and the steady supply of freshwater to areas downstream. The threat is highlighted by the Intergovernmental Panel on Climate Change (IPCC), which notes that "mountainous areas are experiencing faster warming compared to global averages, leading to widespread impacts on water resources" (Intergovernmental Panel on Climate Change (IPCC) 2023). Enhanced melting of glaciers and snowpacks, which are vital factors of upland water supply, is one of the most urgent and evident effects (Immerzeel et al. 2020). The melting of glaciers by rising global temperatures has dropped the amount of meltwater available to support river systems downstream (Immerzeel et al. 2020). This phenomenon affects the amount of water that flows through mountainous areas, but it also messes with the timing and thickness of water releases, which causes difficulties for communities downstream that depend on this controlled inflow for a variety of uses, such as hydropower generation and agriculture, this interrupts hydrological cycles, increases evaporation, and changes rush patterns which in turn contribute to water shortages in these places (Intergovernmental Panel on Climate Change (IPCC) 2023). The irregular nature of rainfall worsens the situation, resulting in landslides and unforeseen floods in hilly areas, which can have an impact on both ecosystems and human population (Kelkar and Kumthekar 2019). The UNEP underlines these challenges, stating that "changes in rain patterns and increasing temperatures are significantly impacting water scarcity and quality in mountainous areas."

10.2 Microbial Dynamics and Mountainous Water Resources

10.2.1 Inquisition of the Significance of Microbial Ecosystems in Mountainous Water Systems

In habitats of high-altitude mountains, microbial communities are the first to settle. They have a major role in carbon and nitrogen fixation, elemental transformation, and the development of pioneer plants and mineral soil fertility (Ciccazzo et al. 2015). When the ice melts, water is released and microbially mediated biogeochemical processes are facilitated, cryoconite holes are tiny, water-filled depressions on glacier surfaces where a large amount of biological activity takes place (Ciccazzo et al. 2015; Hotaling, Hood and Hamilton 2017; Shamurailatpam et al. 2023). Microbial communities are abundant in mountain environments with significant organic C storage, but they are constrained by relative C and P (Pan et al. 2021). This offers crucial information about the relationship between microbial metabolisms and environmental gradients in ecosystems found in arid mountains (Pang 2022).

In these ecosystems, microorganisms are essential for the cycling of nutrients. Decomposition of organic matter, nitrogen fixation, carbon conversion, and nutrient recycling are some of their roles, all of which enhance the ecosystem's general health and production (Ciccazzo et al. 2015). Within these habitats, microbes play a significant role in the biological diversity that makes them essential elements of the natural heritage (Wang et al. 2022). Although representatives of other phyla, such as Bacteroidetes, are also found, *Actinobacteria* and *Proteobacteria* are the most often detected groups of heterotrophic bacteria (Berg et al. 2009). Allochthonous carbon sources sustain most of the supraglacial heterotrophic activity (Hotaling et al. 2017).

10.2.2 Interaction Between Microbial Communities and Climatic Variables

Temperature and nutrient enrichment have direct effects on bacterial variety and oblique results via number one productiveness in subtropical and subarctic environments, as verified by using a macroecological test performed on mountain slopes eight (Wang et al. 2022). Abnormally low temperatures cause harm to microbial cells' membranes and cell partitions, which leads to mobile lysis, on the opposite hand, due to the fact snow cover ensures steady, mildly frigid temperatures, bacterial cells are shielded from freezing (Immerzeel et al. 2020). Cell loss of life is the result of decreased water hobby, an ionic imbalance, and desiccation. Proteins go through bloodless denaturation while the temperature drops under −15 °C. Microorganisms interpret low temperatures as heightened membrane tension, and inside the worst case scenario, as an overexpression of genes for bloodless-surprise proteins, fatty acid desaturase (which increases membrane fluidity, increasing unsaturation of the hydrophobic tails of the lipidic double layer), bloodless-lively enzymes, superoxide dismutase, catalase, and several different proteins or enzymes worried in transcription, translation, and replication and several other techniques (Viviroli et al. 2007). The capability of psychrophilic microorganism to create proteins that shield them against ice crystal formation in the cytosol and nearby extracellular surroundings is an exciting edition. Exopolymers (EPS), which are overall made from high molecular weight pentose sugars, play an essential role in the mechanism of attachment to mineral particles as well as in protection toward dehydration and publicity to excessive and low temperatures (Bhagat et al. 2021). Mucous slime, that is secreted by EPS and offers a conducive environment for existence, is considerable inside the Arctic and Antarctic (Ciccazzo et al. 2015).

10.2.3 Essence of Microbial Retorts for Ecological Stability in Mountainous Regions

To maintain hilly water zones ecologically strong, microorganisms are vital. It has been determined that the microbial populations in high alpine lakes are extraordinarily extraordinary and reservoirs of evolutionary ability, making them an enormous part of the natural history (Casamayor 2017). Native microbiomes determined within the Himalayas were recounted for his or her capability benefits to agriculture and environmental sustainability. For plants developing in bloodless climates and hilly areas, in addition to for the bioremediation of pollution, they were proposed as biopesticides and fertilizers (Suyal et al. 2021). Research on the range and capacity of microbes in low temperature environments has centered on the Indian Himalayan Region if you want to expand microbe-based solutions for environmental and agricultural applications (Trivedi et al. 2012).

Microbial groups play a key position in biogeochemical procedures such strength and fabric transfer, organic count breakdown, and nutrient recycling. Microbial groups are crucial to the renovation of thriving aquatic habitats because of those mechanisms. Warming and eutrophication, but, may endanger the maximum bizarre microbial species that have located the best circumstances for survival in these (ultra)oligotrophic and diluted environments sixteen (Grossart et al. 2020). The microbial communities determined in mountainous areas and mountain water structures are crucial for the maintenance of ecological stability because of their roles in primary colonization, soil formation, plant boom, nutrient biking, biodiversity conservation, and edition to international activities sixteen (Pang et al. 2022).

By generating plant growth regulators and soluble vitamins, they were proven to sell plant development (Pandey and Yarzábal 2019). They can also bioremediate contaminants inclusive of insecticides and heavy metals, which facilitates to maintain the sustainability of the surroundings (Sharma et al. 2020).

10.3 Climatic Challenges and Extreme Climatic Affairs

The peculiarities of high-altitude habitats gift plenty of climate difficulties and cruel occasions for microorganisms living in mountain water ecosystems. The dynamics of microbial groups, ecological tactics, and the steadiness of ecosystems can all be considerably affected by these demanding situations.

10.3.1 Situations Leading to Changes in Ecosystem and Microbial Group Dynamics

10.3.1.1 Temperature Fluctuations

The temperature varies significantly between day and night, these variations can affect cell features, enzyme hobby, and metabolic costs (Harwati 2007).

10.3.1.2 UV Radiation Exposure

Radiation can harm an organism's DNA, proteins, and mobile membranes, which can influence the organism's capability to live to tell the tale and reproduce (Fernández Zenoff et al. 2006).

10.3.1.3 Oxygen Levels

Temperature, water pace, and altitude versions can all have a large impact at the oxygen contents in alpine water our bodies and can affect cellular processes (Lee 2011).

10.3.1.4 Extreme Weather Events

Extreme climate occasions like storms, extreme precipitation, and flash floods have the capacity to upend microbial populations, adjust habitat structures, and influence nutrient dynamics (Alexander 2013).

10.3.1.5 Nutrient Limitation

Nutrients poor conditions are not unusual in mountain water habitats, particularly in alpine and glacier settings. For microorganisms to continue developing and metabolizing, they want to correctly use scarce resources like carbon, nitrogen, and phosphorus (Anesio and Laybourn-Parry 2012).

10.3.1.6 Seasonal Changes

Seasonal versions in temperature, precipitation, and aid availability are discovered in mountainous environments. For microorganisms to maximize their metabolic techniques, reproductive processes, and survival strategies, they need to alter to those seasonal versions (Wilhelm et al. 2013).

10.3.2 Investigation of Climatic Challenges Encountered by Water Resources in Mountainous Regions

Various climate issues that affect the sustainability and availability of water supplies are experienced by mountainous areas (Beniston 2003). The influence of weather on solar radiation estimation, small and nonrepresentative hydrometeorological monitoring networks, and high geographical variability in conditions and processes are some of these problems (Luttenauer et al. 2023). In hilly areas, the way hillslopes are oriented can also influence how much solar radiation reaches the soil or canopy, which can influence water drainage and evapotranspiration. There are significant difficulties in assessing the current and future availability of water due to the complicated hydro-climatic environment and the paucity of cooperative efforts in these regions (Vishwakarma et al. 2022). Ensuring regional water security requires an understanding of the spatiotemporal dynamics of the hydrological cycle in mountainous locations (Grießinger et al. 2022).

Warming would cause a systematic shift in the seasonality of rivers that are dominated by snowmelt, and snow-dominated areas are more susceptible to temperature fluctuations, early snowmelt may result in localized spring flooding more frequently; and in locations dominated by nival regimes today, summer irrigation water shortages may arise (Immerzeel et al. 2020). Changes in seasonality and reductions in glacial melt may result in a systematic decline in water availability as well as a diminished glacier runoff buffer during the dry season in areas where there is a strong reliance on glacier runoff such as the Tropical Andes (Viviroli et al. 2011).

Depletion of snow and ice-covered areas raises the amount of shortwave radiation absorbed because it lowers surface albedo which has a significant impact on temperature trends in mid elevations that are close to the yearly 0 °C isotherm and is most pronounced in the winter and spring in mid latitudes due to its close relationship to the seasonal snowfall this creates a cloud-radiation feedback cycle, increasing incoming solar radiation (Walsh and Chapman 1998). Furthermore, mountains have a greater direct exposure to the middle troposphere, where an improved hydrologic cycle will increase latent heat release and promote warming in high mountain regions, particularly in tropical latitudes (Viviroli et al. 2011).

10.3.3 Influence of Extreme Weather Occurrences, Including Precipitation and Heat Waves, on Microbial Communities

Reduced precipitation in montane rainforests led to a change in the species composition of testate amoebae and a drop in microbial respiration and biomass (Krashevska 2012). The number of microorganisms in watercourses, including bacterium and parasite infection, significantly increases during periods of intense rainfall and runoff (Kistemann 2002). Higher loads of parasite protozoa and fecal indicator bacteria were found in surface waters in catchment areas exposed to varying fecal sources due to increasing precipitation (Tryland et al. 2011).

Depending on the particular environment, microorganisms' reaction to heat waves in mountain water can differ. A heat wave altered the makeup of the phytoplankton population in deep prealpine Lake Mondsee, Austria, but had no overall influence on phytoplankton (Bergkemper and Weisse 2017). Heat waves and significant fluctuations in temperatures may cause changes in the variety and organization of bacterial communities during this time as well as an increase in the decomposition of aquatic litter. In order to investigate the effects of both constant (+4 °C) and variable (randomly +0~8 °C) warming treatments on the submerged macrophyte litter decomposition process (Pan et al. 2021).

Vad et al. 2023 investigated whether the geographical insurance offered by dispersal mitigates the detrimental impacts of heatwaves and how multitrophic plankton populations react to them. Researchers initially created plankton populations from a geographically remote mountain lake in a mesocosm experiment. Next they investigated the possibility that the experimental communities' resilience and recovery during and after the heatwave manipulation were influenced by an early dispersal event from a varied regional species pool (Vad et al. 2023). For phytoplankton and zooplankton, dispersion was adjusted independently in order to separate the impacts of dispersal of various trophic levels. They initially hypothesized that the experimental heatwave would alter the balance of biomass in the ecosystem. Second, zooplankton and other higher trophic levels were expected to be more negatively impacted than lower-trophic level organisms due to their distinct metabolic restrictions and sensitivity to resource availability, which would lead to a weaker top-down regulation. Third, it was predicted that greater trophic level animals would benefit more from this spatial insurance effect, which would strengthen community resistance and recovery when connected to a regional species pool (Vad et al. 2023).

10.3.4 Interaction Between Microbial Dynamics, Climate-Related Issues, and Ecosystem Health

In mountain water resources, microbial dynamics, climate-associated challenges, and environment health are intertwined (Immerzeel et al. 2020). The degradation of mountain ecosystems because of climate change and human pastime has reduced those ecosystems' resilience and raised the susceptibility of groups downstream to climate-associated calamities (Rashid et al. 2022). Water sustainability and atmosphere fitness have been jeopardized by means of the interplay between human activity and herbal structures in watersheds, which have changed hydrological processes (He and Winde 2018). The motion of microbiological pollution in montane areas is substantially prompted by means of hydrological connection, and regulating microbiological pollution in mountain catchments calls for a recognition of seasonality (Tetzlaff and Birkel 2010).

10.3.4.1 Microbial Dynamics

In alpine water habitats, microbial groups are diverse and essential to the cycling of nutrients and different ecosystem techniques (Yi et al. 2020). They enjoy dynamic adjustments which are impacted by using predators, temperature, and the provision of nutrients.

10.3.4.2 Climate-Related Issues

At massive local scales, weather change impacts the microbial functional diversity in streams. Risks which includes debris flows, avalanches, landslides, river floods, and lake outburst floods may additionally upward thrust as a result of climate exchange that exacerbate melting of glaciers (Kaushik et al. 2021; Wyss et al. 2022).

10.3.4.3 Interaction

The nation of the alpine water environment may be substantially impacted with the aid of the interplay between microbial dynamics and issues associated with climate exchange. For example, nutrient cycling may be suffering from changes in microbial groups brought approximately by way of climate alternate, that may have an impact at the environment's preferred fitness and capability (Yi et al. 2020).

10.4 Mountainous Ecological Resilience

10.4.1 Exploring the Notion of Ecological Resilience in Mountain Regions

The ability of ecological systems to endure change, or to continue growing in constantly shifting surroundings, is referred to as their resilience (Wyss et al. 2022). The concept of ecological resilience in these locations is critical for understanding mountain ecosystems' ability to withstand, adapt to, and recover from disturbances such as climate change and human activity (Folke 2004). The factors listed below have an impact on mountains' ecological resilience:

10.4.1.1 Sensitivity to Environmental Changes

Mountain ecosystems are extremely susceptible to environmental changes due to their steep slopes, height variation, and complex interactions between biotic and abiotic elements (Tito et al. 2020).

10.4.1.2 Biodiversity as a Resilience Indicator

The ecological resilience of mountainous regions is significantly influenced by biodiversity, diverse ecosystems are stronger and more resilient to change (Xu and Grumbine 2014). The ecosystem's stability and functionality are facilitated by the complex web of interactions between species (Wyss et al. 2022).

10.4.1.3 Altitude Gradients and Microclimates

Different microclimates are created in mountainous places due to their topography, which is frequently defined by altitude gradients. Different elevation zones may react to climatic changes differently, necessitating management and conservation plans specific to each microclimate (Körner 2007).

10.4.1.4 Soil Stability and Water Dynamics

In mountainous areas, soil stability is essential to ecological resilience (Immerzeel et al. 2020). Landslides and soil erosion are common in these places due to their steep slopes, alpine plants are among the several types of vegetation cover that are essential for stabilizing soil and reducing erosion (Pohl et al. 2009).

10.4.1.5 Interconnectedness of Ecosystem Components

The complex balance that forms between species, soils, water supplies, and climate determines the resilience of mountain ecosystems (Nettier et al. 2017). Understanding these relationships is essential to creating management and conservation plans that work in mountainous areas (Beniston et al. 1997).

10.4.2 Contribution of Microbial Dynamics to Ecological Resilience

Ecological resilience is largely dependent on the important and complex role that microbial dynamics play in ecosystem function thus understanding the complex interplay between microbial dynamics and ecological resilience is critical for effective conservation and management (Philippot et al. 2021).

10.4.3 Case Studies Exhibiting Thriving Ecological Resilience in Mountains

10.4.3.1 The Bhutanese Himalayas

In the Bhutanese Himalayas, robust ecosystems and abundant biodiversity are the results of the nation's dedication to comprehensive conservation and sustainable development, this harmonious cohabitation between human activity and the environment has been achieved through Bhutan's unique strategy, which includes the promotion of traditional ecological knowledge, community-based natural resource management, and the integration of conservation into national policies (BMCI and ICIMOD 2016).

10.4.3.2 The Swiss Alps

The Swiss Alps provide another compelling example of ecological resilience in mountainous areas, with strict regulations on tourism, sustainable agriculture, and reforestation initiatives helping to preserve alpine ecosystems, protect water resources, and ensure the survival of diverse plant and animal species (FADA 2020).

10.5 Strategies for Sustainable Management

10.5.1 Significance of Sustainable Management Tactics to Safeguard Water Resources in the Mountains

The significance of sustainable management measures for safeguarding water resources in mountainous areas is emphasized, considering their important function as water

Table 10.1 Various methods by which microbial dynamics contribute to ecological resilience.

Microbial contribution	Description	Example	References
Nutrient cycling and decomposition	Microorganisms are essential in the decomposition of organic materials, which releases nutrients back into the ecosystem.	Decomposition of leaf litter by fungi and bacteria, enriching soil with nutrients.	(Allison and Martiny 2008)
Soil health and stability	Microbial communities improve water retention and contribute to the structure of the soil by reducing erosion.	Mycorrhizal fungi forming symbiotic associations with plant roots, improving soil stability.	(Bardgett and Van Der Putten 2014)
Plant–microbe interactions	Plant health is enhanced by nutrient uptake through symbiotic connections between microorganisms and plants.	Nitrogen-fixing bacteria forming nodules on legume roots, facilitating nitrogen uptake.	(Van Der Heijden et al. 2008)
Microbial response to environmental changes	Microbial communities represent the health of an ecosystem and are sensitive indicators of environmental changes.	Changes in microbial community composition in response to increased temperature.	(Strickland and Rousk 2010)
Microbial diversity and functional redundancy	The stability of ecosystems is ensured by the functional redundancy provided by diverse microbial communities.	Multiple microbial species with similar functions maintaining nutrient cycling in diverse ecosystems.	(Allison and Martiny 2008)
Microbial adaptations to extreme environments	In mountainous areas, microbes have adapted to extreme environments, which have increased ecosystem resilience.	Psychrophilic bacteria thriving in cold mountain soils, maintaining nutrient cycling.	(Schmidt et al. 2011)
Microbial facilitation of plant stress tolerance	Microbes enhance the resilience by assisting plants in coping with stressors.	Endophytic fungi promoting plant drought tolerance by improving water use efficiency.	(Porras-Alfaro and Bayman 2011)

catchments and sources for downstream regions. Using sustainable management practices is essential to preserving the environment's overall health, quantity, and quality. The following (Table 10.1) are salient points highlighting the significance of these strategies:

10.5.1.1 Ecosystem Health and Biodiversity Preservation

Sustainable management approaches help to maintain the health and biodiversity of mountain ecosystems, which is critical for the resilience of flora and fauna suited to these conditions (Biswas and Biswas 2021).

10.5.1.2 Downstream Water Supply

Mountains serve as natural water towers, supplying water for everyday necessities, industry, and agriculture in downstream regions. Sustainable management guarantees downstream areas a steady and dependable water supply (Nepal et al. 2014).

10.5.1.3 Climate Regulation and Adaptation

Mountain ecosystems are essential for controlling the climate. Reducing glacier retreat and changing precipitation patterns are two effects of climate change that can be adapted to with the help of sustainable water management (Elsanabary and Gan 2014).

10.5.1.4 Prevention of Soil Erosion

In mountainous areas, sustainable management techniques like afforestation and erosion control help stop soil erosion, protecting water quality, and lowering sedimentation (Wang et al. 2024).

10.5.1.5 Hydroelectric Power Generation

Numerous hilly areas are used to generate hydroelectric power. An equilibrium between energy production and environmental preservation is guaranteed by sustainable water management (Nautiyal and Goel 2020).

10.5.1.6 Balancing Water Use for Agriculture

By facilitating effective irrigation and lowering the possibility of agricultural water scarcity, sustainable water management techniques meet the demands of agriculture in hilly regions (Shukla et al. 2021).

10.5.1.7 Promotion of Transboundary Water Cooperation

Transboundary collaboration is necessary for sustainable water management in many mountainous locations that span multiple countries. Sustainable methods reduce conflict over shared water resources and promote cooperation (Rasul 2021).

10.5.2 Discussion on Contemporary Challenges and Voids in Sustainable Management for Mountains

Contemporary challenges, voids in sustainable mountain management and a variety of environmental, social, and economic reasons present intricate obstacles that that need for prompt attention and calculated actions (Mengist et al. 2020). A significant obstacle is that these ecosystems are more vulnerable to the effects of climate change, such as changing precipitation patterns, a rise in the frequency of extreme weather events, and glacier retreat. The vulnerability of mountainous areas is increased by these changes, which also have an impact on biodiversity, water availability, and ecosystem services (Mengist et al. 2020).

Another major concern is unsustainable tourism practices, which lead to ecosystem destruction, pollution, and cultural disruption in mountainous places (Gössling et al. 2020). Mountain ecosystems are further threatened by rapid urbanization and changes in land use, which result in habitat fragmentation, a decline in biodiversity, and changes in land use patterns. These issues are made worse by inadequate infrastructure planning, which can lead to habitat destruction and soil erosion in the case of poorly designed projects like hydropower dams and highways (Liu et al. 2023).

The frequency of deforestation and illegal logging endangers mountain biodiversity and water quality, and weak law enforcement and governance systems enable these activities to continue (Sloan and Sayer 2015). One enduring gap that prevents local populations from being fully empowered in conservation initiatives is the lack of community involvement in decision-making processes (Reed 2008).

There are gaps in our knowledge of the complex interactions that exist within mountain ecosystems due to a lack of research on ecological dynamics, the effects of climate change, and socioeconomic issues (Viviroli et al. 2007). Cultural heritage protection is frequently disregarded, which leads to the decline of customary knowledge and behaviors (Baival and Fernández-Giménez 2012). The difficulties are exacerbated by sector and fragmented policies, which frequently overlook the interdependence of mountain ecosystems (Byers 2018). Insufficient finance channels impede the execution of sustainable management activities, restricting progress in conservation, climate resilience, and sustainable livelihood programs (Naeem et al. 2019).

10.5.3 Investigating Efficacious Strategies and Approaches for the Sustainable Management of Water Resources in Mountainous Regions

Investigating effective strategies and approaches for the sustainable management of water resources in mountainous regions necessitates a thorough awareness of the specific problems that these terrains present to guarantee the water ecosystems' long-term resilience in the highest regions.

10.5.3.1 Watershed Management Programs

Integrated watershed management strategies are essential in mountainous areas because they view the entire watershed. Such projects seek to retain natural vegetation, minimize soil erosion, and improve water retention, thereby contributing to long-term water supply (Wang 2019).

10.5.3.2 Afforestation and Reforestation

In mountainous areas, the main goals of reforestation and planting are to improve water infiltration, stop soil erosion, and restore damaged landscapes. Trees ensure sustainable water supplies by stabilizing slopes, regulating water flow, and enhancing biodiversity (Filoso 2017).

10.5.3.3 Sustainable Agricultural Practices

Organic farming, less reliance on agrochemicals, and water-efficient farming methods are prioritized in mountainous areas by sustainable agricultural practices. Resilient farming systems are encouraged, soil erosion is reduced, and water quality is preserved. (Allan 2007)

10.5.3.4 Community-Based Water Management

Local communities are included in water resource decision-making processes through community-based water management. By incorporating traditional knowledge and encouraging a sense of ownership, this strategy guarantees sustainable water practices that are in line with regional requirements (Ikhlas and Ramadan 2024).

10.5.3.5 Smart Technologies for Monitoring

Adoption of smart technology offers real-time data on water quality, flow, and usage, such as sensor-based monitoring systems. Making well-informed decisions, identifying problems early, and launching appropriate remedies are all made possible by this data (Palermo 2022).

10.5.3.6 Indigenous Water Governance

It is necessary to incorporate traditional knowledge and practices into contemporary approaches to water management to recognize and integrate indigenous water governance systems (Evans and Sadler 2008).

10.6 Vulnerability and Adaptability

10.6.1 Investigation of Mountainous Water Resources' Vulnerability to Climatic Challenges

The essential significance that mountain water resources are particularly vulnerable to the effects of climate change (Immerzeel et al. 2020). The following are some important details that demonstrate how vulnerable mountainous water resources are to climate change:

10.6.1.1 Melting of Glaciers and Snowpacks

Warm summertime air exacerbates the melting process by melting glaciers' surfaces and producing meltwater that pierces the ice and descends to the glacier bed leading to Calving, or the breaking off enormous pieces of ice into the ocean, which rises and mixes the warm ocean water upward because it is lighter than the surrounding seawater. (NASA 2020).

10.6.1.2 Altered Precipitation Patterns

The atmosphere's increasing moisture content may cause more intense rainstorm events, these extremes in rainfall are further amplified by warming-related reductions in the snow percentage, which changes the hydrological balance in mountainous regions (Pepin et al. 2022).

10.6.1.3 Increased Evaporation Rates

Higher temperatures speed up the melting of snow and ice, which might temporarily boost the amount of water available before finally decreasing streamflow and water supplies as the snowpack decreases (Konapala et al. 2020).

10.6.1.4 Changing Ecosystems

Mountainous areas have ecosystems that are suited to climatic circumstances. These ecosystems may see major changes that impact the quantity and quality of water supplies as temperatures increase and precipitation patterns alter (Garcia 2012).

10.6.2 Analysis of Strategies for Adaptation Employed by the Ecological System in Mountains

Mountain ecosystems have evolved several responses to climate change, combining human intervention with natural processes to increase resilience. Here are a few of the tactics used:

10.6.2.1 Elevation Migration

To avoid the rising heat, some animals are shifting to higher altitudes (Spence and Tingley 2020). Though it can also cause competition and stress on the limited habitats at higher elevations, this natural adaption approach aids in the preservation of biodiversity (Parmesan and Yohe 2003).

10.6.2.2 Phenotypic Plasticity

This speaks to an organism's capacity to adapt its phenotypic in response to environmental changes. A great deal of phenotypic flexibility is exhibited by mountain species, which enables them to adjust to shifting climate conditions without having to migrate (Parmesan and Yohe 2003).

10.6.2.3 Microclimate Exploitation

Certain species are using the microclimates found in hilly areas. Species that are under stress due to warming temperatures may find shelter in microhabitats that provide cooler and moister circumstances (Scherrer and Körner 2010).

10.6.2.4 Conservation of Genetic Diversity

The genetic diversity of mountain species must be preserved if they are to remain adaptable to shifting environmental conditions (Sgrò et al. 2011). This may entail seed banks and other ex situ conservation methods in addition to the preservation of important ecosystems (Frankham et al. 2010).

10.6.2.5 Adaptive Management Practices in Agriculture and Forestry

Maintaining agricultural viability, water supplies, and soil health are all essential for the resilience of mountain ecosystems and the populations who depend on them. These may be achieved by putting into practice sustainable land management techniques that are flexible enough to adjust to changing climatic circumstances (Lin 2011).

10.6.2.6 Climate-Resilient Infrastructure

Building infrastructure that can withstand the effects of climate change, such as floods and landslides, is essential to preserving access and services in mountain communities and promoting ecological and human resilience (Khan et al. 2021).

10.6.3 Comprehending the Vulnerability for Acquiring Effective Adaptation Strategies

A wide range of elements, roughly classified into environmental, socioeconomic, and policy-related dimensions (Table 10.2), impact the successful implementation of adaptation methods against climate change in mountainous regions.

Table 10.2 List of the main variables that affect the implementation of adaptation strategies that are used to combat climate change in mountainous areas.

S. No.	Category	Factors	Description	Reference
01.	**Environmental factors**	Ecosystem sensitivity	Priorities and requirements for adaptation are impacted by the mountain ecosystems' innate susceptibility to the effects of climate change, including altered precipitation patterns, glacier retreat, and biodiversity loss.	(Beniston 2003)
		Geographical isolation	Because of their remoteness, mountains might hinder species' ability to migrate naturally in response to climate change, requiring human assistance in the form of adaptation strategies.	(Scott 2003)
02.	**Socioeconomic factors**	Livelihood dependencies	A lot of mountain towns rely primarily on climate-sensitive industries including forestry, agriculture, and tourism, which affects the acceptability and viability of adaption measures.	(McDowell et al. 2014)
		Infrastructure and access	A lack of market, knowledge, or technological access may make it more difficult to put adaptation plans into action.	(Agrawal 2008)
		Cultural factors	While cultural values and traditional knowledge can be very important in determining adaptation techniques, they can also prevent new practices from being adopted.	(Xu and Grumbine 2014)
03.	**Policy and institutional factors**	Governance and institutional capacity	It is crucial that regional, national, and local government structures and institutions effectively support adaptation programs.	(Biesbroek et al. 2010)
		Policy integration and coordination	Implementation is impacted by how climate change adaptation is included into larger development planning and how various sectors and levels of government collaborate with one another.	(Winkler 2023)
		Financial resources	The availability and accessibility of funding to assist adaption measures are essential for their successful execution.	(Ayers and Huq 2009)

Table 10.2 (Continued)

S. No.	Category	Factors	Description	Reference
04.	**Technological and information factors**	Access to adaptation technologies	It is crucial to have access to the availability of technology that help with adaptation, such as crop types resistant to climate change or irrigation techniques that use less water.	(McCarthy et al. 2001)
		Information and knowledge	Planning and putting into practice successful strategies depend on having access to reliable information and understanding about the effects of climate change and available choices for adaptation.	(Moser and Ekstrom 2010)

10.7 Case Reports and Best Practices

10.7.1 Case Studies Exhibiting Mountainous Region's Ecological Resilience, Microbial Dynamic, and Sustainable Management

A wealth of case studies from mountainous places around the globe provide illuminating examples of microbial dynamics, ecological resilience, and sustainable management techniques. These studies emphasize the need for incorporating sustainable practices into management techniques, the relevance of microbial populations in ecosystem function, and how ecosystems respond to external stressors.

10.7.1.1 The Rocky Mountains—Microbial Dynamics in Alpine Soils

King and Freeman (2010) conducted a study in the Rocky Mountains on the microbial dynamics of alpine soils has demonstrated the ways in which microbial communities adjust to changing climatic conditions. These results highlight the significance of microbial activity and diversity for the health of high-altitude ecosystems and nutrient cycling.

10.7.1.2 The Andes – Water Management and Sustainable Practices

Valdivia et al. (2010) emphasize on sustainability of native techniques for managing water in the Andes, such building terraces and irrigation channels. These techniques demonstrate the fusion of traditional knowledge with sustainable land management by promoting agricultural production, reducing erosion, and conserving soil.

10.7.1.3 The Tibetan Plateau – Grazing Practices and Ecosystem Health

Wu et al. (2014) in their study on Tibetan Plateau's sustainable grazing methods show how to strike a balance between ecosystem health and subsistence requirements. In order to avoid overgrazing, maintain the integrity of the grasslands, and sustain the livelihoods of the surrounding populations, rotational grazing and the upholding of traditional pastoralism have proven beneficial.

10.7.2 Insight from Successful Initiatives and Identification of the Most Promising Approaches

In mountainous areas, effective programs, and viable ways to mitigating climate change focus on integrated, community-based approaches that improve ecological resilience and socioeconomic development.

10.7.3 Application-Based Findings for a Global and Diverse Form of Mountain Ecosystems

Application-based findings from various mountain ecosystems throughout the world highlight the need for customized, context-specific ways to controlling the effects of climate change. Following are some application-based learnings from global mountain ecosystems:

10.7.3.1 Enhanced Ecosystem Services through Biodiversity Conservation

Initiatives to conserve biodiversity have been found to enhance ecosystem services, including soil stability, carbon sequestration, and water management, in mountainous locations ranging from the Andes to the Himalayas (Table 10.3). Creating protected places, replanting native

Table 10.3 Overview of successful programs, promoting mountain ecosystems' ability for adaptation and their sustainable management.

S.No.	Successful initiatives	Description	Reference
01.	**Community-based adaptation in the himalayas**	In the Himalayas, the International Centre for Integrated Mountain Development (ICIMOD) has carried out several community-based adaptation initiatives that use traditional knowledge and methods, such agroforestry and rainwater collection, to increase resistance to the effects of climate change.	(Acharya and Sellmyer 2016)
02.	**Ecosystem-based adaptation in the Andes**	To offer services that aid in community adaptation to climate change, ecosystems in the Andes are restored and conserved using the ecosystem-based adaptation (EbA) strategy.	(Elgon 2022)
03.	**Climate-resilient agriculture in the African Highlands**	Promoting climate-resilient agricultural techniques including crop diversification, conservation agriculture, and integrated pest control has shown to be successful for the African Highlands Initiative.	(Lin 2011)
04.	**Renewable energy solutions in mountainous regions**	These programs provide energy availability for isolated mountain villages while simultaneously lowering greenhouse gas emissions, promoting the socioeconomic growth of such areas.	(Mainali and Silveira 2015)
05.	**Transboundary conservation initiatives**	Transboundary conservation measures are crucial in mountainous regions because the effects of climate change frequently transcend country boundaries.	(Della Corte et al. 2019)

plants, and encouraging wildlife corridors are some of these initiatives (Bellard et al. 2012; Negi et al. 2017).

10.7.3.2 Sustainable Agricultural Practices

Agroforestry, intercropping, and organic farming are examples of climate-resilient agricultural techniques that have proven successful in hilly areas like East Africa and the Alps. These methods improve livelihoods and food security while also helping to preserve soil and water resources (Lin 2011; Brang et al. 2014).

10.7.3.3 Community-based Natural Resource Management (CBNRM)

From the Hindu Kush Himalaya to the Rocky Mountains, mountain ecosystems throughout the world have shown success when local populations are given the tools, they need to manage natural resources. Community-led conservation and management practices (CBNRM) practices promote adaptive ability to climate change, strengthen local governance, and enable sustainable resource use (Ostrom 2009; Xu and Grumbine 2014).

10.7.3.4 Innovative Water Management Solutions

Mountain people from the Andes to the Tibetan Plateau have used creative water management techniques in response to shifting precipitation patterns and receding glaciers. To guarantee water supply for home and agricultural use, they include creating community-managed water storage facilities, installing modern irrigation systems, and resurrecting historic methods of gathering water (Walsh and Chapman 1998; Díaz et al. 2015; Antonelli et al. 2018).

10.8 Potential for Research and Future Routes

10.8.1 Understanding Lacunae of Ecological Resilience, Microbial Dynamics in Water Resources of Mountains

Understanding the lacunae of ecological resilience and microbial dynamics in water resources of mountains is pivotal for effective conservation and management strategies; however, there is a lack of comprehensive understanding of microbial diversity and performing in mountain water resources (Viviroli et al. 2007). It has been proved that microbial communities in mountain lakes, aqueducts, and canals, there is still limited knowledge about the specific microbial taxa present and their ecological roles (Díaz et al. 2015).

Climate change is altering the hydrological cycles of mountain water resources, including changes in temperature, rainfall patterns, and glacial water melt flow (Kundzewicz et al. 2008). All presently available climate models prognosticate a trend of global warming that increases in surface temperatures has far reaching consequences for the hydrological cycle, particularly in regions where water flow is presently dominated by melting snow or ice (Viviroli et al. 2007; Antonelli et al. 2018; Immerzeel et al. 2020; Elgon 2022). However, there is limited knowledge about how microbial communities respond to and alleviate pollution in mountain waters (Schirpke 2017).

Research gaps exist in understanding the role of microbial communities in interlinking biogeochemical processes across ecosystem boundaries and in assessing the adaptability of linked river systems to environmental changes. Addressing these issues requires an interdisciplinary approach that integrates microbial ecology, hydrology, and ecosystem sciences. Cooperative initiatives are essential for advancing our understanding of microbial dynamics and ecological adaptability in mountain water reservoirs and for developing science-based conservation practices and operation strategies to cover these vital ecosystems (Wilhelm et al. 2013).

10.8.2 Exploration of Potential Research Opportunities and Avenues

Research opportunities in mountain river ecosystems encompass a broad range of interdisciplinary investigations, from understanding the hydrological process shaping these environments to exploring the ecological dynamics of their aquatic habitats the links between the evolution and distribution of biodiversity and the formation of mountains also remain poorly understood (Rashid et al. 2022). A better understanding of biosphere interactions is needed to understand the patterns and evolution of mountain biodiversity and how they affect river ecosystems (Antonelli et al. 2018). It is well known that mountain are hotspots of water supply and unprecedented human alterations of the Earth's land surface are threatening ecosystem services supply and ultimately people's well-being (Verburg 2015). A recent study shows that the most important water reservoirs of the world, which are in mountain areas, are also among the most vulnerable and their impact on human well-being is of paramount importance (Immerzeel et al. 2020).

10.8.3 Significance of Ongoing Research in Managing Future Strategies

Given the problems faced by climate change, including declining snowpacks, changed precipitation patterns, and growing water demand, ongoing research is essential to controlling future for mountain water supplies. The management of water resources must change because of these considerations, especially in mountainous areas where water supplies are essential for towns downstream (Beniston 2003). According to a study, there is a necessity to create and execute climate adaptation plans, the research highlights the necessity of reassessing water management techniques and the significance of scientific and water managers working together (Ciccazzo et al. 2015). Therefore, to strengthen the resilience of the water supply, it is necessary to investigate solutions for climate adaptation, such as those of managed aquifer recharge and forecast-informed reservoir management (Chao 2021). The impact of climate change on alpine water resources worldwide is further studies by researches that points out that to achieve long-term, ongoing interaction, it is necessary for us to integrate different disciplines and enhance the knowledge flow between managers and researchers (Konapala et al. 2020). The research also emphasizes the necessity of more accurate scenario projections and in-depth regional research for the efficient management of mountain water resources (Drenkhan et al. 2023).

10.9 Conclusion

Climate change provides a serious challenge to highland water reservoirs, affecting volume, timing, and quality. The rapid melting of glaciers and changes in rainfall patterns undermine the delicate balance of these ecosystems, affecting downstream occupants and ecosystems.

The ramifications go beyond the immediate mountainous areas, affecting water availability and quality in downstream places. Addressing these difficulties demands a holistic and flexible approach that considers both the ecological integrity of mountainous regions and the conditions of populations that calculate on these vital water resources. Sustainable planning for land use, reforestation programs, and soil erosion control are critical for limiting the effects of climate change on highland water resources is also required. Afforestation and reforestation conditioning in mountainous areas can help to palliate the goods of climate change by stabilizing soil, lowering the threat of landslides, and improving water retention. Investments in present-day structure related to water, such as better containment facilities and better irrigation systems, are critical for improving water resilience. Additionally, since highland ecosystems are globally present, it is necessary to encourage international cooperation. Developing adaptability requires international exchanges of information, technology, and resources.

To sum up, it is critical to take a comprehensive approach to comprehending microbiological dynamics, ecological resilience, and sustainable management of mountain water resources to guarantee their long-term viability and health. The interdependence of microbial populations and ecological systems highlights the necessity of integrated management solutions. In addition, studies emphasize how crucial it is to take socioeconomic variables into account while managing water resources to meet sustainability objectives. In conclusion, considering the problems posed by climate change, cooperation, and commitment to protecting vital ecosystems are very important. Effective conservation initiatives necessitate collaboration and collaborations across several sectors.

References

Acharya, S. and Sellmyer, A. (2016). Anjal prakash-p 30; Brian D Collins-p 22; Corinna Wallrapp-pp 19, 42; Dependra tandukar-p 17; Dhuwa-p 48; eSRI-p 52; Faisal Mueen Qamer-p 3; Ghanashyam Sharma-p 10; Haiya Zhang-p 45, *Jitendra Bajracharya-Cover*, 57. www.icimod.org/himaldoc (accessed 6 October 2024).

Agrawal, A. (2008). *The Role of Local Institutions in Adaptation to Climate Change*, 734–764.

Alexander, J. (2013). A metagenomic snapshot of taxonomic and functional diversity in an Alpine glacier cryoconite ecosystem. *Environ. Res. Lett* 8: 35003–35014. https://doi.org/10.1088/1748-9326/8/3/035003.

Allan, C. (2007). Can adaptive management help us embrace the Murray-Darling Basin's wicked problems? *Adaptive and Integrated Water Management*, pp. 61–73. https://doi.org/10.1007/978-3-540-75941-6_4.

Allison, S.D. and Martiny, J.B.H. (2008). Colloquium paper: resistance, resilience, and redundancy in microbial communities. *Proceedings of the National Academy of Sciences of the United States of America* 105 (Suppl 1): 11512–11519. https://doi.org/10.1073/PNAS.0801925105.

Anesio, A.M. and Laybourn-Parry, J. (2012). Glaciers and ice sheets as a biome. *Trends in Ecology & Evolution* 27 (4): 219–225. https://doi.org/10.1016/J.TREE.2011.09.012.

Antonelli, A., Kissling, D.W., Flantua, S.G.A. et al. (2018). Geological and climatic influences on mountain biodiversity. *Nature Geoscience* 11 (10): 718–725. Available at: https://doi.org/10.1038/s41561-018-0236-z.

Ayers, J.M. and Huq, S. (2009). The value of linking mitigation and adaptation: a case study of bangladesh. *Environmental Management* 43 (5): 753–764. https://doi.org/10.1007/s00267-008-9223-2.

Baival, B. and Fernández-Giménez, M.E. (2012). Meaningful learning for resilience-building among Mongolian pastoralists. *Nomadic Peoples* 16 (2): 53–77. https://doi.org/10.3167/NP.2012.160205.

Bardgett, R.D. and Van Der Putten, W.H. (2014). Belowground biodiversity and ecosystem functioning. *Nature* 515 (7528): 505–511. 10.1038/nature13855.

Bellard, C., Bertelsmeier, C., Leadley, P. et al. (2012). Impacts of climate change on the future of biodiversity. *Ecology Letters* 15 (4): 365–377. https://doi.org/10.1111/j.1461-0248.2011.01736.x.

Beniston, M. (2003). Climatic change in mountain regions: a review of possible impacts. *Climatic Change* 59 (1–2): 5–31. https://doi.org/10.1023/A:1024458411589.

Beniston, M. (2006). Mountain weather and climate: a general overview and a focus on climatic change in the Alps. *Hydrobiologia* 562 (1): 3–16. https://doi.org/10.1007/s10750-005-1802-0.

Beniston, M., Diaz, H.F., and Bradley, R.S. (1997). Climatic change at high elevation sites: an overview. *Climatic Change* 36 (3–4): 233–251. https://doi.org/10.1023/a:1005380714349.

Berg, K.A., Lyra, C., Sivonen, K. et al. (2009). High diversity of cultivable heterotrophic bacteria in association with cyanobacterial water blooms. *ISME Journal* 3 (3): 314–325. https://doi.org/10.1038/ismej.2008.110.

Bergkemper, V. and Weisse, T. (2017). Phytoplankton response to the summer 2015 heat wave – a case study from prealpine Lake Mondsee, Austria. *Inland Waters* 7 (1): 88–99. https://doi.org/10.1080/20442041.2017.1294352.

Bhagat, N., Raghav, M., Dubey, S., and Bedi, N. (2021). Bacterial exopolysaccharides: Insight into their role in plant abiotic stress tolerance. *Journal of Microbiology and Biotechnology* 31 (8): 1045–1059. https://doi.org/10.4014/jmb.2105.05009.

Biesbroek, G.R., Swart, R.J., Carter, T.R. et al. (2010). Europe adapts to climate change: comparing national adaptation strategies. *Global Environmental Change* 20 (3): 440–450. https://doi.org/10.1016/j.gloenvcha.2010.03.005.

Biggs, J., von Fumetti, S., and Kelly-Quinn, M. (2017). The importance of small waterbodies for biodiversity and ecosystem services: implications for policy makers. *Hydrobiologia* 793 (1): 3–39. https://doi.org/10.1007/s10750-016-3007-0.

Bik, H.M. (2019). Microbial community succession and nutrient cycling responses following perturbations of experimental saltwater aquaria. *mSphere* 4 (1): https://doi.org/10.1128/msphere.00043-19.

Biswas, P.L. and Biswas, S.R. (2021). Mangrove forests: ecology, management, and threats. In: *Life on Land*, 627–640. Cham: Springer International Publishing.

BMCI & ICIMOD. (2016). *Bhutan Climate Change-Handbook*. lib.icimod.org/record/32399/files/icimodBhutanClimate016.pdf.

Brang, P., Spathelf, P., Larsen, J.B. et al. (2014). Suitability of close-to-nature silviculture for adapting temperate European forests to climate change. *Forestry* 87 (4): 492–503. https://doi.org/10.1093/forestry/cpu018.

Byers, E. (2018). Global exposure and vulnerability to multi-sector development and climate change hotspots. *Environmental Research Letters* 13 (5): https://doi.org/10.1088/1748-9326/AABF45.

Casamayor, E.O. (2017). Towards a microbial conservation perspective in high mountain lakes. *Advances in Global Change Research* 62: 157–180. https://doi.org/10.1007/978-3-319-55982-7_7/FIGURES/11.

Chao, J. (2021). Managing water resources in a low-to-no-snow future. *BerkeleyLab* 1–6.

Ciccazzo, S., Esposito, A., Borruso, L., and Brusetti, L. (2015). Microbial communities and primary succession in high altitude mountain environments. *Annals of Microbiology* 66 (1): 43–60. https://doi.org/10.1007/S13213-015-1130-1.

Delgado-Baquerizo, M. (2018). A global atlas of the dominant bacteria found in soil. *Science* 359 (6373): 320–325. https://doi.org/10.1126/science.aap9516.

Della Corte, V., Gaudio, G.D., Sepe, F., and Sciarelli, F. (2019). Sustainable tourism in the open innovation realm: a bibliometric analysis. *Sustainability* 11 (21): 1–18. https://doi.org/10.3390/su11216114.

Díaz, S., Demissew, S., Carabias, J. et al. (2015). The IPBES conceptual framework - connecting nature and people. *Current Opinion in Environmental Sustainability* 14: 1–16. https://doi.org/10.1016/j.cosust.2014.11.002.

Drenkhan, F., Buytaert, W., Huggle, C. et al. (2023). Looking beyond glaciers to understand mountain water security. *Nature Sustainability* 6 (2): 130–138. https://doi.org/10.1038/s41893-022-00996-4.

Dubey, A. (2019). Soil microbiome: a key player for conservation of soil health under changing climate. *Biodiversity and Conservation* 28 (8–9): 2405–2429. https://doi.org/10.1007/s10531-019-01760-5.

Elgon, M. (2022). Using ecosystem-based adaptation to enhance livelihoods and improve resilience in mountain ecosystems, 1–10.

Elsanabary, M.H. and Gan, T.Y. (2014). Wavelet analysis of seasonal rainfall variability of the upper Blue Nile basin, its teleconnection to global sea surface temperature, and its forecasting by an artificial neural network. *Monthly Weather Review* 142 (5): 1771–1791. https://doi.org/10.1175/MWR-D-13-00085.1.

Evans, R.G. and Sadler, E.J. (2008). Methods and technologies to improve efficiency of water use. *Water Resources Research* 44 (7): https://doi.org/10.1029/2007WR006200.

FADA (2020). Swiss expertise supports the sustainable development of mountain regions Swiss expertise devoted to mountain environments, pp. 1–6.

Fernández Zenoff, V., Siñeriz, F., and Farías, M.E. (2006). Diverse responses to UV-B radiation and repair mechanisms of bacteria isolated from high-altitude aquatic environments. *Applied and Environmental Microbiology* 72 (12): 7857–7863. https://doi.org/10.1128/AEM.01333-06.

Fierer, N. (2012). Cross-biome metagenomic analyses of soil microbial communities and their functional attributes. *Proceedings of the National Academy of Sciences of the United States of America* 109 (52): 21390–21395. https://doi.org/10.1073/pnas.1215210110.

Filoso, S. (2017). Impacts of forest restoration on water yield: A systematic review. *PLOS ONE* 12 (8): e0183210. https://doi.org/10.1371/JOURNAL.PONE.0183210.

Folke, C. (2004). Regime shifts, resilience, and biodiversity in ecosystem management. *Annual Review of Ecology, Evolution, and Systematics* 35: 557–581. https://doi.org/10.1146/annurev.ecolsys.35.021103.105711.

Frankham, R., Ballou, J.D., and Briscoe, D.A. (2010). *Introduction to Conservation Genetics*, 1–4. Cambridge: Cambridge University Press https://doi.org/10.1017/cbo9780511809002.

Garcia, J. (2012). Climate change in mountain ecosystems. *Mountain Research Initiative Newsletter* 7: 48–49.

Gössling, S., Scott, D., and Hall, C.M. (2020). Pandemics, tourism and global change: a rapid assessment of COVID-19. *Journal of Sustainable Tourism* 1–20. https://doi.org/10.1080/09669582.2020.1758708.

Grießinger, J., Meier, W.J.H., and Hochreuther, P. (2022). Decreasing water availability as a threat for traditional irrigation-based land-use systems in the Mustang Himalaya/Nepal, pp. 253–266. https://doi.org/10.1007/978-3-030-70238-0_8.

Grossart, H.P., Massana, R., McMahon, K.D., and Walsh, D.A. (2020) 'Linking metagenomics to aquatic microbial ecology and biogeochemical cycles', *Limnology and Oceanography*, 65(S1), pp. S2–S20. https://doi.org/10.1002/lno.11382.

Halofsky, J.E., Peterson, D.L., Karen Dante-Wood, S., and Hoang, L. (2018). Toward climate-smart resource management in the northern rockies. *Climate Change and Rocky Mountain Ecosystem* 221–228. https://doi.org/10.1007/978-3-319-56928-4_12.

Harwati, T.U. (2007). Characterization of diverse hydrocarbon-degrading bacteria isolated from Indonesian Seawater. *Microbes and Environments* 22 (4): 412–415. https://doi.org/10.1264/JSME2.22.412.

He, C. and Winde, F. (2018). Water sustainability and watershed ecosystem health. *Ecosystem Health and Sustainability* 4 (10): 241–242. https://doi.org/10.1080/20964129.2018.1538666.

Hotaling, S., Hood, E., and Hamilton, T.L. (2017). Microbial ecology of mountain glacier ecosystems: biodiversity, ecological connections and implications of a warming climate. *Environmental Microbiology* 19 (8): 2935–2948. https://doi.org/10.1111/1462-2920.13766.

Ikhlas, N. and Ramadan, B.S. (2024). Community-based watershed management (CBWM) for climate change adaptation and mitigation: research trends, gaps, and factors assessment. *Journal of Cleaner Production* 434: 140031. https://doi.org/10.1016/J.JCLEPRO.2023.140031.

Immerzeel, W.W., Lutz, A.F., Andrade, M. et al. (2020). Importance and vulnerability of the world's water towers. *Nature* 577 (7790): 364–369. https://doi.org/10.1038/s41586-019-1822-y.

Intergovernmental Panel on Climate Change (IPCC) (2023). *Mountains, Climate Change 2022 – Impacts, Adaptation and Vulnerability*. https://doi.org/10.1017/9781009325844.022

Kaushik, H., Ramanathan, A.L., Soheb, M. et al. (2021). Climate change-induced high-altitude lake: hydrochemistry and area changes of a moraine-dammed lake in Leh-Ladakh. *Acta Geophysica* 69 (6): 2377–2391. https://doi.org/10.1007/s11600-021-00670-x.

Kelkar, S. and Kumthekar, M.B. (2019). Need of sustainable planning and development against natural disaster resulting because of climate change and human intervention. *SSRN Electronic Journal* [Preprint]. https://doi.org/10.2139/ssrn.3366750.

Khan, A.S., Ramachandran, A., and Palanivelu, K. (2021). Urban climate change adaptation framework for Chennai city, India. In: *Handbook of Climate Change Management: Research, Leadership Transformation*, vol. 4 (ed. J.M. Luetz and D. Ayal), 3431–3453. Cham: Springer International Publishing https://doi.org/10.1007/978-3-030-57281-5_32.

King, A.J. and Freeman, K.R. (2010). Biogeography and habitat modelling of high-alpine bacteria. *Nature Communication* **1**: 53. https://doi.org/10.1038/ncomms1055.

Kistemann, T. (2002). Microbial load of drinking water reservoir tributaries during extreme rainfall and runoff. *Applied and Environmental Microbiology* 68 (5): 2188–2197. https://doi.org/10.1128/AEM.68.5.2188-2197.2002/ASSET/29466246-C9D2-4F53-9C1B-9D3B38CB2B6B/ASSETS/GRAPHIC/AM0521244004.JPEG.

Konapala, G. (2020). Climate change will affect global water availability through compounding changes in seasonal precipitation and evaporation. *Nature Communications* 11 (1): 1–34. https://doi.org/10.1038/s41467-020-16757-w.

Körner, C. (2007). The use of "altitude" in ecological research. *Trends in Ecology and Evolution* 22 (11): 569–574. https://doi.org/10.1016/J.TREE.2007.09.006.

Krashevska, V. (2012). Consequences of exclusion of precipitation on microorganisms and microbial consumers in montane tropical rainforests. *Oecologia* 170 (4): 1067–1076. https://doi.org/10.1007/S00442-012-2360-6/FIGURES/2.

Kundzewicz, Z.W., Mata, L.J., Arnell, N.W. et al. (2008). The implications of projected climate change for freshwater resources and their management. *Hydrological Sciences Journal* 53 (1): 3–10. https://doi.org/10.1623/hysj.53.1.3.

Lee, C.K. (2011). The Inter-Valley Soil Comparative Survey: the ecology of Dry Valley edaphic microbial communities. *The ISME Journal 2012 6:5* 6 (5): 1046–1057. https://doi.org/10.1038/ismej.2011.170.

Lin, B.B. (2011). Resilience in agriculture through crop diversification: adaptive management for environmental change. *BioScience* 61 (3): 183–193. https://doi.org/10.1525/bio.2011.61.3.4.

Liu B, Wang Y, Zhang H, Zhou Y, Zhang C, Yang N, Wang W. (2023). The Variations of microbial diversity and community structure along different stream orders in Wuyi Mountains. https://doi.org/10.1007/s00248-023-02240-8.

Luttenauer, D., Weill, S. and Ackerer, P. (2023). Impact of mountain topography on potential evapotranspiration and water drainage. https://doi.org/10.5194/EGUSPHERE-EGU23-982.

Mainali, B. and Silveira, S. (2015). Using a sustainability index to assess energy technologies for rural electrification. *Renewable and Sustainable Energy Reviews* 41: 1351–1365. https://doi.org/10.1016/j.rser.2014.09.018.

McCarthy, Canzianii, Learyi, Dokken, W. (2001) Climate change: impacts, adaptation & vulnerability. *IPCC Working Group*, 10032. http://www.ipcc.ch/ipccreports/tar/wg2/index.htm (accessed 7 October 2024).

McDowell, G., Stephenson, E., and Ford, J. (2014). Adaptation to climate change in glaciated mountain regions. *Climatic Change* 126 (1–2): 77–91. https://doi.org/10.1007/s10584-014-1215-z.

Mengist, W., Soromessa, T., and Legese, G. (2020). Ecosystem services research in mountainous regions: a systematic literature review on current knowledge and research gaps. *Science of the Total Environment* 702: 134581. https://doi.org/10.1016/j.scitotenv.2019.134581.

Moser, S.C. and Ekstrom, J.A. (2010). A framework to diagnose barriers to climate change adaptation. *Proceedings of the National Academy of Sciences of the United States of America* 107 (51): 22026–22031. 10.1073/pnas.1007887107.

Naeem, F., Phiri, P., Rathod, S., and Ayub M. (2019). Cultural adaptation of cognitive–behavioural therapy. *BJPsych Advances* 25 (6): 387–395. https://doi.org/10.1192/BJA.2019.15.

NASA (2020). The anatomy of glacial ice loss – climate change: vital signs of the planet. *Nasa* [Preprint]. https://climate.nasa.gov/news/3038/the-anatomy-of-glacial-ice-loss/ (accessed 7 October 2024).

Nautiyal, H. and Goel, V. (2020). Sustainability assessment of hydropower projects. *Journal of Cleaner Production* 265: https://doi.org/10.1016/J.JCLEPRO.2020.121661.

Negi, V.S. and Smith, E. (2017). Climate change impact in the Western Himalaya: people's perception and adaptive strategies. *Journal of Mountain Science* 14 (2): 403–416. https://doi.org/10.1007/s11629-015-3814-1.

Nepal, S., Flügel, W.A., and Shrestha, A.B. (2014). Upstream-downstream linkages of hydrological processes in the Himalayan region. *Ecological Processes* 3 (1): 1–16. https://doi.org/10.1186/S13717-014-0019-4/FIGURES/4.

Nettier, B., Dobremez, L., Lavorel, S. et al. (2017). Resilience as a framework for analyzing the adaptation of mountain summer pasture systems to climate change. *Ecology and Society* 22 (4): https://doi.org/10.5751/ES-09625-220425.

Ostrom, E. (2009). A general framework for analyzing sustainability of social-ecological systems. *Science* 325 (5939): 419–422. https://doi.org/10.1126/science.1172133.

Palermo, S.A. (2022). Smart technologies for water resource management: an overview. *Sensors* 22 (16): 6225. https://doi.org/10.3390/S22166225.

Pan, M., Wang, T., Hu, B. et al. (2021). Mesocosm experiments reveal global warming accelerates macrophytes litter decomposition and alters decomposition-related bacteria community structure. *Water* 13 (14), p. 1940: https://doi.org/10.3390/W13141940/S1.

Pandey, A. and Yarzábal, L.A. (2019). Bioprospecting cold-adapted plant growth promoting microorganisms from mountain environments. *Applied Microbiology and Biotechnology* 103 (2): 643–657. https://doi.org/10.1007/S00253-018-9515-2.

Pang, D. (2022). Features and driving factors of microbial metabolic limitation in mountain ecosystems in arid areas: A case study on the Helan Mountains, Northwest China. *Frontiers in Forests and Global Change* 5: 1–38. https://doi.org/10.3389/ffgc.2022.1024656.

Parmesan, C. and Yohe, G. (2003). A globally coherent fingerprint of climate change impacts across natural systems. *Nature* 421 (6918): 37–42. https://doi.org/10.1038/nature01286.

Pepin, N., Alder, C., Kotlarski, S. et al. (2022). Mountains undergo enhanced impacts of climate change. *Eos Earth & Space Science News* 103: 1–7. https://doi.org/10.1029/2022eo185278.

Philippot, L., Griffiths, B.S., and Langenheder, S. (2021). Microbial community resilience across ecosystems and multiple disturbances. *Microbiology and Molecular Biology Reviews* 85 (2): https://doi.org/10.1128/mmbr.00026-20.

Pohl, M., Alig, D., Körner, C. et al. (2009). Higher plant diversity enhances soil stability in disturbed alpine ecosystems. *Plant and Soil* 324 (1): 91–102. https://doi.org/10.1007/s11104-009-9906-3.

Porras-Alfaro, A. and Bayman, P. (2011). Hidden fungi, emergent properties: endophytes and microbiomes. *Annual Review of Phytopathology* 49: 291–315. https://doi.org/10.1146/ANNUREV-PHYTO-080508-081831.

Rashid, I., Mauri, P., Artur, G. et al. (2022). Editorial: ecosystem and hydrological responses in mountain environments to the changing climate. *Frontiers in Environmental Science* 10: 880386. https://doi.org/10.3389/FENVS.2022.880386/BIBTEX.

Rasul, G. (2021). Beyond hydropower: towards an integrated solution for water, energy and food security in South Asia. *International Journal of Water Resources Development* 37 (3): 466–490. https://doi.org/10.1080/07900627.2019.1579705.

Reed, M.S. (2008). Stakeholder participation for environmental management: a literature review. *Biological Conservation* 141 (10): 2417–2431. https://doi.org/10.1016/J.BIOCON.2008.07.014.

Scherrer, D. and Körner, C. (2010). Infra-red thermometry of alpine landscapes challenges climatic warming projections. *Global Change Biology* 16 (9): 2602–2613. https://doi.org/10.1111/j.1365-2486.2009.02122.x.

Schirpke, U. (2017). Future impacts of changing land-use and climate on ecosystem services of mountain grassland and their resilience. *Ecosystem Services* 26: 79–94. https://doi.org/10.1016/j.ecoser.2017.06.008.

Schmidt, G.A., Jungclaus, J.H., Ammann, C.M. et al. (2011). Climate forcing reconstructions for use in PMIP simulations of the last millennium (v1.0). *Geoscientific Model Development* 4 (1): 33–45. https://doi.org/10.5194/GMD-4-33-2011.

Scott, D. (2003). Climate change and tourism in the mountain regions of North America. *Conference on Climate Change and Tourism* (April), pp. 9–11. https://www.researchgate.net/

publication/267198542_Climate_Change_and_Tourism_in_the_Mountain_Regions_of_North_America (accessed 7 October 2024).

Sgrò, C.M., Lowe, A.J., and Hoffmann, A.A. (2011). Building evolutionary resilience for conserving biodiversity under climate change. *Evolutionary Applications* 4 (2): 326–337. https://doi.org/10.1111/j.1752-4571.2010.00157.x.

Shamurailatpam, M.S., Telling, J., Wadham, J.L. et al. (2023). Factors controlling the net ecosystem production of cryoconite on Western Himalayan glaciers. *Biogeochemistry* 162 (2): 201–220. https://doi.org/10.1007/s10533-022-00998-6.

Sharma, V., Dahiya, D. and Vasanth, D. (2020). Characteristics of Microbial Community and Enzyme Activities in Higher Altitude Regions, pp. 201–226. https://doi.org/10.1007/978-981-15-1902-4_11.

Shukla, S., Jain, S.K., and Kansal, M.L. (2021). Hydrological modelling of a snow/glacier-fed western Himalayan basin to simulate the current and future streamflows under changing climate scenarios. *Science of the Total Environment* 795: https://doi.org/10.1016/J.SCITOTENV.2021.148871.

Sloan, S. and Sayer, J.A. (2015). Forest Resources Assessment of 2015 shows positive global trends but forest loss and degradation persist in poor tropical countries. *Forest Ecology and Management* 352: 134–145. https://doi.org/10.1016/J.FORECO.2015.06.013.

Spence, A.R. and Tingley, M.W. (2020). The challenge of novel abiotic conditions for species undergoing climate-induced range shifts. *Ecography* 43 (11): 1571–1590. https://doi.org/10.1111/ecog.05170.

Strickland, M.S. and Rousk, J. (2010). Considering fungal: Bacterial dominance in soils – methods, controls, and ecosystem implications. *Soil Biology and Biochemistry* 42 (9): 1385–1395. https://doi.org/10.1016/J.SOILBIO.2010.05.007.

Suyal, D.C., Joshi, D., Kumar, S. et al. (2021). Himalayan microbiomes for agro-environmental sustainability: current perspectives and future challenges. *Microbial Ecology* 84 (3): 643–675. https://doi.org/10.1007/S00248-021-01849-X.

Tetzlaff, D. and Birkel, C. (2010). Hydrological connectivity and microbiological fluxes in montane catchments: the role of seasonality and climatic variability. *Hydrological Processes* 24 (9): 1231–1235. https://doi.org/10.1002/hyp.7680.

Tito, R., Vasconcelos, H.L., and Feeley, K.J. (2020). Mountain ecosystems as natural laboratories for climate change experiments. *Frontiers in Forests and Global Change* 3 (March): 1–8. https://doi.org/10.3389/ffgc.2020.00038.

Trivedi, P., Pandey, A., and Palni, L.M.S. (2012). Bacterial inoculants for field applications under mountain ecosystem: present initiatives and future prospects. *Bacteria in Agrobiology: Plant Probiotics* 9783642275 (1): 15–44. https://doi.org/10.1007/978-3-642-27515-9_2.

Tryland, I., Robertson, L., Blankenberg, A.B. et al. (2011). Impact of rainfall on microbial contamination of surface water. *International Journal of Climate Change Strategies and Management* 3 (4): 361–373. https://doi.org/10.1108/17568691111175650.

Vad, C.F., Anett, H.-E., Pavel K. et al. (2023). Spatial insurance against a heatwave differs between trophic levels in experimental aquatic communities. *Global Change Biology* 29 (11): 3054–3071. https://doi.org/10.1111/GCB.16692.

Valdivia, C., Seth, A., Gilles, J.L. et al. (2010). Adapting to climate change in Andean ecosystems: landscapes, capitals, and perceptions shaping rural livelihood strategies and linking knowledge systems. *Annals of the Association of American Geographers* 100 (4): 818–834. https://doi.org/10.1080/00045608.2010.500198.

Van Der Heijden, M.G.A., Bardgett, R.D., and Van Straalen, N.M. (2008). The unseen majority: soil microbes as drivers of plant diversity and productivity in terrestrial ecosystems. *Ecology letters* 11 (3): 296–310. https://doi.org/10.1111/J.1461-0248.2007.01139.X.

Verburg, P.H. (2015). Land system science and sustainable development of the earth system: a global land project perspective. *Anthropocene* 12 (December): 29–41. https://doi.org/10.1016/j.ancene.2015.09.004.

Vishwakarma, B.D., Kumar, P., Srivastava, S. et al. (2022). Challenges in understanding the variability of the cryosphere in the Himalaya and its impact on regional water resources. *Frontiers in Water* 4: 1–92. https://doi.org/10.3389/frwa.2022.909246.

Viviroli, D., Dürr, H.H., Messerli, B. et al. (2007). Mountains of the world, water towers for humanity: typology, mapping, and global significance. *Water Resources Research* 43 (7): 1–13. https://doi.org/10.1029/2006WR005653.

Walsh, J.E. and Chapman, W.L. (1998). Arctic cloud-radiation-temperature associations in observational data and atmospheric reanalyses. *Journal of Climate* 11 (11): 3030–3045. 10.1175/1520-0442(1998)011<3030:ACRTAI>2.0.CO;2.

Wang, Y. (2019). Drivers of change to mountain sustainability in the Hindu Kush Himalaya. *The Hindu Kush Himalaya Assessment: Mountains, Climate Change, Sustainability and People* 17–56. https://doi.org/10.1007/978-3-319-92288-1_2.

Wang, J., Hu, A., Meng, F. et al. (2022). Embracing mountain microbiome and ecosystem functions under global change. *New Phytologist* 234 (6): 1987–2002. https://doi.org/10.1111/nph.18051.

Wang, Y., Sun, X., Miao, L. et al. (2024). State-of-the-art review of soil erosion control by MICP and EICP techniques: problems, applications, and prospects. *Science of The Total Environment* 912: 169016. https://doi.org/10.1016/J.SCITOTENV.2023.169016.

Wilhelm, L., Singer, G.A., Fasching, C. et al. (2013). Microbial biodiversity in glacier-fed streams. *The ISME Journal* 7 (8): 1651. https://doi.org/10.1038/ISMEJ.2013.44.

Winkler, R. (2023). On the relationship between adaptation and mitigation. *SSRN Electronic Journal* 745–777. https://doi.org/10.2139/ssrn.4422846.

Wu, N., Ismail, M., Joshi, S. et al. (2014). Livelihood diversification as an adaptation approach to change in the pastoral Hindu-Kush Himalayan region. *Journal of Mountain Science* 11 (5): 1342–1355. https://doi.org/10.1007/s11629-014-3038-9.

Wyss, R., Luthe, T., Pedoth, L. et al. (2022). *Mountain resilience: a systematic literature review and paths to the future* 42 (2): A23–A36. https://doi.org/10.1659/MRD-JOURNAL-D-21-00044.1.

Xu, J. and Grumbine, R.E. (2014). Building ecosystem resilience for climate change adaptation in the Asian highlands. *Wiley Interdisciplinary Reviews: Climate Change* 5 (6): 709–718. https://doi.org/10.1002/wcc.302.

Yarzábal, L.A., Salazar, L.M.B., and Batista-García, R.A. (2021). Climate change, melting cryosphere and frozen pathogens: should we worry. . .? *Environmental Sustainability* 4 (3): 489–501. https://doi.org/10.1007/s42398-021-00184-8.

Yi, J., Lo, L.S.H., and Cheng, J. (2020). Dynamics of microbial community structure and ecological functions in Estuarine Intertidal sediments. *Frontiers in Marine Science* 7: 585970. https://doi.org/10.3389/FMARS.2020.585970/BIBTEX.

11

Modeling the Glacier Energy and Mass Budget of the Phuche Glacier in the Cold-arid Trans-Himalayan Region, Ladakh Range

Jaydeo Kumar Dharpure[1,3], Ajanta Goswami[1,2], Akansha Patel[1], Sanjay Kumar Jain[3], and Anil Vishnupant Kulkarni[4]

[1] *Centre of Excellence in Disaster Mitigation and Management, Indian Institute of Technology Roorkee, Roorkee, Uttarakhand, India*
[2] *Department of Earth Sciences, Indian Institute of Technology Roorkee, Roorkee, Uttarakhand, India*
[3] *Water Resources System Division, National Institute of Hydrology Roorkee, Roorkee, Uttarakhand, India*
[4] *Divecha Center for Climate Change, Indian Institute of Science, Bangalore, Karnataka, India*

11.1 Introduction

Glaciers are retreating all around the world's mountain regions in the effect of global climatic fluctuations (Marzeion et al. 2015; Ebrahimi and Marshall 2016; Gao et al. 2019). In response to this, the reshaping of the freshwater stored in glaciers may occur, which influences the regional water supply (Radić and Hock 2011) and also changes the timing of streamflow around the year (Dadic et al. 2008). On the other hand, the snow and glaciers of the Himalayan region are a significant support for the upstream and downstream population. Many studies have reported a mass loss (Banerjee 2017; Murtaza and Romshoo 2017; Bandyopadhyay et al. 2019; Patel et al. 2021a) and changes in the pattern of snow cover or snowfall (Ahmad et al. 2018; Basnett and Kulkarni 2019; Dharpure et al. 2020; Dharpure et al. 2021) over the Himalayan region. The variation in snow and glaciers extends causes a significant influence on the basin which are mainly dependent upon meltwater originating from snow and glaciers during the summer season.

However, the weather data from high elevations in the Himalayas and trans-Himalaya are seldom available. Due to the scarcity of such data sets, climatic forcing on glacier changes is often achieved by extrapolating of available low elevation climate data using elevation-dependent or standard relationships. Besides, the complex topography of this high mountain region ensures many topoclimatic zones with varying characteristics (Khan et al. 2017; Yadav et al. 2020). A significant difference in near-surface temperature lapse rates in different climate regimes of the Himalayas with distinct seasonal variations was reported across the region (Kattel et al. 2013; 2015; Heynen et al. 2016; Du et al. 2018; Thayyen and Dimri 2018; Pratap et al. 2019). Therefore, a reliable estimation of glacier mass variation and their response to future climatic change over the region is required,

Water Resources Management in Mountain Regions, First Edition. Edited by Virendra Bahadur Singh, Sughosh Madhav, Rakesh Kumar Gupta, Prerna Diwan, and Amit Kumar.

which causes a significant impact on regional water availability and also help in understanding the physical processes of the glacier and climate variability (Oerlemans et al. 1998; Oerlemans and Klok 2002).

Earlier, the comprehensive surface energy balance (SEB) model was used to quantify the glacier-climate interaction (Hoinkes 1954); though, this method was substantially improved afterward. Then, the improved SEB method was applied all around the world (Ayad Ali Faris Beg et al. 2016; Pellicciotti et al. 2005). Previously, many studies were carried out on the glacier SEB in high mountain Asia (Acharya and Kayastha 2018; Patel et al. 2021b; Srivastava and Azam 2022). However, studies on glacier SEB in the Ladakh range, the Karakoram region, and the eastern Himalayas are not yet available. On the other hand, a study conducted by Wani et al. (2021) on the upper Ganglass catchment in the Ladakh range uses the meteorological observation collected through automatic weather station (AWS) installed at a lower elevation (4727 m a.s.l.) to quantify point-based SEB. However, considering low altitude station data does not provide a complete insight into the glaciers lying above the ~5300 m a.s.l. because this region mainly contains glaciers above this elevation (Schmidt and Nüsser 2017).

Therefore, the SEB components of the Phuche glacier in the upper Ganglass catchment, Ladakh range, have been analyzed, and the glacier surface mass balance was estimated. The study includes the meteorological observation recorded at 5600 m a.s.l. in order to emphasize the lack of physical-based energy balance measurement over this range. This study also bridges the gap of long-term mass balance study and their interaction with climatic response over the cold-arid region of the Himalayas. The monthly, annual, and seasonal variation of SEB components at a point location were analyzed to understand the distribution of each component's characteristic and their controlling processes in glacier mass balance estimation. This research also aims to present 5-year consecutive observations for the research community that requires field observation data for model calibration/validation.

11.2 Study Area

The study site is situated in the Ladakh range of the trans-Himalaya at an elevation of 5600 m a.s.l. (34° 16′ 39.40″N, 77° 33′ 35.34″E) in the Leh city, Union Territory of Ladakh, India (Figure 11.1). The Ladakh range extends from the Indus valley in the south to the Nubra-Shyok valleys in the north, sandwiched between the Zanskar group of mountains to the south and the Karakoram in the northwards side. About 79% of the glaciers in the Ladakh range are <0.75 km^2, and approximately 4% of the glacier have >2 km^2, mainly occupying the high elevation region above 5300 m a.s.l. (Schmidt and Nüsser 2017). Glaciers in the region are experiencing mass loss, and individual glacier cover area has reduced by 17–22% since 1969 (Schmidt and Nüsser 2017). Due to scarce precipitation and higher summer temperature, this region of the trans-Himalaya was mainly known as a cold-arid desert. The region was characterized by strong land-atmosphere interactions, rarefied atmosphere, and strong incoming solar radiation (Wani et al. 2020). This cold-arid region covers a significant portion of the upper Indus River basin with a large snow and glacier extent of the Suru, Zanskar, and Shyok river basins. It contains steep mountain slopes as well as valley bottom filled with glacio-fluvial deposit (Jowhar 2001; Wani et al. 2020).

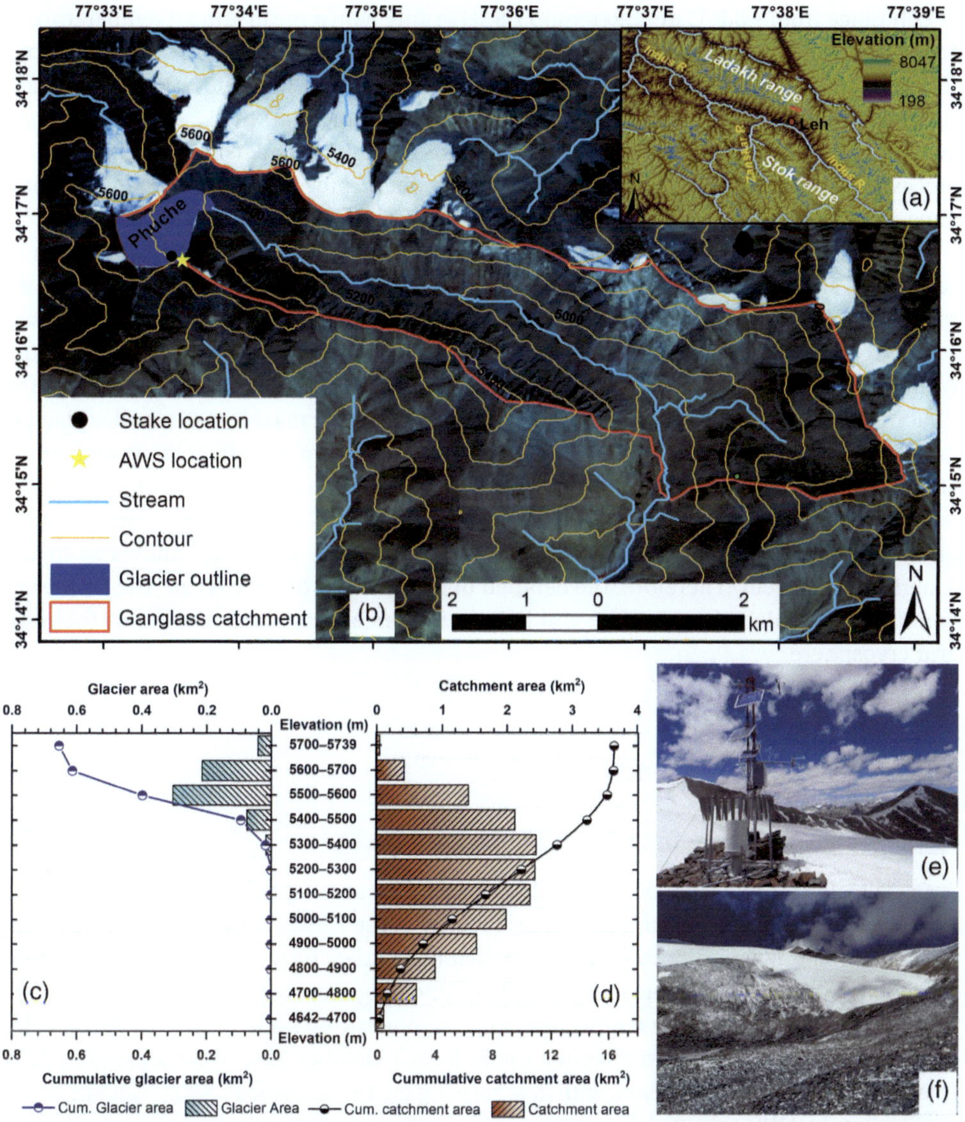

Figure 11.1 Location map of the (a) cold-arid region with glacier boundaries, different ranges, and AWS location and (b) upper Ganglass catchment with the Phuche glacier, AWS, and stake location overlay on satellite composite image. *Source:* U.S. Department of the Interior / https://earthexplorer.usgs.gov/ / last accessed on October 03, 2024. (c, d) represent the hypsometry distribution of Phuche glacier and upper Ganglass catchment and (e, f) indicate the field photographs of AWS location and glacier terminus during 9–15 September 2017.

The long-term meteorological variables were recorded at Leh station (3500 m a.s.l.), indicating high inter annual variability with total less than 100 mm mean annual precipitation during 1882–1968. However, the average summer rainfall was 37.5 mm between July and September while the winter average precipitation was 27.3 mm from January to March (Archer and Fowler 2004; Nüsser et al. 2012). Mean monthly temperatures in Leh range from −8.2°C in January to 17.8 °C in July with a mean annual temperature of 5.8°C during the period from 1951–1980, which is unusually high for these elevations (Brazel and Marcus 1991). Recent studies have shown that high elevation ridges have significant permafrost cover in the Ladakh region (Wani et al. 2021, 2020). Therefore, meteorological conditions at higher elevation are crucial for understanding the physical processes, moisture-temperature interaction, cryospheric process and for establishing their impact on water availability.

The AWS site is situated approximately 35 km from Leh city, and the base camp to access this site is set up in South Pullu at 4700 m a.s.l., which is 25 km from Leh town en route to Khardung-La (Pass). The study site lies within the snow cover area and closes to the high elevation margins of the Phuche glacier. The Phuche glacier is a small cirque-type glacier that is 1.3 km long and covering an area of 0.58 km^2. The elevation of the Phuche glacier ranged from 5400 to 5745 m a.s.l. The drastic mass loss of smaller glaciers may create a warm hotspot that can imbalance the ecosystem and affect the water supply of this region. The meltwater of this glacier is a part of the headwater zone of Leh stream, which is the main water source for the Ganglass village and Leh city downstream. These headwater zones are also considered the recharging zone of the groundwater in the Ganglass catchment. Therefore, the detailed understanding of the climate and SEB components at this high elevation site will help in understanding the processes of the cold-arid cryosphere.

11.3 Data and Methods

11.3.1 Meteorological Data Collection at 5600 m a.s.l.

A Campbell Scientific CR1000 AWS has been installed at a high altitude (5600 m a.s.l. on the ridge of Phuche glacier) in the cold-arid region of the Ladakh range for weather monitoring. The station is equipped with sensors that record data on a half-hourly basis, including global radiation [incoming-shortwave (SWI) and outgoing-shortwave radiation (SWO)], albedo, net radiation (R_n), air temperature (T_a), barometric pressure (P_{atm}), relative humidity (RH), wind speed (u), and direction (WD) from 1 October 2012 to 30 September 2017 (covering five hydrological years, with data missing for 14–31 October 2012). Any measurement errors in AWS-collected data were rectified before analyzing the components, if available. These errors are associated due to ill-functioning of the sensor, and they vary with the sensor accordingly. Then, the corrected variables were analyzed on an hourly, daily, and monthly timescales.

For winter precipitation, we attempted to measure it using the Geonor precipitation gauge; however, it failed to achieve good results due to technical snags and harsh weather conditions. Therefore, we have utilized the empirical relationship of snow thickness and density to calculate the accumulated snow water equivalent (SWE). These components were measured through a field survey conducted in May. Snow thickness was monitored through a snow stake installed over the glacier, and their snow density was estimated by digging the snow pit near the AWS site.

11.3.2 Data Preprocessing

After eliminating the measurement error, the abrupt spike in the AWS dataset was removed and filled through linear interpolation. Apart, the half-hourly shortwave radiations (SWI and SWO) measured using Kipp and Zonen (CMP6) were carefully processed. The computed R_n and total energy were dominant by the change in shortwave radiation. For this, we have considered a cutoff of $5\,\mathrm{W/m^2}$ for nighttime data of incoming and outgoing solar radiation, which is set to be zero. The collected data indicate that SWO sometimes shows a higher value than SWI during morning and evening. This occurs because the solar angle at that time was low due to the poor cosine response of the upward-looking sensor (SWI) (Nicholson et al. 2013). However, the SWO sensor mostly receives isotropic radiation; therefore, SWO was relatively less sensitive than SWI toward uncertainties caused due to poor cosine response (van den Broeke et al. 2004). In this study, we reanalyzed the SWI from raw SWO and with estimated accumulate albedo (α_{acc}), as given by Eq. (11.1).

$$\alpha_{\text{acc}} = \frac{\sum_{i=1}^{24} \text{SWO}_i(\text{raw})}{\sum_{i=1}^{24} \text{SWI}_i(\text{raw})} \tag{11.1}$$

The reflected radiation (SWO) was sometimes higher compared to incoming radiation during the daytime. These higher values may have occurred because the incoming radiation sensor was covered with snow due to heavy snowfall. In this case, incoming radiation was estimated through reflected radiation which is divided by considering the maximum realistic value of snow albedo (0.9) (Oerlemans and Klok 2002; Favier et al. 2004).

The specific humidity was estimated using the T_a and RH measurement as described in Eq. (11.2):

$$\text{Specific humidity}(\text{g/kg}) = 1000\left[0.622\frac{e_a}{P_{\text{atm}}}\right] \tag{11.2}$$

where saturated vapor pressure $(e_s) = 6.112 \times \exp^{\left(\frac{17.67 \times T_a}{243.5 + T_a}\right)}$ and $(e_a) = (e_s \times \text{RH})/100$. The surface temperature (T_s) is a significant variable in order to quantify the turbulent heat fluxes on the glacier surface. The half-hourly observation of solar radiation, T_a, RH and u were used for estimating the T_s, as discussed by Fujita and Ageta (2000). The u was measured at 6.5 m height above the surface, it is required to convert at 2 m for energy fluxes calculation. This height (6.5 m) was converted at 2 m using the logarithmic wind profile function, explained by Bastiaanssen et al. (1998).

11.3.3 Surface Energy Balance (SEB) Modeling

The modeling of SEB components was carried out by the calculation of energy fluxes directed toward the glacier surface, which is considered positive, and when it goes upward from the surface, it becomes negative. Assuming no significant horizontal energy transfer over a unit glacier volume and for a unit of time, the SEB calculated at the snow surface can be expressed by Eq. (11.3) (Oke 1987):

$$F_s = R_n + H_s + H_l + H_g + P \tag{11.3}$$

where R_n is the combination of net shortwave and net longwave radiations, H_s and H_l are the sensible and latent energy fluxes, respectively. H_g is the ground heat flux, and P is the supplied heat from precipitation, which is small enough to be neglected. F_s is the total energy available at the glacier surface. The radiation and energy fluxes were measured in W/m^2.

Measurement of energy fluxes at the glacier surface was carried out daily during the summer season, between May and September, to understand the melting pattern around the selected year. H_s and H_l were carried out by the bulk aerodynamics methods (Azam et al. 2014a; Oke 1987) and calculated by Eqs. (11.4) and (11.5):

$$H_l = \rho_a \frac{L_s k^2 u(q - q_s)}{\left(\ln \frac{z}{Z_{0m}}\right)\left(\ln \frac{z}{Z_{0q}}\right)} (\varnothing_m \varnothing_v)^{-1} \tag{11.4}$$

$$H_s = \rho_a \frac{C_p k^2 u(T_a - T_s)}{\left(\ln \frac{z}{Z_{0m}}\right)\left(\ln \frac{z}{Z_{0T}}\right)} (\varnothing_m \varnothing_h)^{-1} \tag{11.5}$$

where ρ_a is the air density (in kg/m^3) from the following equation [$\rho_a = \frac{P_{\text{atm}}}{R_a T_a}$, where R_a = specific gas constant (for dry air) and P_{atm} = Atmospheric pressure (in hPa) at AWS]. L_s is the latent heat of sublimation (2.834×10^6 J/kg), z_{0T}, z_{0q}, and z_{0m} are the surface roughness parameters for temperature, humidity, and momentum, respectively. C_p is the specific heat capacity (for air at constant pressure) ($C_p = C_{\text{pd}}(1 + 0.84q)$ with $C_{\text{pd}} = 1005$ J/kg/K, the specific heat capacity (for dry air at constant pressure)). k is the von Korman constant ($k = 0.4$), q and q_s are the mean specific humidity of the air and surface, respectively. The nondimensional stability functions of momentum ($\varnothing_m$), heat ($\varnothing_h$), and moisture ($\varnothing_v$) are given by R_{ib}. Where R_{ib} is the bulk Richardson number that described the stability of the surface layer that relates the effects of buoyancy to mechanical forces (Moore 1983), expressed by Eq. (11.6).

$$R_{ib} = \frac{g(T_a - T_s)(z - Z_{0m})^2}{T_a u^2 (z - Z_{oT})} \tag{11.6}$$

where g is the acceleration of gravity (g = 9.8 m/s^2). Stability of the surface boundary layer is measured by bulk Richardson number R_{ib}. Sometimes very high and low values of R_{ib} were observed. These high and low values gave rise to a huge fluctuation in H_s and H_l values. Therefore, the lower and upper limits were fixed by calibrating the R_{ib} in order to remove all the possible outliers (extreme and sudden high or low values of H_s and H_l) in the values, as explained in Eqs. (11.7) and (11.8). An agreement with $-0.40 < R_{\text{ib}} < 0.23$ was achieved and selected for final H_s and H_l calculations (Denby and Smeets 2000; Favier et al. 2011).

$$\text{if } R_{\text{ib}} > 0 \left(\text{stable}\right): \left(\varnothing_m \varnothing_h\right)^{-1} = \left(\varnothing_m \varnothing_v\right)^{-1} = \left(1 - 5R_{\text{ib}}\right)^2 \tag{11.7}$$

$$\text{if } R_{\text{ib}} < 0 \left(\text{unstable}\right): \left(\varnothing_m \varnothing_h\right)^{-1} = \left(\varnothing_m \varnothing_v\right)^{-1} = \left(1 - 16R_{\text{ib}}\right)^{0.75} \tag{11.8}$$

11.3.4 Calculation of Total Ablation (Melt)

An energy balance model based on the physical condition was applied to quantify the surface melt by assessing the radiation and energy fluxes that transfer to and from the glacier surface. When the surface temperature is equal to 0 °C, any surplus energy is assumed to transform the glacier surface layer into the melt. Another way of glacier mass loss/gain is also attained by the sublimation/resublimation process. The melt occurs when the positive surface energy is available, whereas the negative sign indicates the loss of energy at the glacier surface (Hock 2005), as follows by Eq. (11.9):

$$\text{Melt}\left(\text{m w.e.}\right) = \frac{F_s}{L_f \times \rho_w} \tag{11.9}$$

where $\rho_w = 1000$ kg/m^3 (water density) and L_f = latent heat of fusion/melt ($L_f = 3.34 \times 10^5$ J/kg). F_s = total energy is available for glacier melt, and it occurs when the surface temperature is at 0 °C. Sublimation is derived from the H_l when it is negative (energy going away from the surface), by Eq. (11.10) as given below:

$$\text{Sublimation}\left(\text{m w.e.}\right) = \frac{H_l}{L_s \times \rho_w} \tag{11.10}$$

where L_s = latent heat of sublimation ($L_s = 2.834 \times 10^6$ kg/m^3) and H_l is the latent heat. To estimate the total ablation, the estimated melt is added with the sublimation and subtracted with the resublimation of the study site.

11.4 Results

11.4.1 Meteorological Variables

11.4.1.1 Air Temperature and Relative Humidity

The minimum, maximum, and mean T_a were calculated using half-hourly dataset from AWS to capture the seasonal and annual variations during the observational period (2012–2017). During this period, the daily mean minimum and maximum T_a were measured as 14.2 and −29.8 °C, respectively, with a mean value of −7.8 °C. The interannual daily temperature variation was higher in 2016/2017 and the lower value was attained in 2014/2015. This finding was consistent with the observation of Soheb et al. (2020) over the Stok glacier, Ladakh range. The mean monthly temperature suggested that prolonged subzero mean temperatures were noted over the location, varying from September to May (Figure 11.2a).

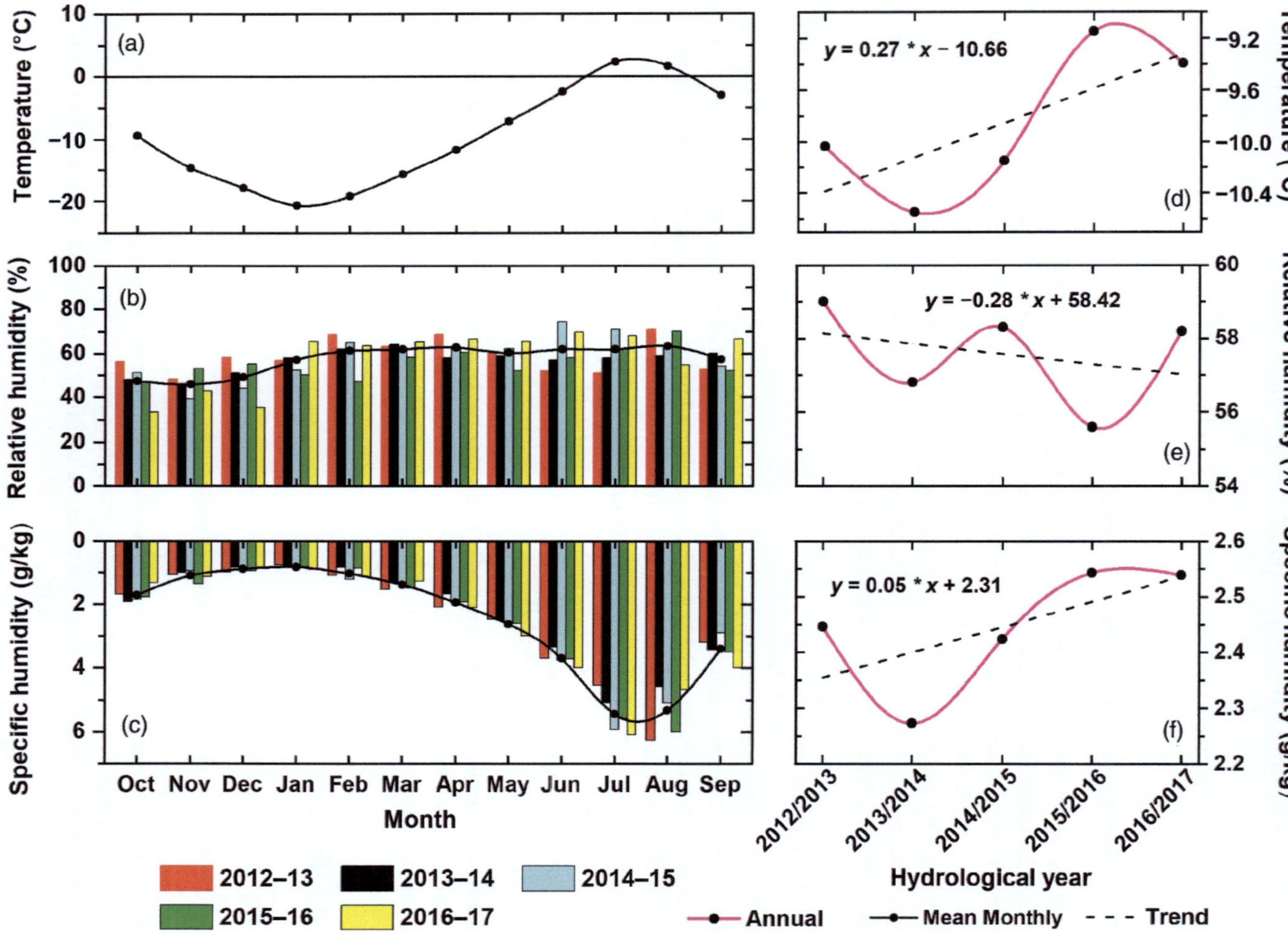

Figure 11.2 Mean monthly and annual distribution of (a) air temperature, (b) relative humidity, and (c) specific humidity for the study period 2012–2017.

The positive mean temperature was found in July and August; however, November to April experienced negative temperatures (<−10 °C). A similar pattern of temperature was observed by Soheb et al. (2020) over the Stok glacier, Ladakh range and also reported by Azam et al. (2014b) for the Chhota Shigri glacier, western Himalayas. Results also highlighted that January was the coldest month (−20.7 °C), whereas July was the warmest month (2.4 °C). The linear trend of mean monthly T_a was increasing for each month except March and June. On the other hand, the mean annual temperature was maximum (−8.9 °C) in 2015/2016 and minimum (−10.5 °C) in 2012/2013 with a mean value of −9.8 °C. Annual temperature range extended from 14.2 to −29.8 °C during the observation period. It was also found that the period from 2012 to 2015 experienced lower temperature from the mean, and it increased by about 13.0% in 2015/2016 followed by 8.1% in 2016/2017 relative to the previous period. Overall, the mean annual temperature shows an increasing trend (0.27 °C/yr) during the observational period. Bhutiyani et al. (2007) have demonstrated that the mean maximum, minimum and mean annual temperature experienced significant increasing trend for Shimla (1901–2002), Srinagar (1901–2002), and Leh (1901–1989). A study conducted by Chevuturi et al. (2018) observed that temperature shows slightly increasing trend from time period 1901–1979, afterwards temperature shows decreasing trend during 1979–1991 and then for short time interval (1991–2013), the temperature again increases rapidly. This variation in temperature possibly effects of foreshadowing over the region.

In addition to temperature, RH is one of the influential variables at higher elevations in determining the arid characteristics of the region. In the daily analysis, the minimum and maximum value of RH varies from 0.2 to 100%, with a mean annual value of 57.6%. During the observational period, the RH range of 50–80% covers the highest percentage of days (55%), followed by 35% of days with RH <50%, and only 10% of the days with RH >80%. The seasonal variation of RH was characterized by the mean monthly value of RH, showing maximum in August (63.5%) and minimum in November (45.9%) (Figure 11.2b). However, a consistent high mean monthly RH values were observed from February to August, ranging from 60.3 to 63.5%. Maximum humidity highly associated with low air temperature and high rainfall or cloudy conditions and vice versa (Pratap Singh 2005). The yearly trend obtained from linear regression indicates a decline (−0.28 %/yr) over the entire observational period. This decreasing trend in RH will further be responsible for the changing rainfall pattern and cloudy conditions (Singh et al. 2006). Based on annual T_a and RH, our findings show that these two variables follow an inverse relationship with each other; if T_a increases, then RH will likely be decreased, or vice versa.

Subsequently, the mean monthly pattern of specific humidity was calculated to better understand the total water vapor content in the air (Figure 11.2c). Results indicate that the July and August months have the highest water vapor content in the air with a specific humidity of 5.44 and 5.32 g/kg, respectively, while December and January recorded the lowest value at 0.89 and 0.83 g/kg, respectively, with a mean value of 2.45 g/kg. The winter months (November–March) received an average water vapor content of 1.04 g/kg, significantly lower than the 3.45 g/kg during the summer months (April–October). This indicates that the higher moisture influx was concentrated during the summer compared to the winter season. Thayyen and Dimri (2018) observed mean monthly specific humidity ranging from 2.92 to 11.4 g/kg at 3763 m a.s.l. over the Dingad catchment, western Himalaya. They also revealed that the winter months (NDJFM) exhibits lower water vapor in the atmosphere than monsoon season. Additionally, the higher elevation (3763 m a.s.l.) experienced lower specific humidity than lower elevation (2540 m a.s.l.) throughout the year. The highest mean annual specific humidity was observed at 2.54 g/kg in 2015/2016 and 2016/2017, while the lowest value was recorded at 2.27 g/kg in 2013/2014. Combining our results with specific humidity and T_a, both follows a similar pattern; higher temperature during the summer is responsible for increase in water content in the atmosphere, and vice-versa.

11.4.1.2 Wind Speed and Direction

The half-hourly observation of u and WD was analyzed to understand the wind regimes that regulate moisture transportation, cloud formation, the occurrence of precipitation, and surface energy balance. The analysis was performed for both day (07 : 00–18 : 00 hours) and night (19 : 00–06 : 00 hours) as well as for the summer and winter periods. It was noted that the orographic processes influence wind regime with anabatic flows attaining a higher speed during daytime than nighttime. A similar pattern of wind regime and their observation was matched with the finding of Singh et al. (2007) over the Gangotri glacier, central Himalayas.

The daytime wind is dominated by south-south-east (SSE) mainly through the local valley (upper Ganglass), and during nighttime southern wind dominates, blowing directly from the main Indus valley (Figure 11.3). The seasonal analysis indicates that the southern winds dominate the study area during summer and winter periods; however, the u was higher in the winter relative to the summer season. This finding was consistent with Azam et al. (2014b) and Soheb et al. (2020). During the five observational periods, the monthly maximum u ranged from 14.3 to 19.2 m/s, reaching the "near gale" class in the Beaufort scale. And the average u was varied from 3.5 to 4.9 m/s, which is of the "gentle breeze" category. Apart, "light breeze," "Gentle breeze," "moderate breeze," and "fresh breeze" conditions prevail for about 26%, 45%, 21%, and 5% of days, respectively, while the calm and light air classes of Beaufort scale do not cover the study site in annual scale. The annual distribution of u indicates an increasing trend at the rate of 0.24 m/s/yr for the entire observational period (Figure 11.3). This suggests that the rising pattern of u may cause a significant impact on turbulent heat fluxes (sensible and latent heat) over the region (Acharya and Kayastha 2018).

11.4.1.3 Radiative and Heat Fluxes

The monthly, seasonal, and annual distribution of shortwave radiation (SWI and SWO), albedo, and its net shortwave radiation (SWN: SWI–SWO) were quantified for the observational period to understand the variation of radiative forces on surface melt. The daily variation of SWI shows higher variability compared to SWO in all hydrological years. The variability in SWO was higher in 2016/2017 and lower in 2012/2013. This higher variability in SWI is mainly influenced by the presence of cloud cover and terrain properties. In the monthly analysis, SWI was observed highest in May (356.1 W/m^2) and lowest in December (170.0 W/m^2). SWO was maximum in May (238.7 W/m^2), and a minimum value was recorded in August (76.6 W/m^2). The annual mean SWI during the observation years ranges between 237.5 and 300.6 W/m^2 with a mean of 262.9 W/m^2, while the annual mean SWO

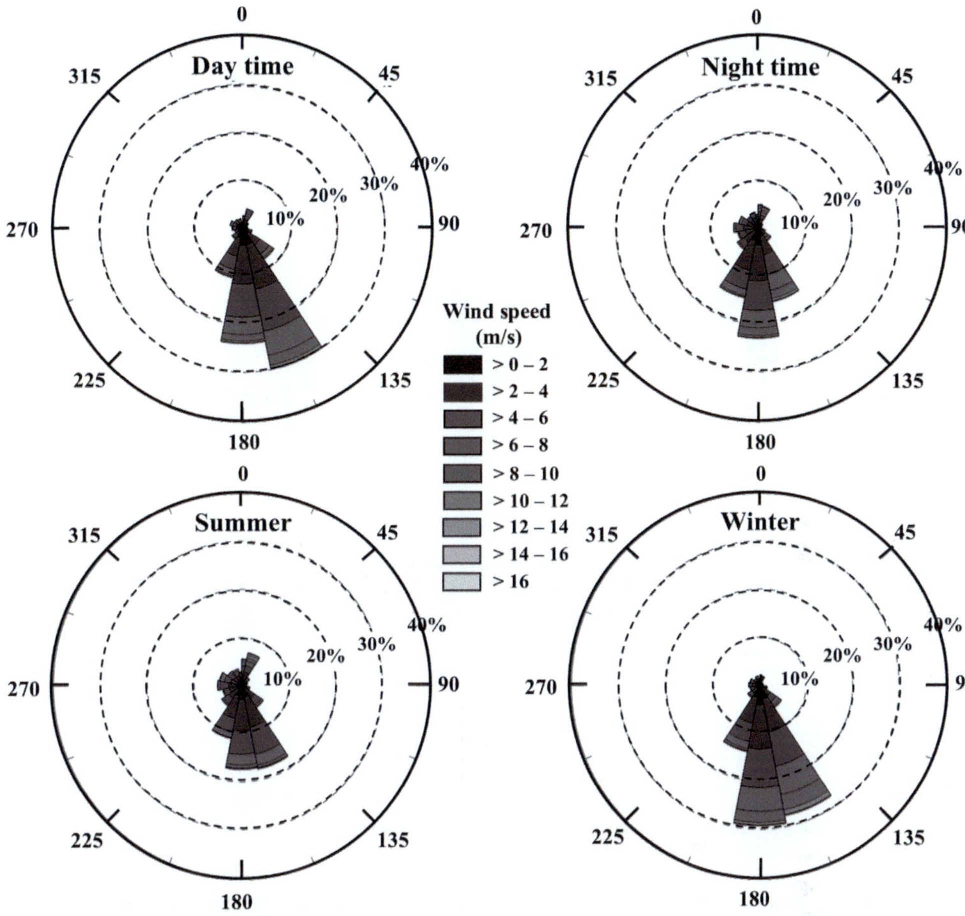

Figure 11.3 Wind speed and direction from half-hourly data for daytime (07 : 00–18 : 00 hours) and nighttime (19 : 00–06 : 00 hours) as well as for summer (April–September) and winter (October–March) seasons.

ranges between 129.6 and 158.6 W/m^2, with a mean of 141.2 W/m^2. This indicates that the higher amount of SWI was trapped in the surface and a less amount of radiation reflects back during the summer season. This observation was consistent with other authors findings (Favier et al. 2011; Azam et al. 2014b; Patel et al. 2021b). The higher absorption of radiation is mainly owing to reduced albedo due to snow aging and even exposure of the ground surface. Our findings also include that the albedo was higher in April (0.72) and lower in August (0.28) (Figure 11.4).

The SWN was calculated from corresponding SWI and SWO, showing that the daily value at the study site ranged from 9.0 to 416.3 W/m^2 with an average of 122.2 W/m^2. The monthly distribution of SWN demonstrates a higher value in August followed by July and September, while a lower value was nearly consistent from January to March. The annual SWN was minimum in 2014/2015 (98.6 W/m^2) and maximum in 2015/2016 (153.9 W/m^2), with an average value of 122.2 W/m^2. However, SWI and SWO were highest in 2016/2017,

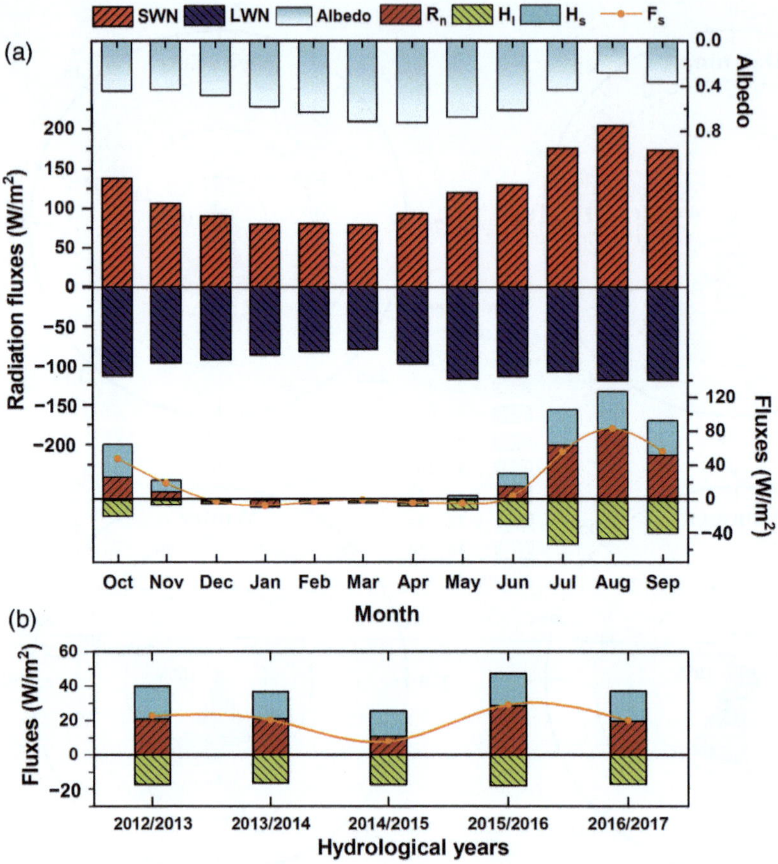

Figure 11.4 (a) Mean monthly variation of net shortwave (SWN), net longwave (LWN: R_n - SWN), albedo, net radiation (R_n), sensible (H_s) and latent (H_l) heat fluxes; (b) interannual distribution of net radiation (R_n), turbulent heat fluxes (H_l and H_s), and total energy available for melt (F_s) over the study site during 2012–2017.

whereas the lowest values were attained in 2014/2015 and 2012/2013, respectively. The linear trend of SWI, SWO, and SWN was increasing with a rate of 18.5, 7.0, and 11.0 W/m^2/yr, whereas albedo shows a declining trend (−0.90%/yr). This indicates that this region may experience less snow accumulation or snowfall amount during the study period. Additionally, the net longwave radiation (LWN) was measured by subtracting the SWN from R_n, indicating a similar distribution as SWN in mean monthly analysis. However, the contribution of LWN in R_n was higher during winter than in the summer season, while SWN was at its maximum in the summer season relative to winter. This means that LWN was inversely proportional to SWN in terms of their contribution toward R_n in seasonal analysis.

After analyzing the variation of radiation fluxes, R_n was measured over the study site, which indicates that the mean daily R_n ranges from −88.3 to 184.6 W/m^2 with a mean of 20.1 W/m^2. The R_n value starts positive from May onwards and attains its maximum value during August (84.5 W/m^2) and further decreases while moving forward till November. The other remaining months show negative R_n with a maximum in January ($-6.5\,W/m^2$) and a minimum in March ($-0.5\,W/m^2$) with an average of $-3.1\,W/m^2$ (Figure 11.4a). This positive R_n indicates that these particular months may be able to contribute surface melt over the region. The mean annual R_n shows an increasing trend with a rate of $2.4\,W/m^2/yr$ for the study period, with maximum value in 2015/2016 followed by 2013/2014 and the minimum value was received in 2014/2015 (Figure 11.4b). By combining T_a, albedo, and R_n for the period 2015/2016, we can conclude that due to an increase in T_a (-9.1°C) it affects the albedo (0.49) and causes an adverse effect on R_n ($108.0\,W/m^2$). This indicates that this region is highly driven by temperature variation and radiation fluxes. Similar observations were explained by several authors (Meena et al. 2015; Shafiq et al. 2016; Wani et al. 2021).

The turbulent heat fluxes (H_s and H_l) were measured from hourly to monthly and yearly scales (Figure 11.4a). The heat fluxes illustrate that the mean monthly distribution of both the fluxes follows an opposite pattern; however, a higher value is attained during May–November. The maximum value of H_s was observed in August, and the minimum was attained in January. In contrast, the H_l was highest in January, and the lowest value was observed in July. The estimated patterns of H_s and H_l were matched with the observations of T_a and u of the study site. Our finding revealed that the T_a was higher in July followed by August, and even u was lower during these months. The obtained monthly pattern of H_s and $H_l$were well matched with the observations of many authors carried out over the Himalayan region (Amatya et al. 2015; Acharya and Kayastha 2018; Patel et al. 2021b). Other than this, the yearly analysis suggests that H_l shows a declining linear trend with a rate of $-0.1\,W/m^2/yr$, while H_s indicates no change over the study period. This means the turbulent heat flux was not significantly changed as compared to radiation components. The value of H_l was negative or nearly zero throughout the temporal analysis. This means the study site was mainly influenced by the process of sublimation rather than resublimation. Wani et al. (2021) have also highlighted a similar observation.

By the summation of radiations and energy fluxes over the study site, the obtained total surface melt (F_s) was used to quantify the SEB from 2012 to 2017. Only positive F_s was considered during the summer months (May–September). The average annual F_s was $56.2\,W/m^2$ with a maximum ($112.8\,W/m^2$) in 2015/2016, and a minimum ($11.2\,W/m^2$) was observed in 2014/2015. In the summer, the mean value of R_n, H_s, and H_l were 43.7, 29.7, and $-34.8\,W/m^2$, respectively, which indicate that the R_n and H_s were the main contributor in the total energy available for melt. The contribution of R_n was higher as compared to other fluxes for the study site. A similar observation was illustrated by many authors (Aizen et al. 2002; Acharya and Kayastha 2018; Soheb et al. 2018). However, the total surface melt and contribution of each variable were changing during the winter season, as the region was experiencing very low temperature, resulting in a negative value.

11.4.1.4 Atmospheric Pressure

The P_{atm} at the site location was recorded from 1 October 2012 to 30 September 2015, which is used for H_s and H_l calculations. The mean daily P_{atm} ranged from 496.1 to 521.9 hPa with a mean value of 512 hPa. In a daily variation, the year 2012/2013 shows some data gap; however, complete data were missing during 2015/2016 and 2016/2017 over the study site. Therefore, the hourly average value of P_{atm} was used for estimating the turbulent heat fluxes in 2015/2016 and 2016/2017. The mean monthly analysis indicates that the P_{atm}

starts decreasing from October onward and attains minimum value in February (502.6 hPa), then suddenly it starts increasing till September with a maximum value in August (512.7 hPa). Our result also illustrates that the coefficient of variation (CV) was highly variable in February (0.63) followed by January (0.58) and lower value attained in September (0.07) followed by August (0.08). This indicates that the CV was higher in the months where P_{atm} was lower (January and February) than in other months. This resultant observation was well-matched with the u of the study site, which was highest in February (4.9 m/s) and lowest in August (3.5 m/s). In the yearly analysis, the mean P_{atm} during 2012/2013, 2013/2014, and 2014/2015 were 511.2, 511.1, and 512.0 hPa, respectively. The yearly variation of P_{atm} was not much change relative to monthly variation.

11.4.1.5 Accumulated Winter Precipitation

The continuous monitoring of precipitation at this higher elevation is difficult; therefore, the accumulated snow water equivalent-based measurement was performed in May every year from 2012 to 2017 (Table 11.1). In the snow water equivalent measurement, winter snow accumulation was measured through stake measurement, and it was multiplied by snowpack density estimated by digging snow pit for snow density measurement close to the AWS site. The estimated winter precipitation was minimum (589 mm w.e.) in 2013/2014 and maximum (980 mm w.e.) in 2016/2017 with a mean value of 734 mm w.e. Our findings also indicated that the winter precipitation shows an increasing trend at the rate of 98.2 mm w.e./yr. However, the mean winter precipitation recorded at Leh station was 47.5 mm w.e. (Thayyen et al. 2020). This indicates that the steep precipitation gradient may exist between the valley bottom and mountain ridge, which play a critical role in the hydrological and cryospheric system response of the region. Even though this region is devoid of significant monsoon incursions, occasional high-intensity rainfall events in summer, which was reported by Thayyen et al. (2013). However, there was less evidence suggesting that such extreme events impacted the AWS site at 5600 m a.s.l. A high precipitation gradient between these two closely placed valley bottom station and overlooking ridge station at 5600 m a.s.l. suggest that the cooler temperature along the mountain ridge is mainly controlled by the mountain slope, and it plays a significant role in assessing the precipitation amount at the higher altitude of the cold-arid region. The mean annual linear trend shows an increasing trend with a rate of 98.18 mm w.e./yr during the

Table 11.1 Accumulated winter precipitation at 5600 m a.s.l. from 2012 to 2017.

Year	Date of measurement	Snowpack thickness (m)	Snowpack density (kg/m^3)	Accumulated precipitation (mm w.e.)
2012/2013	19 May 2013	1.6	380.0	611.8
2013/2014	22 May 2014	1.8	320.0	588.8
2014/2015	25 May 2015	1.7	390.0	655.2
2015/2016	16 May 2016	1.9	430.0	834.2
2016/2017	30 May 2017	2.0	490.0	980.0
Mean				734.0

studied period. However, Bhutiyani et al. (2010) recorded ground based precipitation data at 3504 m a.s.l., illustrating a decreasing trend of total annual, winter, and monsoon precipitation during 1901–1989. On the other hand, low precipitation in Leh was observed before 1970 which subsequently increased during 1970–1995 and then again showing declining trend after 1995 till 2012 Chevuturi et al. (2018). In general, temperature and precipitation show inverse relationship between each other however this study exhibits increasing trend in both precipitation and temperature. This means higher temperatures rise may hold more water in the atmosphere result more evaporation which in turn increases overall precipitation but will not necessarily increases the amount of water available for drinking, irrigation and other purposes. Blunden and Boyer (2020) suggested an inverse relationship exists between temperature and precipitation in this region. However, this study shows a similar increasing trend in precipitation and temperature.

11.4.2 Diurnal Variability

The mean diurnal variation (half-hourly values) of T_a, T_s, RH, u, albedo, SWI, SWO, and R_n were quantified for the study period, indicating that the distribution of T_a, T_s, SWI, SWO, albedo, and R_n follow a similar pattern (Figure 11.5). Diurnal temperature variation shows that maximum temperature was often observed around 14 : 00 hours, and the minimum was around 06 : 00 hours. Results demonstrated that all the radiations intensity increases with the increase in daylight hours, and it reaches a maximum at noon (12 : 00 hours) then starts decreasing till 19 : 00 hours. It was also noted that the radiation was negative or nearly zero between 19 : 00 hours and 06 : 00 hours. This occurred due to the Earth's rotation, resulting in a change in solar angle with daylight hours. These results were in agreement with Al-Hilphy et al. (2014).

In the diurnal analysis, T_a and all radiative fluxes follow a similar pattern; however, radiation attains its peak at noon, two hours before T_a. This finding was consistent with Azam et al. (2014b). The daily variation of SWI, SWO, albedo, and R_n were also related with the change in cloud cover amount and their formation. A similar observation was illustrated by many authors (Cawkwell and Bamber 2002; Qian et al. 2007). The RH of the study site shows an almost consistent pattern throughout the day, similar results were observed by Soheb et al. (2018). On the other hand, u was maximum in the afternoon time (16 : 00 hours), and then it decreases as time progresses. The higher u in the afternoon is a common phenomenon for the valley region, as illustrated by numerous authors (Klok et al. 2005; Schaefer et al. 2020). As a consequence of an increasing u the T_a of the region was deficit during the mid-afternoon.

11.4.3 Sublimation/Resublimation

The sublimation/resublimation of the study site was calculated using the daily computed H_l. The positive value of H_l resulted in the resublimation process, while sublimation occurred when the H_l was negative. The sublimation process occurred throughout the year, with a higher value observed in summer and a lower was received in the winter. In contrast, the resublimation process only occurs in the winter season. This observation

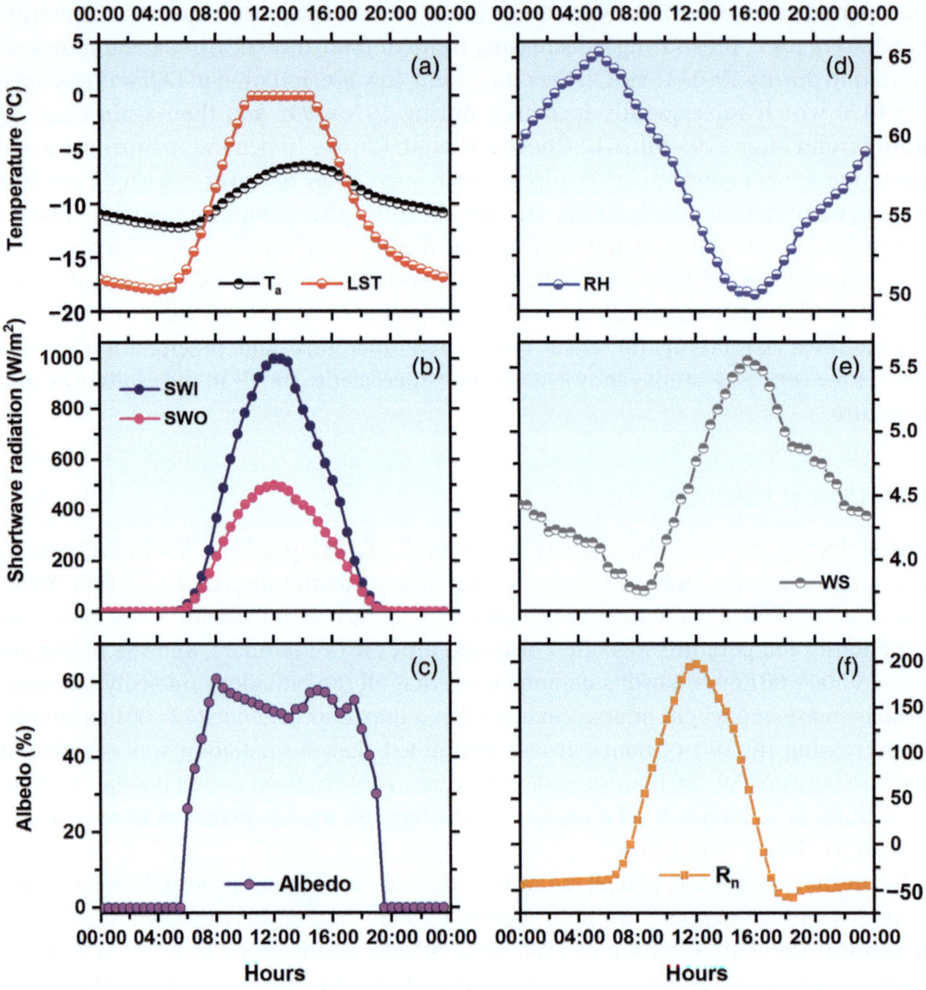

Figure 11.5 Mean diurnal variation of climatic and meteorological variables (a) Temperature, (b) Shortwave radiation, (c) Albedo, (d) Relative humidity, (e) wind speed, and (f) Net radiation at 5600 m a.s.l. elevation for the period 2012–2017.

shows that the region was highly exposed to radiation and causes a significant loss on the surface (Wani et al. 2021). During the observational period, the mean surface mass loss was 162.2 mm w.e. through sublimation, while a negligible amount of mass gain (1.3 mm w.e.) was observed through resublimation. Both sublimation and resublimation were found maximum in 2015/2016, while 2013/2014 and 2014/2015 were experienced the minimum sublimation and resublimation, respectively (Figure 11.6). The linear trend of sublimation and resublimation were increasing with a rate of 3.32 mm w.e./yr and 0.8 mm w.e./yr, respectively. This indicates that both the processes witnessed an increasing trend for the selected period; however, the rate of resublimation was relatively less increasing than sublimation, which means the glacier loses its mass at a higher rate than mass gain.

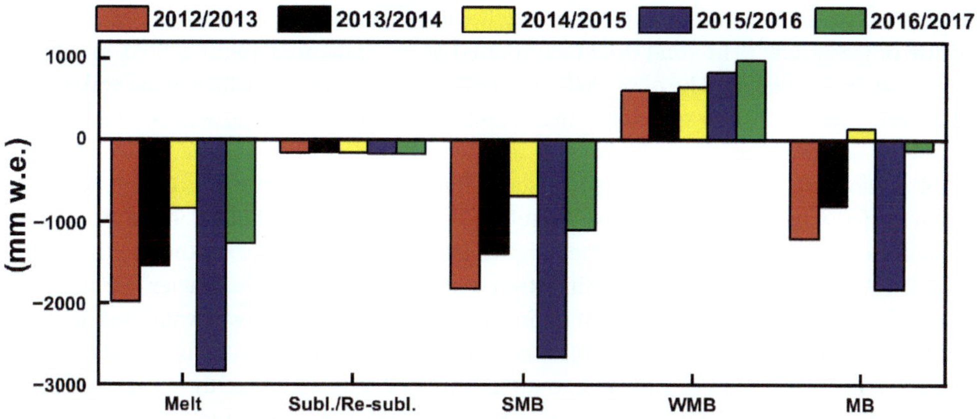

Figure 11.6 Interannual variation of melt, sublimation/resublimation, summer mass balance (SMB), winter mass balance (WMB), and overall annual mass balance of the study site during 2012–2017.

11.4.4 Mass Balance

The year-wise mass balance was measured by adding the summer mass balance (SMB) and winter mass balance (WMB) (Figure 11.6). Results reveal that the selected glacier was losing its mass for the entire observational period except for the year 2014/2015, which experienced a slightly positive mass balance (142.8 mm w.e.). This may have occurred because this year had lower temperatures and higher precipitation compared to other years. The glacier experienced a higher mass loss in 2015/2016 (−1826.5 mm w.e.) followed by 2012/2013 (−1208.2 mm w.e.), while the loss was comparatively less (−122.7 mm w.e.) during 2016/2017. This highest mass loss may have occurred due to higher annual temperatures. Besides, this region receives a higher amount of solid precipitation during winter but is not able to compensate for the melting caused during the summer season. This finding was well-matched with the observation highlighted by Soheb et al. (2020) for the Stok glacier, Ladakh range. Similarly, the year 2013/2014 received comparatively lower winter precipitation (588 mm w.e.) and relatively lower summer temperature, which may change the albedo and reduces the excessive melting; therefore, the study site witnessed moderate mass loss. Overall, the obtained mass balance result concludes that the precipitation occurring during the winter season and temperature variation in the summer season were the main contributors that influenced the glacier surface and regime of the mass loss/gain for the observational period.

11.4.5 Validation

The observed glacier melting from the SEB components was validated with the total melting measured on stake PCA4. This stake was mounted on the glacier surface near the AWS location. For comparing the SEB-based mass balance, the observation of the summer months of 2017 was used. The first observation of the snow stake started from 30 May 2017 to 15 September 2017 with an interval of 10–15 days. The density of accumulated snow was measured on 30 May 2017, which varied from 0.46 to 0.62 g/cm^3 with a mean value of 0.54 g/cm^3. The total thickness melted at the stake was about 2080.0 mm during the observation period. The observed total melting at the glacier surface using stake measurement was 1220.6 mm w.e. while the SEB-measured total melting was 1447.5 mm w.e., showing a difference of 226.9 mm w.e. for the selected period. On the other hand, the sublimation was observed on the snow/ice surface, accounting for a loss of 168.1 mm w.e. (13.2% of the total melt), and resublimation was observed at about 0.9 mm w.e. Result indicates that sublimation consumes higher latent heat compared to the melting of the glacier; therefore, this higher sublimation serves to perverse the glacier surface from excessive melting. The computed ablation using the SEB model was validated with the stake PCA4 measured ablation, suggesting a higher correlation of 0.87 for the period. This higher correlation between these two ablation rates indicates that the used model was robust, and their results provided a reliable mass balance across the glacier. The computed ablation (SEB) was overestimated by 18.6% than the measured ablation (Stake-based). Similarly, many authors have reported an overestimation of SEB against the stake measured ablation (Hay and Fitzharris 1988; Acharya and Kayastha 2018; Liang et al. 2018).

11.4.6 Comparison Between 2014/2015 and 2015/2016

The maximum (2015/2016) and minimum (2014/2015) melting years were considered to quantify the driving factors that cause the higher/lower glacier melting over a certain period of time (Figure 11.7). Results demonstrate that the RH and specific humidity almost follow a similar pattern in both years, while T_a and T_s experienced higher value in 2015/2016 relative to 2014/2015. The monthly distribution of RH suggests a higher value in 2014/2015 than in 2015/2016. Apart from this, it was also observed that the T_s was consistently zero with higher specific humidity in the summer season for the period 2015/2016. This finding also indicates that in 2015/2016, the Phuche glacier witnessed positive or zero temperatures (T_a and T_s) for a longer period (June to September) than 2014/2015 (3 months, i.e. July to September). On the other hand, the R_n and turbulent heat fluxes (H_s and H_l) were higher in 2015/2016 in comparison to 2014/2015 by a mean difference of 42.17 W/m^2, 9.0 W/m^2, and −1.72 W/m^2, respectively. Our findings illustrate that in 2015/2016, the melting period was longer, starting from May to the end of September and attaining maximum value during July. The maximum change in these two consecutive years was mainly occurred in the R_n component due to a change in albedo. This variation in albedo was predominately driven by the change in temperature. The increase in R_n further contribute to the variation of total surface energy available for melt (F_s). The monthly variation of F_s was nearly similar with the R_n, while the loss caused by sublimation was almost consistent in both the years. Overall, the comparison of these two contrasting years suggests that the maximum melting in 2015/2016 was mainly dominated by the change in albedo cause due to the higher positive temperature for a longer period.

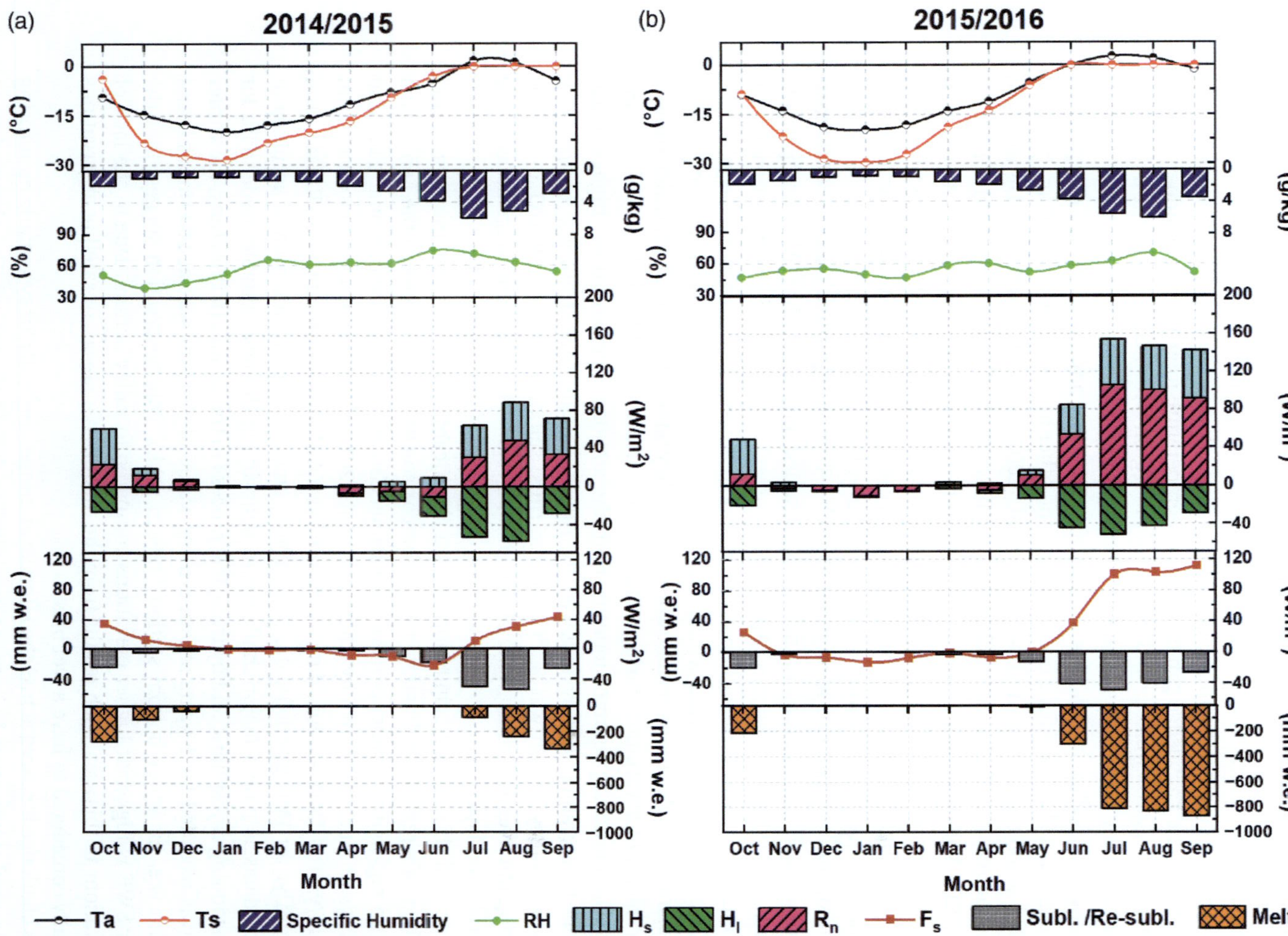

Figure 11.7 Comparison of meteorological, radiation, and energy fluxes components for the two extreme periods (a) 2014/2015 (minimum mass loss) and (b) 2015/2016 (maximum mass loss) during the observational period.

11.5 Discussion

11.5.1 Importance of Meteorological Variables at 5600 m a.s.l.

This paper presents a detailed investigation of meteorological variables measured at 5600 m a.s.l. in the cold-arid trans-Himalayas. This detailed information of higher elevation is crucial for filling the knowledge gap of moisture-temperature interplay in the upper reaches. Present knowledge on weather and climate of the region is formed based on the data generated at the valley bottom station at Leh (3256 m a.s.l.). Leh station in the valley and the ridge station at 5600 m a.s.l. is separated by around 13 km and lie in the line of sight. Leh station data is widely used for climate change and other hydro-meteorological analysis as it provided a long-term record (Bhutiyani et al. 2007, 2010; Thayyen and Dimri 2014; Mukhopadhyay and Khan 2015; Schmidt and Nüsser 2017; Chevuturi et al. 2018; Thayyen et al. 2020). Valley bottom station at Leh (3256 m a.s.l) experienced an annual range of temperature from −22 to 35.4°C, while the mountain ridge station at 5600 m a.s.l. experienced an annual range of −29.8 to 14.2°C, suggesting a significantly higher temperature lapse rate during summer months (8.5–9.0°C/km) (Thayyen and Dimri 2014). Due to this steep temperature lapse rate in summer (March—October), positive mean monthly temperature at 5600 m a.s.l. elevation mainly constrained only 2 months (JA) as compared to nine months at the valley bottom (April to November). Significant interannual variation in mean monthly temperature was observed in June followed by August and September (Cv, 0.87, 0.43, and 0.38), whereas winter months experienced relatively stable temperature regime. Interestingly, this pattern is reversed for valley bottom station at Leh, which showed higher interannual variation during winter months (November–March: Cv, 0.91–0.31) and little deviation during summer months. Both the stations experienced the highest mean monthly temperature in July followed by August. Mean annual average temperature (MAAT) at valley station was 8.7°C and the ridge station recorded a MAAT of −9.8°C reflecting the overall characteristic of the climate regime of valley and ridge areas of the cold-arid Himalaya.

Other than this, total precipitation plays a significant role in defining the climate and hydrology of a region. To illustrate this, a comparison was drawn between the Leh station data and regional assessments made by various climate models, observational data, and reanalysis products as shown by Chevuturi et al. (2018). Regional assessment varies from ~60 mm/yr for IMD station data, ~150 mm/yr (APHRO), ~200 mm/yr (TRMM), and ~300 mm/yr for IMD gridded data. The varying amount of precipitation and accuracy of gridded datasets pose a major challenge in quantifying the best suitable model data for understanding cryospheric and hydrological process in the region. In this study, the winter SWE was recorded for 5 years shows a mean value of 741 mm w.e. at 5600 m a.s.l. and 47.5 mm w.e. for valley bottom station (3256 m a.s.l.). This observation suggest that the winter SWE is higher at higher elevations compared to valley bottom. Due to the lack of observations in this area, such results from the models, reanalysis, or satellite observations generally considered to have a wet bias.

This challenge is shared across the Himalayan ranges and calls for systematic monitoring of valley and ridges areas to improve precipitation data in the region. The observed steep winter precipitation gradient is indicative of steep hydrological gradient existing in the system, providing critical information for assessing the high elevation cryospheric systems of the region. The study provides valuable information on prevailing weather conditions at a high elevation site in the Upper Indus basin, filling a critical data gap in the region. This data can be used to improve the evaluation of climate models/reanalysis data and cryospheric and hydrological modeling in the region.

11.5.2 Comparison of Measured Mass Balance with Other Studies

The impact of climatic variables on the glacier mass balance has been previously analyzed by many authors over the Himalayan glaciers (Yi and Sun 2014; Mandal et al. 2015; Soheb et al. 2020) at regional and basin scales; however, it is still not well understood for glaciers in cold-arid region in the Ladakh range. A study conducted by Wani et al. (2021) suggested that temperature, delayed snow occurrence, and elongated snow cover period in the upper Ganglass catchment, Ladakh, are influential drivers that controlled the SEB of the region. Therefore, we have addressed this topic by analyzing the surface melting through meteorological variables collected by AWS on the Phuche glacier (Ladakh range) and by field-based snow depth measurement for SEB modeling in detail. Based on data availability, we considered an extended study period (from 2012 to 2017) to compare the computed surface melting. Our findings add value to previous studies, suggesting that surface melting in this region is closely related to summer temperature and changes in albedo. A glaciological based mass balance study was conducted by Soheb et al. (2020) over the Stok glacier of Ladakh range, indicated that the glacier experienced a mass loss throughout the study period from 2014–2019, except in 2018/2019.

Soheb et al. (2020) observed the higher mass loss (-730.0 ± 380.0 mm w.e.) in 2015/2016 compared to other years; however, our study shows an even higher mass loss of -1826.5 mm w.e. in the same period. This variation in the analysis may have occurred because we used meteorological data and energy fluxes to measure mass balance at the glacier surface, which may provide the more efficient explanation of condition for glaciers in the cold and arid region. Furthermore, this discrepancy may also be attributed to the fact that the measured mass balance was observed at the ridge of the glacier surface, which can receive higher energy from the surrounding debris and rock, leading to reduced surface albedo and consequently more melting. The higher mass loss in 2015/2016 may be occurred due to significant changes in climatic variables compared to other years during the study period. Additionally, Allen et al. (2016) have indicated that 2016 marked the 37th consecutive year of alpine glacier retreat across the world based on preliminary data.

11.5.3 Comparison of SEB Components with Other Studies

This section compares the observed SEB components and their mass balance with previously published literature on glacier mass balance in the Himalayan region. According to previous studies (Zhang et al. 2013; Azam et al. 2014a; Acharya and Kayastha 2018; Patel et al. 2021a), R_n shows the highest contribution in F_s which was mainly controlled by the albedo and terrain properties. The present result also illustrates that the R_n and H_s were the main contributor in the total energy in the F_s, whereas the H_l was the greatest energy loss which produces a loss in double through latent heat and by process of sublimation. A similar contribution of H_l was revealed by Mölg and Hardy (2004). Numerous authors have also highlighted that the variation in R_n was highly dominant by the SWN under the clear sky condition during the summer season (Kayastha et al. 1999; Acharya and Kayastha 2018). Our findings also suggest that SWN obtained from this study shows an inverse relationship with albedo.

Over the study area, the precipitation during the summer season often occurs in liquid form; therefore, the albedo value becomes constant or slightly reduced over time. However, when the precipitation occurs in solid form during the winter season, the albedo abruptly changes and control the SWN for melting. Several authors have experienced similar patterns on high mountain glaciers (Klok and Oerlemans 2002; Paul and Kotlarski 2010). Our results also indicate that the region experienced negative H_l throughout the year, further exaggerated the amount of sublimation over the glacier surface. This observation was previously highlighted by Sun et al. (2014) over the Laohugou glacier No.12, western Qilsian mountains. They indicated that H_l was negative in the entire summer season (June–September), with slightly positive values observed on 2 days in July (2 and 3) for the year 2011. Similarly, Aizen et al. (2002) and Li et al. (2011) found that H_l was negative during their observational period for Xixibangma Glacier, and Keqicar Glacier, located in the South-central Tibetan Plateau and Southwest of the Tianshan, respectively. Our results also conclude that the region was predominately influenced by sublimation rather than the resublimation process. Similar experiences have been observed by numerous authors over High-Mountain Asia glaciers, which are mainly dominated by the process of sublimation (Aizen et al. 2002; Zhang et al. 2013; Sun et al. 2014).

11.5.4 Assessing the Implication of Mass Balance on the Socioeconomic Aspect

By gathering all the information related to the mass balance of this catchment and even in the Himalayan region, we can conclude that the glaciers are losing their mass at a significant rate, which varies from location to location. In the broader context, the study region falls under the Indus basin, predominately driven by summer snow and glacier melt. With this higher mass loss in the glaciers of the Indus basin, the streamflow is mainly influenced, which further increase the probability of extreme events occurring if any abrupt changes may happen in precipitation. Many studies have been indicated that the higher melting of the glacier can possibly be a reason for the occurrence of extreme events (Bhambri et al. 2018; Sattar et al. 2020). On the other hand, in the long-term aspect, the changing pattern of mass balance over the Himalayan glacier is able to create a water stress-like condition in the near future. This water stress largely impacts the livelihood of upstream and downstream populations, tourism of mountainous regions, and apple farming. Many reports and articles have previously drawn the attention of the research community worldwide to the mass balance of Himalayan glaciers and their implication for billions of people (Bajracharya and Shrestha 2011; Mukherji et al. 2015; Shrestha et al. 2015). Therefore, necessary preventative measure needs to be designed by decision-makers to sustain, restore, and balance the ecosystem.

11.6 Conclusion

The continuous glacier melting of the cold-arid trans-Himalayan region causes a significant concern toward the water supply and their associated future streamflow. Therefore, the energy exchange between the glacier surface and the atmosphere is needed to be studied. In the cold-arid region of the Ladakh range, the first-ever longest meteorological dataset was recorded at 5600 m a.s.l. from 2012 to 2017. The present study aimed to provide insight into the SEB components and their computed mass balance. Windy weather with low T_a and RH was found in the winter period, while warm T_a and higher RH with slightly lower u were noticed during the summer season (from June to September). However, the annual T_a and u show an increasing linear trend with a declining RH over the study period.

An energy balance model, relaying on the physical conditions of the region, was applied to comprehend the melting processes in the Phuche glacier based on their forcing variables (climatic and energy fluxes). The SWN was higher in summer relative to the winter season, and the annual distribution of SWN was minimum in 2014/2015 (98.6 W/m^2), while maximum in 2015/2016 (153.9 W/m^2) with an average value of 122.2 Wm^{-2}. Similarly, the highest value of R_n was observed in 2015/2016 followed by 2013/2014, while lowest in 2014/2015. On the other hand, the snow depth estimated winter precipitation was minimum (589 mm w.e.) in 2013/2014 and maximum (980 mm w.e.) in 2016/2017 with a mean value of 734 mm w.e. Our results conclude that the SWN, R_n, and sublimation/resublimation shows an increasing trend with varying rates, while albedo and H_l indicate a declining trend, and no trend was observed in H_s. In the summer season, R_n and H_s were the main contributors to the total energy available for melt (F_s).

Our findings suggest that the region witnessed a higher loss in 2015/2016 followed by 2012/2013, while the loss was comparatively less during 2016/2017. Among five consecutive years of observation, this study highlights the importance of summer temperature and winter season precipitation on glacier mass balance. Winter precipitation plays an important role in maintaining the albedo and compensating for summer melting. The computed SEB ablation was validated with stake measured total melting, indicating a correlation of 0.87 between them.

Furthermore, a comparison of estimated SEB with other glaciers of the same catchment and the Himalayan region was carried out, suggesting that SWN is the major energy source that controlled the surface melt. Our results indicated that the region was predominately influenced by sublimation than the resublimation process, which was well-matched with the other observations carried out over the High Mountain Asia. The comparison and validation of mass balance suggest that the obtained results are reliable as well as able to predict the future water supply and their management. In the future, a more detailed investigation of glacier mass loss and its implication on socioeconomic aspects is needed. Also, a necessary preventative measure and appropriate law will be designed by the decision-maker to sustain and restore the ecosystem.

Acknowledgment

We thank the Indian Institute of Technology (IIT) Roorkee, India, for providing all the necessary facilities, and National Institute of Hydrology (NIH) Roorkee, India, providing the in situ data and their support. Our acknowledgment extends to anonymous reviewers for their comments and suggestions have significantly improved the paper. We sincere thanks to Late Dr. Renoj J. Thayyen for his invaluable contribution in the chapter, including the establishment of the automatic weather station and organizing fieldwork for data collection and analysis. Special thanks go to the field assistants Mr. Kaustubh and Mr. Stabdan, and the porters for carrying the field instruments and providing necessary supports even in the harsh climatic conditions.

Funding

This study was funded by the Science Engineering Research Board (SERB), Department of Science & Technology, India, with the project no. SR/DGH/PK-1/2009 and EMR/2015/000887.

References

Acharya, A. and Kayastha, R.B. (2018). Mass and energy balance estimation of Yala Glacier (2011–2017), Langtang Valley, Nepal. *Water* 11: 6. https://doi.org/10.3390/w11010006.

Ahmad, S., Israr, M., Liu, S. et al. (2018). Spatio-temporal trends in snow extent and their linkage to hydro-climatological and topographical factors in the Chitral River Basin (Hindukush, Pakistan). *Geocarto International* 35: 1–24. https://doi.org/10.1080/10106049.2018.1524517.

Aizen, V.B., Aizen, E.M., and Nikitin, S.A. (2002). Glacier regime on the northern slope of the Himalaya (Xixibangma glaciers). *Quaternary International* 97–98: 27–39. https://doi.org/10.1016/S1040-6182(02)00049-6.

Al-Hilphy, A.R.S., Hassan, A., Majeed, G.H., and Yaqoob, G.Y. (2014). A practical study for incident solar radiation intensity in Basrah province and air temperature. *IOSR Journal of Engineering* 4: 13–21. https://doi.org/10.9790/3021-04531321.

Allen, S.K., Rastner, P., Arora, M. et al. (2016). Lake outburst and debris flow disaster at Kedarnath, June 2013: hydrometeorological triggering and topographic predisposition. *Landslides* 13: 1479–1491. https://doi.org/10.1007/s10346-015-0584-3.

Amatya, P.M., Ma, Y., Han, C. et al. (2015). Recent trends (2003–2013) of land surface heat fluxes on the southern side of the central Himalayas, Nepal. *Nature* 11: 957–970. https://doi.org/10.1038/175238c0.

Archer, D.R. and Fowler, H.J. (2004). Spatial and temporal variations in precipitation in the Upper Indus Basin, global teleconnections and hydrological implications. *Hydrology and Earth System Sciences* 8: 47–61.

Azam, M.F., Wagnon, P., Patrick, C. et al. (2014a). Reconstruction of the annual mass balance of Chhota Shigri glacier, Western Himalaya, India, since 1969. *Annals of Glaciology* 55: 69–80. https://doi.org/10.3189/2014AoG66A104.

Azam, M.F., Wagnon, P., Vincent, C. et al. (2014b). Processes governing the mass balance of Chhota Shigri Glacier (western Himalaya, India) assessed by point-scale surface energy balance measurements. *The Cryosphere* 8: 2195–2217. https://doi.org/10.5194/tc-8-2195-2014.

Bajracharya, S.R. and Shrestha, B. (2011). *The Status of Glaciers in the Hindu Kush–Himalayan Region*. Kathmandu: ICIMOD https://doi.org/www.icimod.org/publications.

Bandyopadhyay, D., Singh, G., and Kulkarni, A.V. (2019). Spatial distribution of decadal ice-thickness change and glacier stored water loss in the Upper Ganga basin, India during 2000–2014. *Scientific Reports* 9: 1–9. https://doi.org/10.1038/s41598-019-53055-y.

Banerjee, A. (2017). Brief communication: Thinning of debris-covered and debris-free glaciers in a warming climate. *The Cryosphere* 11: 133–138. https://doi.org/10.5194/tc-11-133-2017.

Basnett, S. and Kulkarni, A.V. (2019). Snow cover changes observed over Sikkim Himalaya. In: *Environmental Change in the Himalayan Region*. Springer Nature Switzerland (ed. A. Saikia and P. Thapa), 219–232. Cham: Springer International Publishing https://doi.org/10.1007/978-3-030-03362-0.

Bastiaanssen, W.G.M., Pelgrum, H., Wang, J. et al. (1998). A remote sensing surface energy balance algorithm for land (SEBAL), Part 1: formulation. *Journal of Hydrology* 212–213: 213–229. https://doi.org/10.1016/S0022-1694(98)00254-6.

Beg, A.A.F., Al-Sulttani, A.H., Ochtyra, A. et al. (2016). Estimation of evapotranspiration using SEBAL algorithm and Landsat-8 data—a case study: Tatra mountains region. *Journal of Geological Resource and Engineering* 4: 257–270. https://doi.org/10.17265/2328-2193/2016.06.002.

Bhambri, R., Misra, A., Kumar, A. et al. (2018). Glacier lake inventory of Himachal Pradesh. *Himalayan Geology* 39: 1–32.

Bhutiyani, M.R., Kale, V.S., and Pawar, N.J. (2007). Long-term trends in maximum, minimum and mean annual air temperatures across the Northwestern Himalaya during the twentieth century. *Climatic Change* 85: 159–177. https://doi.org/10.1007/s10584-006-9196-1.

Bhutiyani, M.R., Kale, V.S., and Pawar, N.J. (2010). Climate change and the precipitation variations in the northwestern Himalaya: 1866-2006. *International Journal of Climatology* 30: 535–548. https://doi.org/10.1002/joc.1920.

Blunden, J. and Boyer, T. (2020). State of the climate in 2020. *Bulletin of the American Meteorological Society* 102 (8): S1–S475. doi: 10.1175/2021BAMSStateoftheClimate.1.Special.

Brazel, A.J. and Marcus, M.G. (1991). July temperatures in Kashmir and Ladakh, India: comparisons of observations and general circulation model simulations. *Mountain Research and Development* 11: 75–86. https://doi.org/10.2307/3673568.

van den Broeke, M., van As, D., Reijmer, C., and van de Wal, R. (2004). Assessing and improving the quality of unattended radiation observations in Antarctica. *Journal of Atmospheric and Oceanic Technology* 21: 1417–1431. https://doi.org/10.1175/1520-0426(2004)021<1417:AAITQO>2.0.CO;2.

Cawkwell, F.G.L. and Bamber, J.L. (2002). The impact of cloud cover on the net radiation budget of the Greenland ice sheet. *Annals of Glaciology* 34: 141–149. https://doi.org/10.3189/172756402781817789.

Chevuturi, A., Dimri, A.P., and Thayyen, R.J. (2018). Climate change over Leh (Ladakh), India. *Theoretical and Applied Climatology* 131: 531–545. https://doi.org/10.1007/s00704-016-1989-1.

Dadic, R., Corripio, J.G., and Burlando, P. (2008). Mass-balance estimates for Haut Glacier d' Arolla, Switzerland, from 2000 to 2006 using DEMs and distributed mass-balance modeling. *Annals of Glaciology* 49: 22–26. https://doi.org/10.3189/172756408787814816.

Denby, B. and Smeets, C.J.P.P. (2000). Derivation of turbulent flux profiles and roughness lengths from katabatic flow dynamics. *Journal of Applied Meteorology* 39: 1601–1612. https://doi.org/10.1175/1520-0450(2000)039<1601:DOTFPA>2.0.CO;2.

Dharpure, J.K., Goswami, A., Patel, A. et al. (2021). Assessment of snow cover variability and its sensitivity to hydro-meteorological factors in the Karakoram and Himalayan region. *Hydrological Sciences Journal* 66: 2198–2215. https://doi.org/10.1080/02626667.2021.1985125.

Dharpure, J.K., Patel, A., Goswami, A., and Kulkarni, A.V. (2020). Spatiotemporal snow cover characterization and its linkage with climate change over the Chenab river basin, western Himalayas. *GIScience and Remote Sensing* 57: 1–25. https://doi.org/10.1080/15481603.2020.1821150.

Du, M., Zhang, M., Wang, S. et al. (2018). Near-surface air temperature lapse rates in Xinjiang, northwestern China. *Theoretical and Applied Climatology* 131: 1221–1234. https://doi.org/10.1007/s00704-017-2040-x.

Ebrahimi, S. and Marshall, S.J. (2016). Surface energy balance sensitivity to meteorological variability on Haig Glacier, Canadian Rocky Mountains. *The Cryosphere* 10: 2799–2819. https://doi.org/10.5194/tc-10-2799-2016.

Favier, V., Agosta, C., Genthon, C. et al. (2011). Modeling the mass and surface heat budgets in a coastal blue ice area of Adelie Land, Antarctica. *Journal of Geophysical Research: Earth Surface* 116: 1–14. https://doi.org/10.1029/2010JF001939.

Favier, V., Wagnon, P., Chazarin, J.P. et al. (2004). One-year measurements of surface heat budget on the ablation zone of Antizana Glacier 15, Ecuadorian Andes. *Journal of Geophysical Research – Atmospheres* 109: 1–15. https://doi.org/10.1029/2003JD004359.

Fujita, K. and Ageta, Y. (2000). Effect of summer accumulation on glacier mass balance on the Tibetan Plateau revealed by mass-balance model. *Journal of Glaciology* 46: 244–252. https://doi.org/10.3189/172756500781832945.

Gao, J., Yao, T., Masson-Delmotte, V. et al. (2019). Collapsing glaciers threaten Asia's water supplies. *Nature* 565: 19–21. https://doi.org/10.1038/d41586-018-07838-4.

Hay, J.E. and Fitzharris, B.B. (1988). A comparison of the energy-balance and bulk-aerodynamic approaches for estimating glacier melt. *Journal of Glaciology* 34: 145–153.

Heynen, M., Miles, E., Ragettli, S. et al. (2016). Air temperature variability in a high-elevation Himalayan catchment. *Annals of Glaciology* 57: 212–222. https://doi.org/10.3189/2016AoG71A076.

Hock, R. (2005). Glacier melt: a review on processes and their modelling. *Progress in Physical Geography* 3: 362–391.

Hoinkes, H. (1954). Measurement of ablation and heat balance on alpine glaciers. *Journal of Glaciology* 2: 497–501.

Jowhar, T.N. (2001). Geobarometric constraints on the depth of emplacement of granite from the Ladakh batholith, Northwest Himalaya, India. *Journal of Mineralogical and Petrological Sciences* 96: 256–264.

Kattel, D.B., Yao, T., Yang, W. et al. (2015). Comparison of temperature lapse rates from the northern to the southern slopes of the Himalayas. *International Journal of Climatology* 35: 4431–4443. https://doi.org/10.1002/joc.4297.

Kattel, D.B., Yao, T., Yang, K. et al. (2013). Temperature lapse rate in complex mountain terrain on the southern slope of the central Himalayas. *Theoretical and Applied Climatology* 113: 671–682. https://doi.org/10.1007/s00704-012-0816-6.

Kayastha, R.B., Ohata, T., and Ageta, Y. (1999). Application of a mass-balance model to a Himalayan glacier. *Journal of Glaciology* 45: 559–567.

Khan, A.A., Pant, N.C., Sarkar, A. et al. (2017). The Himalayan cryosphere: a critical assessment and evaluation of glacial melt fraction in the Bhagirathi basin. *Geoscience Frontiers* 8: 107–115. https://doi.org/10.1016/j.gsf.2015.12.009.

Klok, E.J., Nolan, M., and Van Den Broeke, M.R. (2005). Analysis of meteorological data and the surface energy balance of McCall Glacier, Alaska, USA. *Journal of Glaciology* 51: 451–461. https://doi.org/10.3189/172756505781829241.

Klok, E.J. and Oerlemans, J. (2002). Model study of the spatial distribution of the energy and mass balance of Morteratschgletscher, Switzerland. *Journal of Glaciology* 48: 505–518. https://doi.org/10.3189/172756502781831133.

Liang, L., Cuo, L., and Liu, Q. (2018). The energy and mass balance of a continental glacier: Dongkemadi Glacier in central Tibetan Plateau. *Scientific Reports* 8: 1–8. https://doi.org/10.1038/s41598-018-31228-5.

Li, J., Liu, S., Zhang, Y., and Shangguan, D. (2011). Surface energy balance of Keqicar Glacier, Tianshan Mountains China during ablation period. *Sciences in Cold and Arid Regions* 3: 0197–0205. https://doi.org/10.3724/SP.J.1226.2011.00197.

Mandal, A., Ramanathan, A., Farooq Azam, M. et al. (2015). Annual and seasonal mass balances of Chhota Shigri Glacier (benchmark glacier, Western Himalaya), India. *EGU General Assembly Conference Abstract* 17: 14078.

Marzeion, B., Leclercq, P.W., Cogley, J.G., and Jarosch, A.H. (2015). Brief communication: global reconstructions of glacier mass change during the 20th century are consistent. *The Cryosphere* 9: 2399–2404. https://doi.org/10.5194/tc-9-2399-2015.

Meena, H.M., Tewari, J.C., and Pandey, C.B. (2015). Influence of weather variation on cropping pattern of Leh district of Ladakh region. *Current World Environment: An International Research Journal of Environmental Sciences* 10: 489–493. https://doi.org/10.12944/cwe.10.2.13.

Mölg, T. and Hardy, D.R. (2004). Ablation and associated energy balance of a horizontal glacier surface on Kilimanjaro. *Journal of Geophysical Research: Atmospheres* 109: 1–13. https://doi.org/10.1029/2003JD004338.

Moore, R.D. (1983). On the use of bulk aerodynamic formulae over melting snow. *Nordic Hydrology* 14: 193–206. https://doi.org/10.2166/nh.1983.016.

Mukherji, A., Molden, D., Nepal, S. et al. (2015). Himalayan waters at the crossroads: issues and challenges. *International Journal of Water Resources Development* 31: 151–160. https://doi.org/10.1080/07900627.2015.1040871.

Mukhopadhyay, B. and Khan, A. (2015). A reevaluation of the snowmelt and glacial melt in river flows within Upper Indus Basin and its significance in a changing climate. *Journal of Hydrology* 527: 119–132. https://doi.org/10.1016/j.jhydrol.2015.04.045.

Murtaza, K.O. and Romshoo, S.A. (2017). Recent glacier changes in the Kashmir alpine Himalayas, India. *Geocarto International* 32: 188–205. https://doi.org/10.1080/10106049.2015.1132482.

Nicholson, L.I., Prinz, R., Mölg, T., and Kaser, G. (2013). Micrometeorological conditions and surface mass and energy fluxes on Lewis Glacier, Mt Kenya, in relation to other tropical glaciers. *The Cryosphere* 7: 1205–1225. https://doi.org/10.5194/tc-7-1205-2013.

Nüsser, M., Schmidt, S., and Dame, J. (2012). Irrigation and development in the upper Indus basin. *Mountain Research and Development* 32: 51–61. https://doi.org/10.1659/MRD-JOURNAL-D-11-00091.1.

Oerlemans, J., Anderson, B., Hubbard, A. et al. (1998). Modelling the response of glaciers to climate warming. *Climate Dynamics* 14: 267–274. https://doi.org/10.1007/s003820050222.

Oerlemans, J. and Klok, E.J. (2002). Energy balance of a glacier surface: analysis of automatic weather station data from the Morteratschgletscher, Switzerland. *Arctic, Antarctic, and Alpine Research* 34: 477. https://doi.org/10.2307/1552206.

Oke, T.R. (1987). *Boundary Layer Climates*, 2nde. Taylor & Francis e-Library.

Patel, A., Goswami, A., Dharpure, J.K. et al. (2021a). Regional mass variations and its sensitivity to climate drivers over glaciers of Karakoram and Himalayas. *GIScience and Remote Sensing* 58: 670–692. https://doi.org/10.1080/15481603.2021.1930730.

Patel, A., Goswami, A., Dharpure, J.K. et al. (2021b). Estimation of mass and energy balance of glaciers using a distributed energy balance model over the Chandra river basin (Western Himalaya). *Hydrological Processes* 35: 1–22. https://doi.org/10.1002/hyp.14058.

Paul, F. and Kotlarski, S. (2010). Forcing a distributed glacier mass balance model with the regional climate model REMO. Part II: downscaling strategy and results for two Swiss glaciers. *Journal of Climate* 23: 1607–1620. https://doi.org/10.1175/2009JCLI3345.1.

Pellicciotti, F., Brock, B., Strasser, U. et al. (2005). An enhanced temperature-index glacier melt model including the shortwave radiation balance: development and testing for Haut Glacier d' Arolla, Switzerland. *Journal of Glaciology* 51: 573–587.

Pratap, B., Sharma, P., Patel, L. et al. (2019). Reconciling high glacier surface melting in summer with air temperature in the semi-arid zone of Western Himalaya. *Water* 11: 1–18. https://doi.org/10.3390/w11081561.

Singh, P., Haritashya, U.K., Ramasastri, K.S., and Kumar, N. (2005). Prevailing weather conditions during summer seasons around Gangotri Glacier. *Current Science* 88: 753–760.

Qian, Y., Wang, W., Leung, L.R., and Kaiser, D.P. (2007). Variability of solar radiation under cloud-free skies in China: the role of aerosols. *Geophysical Research Letters* 34: 1–5. https://doi.org/10.1029/2006GL028800.

Radić, V. and Hock, R. (2011). Regionally differentiated contribution of mountain glaciers and ice caps to future sea-level rise. *Nature Geoscience* 4: 91–94. https://doi.org/10.1038/ngeo1052.

Sattar, A., Goswami, A., Kulkarni, A.V., and Emmer, A. (2020). Lake evolution, hydrodynamic outburst flood modeling and sensitivity analysis in the central himalaya: a case study. *Water* 12: 1–19. https://doi.org/10.3390/w12010237.

Schaefer, M., Fonseca-Gallardo, D., Fariás-Barahona, D., and Casassa, G. (2020). Surface energy fluxes on Chilean glaciers: measurements and models. *The Cryosphere* 14: 2545–2565. https://doi.org/10.5194/tc-14-2545-2020.

Schmidt, S. and Nüsser, M. (2017). Changes of high altitude glaciers in the Trans-Himalaya of Ladakh over the past five decades (1969–2016). *Geosciences* 7: 27. https://doi.org/10.3390/geosciences7020027.

Shafiq, M.U., Bhat, M.S., and Rasool, M. (2016). Variability of precipitation regime in Ladakh region of India from 1901–2000. *Journal of Climatology and Weather Forecasting* 4: 8–11. https://doi.org/10.4172/2332-2594.1000165.

Shrestha, A.B.A., Agrawal, N.N.K., Alfthan, B. et al. (2015). *The Himalayan Climate and Water Atlas: Impact of Climate Change on Water Resources in Five of Asia's Major River Basins*. Kathmandu: ICIMOD, GRID-Arendal and CICERO https://doi.org/10.1659/mrd.mm197.

Singh, P., Arora, M., and Goel, N.K. (2006). Effect of climate change on runoff of a glacierized Himalayan basin. *Hydrological Processes* 20: 1979–1992. https://doi.org/10.1002/hyp.5991.

Singh, P., Haritashya, U.K., and Kumar, N. (2007). Meteorological study for Gangotri Glacier and its comparison with other high altitude meteorological stations in central Himalayan region. *Nordic Hydrology* 38: 59. https://doi.org/10.2166/nh.2007.028.

Soheb, M., Ramanathan, A., Angchuk, T. et al. (2020). Mass-balance observation, reconstruction and sensitivity of Stok glacier, Ladakh region, India, between 1978 and 2019. *Journal of Glaciology* 66: 627–642. https://doi.org/10.1017/jog.2020.34.

Soheb, M., Ramanathan, A., Mandal, A. et al. (2018). Wintertime surface energy balance of a high-altitude seasonal snow surface in Chhota Shigri glacier basin, Western Himalaya. *Geological Society – Special Publications* 462: 155–168. https://doi.org/10.1144/SP462.10.

Srivastava, S. and Azam, M.F. (2022). Mass- and energy-balance modeling and sublimation losses on Dokriani Bamak and Chhota Shigri glaciers in Himalaya since 1979. *Frontiers in Water* 4: 1–21. https://doi.org/10.3389/frwa.2022.874240.

Sun, W., Qin, X., Du, W. et al. (2014). Ablation modeling and surface energy budget in the ablation zone of Laohugou glacier No. 12, western Qilian mountains, China. *Annals of Glaciology* 55: 111–120. https://doi.org/10.3189/2014AoG66A902.

Thayyen, R.J. and Dimri, A.P. (2014). Factors controlling Slope Environmental Lapse Rate (SELR) of temperature in the monsoon and cold-arid glacio-hydrological regimes of the Himalaya. *The Cryosphere Discussions* 8: 5645–5686. https://doi.org/10.5194/tcd-8-5645-2014.

Thayyen, R.J. and Dimri, A.P. (2018). Slope environmental lapse rate (SELR) of temperature in the monsoon regime of the western Himalaya. *Frontiers in Environmental Science* 6: 1–20. https://doi.org/10.3389/fenvs.2018.00042.

Thayyen, R.J., Dimri, A.P., Kumar, P., and Agnihotri, G. (2013). Study of cloudburst and flash floods around Leh, India, during August 4–6, 2010. *Natural Hazards* 65: 2175–2204. https://doi.org/10.1007/s11069-012-0464-2.

Thayyen, R.J., Singh, M.K., and Dimri, A.P. (2020). Visual evidence of constrained area of a cloudburst, 12 June 2018, Tirisha village, Nubra valley, Ladakh, India. *Journal of Climate Change* 6: 47–57. https://doi.org/10.3233/jcc200011.

Wani, J.M., Thayyen, R.J., Gruber, S. et al. (2020). Single-year thermal regime and inferred permafrost occurrence in the upper Ganglass catchment of the cold-arid Himalaya, Ladakh, India. *Science of the Total Environment* 703: 134631. https://doi.org/10.1016/j.scitotenv.2019.134631.

Wani, J.M., Thayyen, R.J., Ojha, C.S.P., and Gruber, S. (2021). The surface energy balance in a cold and arid permafrost environment, Ladakh, Himalayas, India. *The Cryosphere* 15: 2273–2293. https://doi.org/10.5194/tc-15-2273-2021.

Yadav, B.C., Thayyen, R.J., and Jain, K. (2020). Topoclimatic zones and characteristics of the upper Ganga basin. *International Journal of Climatology* 1: https://doi.org/10.1002/joc.6562.

Yi, S. and Sun, W. (2014). Evaluation of glacier changes in high-mountain Asia based on 10 year GRACE RL 05 models. *Journal of Geophysical Research – Solid Earth* 119: 2504–2517. https://doi.org/10.1002/2013JB010860.Received.

Zhang, G., Kang, S., Fujita, K. et al. (2013). Energy and mass balance of Zhadang glacier surface, central Tibetan Plateau. *Journal of Glaciology* 59: 137–148. https://doi.org/10.3189/2013JoG12J152.

12

Estimation of Ice Thickness and Glacier-Stored Water of the Teesta Basin, Eastern Himalayas, Sikkim, India

K. Shruti and M. Geetha Priya

CIIRC, Jyothy Institute of Technology, Bengaluru, India

12.1 Introduction

The frozen component of the Earth's climate system, known as the cryosphere, is distributed across the planet in diverse forms, including seasonal snow, permafrost, sea ice, glaciers, ice caps, lake ice, river ice, ice sheets, ice shelves, and atmospheric ice crystals (Vaughan et al. 2013). Glaciers represent a crucial cryosphere component, covering approximately 10% of the Earth's land surface. They serve as a vital freshwater reservoir by containing substantial water volumes and contribute to the nourishment of major river systems worldwide (Cuffey and Paterson 2010). They act as an important thermoregulator for the Earth's climatic system (Vaughan et al. 2013) and also act as an indicator of climatic changes (Cogley 2012; Dkhar and Tayal 2013). Glaciers are an immense body of ice created through the compression and recrystallization of snow. They advance slowly along a slope, driven by the force of its weight, and endure from one year to the next (Raina 2009). The ice that accumulates to form glaciers can originate from either the deposition of snowfall and ice crystals from the atmosphere or directly from the freezing of liquid on the glacier's surface (Cuffey and Paterson 2010).

The Himalayas constitute the largest glacier storage outside the poles, serving as the main source of freshwater (Tawde et al. 2017; Bolch et al. 2012) for prominent rivers in India, including the Brahmaputra, Indus, and Ganges, along with their tributaries. These rivers, in turn, provide vital water resources for numerous regions. Global warming is causing the Himalayan glaciers to retreat continuously and thus will have a notable change in glacier thickness and volume (Gopika et al. 2023). In recent years, remote sensing methods have become increasingly prevalent in glaciological investigations due to their convenience and ease. This shift is particularly crucial in regions with unpredictable and harsh climatic conditions that can impede traditional field investigations (Wang et al. 2021).

Himalayan glaciers are distinctive in their presence as they are in high-elevation tropical zones, and they have a crucial role in regulating circulation patterns and shaping weather systems of the nearby regions (Beniston 1997). These glaciers, covering an expansive area

Water Resources Management in Mountain Regions, First Edition. Edited by Virendra Bahadur Singh, Sughosh Madhav, Rakesh Kumar Gupta, Prerna Diwan, and Amit Kumar.

of 40 755 square kilometers, serve as crucial freshwater resources for mountain communities, particularly during the summer months (Bolch et al. 2012). This region is the source of many major rivers, including the Brahmaputra, Indus, and the Ganges. Therefore, the Himalayas are recognized as the freshwater tower of South Asia. Glacier retreat affects the water balance of the river systems (Kaser et al. 2010). In India, glaciers are confined to the latitude range of 29.75°–36.15° and the longitude range of 72°E–96°E (Raina et al. 2008; Sangewar and Shukla 2009).

The Himalayan climate is predominantly influenced by significant weather systems: (i) the Indian Summer Monsoon; (ii) the mid-latitude Western Disturbance (Owen et al. 1996). The summer monsoon brings moisture from the Bay of Bengal and the Indian Ocean and delivers an ample amount of precipitation (Chao and Chen 2001; Gadgil 2003; Yancheva et al. 2007). The Himalayas region is classified into three segments, namely the Western Himalaya, the Central Himalaya, and the Eastern Himalaya, depending on the prevailing influence of the two major weather systems (Bajracharya and Shrestha 2011). The Eastern Himalayas region is dominantly influenced by the Indian summer monsoon, and it receives precipitation mainly during the summers (Pandey and Venkataraman 2013).

12.1.1 Glacial Retreat Influence Due to Climate Change

Human activities influence the increase in greenhouse gas concentration (Maurer et al. 2019), which results in global warming (Trenberth 2018). This results in the accumulation of heat all over the world, which results in causing alterations in the Earth's climatic systems (Last et al. 1998). It exerts an extensive negative impact on the environment, socioeconomic aspects, and associated sectors through elevated temperatures, the thawing of snow and ice-covered areas, reduced precipitation, frequent occurrences of floods, droughts, and heatwaves, and a decline in cold waves (Narimisa and Narimisa 2018). The Himalayas act as sensitive hotspots to climate change (Kulkarni et al. 2007).

The escalating temperatures lead to the retreat of glaciers, the melting of snow and ice packs, downstream flooding, coastal erosion, and ultimately, a rise in sea levels. Elevated temperatures contribute to above-average summer melting of mountain glaciers, while reduced snowfall exacerbates the overall shrinkage of glaciers (Bajracharya et al. 2007). India is currently witnessing diverse climate change impacts, including fluctuations in the Indian monsoon, the melting of the Himalayan glaciers, coastal erosion, and an increased frequency of natural disasters (Balasubramanian and Dhulasi Birundha 2012). Any alterations in glacier volume can initiate events such as glacier lake outburst floods, landslides, and ice avalanches, posing significant challenges for mountain communities (Tawde et al. 2017; Prasad et al. 2019).

12.1.2 Use of Remote Sensing in Glacier Studies

Remote sensing entails gathering data regarding objects or regions on the Earth's surface without direct physical interaction. By employing different segments of the electromagnetic spectrum (EMS), remote sensing methods generate images of the Earth's surface (Aggarwal 2004). Monitoring glaciers in the Himalayan region using conventional methods proves to be challenging and labor-intensive because of the rugged mountain terrain and severe climatic conditions. In response to these problems, remote sensing techniques have become crucial for mapping and monitoring glaciers. Satellite images offer the advantage of covering a large number of glaciers per image over extended periods, providing valuable long-term data. With the use of remote sensing techniques, glacier mass balance, volume, and their changes concerning climatic fluctuations can also be monitored (Raup et al. 2014). An accurate assessment of these variables is essential for evaluating the correlation between glaciers and climate change. Various methods are employed for estimating glacier volume, and one such approach involves using scaling relations between glacier area and volume. This method allows for the estimation of glacier-stored water over extensive areas by analyzing satellite images, as exemplified by the work of Bahr (1997). As this method proves inappropriate for individual glaciers, advanced modeling techniques have been introduced by Farinotti et al. (2009). These methods depend on the observed connection between topographical features and glacier shear stress.

12.1.3 Laminar Flow Method

Glacier ice thickness is a crucial parameter for estimating glacier-stored water, as highlighted by Singh et al. (2019). Numerous approaches have been utilized in scholarly works for ice thickness estimation, including those proposed by Bahr (1997); Bahr et al. (2015); Farinotti et al. (2009); Frey et al. (2014); Huss and Farinotti (2012); and McNabb (2012). However, these techniques come with several uncertainties. Addressing this, Gantayat et al. (2014) introduced a method based on laminar flow. This approach estimates the spatial distribution of glacier ice thickness using slope, flow law of glacier ice, and surface velocity. The ice thickness equation is derived from the laminar flow equation of glacier ice and basal shear stress. Glacier surface velocity is estimated using the remote sensing technique proposed by Leprince et al. (2007).

12.1.4 Glacier-stored Water Estimates

Glacier volume serves as a crucial parameter, providing a quantification of the stored water within glaciers. Research on both glacier ice volume and surface area is imperative for gaining insights into the complex interactions between glaciers and climate, as emphasized in studies (Bahr 1997; Farinotti et al. 2009). There is a disparity among the quantities of glaciers, glacier area, and glacier volume. Numerous small glaciers collectively contribute only a minor portion to the overall glacier volume, while a few large glaciers hold a significant portion of the total stored glacier water (Helfricht et al. 2019).

In previous studies, Huss (2008) characterizes total glacier ice volume as the water stored by glaciers within a specific catchment. The ice thickness distribution over the glacier plays a crucial role in shaping the basin's hydrological characteristics. Glacier volume serves as the indicator of the amount of freshwater stored in glacier reserves. Assessing glacier ice thickness not only aids in determining the loss of glacier area but also contributes to predicting future changes in glacier geometry. Ice volume serves as a crucial criterion for estimating future glacier evolution, projecting runoff, conducting sea-level studies, and predicting impacts on the hydrological cycle, as highlighted by Frey et al. (2013). Commonly used methods for estimating glacier volume include volume-area relations (Chen and

Ohmura 1990; Bahr 1997), slope-dependent ice thickness estimations (Haeberli and Hoelzle 1995), and the use of ice thickness models (Farinotti et al. 2009; Huss and Farinotti 2012; Van Pelt et al. 2013).

A new method introduced by Bahr (1997) establishes a connection between unknown glacier ice volume and easily observable surface quantities. This method entails estimating glacier-stored water across extensive regions by leveraging the scaling relationship between surface area and ice volume. The volume-area power-law scaling method offers both practicality and a physically based approach to estimating glacier volume. In comparison to other volume-area scaling methods, like those by Chen and Ohmura (1990) and Macheret and Zhuravlev (1982), the method developed by Bahr (1997) is considered more accurate. Scaling methods are widely employed due to their simplicity and ease of use, further facilitated by the availability of measured and compiled data on glacier areas (Frey et al. 2014). Following the release of the Randolph Glacier Inventory (RGI) in 2012 (Pfeffer et al. 2014), a comprehensive dataset of global glacier coverage has become accessible. This availability of data allows easy calculation of glacier-stored water without the need for data extrapolation. RGI is a valuable resource for researchers and scientists studying glaciers and their associated water storage. Volume estimations for all glaciers and ice caps worldwide, relying on volume-area (V-A) relations, are provided by Marzeion et al. (2012), Grinsted (2013), and Radić et al. (2014). Moreover, Huss and Farinotti (2012) provided the initial assessment of the ice thickness distribution for all glaciers worldwide, utilizing the RGI. These investigations offer significant insights into the volume and distribution of ice thickness across global glaciers, utilizing data accessible in the RGI. The present study focuses on assessing the glacier-stored freshwater for the important glaciers of the Teesta basin in Sikkim, Northeastern India. Most of the Sikkim region is covered in the Teesta and Rangit river basins. The glaciers of the Sikkim region are majorly affected by the easterly disturbances (East Asian monsoon) during the peak ablation season, July–August (Agrawal and Tayal 2018; Hazra and Krishna 2019; Nela et al. 2023). The important glaciers of the Teesta river basin influence the water level in the Teesta river. Landsat 8 datasets are utilized for ice thickness estimations (using the laminar flow method), along with which total glacial stored volume is estimated using the pixel-based volume estimation method. The evaluation of glacier thickness and volume is an important aspect of understanding the rise in water levels (Nela et al. 2023), global climate, and the interaction of the glaciers with the nearby region (Gopika et al. 2023). It also helps in understanding many of the glaciological processes (glacier retreat, runoff mass balance, etc.) which help to understand the global warming effects (Agrawal and Tayal 2018; Immerzeel et al. 2012).

12.2 Study Area

Sikkim, nestled in the northeastern part of India, is bordered by West Bengal to the south, Tibet to the north, Nepal to the west, and Bhutan to the east. Its dimensions stretch approximately 114 kilometers from north to south and 64 kilometers from east to west. Geographically, the state lies within the coordinates of 88°00′51″E to 88°55′25″E in longitude and 27°07′04″N to 28°07′26″N in latitude, covering an area of 7096 km^2. The study region mentioned is significantly affected by the South Asian summer monsoon, where humid air from the Bay of Bengal flows into the Indian subcontinent due to the heating of the Tibetan plateau. This phenomenon leads to precipitation during this period, as supported by various sources discussing the impact and characteristics of the South Asian summer monsoon (Bhatt and Nakamura 2005). During this period, humid air from the Bay of Bengal flows into the Indian subcontinent because of the heating of the Tibetan plateau (Yanai et al. 1992; Benn and Owen 1998), resulting in precipitation. The Himalayas and the Tibetan plateau obstruct the monsoon winds, leading to about 77% of the summer season's precipitation occurring on the southern slopes of the Himalayas. This unique climatic pattern results in a glacier regime known as "summer accumulation," where ablation and accumulation processes occur simultaneously over the summer (Ageta and Higuchi 1984). Approximately 35–40% of the study region remains snow-covered in the summer, and the annual rainfall ranges from 2000 to 4000 mm (Gurung et al. 2011; Kumar et al. 2020). The Gangtok station demonstrates an upward trend in minimum temperature and a downward trend in maximum temperature, with rates of change of 0.036° $C\,yr^{-1}$ and 0.027° $C\,yr^{-1}$, respectively, observed over the years from 1961 to 2017 (Kumar et al. 2020).

The Sikkim Himalaya is home to Mount Kanchenjunga, the world's third-highest peak, and is one of India's biodiversity hotspots, hosting a diverse range of temperate, alpine, tundra, and tropical plant species. The ecological flora and fauna depend largely on glacier meltwater (freshwater). Sikkim state features steep slopes, accounting for over 43% of its total land area, with rugged terrain and escarpments. Hydrologically, streams have sculpted river valleys in the western and southern parts of Sikkim, giving rise to two significant river systems, namely the Teesta and Rangit, which originate from high-altitude glacierized regions. The predominantly rural population (almost 75% of the total) relies heavily on climate-sensitive sectors, such as agriculture and tourism, for livelihoods. The Sikkim region is home to over 100 glaciers, playing a crucial role in sustaining the area. Therefore, studying the climate in this region is significant for understanding future glaciological status (Kumar et al. 2020). The present study specifically focuses on assessing glacier thickness and freshwater storage in six important glaciers within the Teesta basin, a tributary to the Brahmaputra river. The important glaciers, North Lhonak, Lhonak, South Lhonak, Chengsang, and Zemu Onglang Thung, are illustrated in Figure 12.1, and their details are provided in Table 12.1.

Table 12.1 Details of important glaciers.

Glaciers	RGI glacier id	Area (km^2)	Elevation range (m.a.s.l)	Central latitude and longitude	Slope
North Lhonak	RGI60-15.02911	7.5	5418–6754	27°56′41″ and 88°09′13″	12.4
Lhonak	RGI60-15.02908	6.5	5453–6661	27°55′08″ and 88°08′14″	18.5
South Lhonak	RGI60-15.02698	11.7	5202–7409	27°53′52″ and 88°09′18″	23.4
Chengsang	RGI60-15.03150	9.8	5409–6490	27°48′47″ and 88°12′43″	12.6
Zemu	RGI60-15.02703	68.6	4187–7134	27°45′57″ and 88°12′06″	14.8
Onglang Thung	RGI60-15.03200	2.5	4491–6299	27°35′05″ and 88°08′43″	22.7

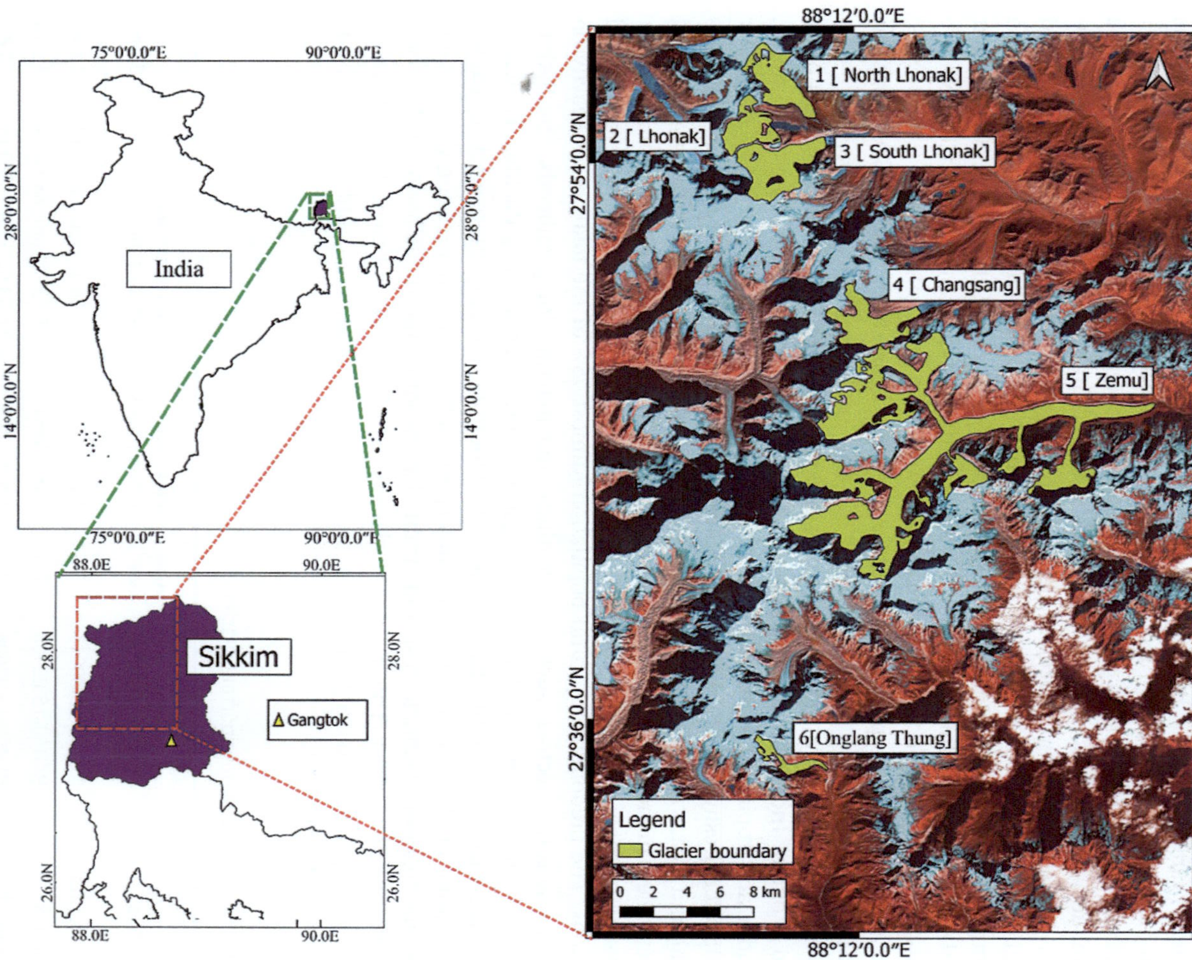

Figure 12.1 Study area map demarcating the important glaciers.

12.3 Data Used

Landsat 8 was launched on 11 February 2013, using an Atlas-V 401 rocket from Vandenberg Air Force Base in California. The launch utilized an extended payload fairing (EPF) produced by United Launch Alliance, LLC. This collaborative project is between USGS (United States Geological Survey) and NASA (National Aeronautics and Space Administration). The satellite consists of two sensors: OLI (Operational Land Imager) and TIRS (Thermal Infrared Sensor), which provide coverage of the Earth. The OLI is equipped with a push-broom sensor featuring a 4-mirror telescope and 12-bit quantization. It gathers data across various spectral bands including Visible (VIS), Near Infrared (NIR), Short Wave Infrared (SWIR), Coastal/Aerosol, and Cirrus, in addition to a panchromatic band. TIRS collects data in the TIR (thermal infrared) region. The sensors capture data in the wavelength range of 0.43–12.51 μm, with a revisit period of 16 days. The TIRS data is aligned with OLI data to generate radiometrically, geometrically, and terrain-corrected quantized (12-bit) data products. Landsat 8 captures 725 scenes per day, with each scene covering an area of 185 kilometers across the track and 180 kilometers along the track. This high acquisition rate increases the likelihood of acquiring cloud-free data. The OLI level 2 collection 2 panchromatic images (15 m) were obtained from (https://earthexplorer.usgs.gov/) and processed for the study. Details of the satellite dataset used are provided in Table 12.2.

The study employs two datasets (master and slave) with a one-year interval to estimate glacier surface ice flow velocity, subsequently deriving ice thickness using the laminar flow model. The Eastern Himalayas region experiences accumulation from July to September, making obtaining cloud-free data challenging. The Sikkim region, characterized by low precipitation during winters, prefers imagery from November to January, as Sikkim glaciers are monsoon-driven, resulting in less/no cloud cover (Basnett et al. 2013).

The proposed study utilizes RGI version 6.0 and ASTER (Advanced Spaceborne Thermal Emission and Reflection Radiometer) GDEM (Global Digital Elevation Model). The RGI v6.0 glacier boundaries were obtained for the South Asia East region under the GLIMS (Global Land Ice Measurement from Space) initiative, which contains glacier outlines larger than 0.01 km^2 (https://www.glims.org/RGI/). The ASTER GDEM version 3 of 30 m spatial resolution (1 arc second) was obtained from https://earthexplorer.usgs.gov/.

12.4 Methodology

A variety of techniques have been developed and utilized in optical image analysis for studying the cryosphere and glaciology. These methods, including Fourier domain cross-correlation (Rolstad et al. 1997), phase correlation (Avouac et al. 2006), normalized

Table 12.2 List of dataset utilized in the study.

Landsat 8 data	Date of acquisition	Path/row
LC08_L1TP_139041_20211212_20211222_02_T1_B8	12 December 2021	Path-139 & Row -41
LC08_L1TP_139041_20221215_20221223_02_T1_B8	15 December 2022	Path-139 & Row -41

cross-correlation (Berthier et al. 2003; Bindschadler et al. 1996; Kääb 2002), least-squares matching (Kaufmann and Ladstädter 2003), and orientation correlation (CCF-O) (Haug et al. 2010), play a crucial role in understanding glacier dynamics.

However, it has been noted that CCF-O and phase correlation, as implemented in COSI-Corr, are considered more robust for global-scale mapping and monitoring of glacier velocities (Heid and Kääb 2012). Numerous image processing software tools are available as open-source options, such as CIAS (Computer Image Analysis Software), IMCORR (Image Correlation), COSI-Corr (Co-registration of Optically Sensed Images and Correlation), and IMGRAFT (Image GeoRectification and Feature Tracking) and utilize these principles, but their application is hindered by time-consuming processing. Nonetheless, COSI-Corr has demonstrated superior efficiency in terms of processing duration and output variables (Jawak et al. 2018). Additionally, COSI-Corr offers various pre- and post-processing tools, providing further advantages (Ayoub et al. 2009).

In this investigation, the COSI-Corr software package (Leprince et al. 2008) was employed to analyze surface displacement. The COSI-Corr module is accessible for free and can be obtained from http://www.tectonics.caltech.edu. COSI-Corr has become widely adopted for monitoring changes in Earth's surface (Ayoub et al. 2009) and for glacier monitoring (Scherler et al. 2008). This software enables accurate co-registration, orthorectification, and correlation at the sub-pixel level of optical images in an intuitive interface. The COSI-Corr algorithm operates on single-band grayscale images and provides two correlation methods for analysis: (i) statistical correlation and (ii) frequency correlation. In this investigation, the choice was made to utilize the panchromatic band of Landsat 8. The frequency correlation, a Fourier-based method, was selected as it is deemed more accurate compared to the statistical correlation (Ayoub et al. 2009). This correlation algorithm is notably sensitive to noise and is best suited for use with high-quality optical images. On the other hand, the statistical correlation aims to maximize the absolute value of the correlation coefficient, presenting a coarser but more robust alternative to the frequency correlation algorithm. It is recommended for correlating optical images characterized by significant noise that may yield unsatisfactory results. For glacier displacement measurement in this study, the frequency correlation algorithm was employed.

Precise co-registration of input images is essential due to the correlation algorithm's sensitivity to misregistration errors. Moreover, the presence of image noise, resulting from sensor noise and variations and stripping in illumination caused by clouds and shadows, can adversely impact displacement results (Scherler et al. 2008). In consideration of computational efficiency, the frequency domain phase correlation algorithm was employed in COSI-Corr. Additionally, the Fourier-based algorithm integrated into COSI-Corr enhances traditional peak correlation (Avouac et al. 2006) methods, achieving accuracy levels as fine as 0.02 pixel.

The frequency-based correlation algorithm used in COSI-Corr necessitates several preliminary parameter configurations, which include:

1) **Window Size**: This specifies the size of the sliding windows.
2) **Robustness Iteration**: This parameter specifies how frequently the frequency mask size should be recalculated during each measurement.

3) **Frequency Mask Size**: The dimension, measured in pixels, of the area to be correlated.
4) **Step**: This parameter sets the distance, in pixels, between consecutive sliding windows.
5) **Mask Threshold**: This involves masking the frequency based on the amplitude of the logarithmic cross-spectrum.

These parameter settings play a pivotal role in optimizing the performance of the frequency correlation algorithm and ensuring accurate displacement measurements.

The methodology involves applying the frequency-based correlation technique to a pair of optical images (referred to as Master and Slave) captured at different time intervals (1 year apart). After the orthorectification process, a frequency correlator was used to correlate the two selected images. This approach is aimed at achieving precise image alignment and correlation, allowing for the accurate analysis of changes between the two time periods. In this study, the window sizes were configured as 32×32, and the step was set to 2×2. Consequently, these parameter settings led to a pixel resolution of 30×30 m. This selection of window size and step size is pivotal in achieving the desired level of precision in the displacement measurements. To enhance the precision of correlation outcomes by minimizing the impact of high-frequency noise, the frequency mask threshold was set at 0.9, and the robustness iteration was established at 2. These parameter settings were consistently applied to the entire dataset to ensure result consistency across the analysis. This approach aims to maintain a balance between noise reduction and result stability throughout the dataset.

The COSI-Corr correlation algorithm generates east-west displacement, north-south displacement, and signal-to-noise (SNR) images. The SNR image serves the purpose of evaluating the quality of the displacement images. Pixels having SNR values greater than or equal to 0.9 were used to calculate displacement (Disp) with Euclidean norms (Eq. 12.1). To determine surface velocity (Vs), the displacement was divided by the time interval (t) between two consecutive images (Eq. 12.2).

$$Disp = \sqrt{(NS)^2 + (EW)^2} \quad (12.1)$$

$$Vs = \frac{Disp}{t} \quad (12.2)$$

The displacement obtained corresponds to the highest correlation values and motion of the features between the two images. As images of two successive years with approximately one-year intervals are taken, the resultant (estimated displacement) corresponds to the annual glacier surface velocity. Using glacier boundaries obtained from RGI 6.0, the ice surface flow velocity of the important glaciers is clipped. The glacier boundaries were modified according to the satellite images and Google Earth imagery.

12.4.1 Ice Thickness Estimation Using Laminar Flow Model

The glacier's ice thickness distribution was calculated from the computed glacier surface ice flow velocity and slope, utilizing the laminar flow model as described in Eqs. 12.3 and 12.4.

$$T = \sqrt[4]{\frac{1.5 * Vs}{Af^3(\rho g \sin a)^3}} \tag{12.3}$$

$$Vs = Vb + \frac{2c}{G+1}\tau b^G T \tag{12.4}$$

where *Vs* is surface velocity, T is ice thickness (m), A is creep parameter (3.24×10^{-24} Pa^{-3} s^{-1}), g is acceleration due to gravity (9.8 m s^{-2}), ρ is ice density (900 kg m^{-3}), *f* is shape factor (0.8), *a* is slope angle, Vb is basal velocity, T is ice thickness, c is the creep parameter (3.24×10^{-24} Pa^{-3} s^{-1} for temperate glaciers), τb is basal stress, and G represents the Glen's flow law exponent, which is taken as 3. The creep parameter varies based on factors such as impurity content of glaciers, temperature, grain size, and fabric, while the shape factor represents the ratio of driving stress to basal stress along the glacier. The ice thickness equation described in Eq. 12.3 was formulated (Gantayat et al. 2014) and derived from the principles of laminar flow within glacier ice.

The freshwater stored volume (Vol) for all the six important glaciers of Sikkim was calculated using the volume-area scaling method using Eq. 12.5.

$$Vol = \Sigma(Tp * Ap) \tag{12.5}$$

where *Tp* represents the thickness values of each pixel and *Ap* represents the pixel area. A detailed step-wise methodology chart is shown in Figure 12.2.

12.5 Results

The ice thickness and total glaciated volume of six glaciers in the Teesta river basin were determined using the laminar flow method, with pixel-based volume estimation at each pixel. Surface velocity was calculated using the normalized cross-correlation of satellite images (Landsat 8) of December 2021–2022 which gave velocity in the range of 12.2–26.5 m/yr, with an average mean velocity of 17.1 m/year. Correlation results showed that Zemu glacier has the highest mean velocity of 26.5 m/year, while North Lhonak has the least maximum of 12.2 m/year.

The distribution of ice thickness across the surveyed Sikkim glaciers, depicted in Figure 12.3, indicates an irregular concentration throughout the region. Ice thickness estimates, derived using the laminar flow model, incorporate considerations of surface velocity and slope. Within the accumulation region, thickness values are notably high and gradually diminish toward the snout. Maximum ice thickness distribution was observed in the central regions of the main trunks of the glaciers.

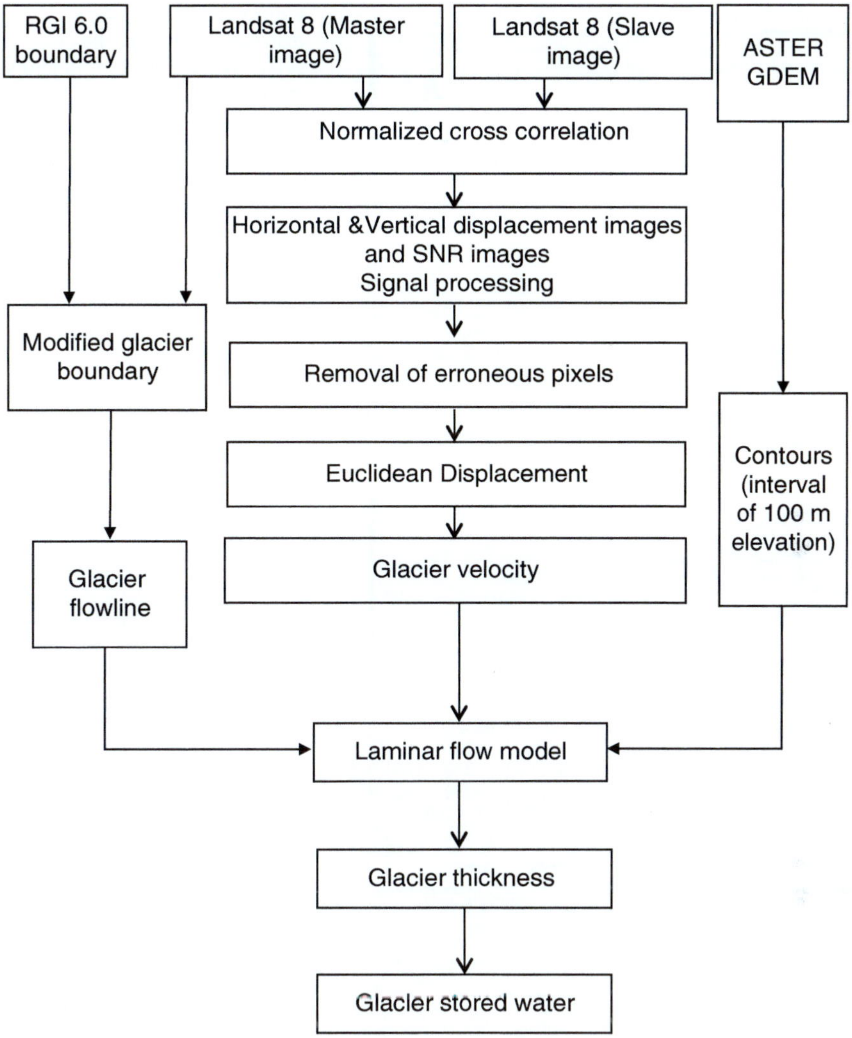

Figure 12.2 Methodology flow chart.

Zemu glacier stands out with the maximum observed ice thickness, reaching 337 meters, highlighting a concentration of ice in its central regions. In contrast, Lhanok Glacier recorded the least maximum thickness at 221 meters. The average ice thickness of Zemu spans from 51 to 162 meters. Notably, all glaciers under study exhibit thicknesses ranging from 100 to more than 300 meters along the central flowline in the accumulation region and over 60 meters near the snout. Furthermore, glaciers with smaller areal extents tend to display lower thickness values compared to those with larger areal extents (Gantayat et al. 2014). Detailed surface ice flow velocity, ice thickness, and total glacier volume for the six glaciers within the Teesta basin, Sikkim, are provided in Table 12.3.

The computation of glacier volume through the pixel-based volume estimation method involves the utilization of pixelated ice thickness data. The stored water within the studied

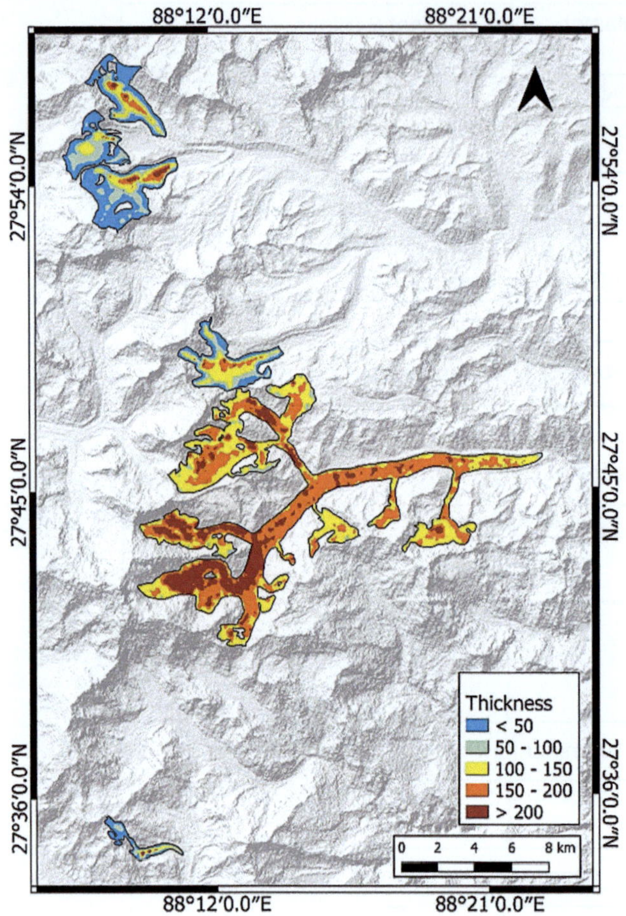

Figure 12.3 Thickness distribution map of the selected six important glaciers in Sikkim.

Table 12.3 Surface ice flow velocity, ice thickness, and volume of the glaciers.

Glacier	Mean velocity (m/year)	Maximum thickness (m)	Average thickness (m)	Total glacier volume (km^3)
North Lhonak	12.2	295	62	0.5
Lhonak	15.9	221	51	0.3
South Lhonak	21	235	61	0.7
Chengsang	13.4	235	79	0.7
Zemu	26.5	337	162	12.2
Onglang Thung	13.7	281	51	0.1

individual glaciers exhibits a range from 0.1 to 12.2 km^3. Cumulatively, the total computed freshwater volume for all six glaciers is approximately 14.5 km^3. The calculation of glacier-stored water reveals that Zemu glacier has the largest stored water of 12.2 km^3, while Onglang Thung exhibits the lowest stored water volume of 0.1 km^3.

Larger glaciers store huge amounts of water compared to smaller glaciers, as the thickness and overall extent vary widely. The big glaciers are significant in total glacier-stored water. More than 80% of water is stored in Zemu glacier (the largest glacier), which occupies an area of more than 75 km^2. Intense global warming leads to glacial retreat, resulting in a decrease in glacier size and total glacier-stored water. Glacier retreat causes changes in the runoff patterns, affecting socioeconomic activities downstream which helps in hydrological planning (Gopika et al. 2023).

12.6 Discussion

Various evaluation parameters, such as ice thickness, area, and volume, serve as crucial indicators for assessing the health of glaciers, offering valuable insights for future policy and decision-making processes. Glacier volume estimates directly relate to the overall quantity of freshwater held within these glacial systems (Huss et al. 2008). The determination of glacier ice thickness holds particular importance for projecting sea-level rise, predicting runoff patterns, and forecasting the future condition of glaciers (H. Frey et al. 2014). This investigation has unveiled significant variations in surface ice flow velocity, ice thickness distribution, and the volume of freshwater stored in individual glaciers within the basin, as outlined in Table 12.3. These fluctuations result from a combination of factors, including local terrain, the presence of debris on the glaciers, slope gradient, and the distinct shapes of the glaciers (Scherler et al. 2008, 2011). Notably, the largest glacier, Zemu, exhibited the highest surface ice flow velocity during the study period, as indicated by individual glacier assessments. This aligns with findings from a study by Sun et al. (2017) for the study period 1999–2003, which demonstrated that larger glaciers tend to move more rapidly than smaller ones. Additionally, a positive correlation between glacier extent and surface velocity was observed in a separate study conducted by Singh et al. (2020) for the study period 2009–2015 (with four image pairs within the study period and having an interval of 14 days between them).

The estimation of glacier ice thickness was conducted using the laminar flow equation, which accounts for both slope gradient and surface velocity of each glacier. A study by Remya et al. (2019) calculated Zemu's ice thickness values to be 439 ± 57 meters based on Landsat 8 data from the USGS during 2015. However, in the year 2022 (our study), these values have decreased to 337 meters (refer to Table 12.3). Also, Bhattacharjee and Pandey (2023) in a separate study for the year 2020, utilized GRD data from the Alaska Satellite Facility (ASF) to assess the Zemu glacier's ice thickness using a similar method. These research findings indicated spatial variability in Zemu glacier's thickness, ranging from 80 ± 9.6 meters to 160 ± 19.2 meters near the snout and from 240 ± 28.8 meters to 320 ± 38.4 meters in the upper accumulation zone. Whereas, present study reveals that the ice thickness values for Zemu glacier range from 60 meters near the snout to more than 300 meters in the upper accumulation zone. This study further demonstrates an association between glacier velocity and depth, showcasing inverse properties. This phenomenon is attributed to drag forces involved in glacier movement and variations in bed topography (Linsbauer et al. 2016; Prasad et al. 2019).

To assess glacier mass loss, the study relies on estimating the entire glacier volume. The study underscores the significance of larger glaciers in gauging the total water volume for a given region, as they retain more freshwater compared to their smaller counterparts. In the Satluj basin, for instance, the majority of larger glaciers store 56% of the glacial water, while approximately 95% of the remaining smaller glaciers collectively store about 0.1 gigatonnes of ice, according to a study by Prasad et al. (2019) for the period 1984–2013. Additionally, Helfricht et al. (2019) highlighted variations in the number, area, and volume of glaciers for the study period 1969–2006. Similar findings have been reported in various studies across different fields, reinforcing the interconnectedness and linear relationship between the glacier volume and area. The results of Gopika's (2020) study on the Bhaga basin for the study duration of 2013–2019 align with other studies, demonstrating a comparable linear relationship between glacier area and volume.

Corroborating evidence comes from volume assessments in the Satluj basin by Prasad et al. (2019) and in the Chandra basin by Pandit and Ramsankaran (2020). The Himalayas' complex topography and diverse climate significantly influence the quantity and seasonal variations in glacier melt. Glacier runoff emerges as a crucial water source for Himalayan rivers, playing a pivotal role in sustaining both hydrological and socioeconomic activities. In the Himalayan region, glacier recession serves as a prominent indicator of climate change. A precise evaluation of stored water becomes imperative for numerous studies about climatology, hydrology, and sea-level dynamics.

12.7 Conclusion

The study employed a laminar flow equation and the pixel-based volume estimation scaling method to calculate the thickness and volume of six glaciers. It is noteworthy that these glaciers exhibit varying thicknesses across the entire region, with maximum thicknesses occurring in the accumulation region and gradually decreasing toward the snout. Obtaining field measurements in such challenging terrains is arduous, underscoring the importance of remote sensing in analyzing glacier parameters. The thickness and volume of glaciers play a crucial role in hydrological modeling for the regions and are integral for future planning and development efforts. The Lhonak glacier exhibited the lowest average and maximum thickness values of 51.3 and 221 meters, respectively, with a mean velocity of 15.9 km^3. In contrast, Zemu glacier, with a maximum ice thickness of 337 meters, stored the highest glacial volume at 12.2 km^3. The observed maximum ice thickness values for the studied glaciers ranged from 221 to 337 meters. The substantial variability in these values is attributed to differences in glacier area and slope variations every 100 meters. Understanding the hydrological dynamics of glaciers through such studies provides valuable insights into their current status, which is crucial for informed decision-making in future planning and development.

Acknowledgments

The authors gratefully acknowledge the support rendered by Prazim Trading and Investment Company Private Limited (PTICL), Bengaluru, and Dr. Krishna Venkatesh, Director, CIIRC, Jyothy Institute of Technology, Bengaluru. Additionally, the authors are thankful to the USGS Earth Explorer for the valuable datasets provided, which greatly contributed to the work.

References

Ageta, Y. and Higuchi, K. (1984). Estimation of mass balance components of a summer-accumulation type glacier in the Nepal Himalaya. *Geografiska Annaler: Series A, Physical Geography* 66 (3): 249–255.

Aggarwal, S. (2004). Principles of remote sensing. *Satellite Remote Sensing and GIS Applications in Agricultural Meteorology* 23 (2): 23–28.

Agrawal, A. and Tayal, S. (2018). Preliminary study on estimation of volume of eastern Himalayan glaciers using remote sensing methods. *Journal of Climate Change* 4 (1): 13–21. https://doi.org/10.3233/jcc-180002.

Avouac, J.-P., Ayoub, F., Leprince, S. et al. (2006). The 2005, Mw 7.6 Kashmir earthquake: sub-pixel correlation of aster images and seismic waveforms analysis. *Earth and Planetary Science Letters* 249 (3–4): 514–528.

Ayoub, François, Sébastien Leprince, and Lionel Keene. 2009. "User's guide to COSI-CORR Co-registration of optically sensed images and correlation." California Institute of Technology: Pasadena, CA, 1–38.

Bahr, D.B. (1997). Width and length scaling of glaciers. *Journal of Glaciology* 43 (145): 557–562.

Bahr, D.B., Tad Pfeffer, W., and Kaser, G. (2015). A review of volume-area scaling of glaciers. *Reviews of Geophysics* 53 (1): 95–140.

Bajracharya, S.R., Mool, P., and Shrestha, B.R. (2007). *Impact of Climate Change on Himalayan Glaciers and Glacial Lakes: Case Studies on GLOF and Associated Hazards in Nepal and Bhutan*, vol. 12. International Centre for Integrated Mountain Development Kathmandu.

Bajracharya, S.R. and Shrestha, B.R. (2011). *The Status of Glaciers in the Hindu Kush-Himalayan Region*. International Centre for Integrated Mountain Development (ICIMOD).

Balasubramanian, M. and Dhulasi Birundha, V. (2012). Climate change and its impact on India. *IUP Journal of Environmental Sciences* VI: 31–46.

Basnett, S., Kulkarni, A.V., and Bolch, T. (2013). The influence of debris cover and glacial lakes on the recession of glaciers in Sikkim Himalaya, India. *Journal of Glaciology* 59 (218): 1035–1046. https://doi.org/10.3189/2013JoG12J184.

Beniston, M. (1997). Variations of snow depth and duration in the swiss alps over the last 50 years: links to changes in large-scale climatic forcings. *Climatic Change* 36 (3–4): 281–300.

Benn, D.I. and Owen, L.A. (1998). The role of the Indian summer monsoon and the mid-latitude westerlies in Himalayan glaciation: review and speculative discussion. *Journal of the Geological Society* 155: 355–363.

Berthier, E., Raup, B., and Scambos, T. (2003). New velocity map and mass-balance estimate of Mertz Glacier, East Antarctica, derived from landsat sequential imagery. *Journal of Glaciology* 49 (167): 503–511.

Bhatt, B.C. and Nakamura, K. (2005). Characteristics of monsoon rainfall around the Himalayas revealed by TRMM precipitation radar. *Monthly Weather Review* 133 (1): 149–165. https://doi.org/10.1175/MWR-2846.1.

Bhattacharjee, S. and Pandey, A.C. (2023). Estimating thickness of zemu glacier, Sikkim (India) using ice-flow velocity approach: a geoinformatics based perspective. *Spatial Information Research* 31: 479–485.

Bindschadler, R., Vornberger, P., Blankenship, D. et al. (1996). Surface velocity and mass balance of ice streams D and E, west Antarctica. *Journal of Glaciology* 42 (142): 461–475.

Bolch, T., Kulkarni, A., Kääb, A. et al. (2012). The state and fate of Himalayan glaciers. *Science* 336 (6079): 310–314. https://doi.org/10.1126/science.1215828.

Chao, W.C. and Chen, B. (2001). The origin of monsoons. *Journal of the Atmospheric Sciences* 58 (22): 3497–3507.

Chen, J. and Ohmura, A. (1990). Estimation of alpine glacier water resources and their change since the 1870s. *IAHS Publ* 193: 127–135.

Cogley, J.G. (2012). Himalayan glaciers in the balance. *Nature* 488 (7412): 468–469.

Cuffey, K.M. and Paterson, W.S.B. (2010). *Physics of Glaciers*. Elsevier.

Dkhar, N.B. and Tayal, S. (2013). Glaciers in ganges basin: an assessment of their dimensions through inventories. *International Journal of Scientific and Engineering Research* 4 (9): 1560–1563.

Farinotti, D., Huss, M., Bauder, A. et al. (2009). A method to estimate the ice volume and ice-thickness distribution of alpine glaciers. *Journal of Glaciology* 55 (191): 422–430.

Frey, H., Machguth, H., Huggel, C. et al. (2013). Ice volumes in the Himalayas and Karakoram: evaluating different assessment methods. *EGU General Assembly Conference Abstracts*, EGU2013-11226.

Frey, H., Machguth, H., Huss, M. et al. (2014). Estimating the volume of glaciers in the Himalayan-Karakoram region using different methods. *Cryosphere* 8 (6): 2313–2333. https://doi.org/10.5194/tc-8-2313-2014.

Gadgil, S. (2003). The Indian monsoon and its variability. *Annual Review of Earth and Planetary Sciences* 31 (1): 429–467.

Gantayat, P., Kulkarni, A.V., and Srinivasan, J. (2014). Estimation of ice thickness using surface velocities and slope: case study at gangotri glacier, India. *Journal of Glaciology* 60 (220): 277–282. https://doi.org/10.3189/2014JoG13J078.

Gopika, J.S. (2020). Estimation of glacier stored water in Bhaga basin, Himalayas. Thesis. Academy of Climate Change Education and Research, Vellanikkara.

Gopika, J.S., Kulkarni, A.V., and Prasad, V. (2023). Glacier volume estimation using laminar-flow and volume–area scaling techniques in the chenab basin. *Journal of the Indian Society of Remote Sensing* 51 (8): 1809–1823. https://doi.org/10.1007/s12524-023-01744-7.

Grinsted, A. (2013). An estimate of global glacier volume. *Cryosphere* 7 (1): 141–151. https://doi.org/10.5194/tc-7-141-2013.

Gurung, D.R., Kulkarni, A.V., Giriraj, A. et al. (2011). Monitoring of seasonal snow cover in Bhutan using remote sensing technique. *Current Science* 1364–1370.

Haeberli, W. and Hoelzle, M. (1995). Application of inventory data for estimating characteristics of and regional climate-change effects on mountain glaciers: a pilot study with the European alps. *Annals of Glaciology* 21: 206–212.

Haug, T., Kääb, A., and Skvarca, P. (2010). Monitoring ice shelf velocities from repeat MODIS and landsat data–a method study on the larsen C ice shelf, Antarctic Peninsula, and 10 other ice shelves around Antarctica. *The Cryosphere* 4 (2): 161–178.

Hazra, Pranata, and Akhouri Pramod Krishna. 2019. Spatio-temporal and surface elevation change assessment of glaciers of Sikkim Himalaya (India) across different size classes using geospatial techniques. *Environmental Earth Sciences* 78(14): 1–20. https://doi.org/10.1007/s12665-019-8390-1.

Heid, T. and Kääb, A. (2012). Evaluation of existing image matching methods for deriving glacier surface displacements globally from optical satellite imagery. *Remote Sensing of Environment* 118: 339–355.

Helfricht, K., Huss, M., Fischer, A., and Otto, J.-C. (2019). Calibrated ice thickness estimate for all glaciers in Austria. *Frontiers in Earth Science* 7: 68.

Huss, M. and Farinotti, D. (2012). Distributed ice thickness and volume of all glaciers around the globe. *Journal of Geophysical Research: Earth Surface* 117 (4): 10–11. https://doi.org/10.1029/2012JF002523.

Huss, M., Farinotti, D., Bauder, A., and Funk, M. (2008). Modelling runoff from highly glacierized alpine drainage basins in a changing climate. *Hydrological Processes* 22 (19): 3888–3902. https://doi.org/10.1002/hyp.7055.

Immerzeel, W.W., van Beek, L.P.H., Konz, M. et al. (2012). Hydrological response to climate change in a glacierized catchment in the Himalayas. *Climatic Change* 110: 721–736. https://doi.org/10.1007/s10584-011-0143-4.

Jawak, S.D., Kumar, S., Luis, A.J. et al. (2018). Evaluation of geospatial tools for generating accurate glacier velocity maps from optical remote sensing data. *Proceedings* 2: 341.

Kääb, A. (2002). Monitoring high-mountain terrain deformation from repeated air-and spaceborne optical data: examples using digital aerial imagery and ASTER data. *ISPRS Journal of Photogrammetry and Remote Sensing* 57 (1–2): 39–52.

Kaser, G., Großhauser, M., and Marzeion, B. (2010). Contribution potential of glaciers to water availability in different climate regimes. *Proceedings of the National Academy of Sciences* 107 (47): 20223–20227.

Kaufmann, Viktor, and Richard Ladstädter. 2003. "Quantitative analysis of rock glacier creep by means of digital photogrammetry using multi-temporal aerial photographs: two case studies in the Austrian alps." In *Proceedings of the 8th International Conference on Permafrost*. AA Balkema Publishers Zurich, Switzerland, 21–25.

Kulkarni, A.V., Bahuguna, I.M., Rathore, B.P. et al. (2007). Glacial retreat in Himalaya using Indian remote sensing satellite data. *Current Science* 69–74.

Kumar, P., Sharma, M.C., Saini, R., and Singh, G.K. (2020). Climatic variability at gangtok and tadong weather observatories in Sikkim, India, during 1961–2017. *Scientific Reports* 10 (1): 1–12. https://doi.org/10.1038/s41598-020-71163-y.

Last, J., Trouton, K., and Pengelly, D. (1998). *Taking Our Breath Away: The Health Effects of Air Pollution and Climate Change*. David Suzuki Foundation.

Leprince, S., Barbot, S., Ayoub, F., and Avouac, J.-P. (2007). Automatic and precise orthorectification, coregistration, and subpixel correlation of satellite images, application to ground deformation measurements. *IEEE Transactions on Geoscience and Remote Sensing* 45 (6): 1529–1558.

Leprince, S., Berthier, E., Ayoub, F. et al. (2008). Monitoring earth surface dynamics with optical imagery. *EOS, Transactions American Geophysical Union* 89 (1): 1–2.

Linsbauer, A., Frey, H., Haeberli, W. et al. (2016). Modelling glacier-bed overdeepenings and possible future lakes for the glaciers in the Himalaya-Karakoram region. *Annals of Glaciology* 57 (71): 119–130. https://doi.org/10.3189/2016AoG71A627.

Macheret, Y.Y. and Zhuravlev, A.B. (1982). Radio echo-sounding of svalbard glaciers. *Journal of Glaciology* 28 (99): 295–314.

Marzeion, B., Jarosch, A.H., and Hofer, M. (2012). Past and future sea-level change from the surface mass balance of glaciers. *The Cryosphere* 6 (6): 1295–1322.

Maurer, J.M., Schaefer, J.M., Rupper, S., and Corley, A. (2019). Acceleration of ice loss across the Himalayas over the past 40 years. *Science Advances* 5: eaav7266.

McNabb, R.W., Hock, R., O'Neel, S. et al. (2012). Using surface velocities to calculate ice thickness and bed topography: a case study at Columbia Glacier, Alaska, USA. *Journal of Glaciology* 58 (212): 1151–1164.

Narimisa, M.R. and Narimisa, M.R. (2018). Climate change and global warming overview; assessing climate change and global warming in local scale. *Revista Publicando* 5 (14): 570–580.

Nela, B.R., Singh, G., and Kulkarni, A.V. (2023). Ice thickness distribution of Himalayan glaciers inferred from DInSAR-based glacier surface velocity. *Environmental Monitoring and Assessment* 195 (1): –15. https://doi.org/10.1007/s10661-022-10658-9.

Owen, L.A., Derbyshire, E., Richardson, S. et al. (1996). The quaternary glacial history of the lahul Himalaya, northern India. *Journal of Quaternary Science: Published for the Quaternary Research Association* 11 (1): 25–42.

Pandey, P. and Venkataraman, G. (2013). Changes in the glaciers of Chandra–Bhaga basin, Himachal Himalaya, India, between 1980 and 2010 measured using remote sensing. *International Journal of Remote Sensing* 34 (15): 5584–5597.

Pandit, A. and Ramsankaran, R. (2020). Modeling ice thickness distribution and storage volume of glaciers in Chandra basin, western Himalayas. *Journal of Mountain Science* 17 (8): 2011–2022. https://doi.org/10.1007/s11629-019-5718-y.

Pfeffer, W.T., Arendt, A.A., Bliss, A. et al. (2014). The randolph glacier inventory: a globally complete inventory of glaciers. *Journal of Glaciology* 60 (221): 537–552.

Prasad, Veena, Kulkarni, A.V., Pradeep, S. et al. 2019. "Large losses in glacier area and water availability by the end of twenty-first century under high emission scenario, Satluj basin, Himalaya." *Current Science* 116: 1721–1730. https://doi.org/10.18520/cs/v116/i10/1721-1730.

Radić, V., Andrew Bliss, A., Beedlow, C. et al. (2014). Regional and global projections of twenty-first century glacier mass changes in response to climate scenarios from global climate models. *Climate Dynamics* 42: 37–58.

Raina, V. (2009). Himalayan glaciers: a state-of-art review of glacial studies, glacial retreat and climate change. Ministry of Environment and Forests, Government of India (MoEF) discussion paper published by MoEF and G.B. Pant Institute of Himalayan Environment & Development, Kosi-Katarmal, Almora, India. https://lib.icimod.org/record/14392

Raina, V. K., D. Srivastava, R. K. Singh, and C. V. Sangewar. 2008. "Glacier regimen, fluctuation, hydrometry and mass transfer studies in the himalayas." *Glacier Atlas of India*, Raina, VK and Srivastava, D., Eds. Geological Society of India, Bangalore, 206–216.

Raup, B.H., Andreassen, L.M., Bolch, T., and Bevan, S. (2014). Remote sensing of glaciers. In: *Remote Sensing of the Cryosphere*, 123–156. Wiley. https://doi.org/10.1002/9781118368909.ch7.

Remya, A.V., Kulkarni, S.P., and Shrestha, D.G. (2019). Volume estimation of existing and potential glacier lakes. *Current Science* 116 (4): 620–627. https://doi.org/10.2307/27137905.

Rolstad, C., Amlien, J., Hagen, J.-O., and Lundén, B. (1997). Visible and near-infrared digital images for determination of ice velocities and surface elevation during a surge on osbornebreen, a tidewater glacier in svalbard. *Annals of Glaciology* 24: 255–261.

Sangewar, C.V. and Shukla, S.P. (2009). Inventory of the Himalayan glaciers (an updated edition). *Special Publication* 34: 169.

Scherler, D., Bookhagen, B., and Strecker, M.R. (2011). Spatially variable response of Himalayan glaciers to climate change affected by debris cover. *Nature Geoscience* 4: 156–159. https://doi.org/10.1038/NGEO1068.

Scherler, D., Leprince, S., and Strecker, M.R. (2008). Glacier-surface velocities in alpine terrain from optical satellite imagery-accuracy improvement and quality assessment. *Remote Sensing of Environment* 112: 3806–3819. https://doi.org/10.1016/j.rse.2008.05.018.

Singh, G., Nela, B.R., Bandyopadhyay, D. et al. (2020). Discovering anomalous dynamics and disintegrating behaviour in glaciers of Chandra-Bhaga sub-basins, part of western Himalaya using DInSAR. *Remote Sensing of Environment* 246: 111885.

Singh, K.K., Negi, H.S., and Singh, D.K. (2019). Assessment of glacier stored water in karakoram Himalaya using satellite remote sensing and field investigation. *Journal of Mountain Science* 16 (4): 836–849.

Sun, Y., Jiang, L., Liu, L. et al. (2017). Spatial-temporal characteristics of glacier velocity in the central karakoram revealed with 1999–2003 landsat-7 ETM+ pan images. *Remote Sensing* 9 (10): 1064. https://doi.org/10.3390/rs9101064.

Tawde, S.A., Kulkarni, A.V., and Bala, G. (2017). An estimate of glacier mass balance for the Chandra basin, western Himalaya, for the period 1984–2012. *Annals of Glaciology* 58 (75): 99–109. https://doi.org/10.1017/aog.2017.18.

Trenberth, K.E. (2018). Climate change caused by human activities is happening and it already has major consequences. *Journal of Energy & Natural Resources Law* 36 (4): 463–481.

Van Pelt, W.J.J., Oerlemans, J., Reijmer, C.H. et al. (2013). An iterative inverse method to estimate basal topography and initialize ice flow models. *The Cryosphere* 7 (3): 987–1006.

Vaughan, D.G., Comiso, J.C., Allison, I. et al. (2013). Observations: cryosphere. *Climate Change* 2103: 317–382.

Wang, Y., Li, J., Lixin, W. et al. (2021). Estimating the changes in glaciers and glacial lakes in the xixabangma massif, central Himalayas, between 1974 and 2018 from multisource remote sensing data. *Remote Sensing* 13 (19): 3903. https://doi.org/10.3390/rs13193903.

Yanai, M., Li, C., and Song, Z. (1992). Seasonal heating of the Tibetan plateau and its effects on the evolution of the Asian summer monsoon. *Journal of the Meteorological Society of Japan* 70 (1): 319–351. https://doi.org/10.2151/jmsj1965.70.1B_319.

Yancheva, G., Nowaczyk, N.R., Mingram, J. et al. (2007). Influence of the intertropical convergence zone on the east Asian monsoon. *Nature* 445 (7123): 74–77.

13

Snowmelt Runoff Estimation Using Landsat 8 Snow Cover Area Products in Beas River Basin, Western Himalayan via SRM Degree-Day Modeling Approach

Gopinadh Rongali[1], Ashok K. Keshari[2], Ashvani K. Gosain[2], Rakesh Khosa[2], and Raghavendra Ashrit[1]

[1] *National Center for Medium Range Weather Forecasting (NCMRWF), Ministry of Earth Sciences (MoES), Noida, India*
[2] *Department of Civil Engineering, Indian Institute of Technology Delhi, New Delhi, India*

13.1 Introduction

Understanding hydrological behavior is an important first step in managing large-scale watershed water resources (Ramahaimandimby et al. 2023). Estimating and predicting snowmelt runoff, as well as the associated stream flow from diverse catchments, is critical for river valley project development and operation, as well as for comprehending various hydrometeorological processes (Zhou et al. 2021). Furthermore, precise estimations of runoff volume are essential for projects related to water resource management, navigation, flood prediction, and warning systems. The snowmelt runoff model (SRM), sometimes referred to as the Martinec model or Martinec-Rango model, is used by researchers to simulate and predict the daily flow of water in mountain basins where the melting of snow significantly contributes to the runoff (Pangali Sharma et al. 2020). An example of a tool that may be used to calculate and forecast the runoff that occurs as a consequence of snow melting in mountainous areas is the SRM. These basins range in height from 305 to 7690 meters and in size from 0.76 to 120 000 km^2. A model run may go on forever and starts with a discharge value that is known or predicted. Snow Cover Area (S), temperature (T), and precipitation (P) measurements are needed inputs to the SRM for various elevation zones. One of the most important technologies for gathering spatial and temporal data on large-scale catchments is remote sensing (RS).

Daoo and Shirvaikar (1978) used an energy balance methodology to compute the runoff resulting from snowmelt in the Beas River basin located in the Himalayan region. The study focused on the three-month period of April, May, and June in the year 1969. The calculated runoff was then compared to the observed discharge of snowmelt from the river. Snowmelt may contribute significantly to a river's total flow, according to

Water Resources Management in Mountain Regions, First Edition. Edited by Virendra Bahadur Singh, Sughosh Madhav, Rakesh Kumar Gupta, Prerna Diwan, and Amit Kumar.

Gupta et al. (1982), and is crucial in predicting snowmelt runoff. Kumar et al. (1993) used satellite data for the purpose of snow cover mapping and snowmelt runoff modeling. Nagler and Rott (1997) developed a method to chart snow in mountainous areas and glaciers using Synthetic Aperture Radar (SAR) data acquired from the European Remote Sensing Satellite (ERS) operating in the C-band. We assessed the SRM approach in two catchment regions situated in the Austrian Alps. In Nagler and colleagues (1998) conducted real-time estimations of snowmelt runoff for basins in the Austrian Zillertal Alps. These basins are connected to a system of reservoirs used for hydropower generation. We generated the projections with a lead time of up to six days. In addition, Nagler and Rott (2000) used a change detection-based automated system for snow mapping using SAR data in the Tuxbach mountain basin in the Eastern Alps, resulting in the creation of melting snow maps. Prasad and Roy (2005) state that accurately calculating the amount of snowmelt runoff is crucial in the Western Himalayan rivers of India. This is because it is necessary to be well-prepared for hydropower production and water management throughout the non-monsoon season. Kumar et al. (2007) used the water balance approach to calculate the mean proportion of runoff from snow and glacier melt in the Beas River's annual flow at Pandoh dam. Singh et al. (2014) conducted a study on a hydrological approach that compares and controls the transfer of hydrologic features from gauged to ungauged catchments. Yuan et al. (2016) established a connection between the amount of rainfall in the mid-Tianshan mountains and the Pacific North America (PNA) pattern. Engelhardt et al. (2017) conducted a simulation of glacial mass balance and discharge for the Chhota Shigri glacier over a period of 60 years. The correlations between the mass balance, weather, discharge, and surface speed of the Chhota Shigri glacier from 2002 to 2019 were determined by Mandal et al. (2020) using in-situ data.

The watershed characteristics – slope, aspect, size, and elevation – that may be obtained using geographic information systems (GIS) methods are also important to the hydrological process (Sreenivasulu and Bhaskar 2010). Presently, advancements in RS and GIS enable us to ascertain albedo (Hofierka and Onačillová 2021), skin temperature (Khan et al. 2022), duration of sunshine, and evapotranspiration (Prasad and Mahadev 2006). Using RS data and the SRM, we attempted to replicate the daily fluctuations of the Beas River up to Pandoh dam. Multiple studies (Jain et al. 2010; Aggarwal et al. 2014; Azam et al. 2019) have shown that the SRM has the ability to accurately and effectively simulate runoff in various Himalayan river basins. This study might provide baseline information for managing the water and land resources in the Beas basin, as well as assist in creating a preliminary database for developing hydropower facilities on the Beas River. The rise in altitude of the basin results in less precipitation contributing to the streamflow, while simultaneously generating a greater amount of snowmelt. The primary source of water flow at altitudes over 3000 meters is the melting of snow. Due to the rugged topography and limited number of ground observation sites, the frequent monitoring and mapping of snow cover and glaciers is consistently a difficult undertaking. This study used the SRM for Windows (WinSRM) in conjunction with Landsat 8 satellite data, RS, and GIS to predict the amount of water runoff caused by the melting of snow. This projection is especially relevant due to the pivotal role played by seasonal snow cover, glaciers, and melt runoff in the Himalayan region.

13.2 Description of the Study Area

13.2.1 The Beas River Basin

The scope of this study focuses on the Beas River basin, extending up to the Pandoh dam, as seen in Figure 13.1. The Beas River basin is located within the latitudes of 31°N and 32°N, and the longitudes of 77°E and 78° E. The Beas River is a major affluent of the Indus. The Rothang Pass, situated at an elevation of 3900 meters, marks the starting point of this route, which extends in a north-south direction until it reaches Larji. The route takes a right turn at Larji and proceeds in a southwest direction, maintaining the same course until it reaches Pandoh dam. The Beas River spans a distance of 116 km until it reaches Pandoh. The Parbati and Sainj Khad rivers, which are supplied by glaciers, are two of the tributaries of the river. The catchment area of the Beas River is 5383 km^2 and reaches as far as the Pandoh dam. The watershed region mostly comprises steeply sloping terrain with exposed bedrock. The river valley is encompassed by towering peaks in the east and north. The elevation of the Pandoh dam is 846 meters, whereas the northeastern boundary of the Parbati sub-height catchment is 6500 meters. The river is extensively blanketed with snow

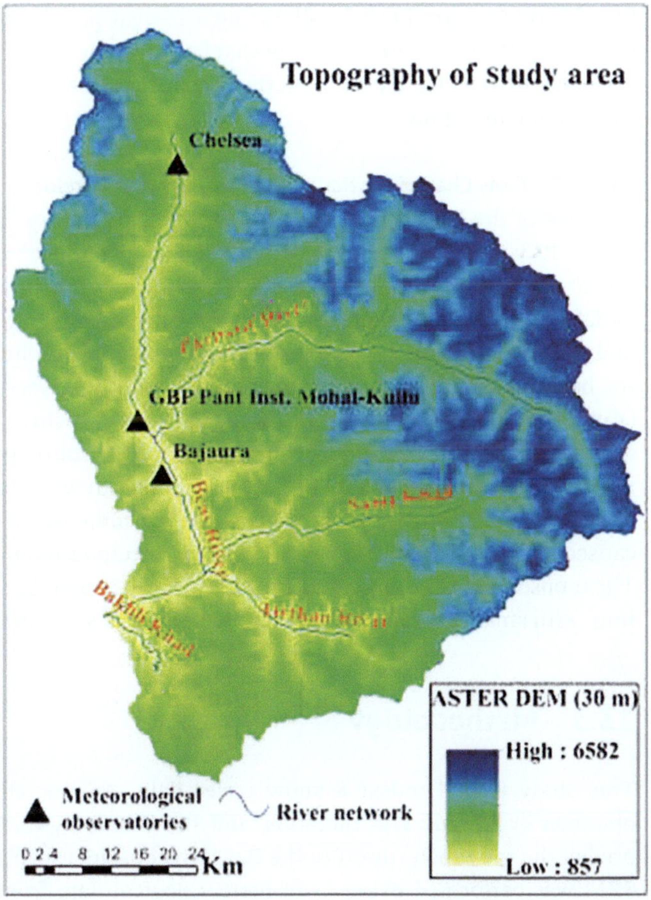

Figure 13.1 The Beas River basin in Himachal Pradesh is displayed in the study area.

throughout the winter season. During the summer months, the Beas River relies mostly on the melting of snow as its main water supply. Several significant tributaries converge upstream of the dam, including the Parbati River at Bhuntar, the Tirthan and Sainj rivers at Larji, the Sabari Nala near Kulu, and the Bakhli Khad next to the Pandoh dam. The Pandoh dam redirects water from the Beas River to the Satluj River, where the Dehar Power Station, situated on the right bank of the Satluj River, harnesses it to produce electricity.

13.2.2 Climate and Water-Related Features of the Beas River Basin

13.2.2.1 Rainfall

The Beas River basin receives a lot of rain from July through September during the monsoon season. The region is influenced by humid breezes originating from the Bay of Bengal and the Arabian Sea. In winter, the upper part of the basin experiences snowfall. The analysis of rainfall data reveals that the area with the largest amount of rainfall is Kothi near Manali, while the area with the lowest amount of rainfall is near Pulga (Singh et al. 1995). The research region has a chilly climate overall. The basin's raingauge station distribution shows that two significant tributaries, the Parvati and Sainj rivers, originate in the eastern portion of the basin, which is devoid of raingauge stations. Based on variations in the basin's average monthly rainfall during a 15-year period. As can be seen, July and August experience the most rainfall, while October through January sees the least amount. Due to variances in aspect and elevation, the climate varies significantly between locations within the catchment region.

13.2.2.2 Flow Characteristics of the Beas River at Pandoh

The flow of the Beas River is determined by two primary factors: the amount of precipitation in its watershed and the rate at which snow and glaciers melt. This river is perennial due to the continuous input of water from baseflow, precipitation, and snowmelt. Seasonal variations in the snowmelt discharge are caused by variations in the basin's temperature and precipitation. The stages are used to measure the discharge at different locations across the basin. With the start of the monsoon season, the average river flow rises significantly (Kumar et al. 2007). During this time, high floods with sharp peaks occur. The amount, allocation, and proportionate occurrence of precipitation in the areas where the river and its tributaries collect water all impact the manner in which floodwater flows in these watercourses. The main water supply during the pre-monsoon season (April–June) is the runoff caused by the melting of snow and glaciers. Precipitation may also influence the situation. The monsoon season, which occurs from July to September, is mostly defined by precipitation, with small contributions from the melting of snow and glaciers.

13.3 Methodology

This study used Landsat 8 snow cover data, ASTER Digital Elevation Model (DEM) obtained by RS and GIS methods, and the Windows version of the SRM (WinSRM) to predict the snowmelt runoff in the Beas River basin, especially up to the Pandoh dam. The SRM was performed using hydrometeorological data from 2013 to 2015. Using Landsat

8 satellite sensing data, a temporal SCA for the Beas River was created between 2013 and 2015. A DEM was used to find aspect maps and elevation zones. The SRM created a temperature index-based model specifically for this experiment to imitate the runoff of melted snow and glaciers. We conducted simulations of snowfall and glacier melt using data obtained from three specific locations on the Meteorological and Oceanographic Satellite Data Archival Centre (MOSDAC) website (www.mosdac.gov.in). Each day, we collect data on temperature and precipitation.

The purpose of the SRM is to replicate and predict the daily flow of water in mountain basins, where the melting of snow has a major role in the amount of water that flows. Researchers have lately used it to forecast the consequences of climate change on seasonal runoff and snow cover. Martinec (1975) introduced the concept of Sediment Rating Curves (SRM) in smaller basins located in Europe. With the progress of satellite remote sensing of snow cover, larger and more extensive basins have also used SRM. The runoff predictions provided by SRM are easily comprehensible. This technique has been used by several scholars to date, examining more than 100 basins across 29 different nations (DeWalle and Rango 2008). The World Meteorological Organization (WMO) has extensively replicated conditions of real-time runoff forecasts and successfully tested SRM in runoff simulations (WMO 1986, 1992). The daily outflow from the basin is multiplied by the predicted outflow from the basin to get the amount of water that is expected to be generated by rain and snowmelt, as determined by

$$Q_{n+1} = \left[c_{Sn} . a_n \left(T_n + \Delta T_n \right) S_n + c_{Rn} P_n \right] \frac{\text{A}.10000}{86400} \left(1 - k_{n+1} \right) + Q_n k_{n+1} \quad (13.1)$$

The variable "a" represents the degree-day factor, measured in centimeters per degree per day, which indicates the amount of snowmelt caused by a one-degree increase in temperature. In the runoff coefficient, snowmelt and rain are denoted by the letters c_S and c_R, respectively. The coefficient describes losses as the ratio of runoff to precipitation, where C represents the losses stated as a ratio of runoff to precipitation. T represents the temperature extrapolation from the station to the average hypsometric elevation of the basin or zone, and it is adjusted based on the temperature lapse rate in degrees Celsius per day. S represents the percentage of the total area covered with snow. P represents the amount of precipitation that leads to runoff, measured in centimeters. The extent and speed of its contribution are governed by a pre-established threshold temperature, TCRIT. If TCRIT identifies the precipitation as new snow, it is retained over the area that was previously devoid of snow until the conditions for melting occur. The variables m and m + 1 represent the consecutive days during a time of real recession flow. The variable n represents the number of days in the discharge computation period, arranged in order. A is the size of the basin or zone in square kilometers. k is the recession coefficient, which indicates a decrease in discharge during a dry time without snowmelt or rainfall. The terms m and m + 1 describe the course of an actual recession flow period. Eq. (13.6) results in a discharge cycle with a time lag of 18 hours between it and the daily temperature cycle. In this situation, the number of degree-days on the nth day is equivalent to the amount of water discharged on that day. A subroutine may cause different durations of delay. Convert cm km^2 d^{-1} to m^3 s^{-1}.

Within a basin or climate, the unique properties of c_R, c_S, the lapse rate necessary for calculating T, TCRIT, k, and the lag time are specific to that area. On the other hand, the variables T, S, and P need to be observed or determined on a daily basis. If the height range of the basin exceeds 500 meters, it is advisable to split the basin into elevation zones about 500 meters apart. The equation of the model is now: The model equation is modified to account for three elevation zones, A, B, and C, within a range of 1500 meters.

$$Q_{n+1} = \left\{ \begin{array}{l} \left[c_{SAn}.a_{An}\left(T_n + \Delta T_{An}\right)S_{An} + c_{RAn}P_{An}\right]\dfrac{A_A.10000}{86400} + \\ \left[c_{SBn}.a_{Bn}\left(T_n + \Delta T_{Bn}\right)S_{Bn} + c_{RBn}P_{Bn}\right]\dfrac{B_B.10000}{86400} + \\ \left[c_{SCn}.a_{Cn}\left(T_n + \Delta T_{Cn}\right)S_{Cn} + c_{RCn}P_{Cn}\right]\dfrac{C_C.10000}{86400} \end{array} \right\}\left(1 - k_{n+1}\right) + Q_n k_{n+1} \tag{13.2}$$

SRM can run in simulation mode without needing to be updated. The discharge data is solely used to assess the simulation's correctness. In ungauged basins, the simulation begins with an anticipated discharge based on a nearby gauged basin. The software WinSRM version 1.12, a SRM, was used for the simulation in the Beas River basin. This basic method has been used to estimate runoff from snowmelt (Figure 13.2). The procedure is called the temperature index of degree-day technique.

There is not much of a meteorological network in the Himalayan mountain ranges for data collection. The most often accessible data are the surface wind speed, relative humidity, and the maximum and lowest daily temperatures. Since temperature indices are often considered

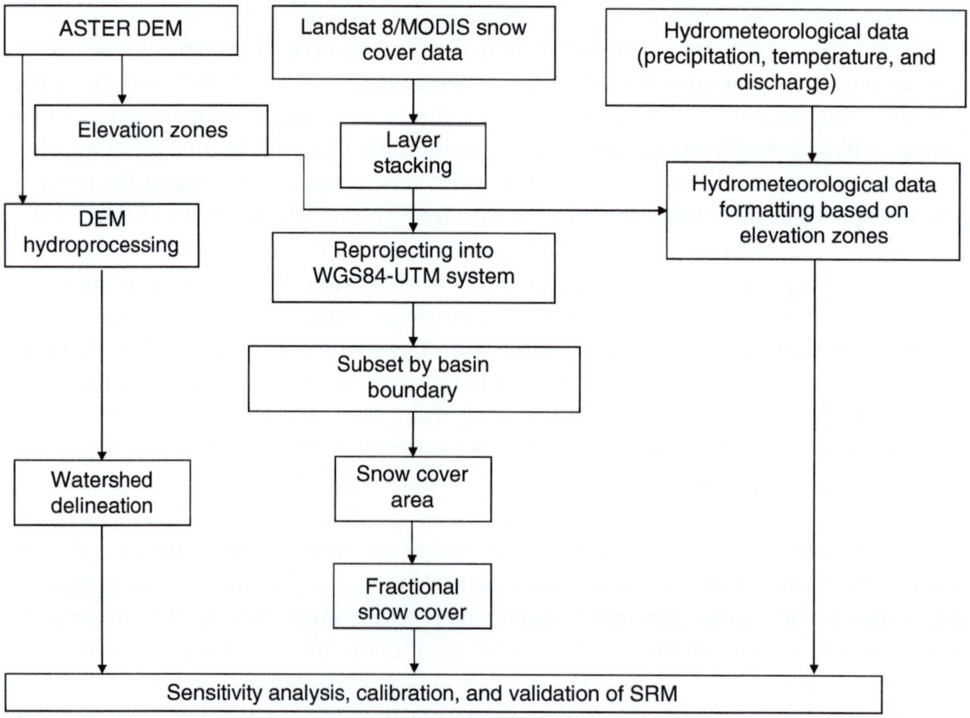

Figure 13.2 Flowchart of the snowmelt runoff model (SRM).

the best predictor of the heat transfer processes involved in snowmelt, they have been used for a long time to estimate snowmelt. Our result suggests that air temperature, recorded in degrees per day, may be used as a replacement for the complex energy balance required to calculate snowmelt. A degree-day is a metric used to quantify the duration of a temperature that deviates by one degree Celsius from a reference temperature during a 24-hour period.

The most straightforward and frequent sentence linking daily snowmelt to the temperature index is,

$$M = D_f\left(T_i - T_b\right) \tag{13.3}$$

M represents the amount of melt produced in a certain time period inside a centimeter of water. Df is the degree-day factor, which measures the rate of melt in centimeters per degree-day. Ti refers to the air temperature index in degrees Celsius, while Tb is the base temperature, often set at 0° Celsius.

R^2 and D_V as defined below are two criteria that are calculated by the model automatically (Martinec et al. 2008):

$$R^2 = 1 - \frac{\sum_{i=1}^{n}\left(Q_i - Q_j\right)^2}{\sum_{i=1}^{n}\left(Q_i - \bar{Q}_j\right)^2} \tag{13.4}$$

$$D_v = \frac{V_R - V_R'}{V_R} \times 100 \tag{13.5}$$

R2 represents the efficiency of the model, measured in units of [–]. Qi represents the daily discharge measured in cubic meters per second [m^3s^{-1}]. Qj represents the simulated daily discharge also measured in cubic meters per second [m^3s^{-1}]. The average daily discharge for either the simulation year or simulation season is denoted as [m^3s^{-1}]. n represents the number of daily discharge values [–]. Dv represents the percentage difference between the total measured and simulated runoff, measured in percent. V_R represents the measured runoff volume in cubic meters [m^3], while V'_R represents the simulated runoff volume also measured in cubic meters [m^3].

13.3.1 Hydrometeorological Analyzes

To forecast a catchment's hydrological behavior, researchers must examine the catchment's previous hydrometeorological data. It is necessary to examine daily and monthly hydrological and meteorological data from several stations within and outside the study zone to determine the SRM model parameters, such as temperature, rainfall, and runoff.

13.3.1.1 Variables

Variables essentially characterize the simulated period's real climatic conditions. This information must be gathered, ideally by in-situ observations, although snow cover may also be determined using RS.

13.3.1.2 Temperature (T)

Data on daily maximum and lowest temperatures were used from 2013 to 2015. The highest and lowest daily temperature data that are available have been used to calculate the average daily temperature. Then, using the temperature lapse rate and superimposing them on top of an ASTER DEM, they are extrapolated to each cell.

The SRM takes either the average temperature for the day or two temperature data points for each day, namely the maximum temperature (TMax) and the minimum temperature (TMin). The SRM uses the lapse rate to extend temperature data to the hypsometric mean heights of all zones. The lapse rate may be determined by analyzing historical data obtained from temperature stations located at different altitudes. Otherwise, we need to assess it within the framework of other basins or climates. SRM simulations often use a lapse rate of 0.65 °C every 100 m. To calculate the altitude adjustment T in Eq. (13.6), adhere to the following procedure:

$$\Delta T = \Upsilon.\left(h_{st} - \bar{h}\right).\frac{1}{100} \tag{13.6}$$

where Υ = lapse rate of temperature [°C per 100 m], h_{st} = the temperature station's elevation [m], and $\bar{h}$ = A zone's hypsometric mean elevation [m]. The degree-day values (T + Δ T) are immediately changed to zero if they become negative, preventing the computation of negative snowmelt.

13.3.1.3 Precipitation (P)

The mean hypsometric elevations of the individual zones are estimated from available precipitation data from three meteorological stations. Rain and snowfall are the two types of precipitation. The crucial temperature determines whether snow or rain is being seen as precipitation (Tcrit). The model generates an output for each zone, whether the runoff is precipitation-runoff, snowmelt-runoff, or a mix of both. The output of the model depends on the critical temperature, the presence of an event (snow or rain), and whether the model is in a snow-free area (SFA) or snow-covered area (SCA).

Estimating average areal precipitation in mountain basins is a very difficult task. The System Response Model (SRM) may receive a basin-wide precipitation input, which can come from a single station or a "synthetic station" produced by merging data from many stations. Alternatively, it can also receive input on a zone-by-zone basis. Precipitation stations situated at low altitudes are susceptible to overestimating the amount of precipitation in basins with a wide range of elevations. For accurate estimation of precipitation data in the relevant zones, it is advisable to use an altitude gradient of 3% or 4% per 100 m. By applying an altitude gradient, it is possible to estimate the averaged data for various height zones and compute the average elevation for each station, assuming there are two stations. It is crucial to acknowledge that the increase in the amount of precipitation as altitude increases does not continue indefinitely but instead reaches a point when it stops, particularly in mountain ranges at very high altitudes.

13.3.1.4 Snow-Covered Area (S)

The model calculates SCA using Normalized Difference Snow Index (NDSI). Small cells that make up catchments each have a unique NDSI value. Typically, the NDSI number falls between −1 and 1. SCA is indicated by a value > = 0.4, while a value of 0.4 indicates a SFA. The model can detect if the cell area is snow-covered or not. The model uses NDSI threshold values, meaning that it identifies all grid cell locations with an NDSI value greater than or equal to 0.4 as snow cover (Kulkarni et al. 2006).

Over the course of the snowmelt season, the amount of seasonal snow cover progressively diminishes. This characteristic is typical in mountain basins. Periodic snow cover mapping may be utilized to interpolate daily data into snow cover loss curves that can be read off and used as a crucial input variable for SRM. You may utilize ground observations (in very small basins), aerial photographs (particularly during a flood emergency), or, best of all, satellites to create a map of the snow cover. The minimum size of an area that may be correctly scanned is dictated by the spatial resolution of the remote sensor (Rango and Martinec 1998).

13.3.2 Model Parameters

In the current work, input parameters (constants) from a similar catchment region of the Beas River were evaluated for snow and glacier melt runoff modeling. These parameters, on the other hand, may be calculated using measurements or approximated using hydrological judgment based on basin features, physical and theoretical relationships, or empirical regression relationships. Future modifications should never go beyond the range of hydrologically and physically acceptable values.

The accuracy with which the model parameters are computed is critical to the SRM's success. However, since snow ground observations are uncommon in the examined area, historical data cannot be used to calibrate the model parameters. We attempt to ascertain the necessary parameters by using hydrometeorological data, empirical regression equations, hypotheses, or the default settings of the model. Initially, the model underwent testing in many designated stages and attained the expected values correctly.

Hydrological modeling requires the ASTER GDEM (Global Digital Elevation Model), multitemporal Landsat 8 satellite images, and hydrometeorological data (temperature and rainfall per day). The SRM model necessitates inputs such as catchment area, elevation zones, aspect (slope direction), SCA, rainfall, temperature, runoff coefficient for snow (Cs), runoff coefficient for rain (Cr), degree-day factor (a), temperature lapse rate (TCRIT), rainfall contributing region (RCA), recession coefficient (k), and time lag (L).

13.3.2.1 Runoff Coefficient, C

The runoff coefficient quantifies the discrepancy between the total water input (including snowmelt and rainfall) and the outflow from the basin, taking into consideration any losses. It ought to match the proportion of observed runoff to measured precipitation over time. Actually, the historical precipitation and runoff ratios are compared to derive the runoff coefficient values. The catch deficiency of precipitation gauges, which mostly impacts snowfall, and a dearth of precipitation data from mountain locations make it difficult to consistently calculate these ratios. The only way water may be lost during the start of the snowmelt season is by evaporation from the snow surface, therefore losses are often not very significant. When some soil is exposed and plant develops, evapotranspiration and interception will cause higher losses. As the snowmelt season comes to an end, the flow of water directly from glaciers and the remaining snowfields may become the dominant source of water in particular basins. This may reduce water losses and increase the runoff coefficient. Furthermore, for snowmelt and rainfall, c is frequently different. The computer software records the snow (C_S) and rain (C_R) data for each elevation zone independently.

It also provides for half-monthly (and, if necessary, daily) revisions to the values. In certain semiarid basins, runoff coefficients may even be lower, especially in the lowest elevation zones. It seems that in the event that a runoff simulation fails at first, the SRM parameter that will change the most often is the runoff coefficient.

The runoff coefficient quantifies the difference between the total water input (snowmelt and rainfall) and the water output from the basin. The runoff coefficient of rain is the ratio of rainfall that transforms into runoff. The Beas River basin has used a runoff coefficient of rainfall ranging from 0.10 to 0.15 years. The snow and glacier runoff coefficients represent the proportion of snowfall that converts into runoff. For the Beas River basin, these coefficients ranged from 0.15 to 0.20. These equations were calibrated using daily runoff observations.

This factor denotes the ratio of precipitation or snowfall that converts into runoff. Ultimately, it should be equal to the proportion of measured runoff to observed precipitation. Martinec et al. (2008) determined the runoff coefficients by analyzing the historical relationship between precipitation and runoff ratios. The daily runoff coefficient was estimated using data from meteorological stations since there were no runoff and precipitation measurements near the research basin's outflow.

13.3.2.2 Degree-Day Factor, a

The number of degree-days T ($^0C^{-d}$) is converted into the daily snowmelt depth M (cm) using the degree-day factor a ($cm^oc^{-1}d^{-1}$): M = a T. The degree-day factor is a constant that varies between 0.55 and 0.75 cm per degree-day. The degree-day factor is found to be 0.65 cm per degree-day.

Both a field-based and a reflectance-based melt factor are calculated by the model (in degree-days). The model in this study uses the reflectance value of each cell to produce the melt factor parameter. The reflectivity of snow and glaciers reduces as their density rises. The reflectance threshold values are affected by the densities of various snow types, such as fresh snow, moist snow, settled snow, depth hoar, wind-packed snow, extremely wet snow, firn, and glacier ice. Martinec (1975) introduced an empirical computation that may be used to determine the melt factor for each specific cell:

$$a = 1.1 * \frac{(density\ of\ snow)}{(density\ of\ water)} \tag{13.7}$$

13.3.2.3 Temperature Lapse Rate, γ

If historical data from temperature stations at different altitudes is available, it may be used to compute the lapse rate. Otherwise, it must be compared to other basins or assessed in light of climatic circumstances. In the current investigation, a lapse rate of 0.71 to 0.87 °C per 100 m was used in SRM simulations. In the Rocky Mountains, somewhat higher, seasonally shifting levels proved to be sufficient.

13.3.2.4 Critical Temperature, T_{CRITT}

Whether snow or rain is present in the observed or predicted precipitation depends on the crucial temperature. Before replicating runoff, models that replicate the accumulation of snow cover depend heavily on traits that occur both during the ablation phase and the accumulation period. SRM just has to be aware of the essential temperature to ascertain whether precipitation immediately results in runoff (rain) or whether it leads to snowfall at TCRITT. In this case, SRM will autonomously retain recently accumulated snow until it dissolves on subsequent days with higher temperatures (Martinec et al. 2008). In this research, the critical temperature is considered to be in °C to ascertain whether precipitation at each elevation zone is rain or snow. In these cases, it is important to reevaluate the critical temperature and temperature lapse rate parameters and make appropriate adjustments to turn snow into rain. Because the temperature utilized is the daily mean, it is difficult to tell the difference between rain and snow. Regardless of the time of day or night, or if it is warmer or colder than usual, precipitation is possible.

13.3.2.5 Rainfall Contributing Area, RCA

There are two potential reactions when we categorize precipitation as rain. Scenario 1 (option 0) assumes that the snow, which is usually dense and dry, will retain any precipitation that occurs during the first phase of snow melting. The rainfall depth decreases when the runoff from the SFA is combined with the runoff from snowmelt, according to the ratio of the snow-free area to the total area. Prior to choosing option 1 in the computer software, the user is required to specify the maturity of the snow cover. In the case of rain falling on the snow cover, the computer software will include the rainfall from the whole zone area into the snowmelt. This is based on the assumption that the snowpack will release an equivalent quantity of water in response to the rain.

13.3.2.6 Recession Coefficient, *k*

The recession coefficient is a vital component of SRM since it indicates the proportion of the daily meltwater flow that immediately contributes to the runoff (1–*k*). The most prevalent approach to ascertain the value of *k* is by analyzing historical discharge data.

The contribution of the discharge from the snowmelt discharge from the day before to the present day is indicated by the recession coefficient. By graphing the daily discharges on a logarithmic scale and determining the slope of the line that best fits the data, we may get the recession coefficient. We do daily calibrations. The range for the Beas River basin is between 0.64 and 1.

A key component of SRM is the recession coefficient, which shows how much meltwater is produced each day and how much of it enters the runoff right away. Examining historical discharge data is often a beneficial approach for determining the value of *k* (Martinec et al. 2008). Hydrometric stations daily compare the logarithmic values of Qn and Qn + 1 on a log–log scale to calculate the slope of the most accurate line, also known as the *k*-value.

13.3.2.7 Time Lag, L

The temporal lag from the hydrographs of past years may be readily established due to the typical daily variations in snowmelt runoff. The Water Regional Authority's previous physiographic studies of the area led to the prediction that the Beas River basin would have a lag period of 18 hours (Northern Khorasan Water Regional Company 2013). Using anticipated hydrological and meteorological parameters, SRM may calculate daily stream flow. Snow cover produced by digital image analysis of Landsat 8 data can also be used.

13.4 Data Used

Hydrological modeling requires the ASTER GDEM, multitemporal Landsat 8 satellite pictures, and hydrometeorological data (temperature and rainfall per day). The primary digital elevation data and RS information used in this study are listed in Table 13.1. The SRM model requires the following inputs: catchment area, elevation zones, aspect (slope direction) of the catchment, territory covered by snow, rainfall, temperature, runoff coefficient, degree-day factor, temperature lapse rate, critical temperature, area where rainfall contributes, recession coefficient, and time lag.

13.4.1 Remote Sensing Datasets

Monthly snow cover area calculation was done using Landsat 8 Operational Land Imager/Thermal Infrared Sensor (OLI/TIRS) satellite images from 2013 to 2015 (Figure 13.3). The sensor consists of a total of 11 bands, which are divided into different resolutions. The OLI multispectral bands (B1 through B7 and B9) have a resolution of 30 m, while the OLI panchromatic band (B8) has a resolution of 15 m. Additionally, there are TIRS bands (B10 and B11) with a resolution of 100 m. The NDSI approach was used to map snow cover. Based on a threshold value, the NDSI maps are further divided into snowy and snow-free zones. The region is considered snowy if the threshold value is larger than 0.4; otherwise, it is considered snow-free. Under mountain shadow conditions, this sort of categorization gives an advantage for SCA estimate. The band combination ratio of (B3–B6) / (B3 + B6) is used to determine NDSI, where B3 is the green band and B6 is the SWIR band.

13.4.2 Topography

ASTER GDEM data has been used to derive catchment borders and other watershed features such as elevation zones, aspect, and slope, in addition to these optical datasets. High-latitude and rocky mountainous regions not covered by SRTM3 may access the ASTER GDEM. From the GDEM of the study area, 12 distinct altitudinal zones were found for thorough analysis of SCA estimation. Table 13.2 lists some of the features of these zones.

13.4.3 Hydrometeorological Data

For the study area, hydrometeorological data on precipitation, maximum, minimum, and average temperatures are necessary. Meteorological data has been gathered for the years 2013–2015. Hydrometeorological data were given by the MOSDAC-Space Applications

Table 13.1 Summary of data used in the study.

Satellite	Sensor	Data type	Time period	Spatial resolution (m)
Landsat 8	OLI/TIRS	Multispectral image	2013–2015	30
ASTER	ASTER	Multispectral image	2008	30

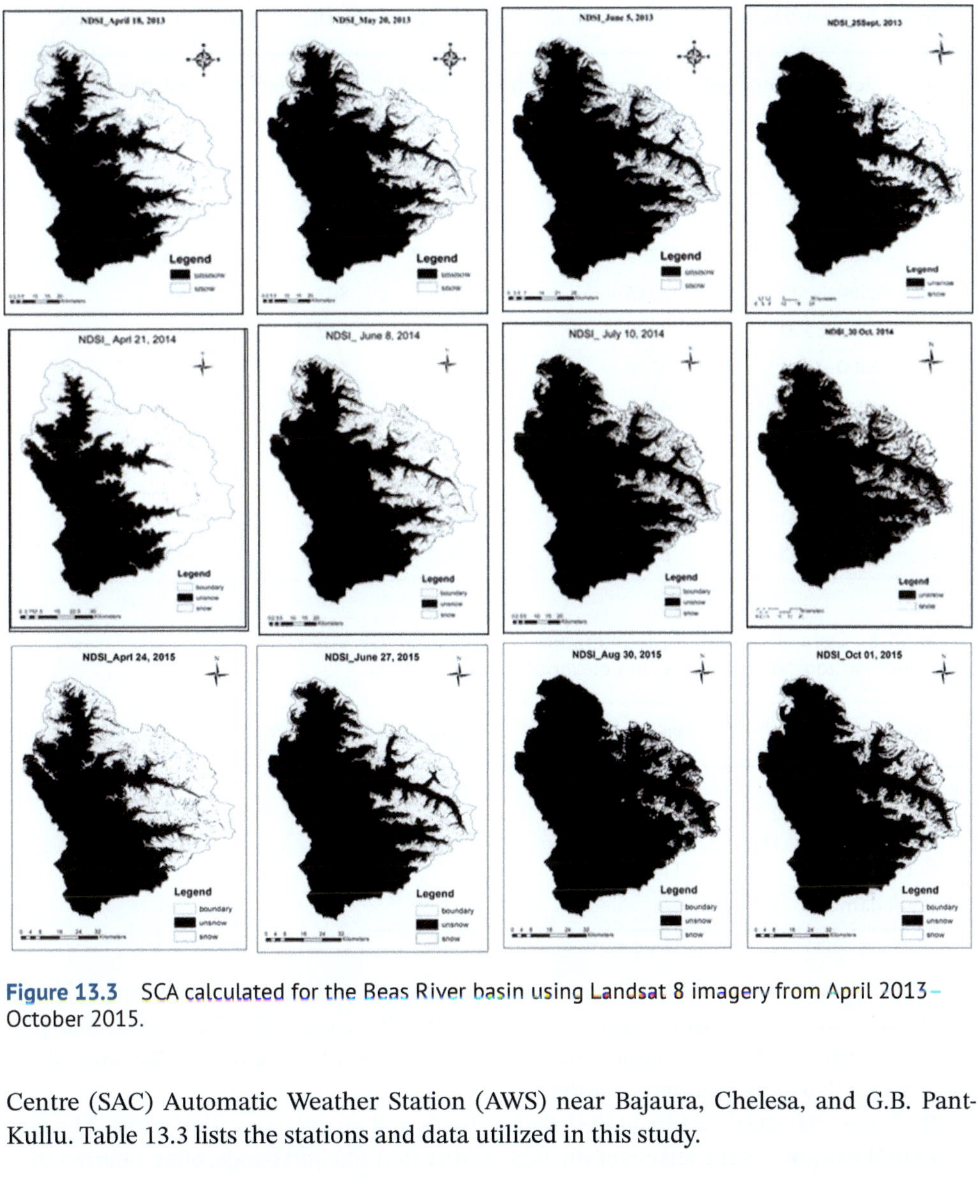

Figure 13.3 SCA calculated for the Beas River basin using Landsat 8 imagery from April 2013–October 2015.

Centre (SAC) Automatic Weather Station (AWS) near Bajaura, Chelesa, and G.B. Pant-Kullu. Table 13.3 lists the stations and data utilized in this study.

13.5 Results and Discussion

13.5.1 Elevation Zonation/Reclassification

We used ArcGIS tools to categorize and segregate the watershed based on different elevations. This was done to model the process of snowmelt runoff utilizing SRM. Using the DEM of the catchment, we identified 12 distinct elevation zones. These zones have different heights and geographic extents. The demarcation of these zones was established by considering the disparity in altitude between the lowest point (857 m) and the highest point

Table 13.2 Features of the 12 elevation zones derived from the DEM of the Beas River basin.

Zone	Elevation range (m)	Mean elevation (m)	Area (%)	Area (km^2)	Cumulative area (km^2)
A	857–1000	957	0.17	9.4	9.4
B	1000–1500	1296	4.69	252.8	262.2
C	1500–2000	1771	10.17	547.7	809.9
D	2000–2500	2259	14.07	757.5	1567.4
E	2500–3000	2745	15	808	2375.4
F	3000–3500	3240	11.39	613.6	2989
G	3500–4000	3754	9.88	532.2	3521.2
H	4000–4500	4260	12.10	651.6	4172.8
I	4500–5000	4745	12.46	671.2	4844
J	5000–5500	5210	8.68	467.4	5311.4
K	5500–6000	5646	1.23	66.5	5377.9
L	6000–6582	6135	0.09	4.9	5382.8

Table 13.3 Data on rainfall (P), temperature (T), discharge (Q), hydrometeorological conditions, and elevation at various sites.

Station	Elevation	Data used
Bajaura	1185 m	P, T
Chelesa	1888 m	P, T
G.B. Pant-Kullu	1175 m	P, T
Pandoh dam	—	Q

(6582 m) inside the watershed. These zones serve as the initial reference locations for the SRM calculations. To estimate snowmelt runoff, we have carefully determined and projected several variables and parameters for each elevation zone individually, since these factors vary depending on the elevation.

The basin characteristics include elevation zones and an area-elevation curve. The current DEM categorizes the terrain of the Beas basin into 12 height bands, often referred to as zones, each covering a distance of 500 m. To calculate the degree-day factor (s) for each elevation zone, we can use the average elevation of each zone and extend the base station temperatures to various heights. The meteorological metrics collected consist of the temperature distribution in each elevation zone, the degree-day factor, and the rainfall distribution using Thiessen polygons. Three hydrometeorological stations gather data on rainfall and temperature. The Thiessen polygon approach is used to convert point observations of rainfall or temperature into spatial data. The SRM model computes the weighted precipitation in each zone by using the elevation zones and the Thiessen polygon.

Using the ASTER GDEM (Figure 13.4), we divided the selected river basin in the research area into 12 elevation bands or zones, each spanning 500 m. Areas A and L were not

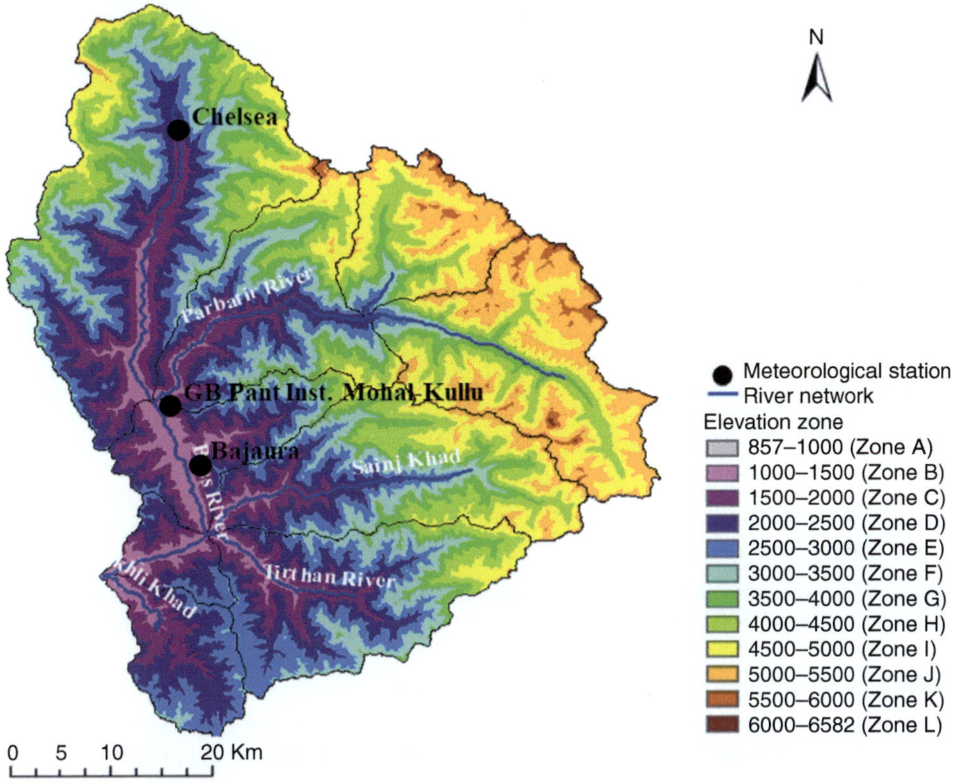

Figure 13.4 The 12 altitudinal zones extracted for this research are shown on the Global Digital Elevation Model (GDEM) of the Beas River basin.

included in this division. We have calculated the average altitude of each height zone by using these bands or zones of elevation. The hypsometric curve in Figure 13.5 illustrates the area distribution for each of the watershed's distinct altitudinal zones, and Table 13.4 lists these zones' attributes (Figure 13.6).

13.5.2 Assessment of the Model Accuracy

Several SRM parameters need to be parameterized to be calibrated and validated. Table 13.5 is a list of the parameters utilized in this SRM simulation study. The period from 2013 to 2014 was the calibration phase (Figure 13.7), while 2014–2015 was the validation phase (Figure 13.8). SRM performs automated calculations and presents the results for volume difference (Dv) and coefficient of determination (R^2), which are well-recognized measures of accuracy. Table 13.6 presents the results of snowmelt runoff estimation using SRM for both the calibration and validation periods. The value of R^2 for the simulation period 2013–2015 ranges from 0.74 (Figure 13.9) to 0.77 (Figure 13.10), according to Table 13.6. The calculated runoff is 27805.12–25087.34 ($10^6\,m^3$), and the average computed runoff is 880.72–794.54 (m^3/s). The measured runoff is roughly 22931.47–23762.97 ($10^6\,m^3$), and the average measured runoff is 724.22–750.48 (m^3/s). The difference in volume ranges from 3.61% in 2013–2014 to 9.76% in

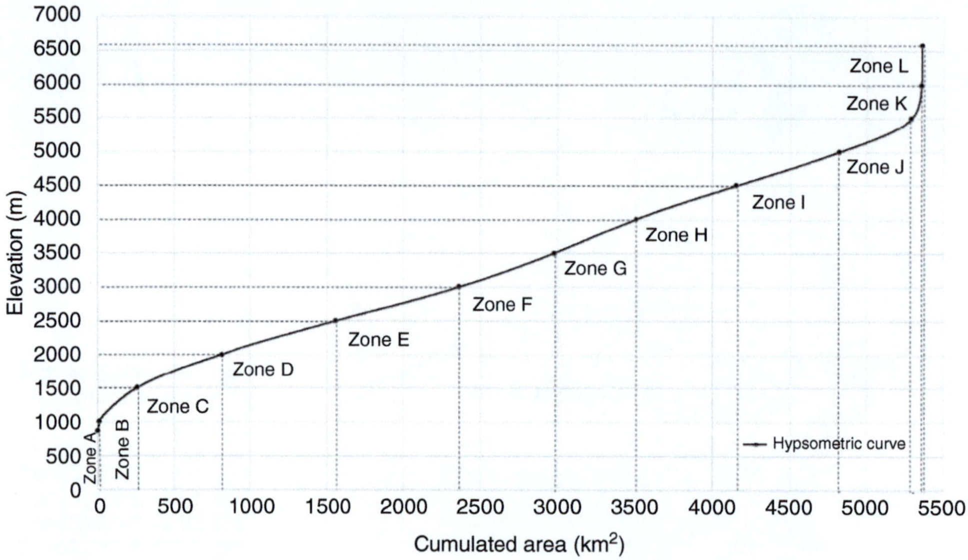

Figure 13.5 Area distribution in the 12 distinct elevation zones is shown on the hypsometric curve of the Beas River basin.

Table 13.4 Percentage of snow cover in elevation zones.

Monthly SCA per elevation zone during April 2013 to October 2015 (%)												
Elev. zone	**A**	**B**	**C**	**D**	**E**	**F**	**G**	**H**	**I**	**J**	**K**	**L**
18 April 2013	0.0	0.0	0.0	0.0	0.4	14.7	59.6	84.5	91.3	92.0	90.8	88.0
20 May	0.0	0.0	0.0	0.0	0.0	0.7	18.3	56.0	82.2	89.9	89.0	85.4
5 June	0.0	0.0	0.0	0.0	0.0	0.1	4.2	32.5	69.3	83.1	84.5	82.6
25 Sept	0.0	0.0	0.0	0.0	0.0	0.0	0.0	13.5	63.4	90.8	92.2	92.7
21 April 2014	0.0	0.0	0.0	0.0	2.7	38.6	95.6	99.3	99.6	99.4	98.3	96.5
8 June	0.0	0.0	0.0	0.0	0.0	0.0	25.1	70.7	91.3	96.5	94.9	91.3
10 July	0.0	0.0	0.0	0.0	0.0	0.0	1.2	16.9	60.2	82.2	80.9	79.9
30 Oct	0.0	0.0	0.0	0.0	0.0	0.0	2.2	22.0	53.6	69.2	68.4	70.3
24 April 2015	0.0	0.0	0.0	0.0	51.0	61.5	73.0	87.2	95.6	97.3	95.1	90.1
27 June	0.0	0.0	0.0	0.0	0.0	18.1	26.0	36.6	86.9	97.3	96.8	93.8
30 Aug	0.0	0.0	0.0	0.0	0.0	0.0	38.1	51.2	71.1	49.9	50.1	49.8
01 Oct	0.0	0.0	0.0	0.0	0.0	0.0	0.0	54.6	58.9	82.4	77.8	67.9

2014–2015. The availability of rainfall data from just three rain gauge stations may be the major explanation for the volume discrepancy between 2013 and 2014, and 2014–2015. The rainfall data available during this era was inadequate to adequately portray the entire quantity of rainfall across the basin, unlike the simulated years 2013–2014 and 2014–2015.

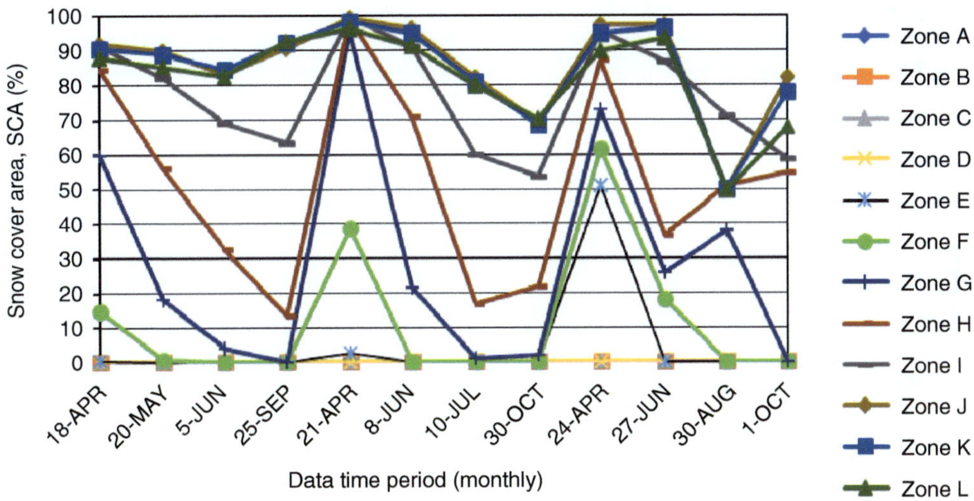

Figure 13.6 Distribution of snow cover in the Beas River basin's 12 various altitudinal zones from 2013 to 2015.

Table 13.5 The range of parameter values used in the SRM model's calibration.

	Range		
Parameters	**Min.**	**Max.**	**Description**
Temperature lapse rate (°C/100 m)	0.71	0.87	Constant for all the zones
Degree-day factor (cm °C^{-1} d^{-1})	0.55	0.75	Varies month wise and zone wise
Critical temperature; TCRIT (°C)	0	2.0	0.0 for Zones 1 to 3, increasing from 0.75 to 2.0 for Zones 4 to 12
Time lag (h)	6	18	6 h for Zones 1 to 3, 12 h for Zones 4 and 5, 15 h for Zones 6 and 7, and 18 h for Zones 8 to 12
Runoff coefficient for snow, CS	0.15	0.20	Varies month wise and zone wise
Runoff coefficient for rain, CR	0.10	0.15	Varies month wise and zone wise
Rainfall contributing area, RCA	0	1	Varies month wise and zone wise (1 for lower zones in all months)
Xcoeff	0.80	1.06	Constant for all the zones
Ycoeff	0.02	0.02	Varies month wise

The spring–summer (April–May) season observed the start of the SRM's rise, chiefly due to an increase in air temperature. The biggest outflow was recorded in August, mostly due to monsoonal rainfall. The period between October and April has the lowest level of snowmelt. Except for the monsoon, the modeling confirms that SRM is the main source of

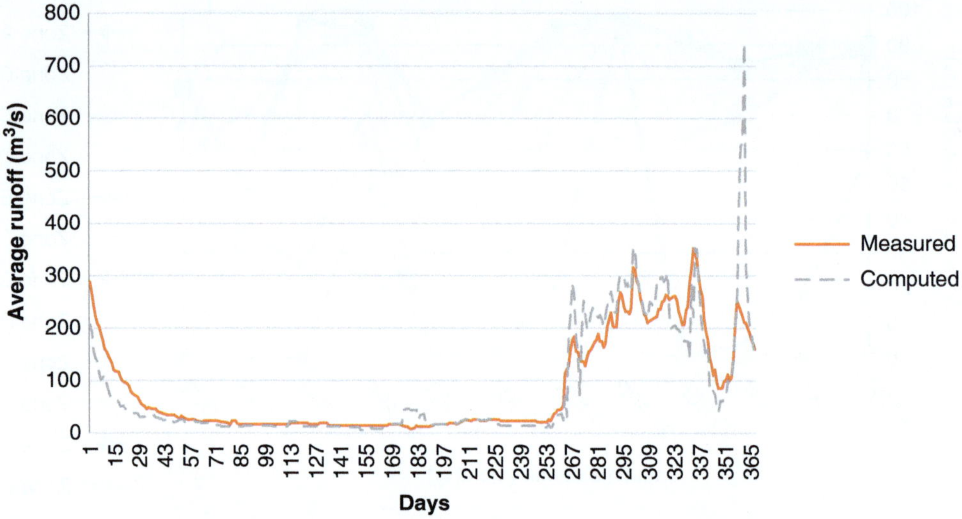

Figure 13.7 Measured runoff versus computed runoff for the period from 2013 to 2014.

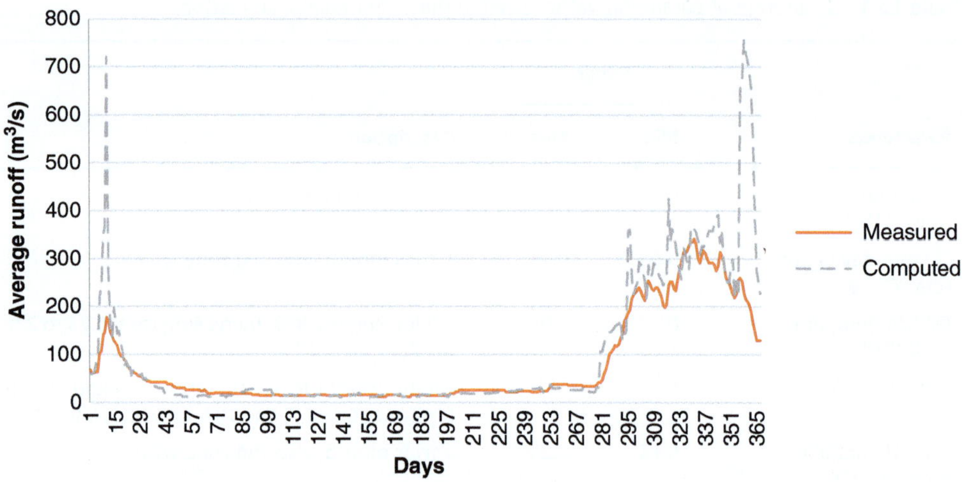

Figure 13.8 Measured runoff versus computed runoff for the period from 2014 to 2015.

Table 13.6 Outcomes of the Beas River basin model simulation from 2013 to 2015, up to the Pandoh dam.

Period	Runoff volume (10^6 m^3)		Average runoff (m^3/s)		Volume difference, D_v (%)	Coefficient of determination, R^2
	Measured	Computed	Measured	Computed		
2013–2014	22 931.47	23 762.97	724.22	750.48	3.61	0.74
2014–2015	25 087.34	27 805.12	880.72	794.54	9.76	0.77

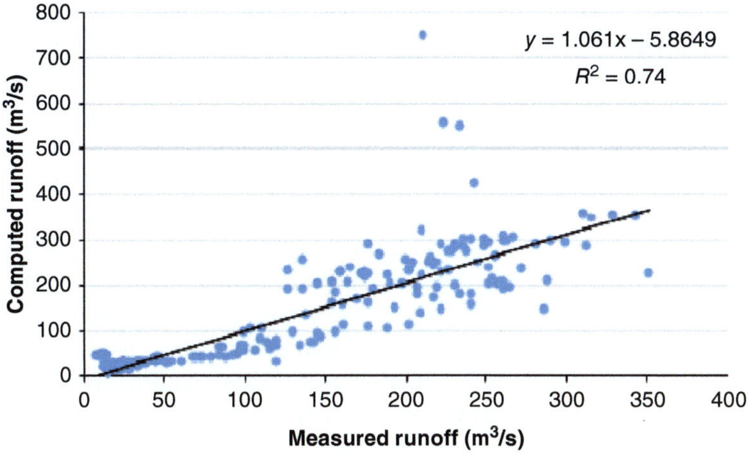

Figure 13.9 Measured versus computed daily discharge to calibrate the model for the 2013–2014 timeframe.

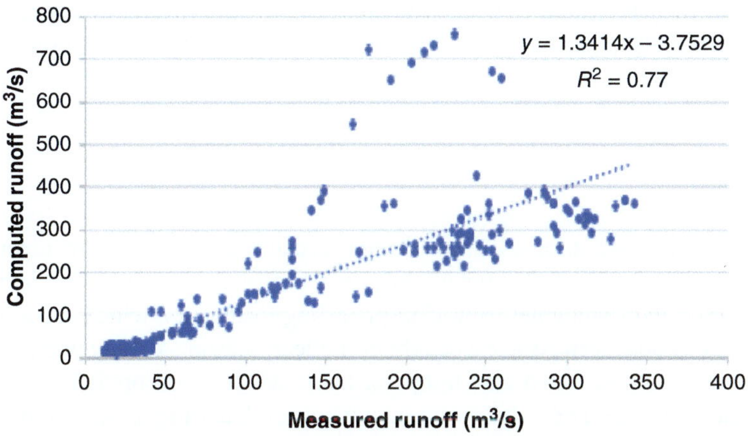

Figure 13.10 Measured versus computed daily discharge to validate the model for the 2014–2015 timeframe.

freshwater in the region all year round. The cumulative impact of rainfall on snow and the usual contribution of rainfall to runoff during the monsoon season is much more than the normal runoff resulting from snow melting, mostly owing to the increase in warmth and precipitation. The snow cover area (SCA) in the basin mostly undergoes melting in the summer, resulting in an increase in both the estimated and observed runoff during this period. Due to the frigid air temperatures and low degree-days in winter, melting is little, particularly in December and February. During the monsoon season, spanning from mid-June to mid-July, there is a substantial amount of precipitation, leading to elevated amounts of both estimated and observed runoff. The SRM methodology relies on the use of snow and rainfall runoff coefficients, together with a degree-day component, as proposed by Martinec and Rango (1986). This implies that the quantity of runoff is directly correlated with the quantity of rainfall, where the runoff coefficients establish the proportion of rainfall that transforms into runoff.

The most significant disparity is in the temporal gap between the predicted and observed peaks of discharge. The recession coefficient, a crucial component of SRM, adjusts and corrects these errors, with (1–*k*) being the fraction of daily meltwater production that immediately converts into runoff (Martinec et al. 2008). The *k*-value was discovered by plotting the values of Qn and Qn + 1 on a graph and analyzing their relationship. We acquired the discharge data for the Pandoh dam from the website (https://cds.climate.copernicus.eu) for the years 2013–2015. This data will be used to determine the k values for the research area. The number of peak flows is larger during the monsoon season owing to multiple severe rainfall events, hence more calibration for *X* and *Y* coefficients is necessary. The estimated and observed runoff peak discharge values vary directly with the runoff coefficient, and the *X* and *Y* coefficients are primarily responsible for the position of flow peaks. Another factor contributing to the SRM's poor peak discharge simulation is the lack of representative rainfall data information in each elevation zone in the basin. A suggestion has been made to compute the mean quantity of snowmelt that enters the basin, taking into consideration any losses, over a significant duration. This calculation might assist in determining the extent to which the snowmelt component contributes to the total outflow of water. SRM separates the constituents of runoff that are caused by precipitation and the thawing of snow.

13.6 Conclusions

From the overall study and observations, it can be concluded at a note that, as previously gone through, in the Himalayan rivers flowing all the 12 months in a year, have a considerable amount of contribution of snowmelt runoff in providing freshwater, which sustains billions of people living in Southern Asia, especially in the lean season. A study was conducted on the integration of the DEM and temporal SCA, which were obtained from satellite-based RS data using GIS. These data were then incorporated into a snowmelt runoff estimation model based on temperature index. This research focuses on estimating snowmelt in the Beas River basin, specifically up to the Pandoh dam. To find out aspect maps and elevation zones, temporal SCA (2013–2015) for the Beas River, which is remotely sensed data and DEM data have been utilized. The SRM is a component of the temperature index-based simulation models used to predict the runoff resulting from snowmelt. We have used hydrometeorological data from the MOSDAC to calculate the amount of snowmelt. Accurate assessment of snowmelt runoff is essential for the effective management of water resources and hydropower generation in the Western Himalayan rivers of India, particularly during the non-monsoon period. This research used the SRM to simulate the behavior of runoff in a snow-covered alpine basin. It does this by merging information from the Landsat 8 database about the extent of snow coverage with daily data on temperature and rainfall in the region.

The temperature index-based SRM WinSRM has shown to be useful in predicting the runoff resulting from snow and glacier melt when data is not available. This is accomplished by using SCA data acquired from Landsat 8 images, ASTER GDEM data, and hydrometeorological data. The elevation zone approach improves the quality of SRM simulations by distributing parameters based on elevation zones and thus enhancing the SRM output. The calibration parameters with the greatest sensitivity in SRM are the recession coefficients, runoff coefficients, and degree-day factor. The snowmelt contribution is especially significant during the period from April to June, when the air temperature consistently rises. The monsoon season, spanning from July to September, is characterized by the highest volume of discharge, resulting from the combined effects of snow and glacier melt as well as rainfall runoff, seen during the research period. Significant melting occurs between the months of October and November. The average coefficient of determination (R^2) for the snowmelt season in the Beas River watershed for the 2013–2015 modeling period was 0.77. The mean volume difference is around plus or minus 6.68%. The mean computed runoff is 772.51 m^3/s, and the mean calculated runoff volume is 25 784.04 million cubic meters.

Mountainous watersheds globally have effectively implemented SRM for several years. By using improved input climatic variables and advanced parameter selection approaches, the Streamflow Routing Model has the capability to forecast streamflow and evaluate the effects of climate change on runoff in mountainous watersheds. This is particularly useful for watersheds with limited data availability in high-altitude regions. While the majority of basins do not have available climatic data, the limited number of SRMs used demonstrated satisfactory model accuracy. The growing availability of RS data enables the use of the model in mountainous basins that are mostly covered in snow, where the gathering of hydrometeorological data is severely limited or nonexistent. It would be beneficial to do more research on the practicality of predicting snowmelt flow in mountainous watersheds with limited hydrometeorological observations. Innovative model parameterization techniques, field observations, GIS tools, and RS data on snow and glacier cover may all be used in this study.

Acknowledgments

The hydrometeorological data for the research was obtained from the MOSDAC website (www.mosdac.gov.in). We are grateful to the Copernicus Climate website (https://cds.climate.copernicus.eu) for providing the Pandoh dam discharge data, and we also acknowledge the United States Geological Survey (USGS) Earth Explorer website for supplying the Landsat 8 and ASTER GDEM data.

References

Aggarwal, S.P., Thakur, P.K., Nikam, B.R., and Garg, V. (2014). Integrated approach for snowmelt run-off estimation using temperature index model, remote sensing and GIS. *Current Science* 106 (3): 397–407.

Azam, M.F., Wagnon, P., Vincent, C. et al. (2019). Snow and ice melt contributions in a highly glacierized catchment of Chhota Shigri glacier (India) over the last five decades. *Journal of Hydrology* 574: 760–773.

Daoo, V.J. and Shirvaikar, V.V. (1978). Snow melt run off estimate from the Beas catchment in the Himalayas using energy balance techniques. *Archiv für Meteorologie, Geophysik und Bioklimatologie, Serie B* 26 (2–3): 183–192.

DeWalle, D. and Rango, A. (ed.) (2008). Snowmelt-runoff model (SRM). In: *Principles of Snow Hydrology*, 306–364. Cambridge: Cambridge University Press.

Engelhardt, M., Ramanathan, A.L., Eidhammer, T. et al. (2017). Modelling 60 years of glacier mass balance and runoff for Chhota Shigri glacier, Western Himalaya, Northern India. *Journal of Glaciology* 63 (240): 618–628.

Gupta, R.P., Duggal, A.J., Rao, S.N. et al. (1982). Snow-cover area vs. snowmelt runoff relation and its dependence on geomorphology—a study from the Beas catchment (Himalayas, India). *Journal of Hydrology* 58 (3–4): 325–339.

Hofierka, J. and Onačillová, K. (2021). Estimating visible band albedo from aerial Orthophotographs in urban areas. *Remote Sensing* 14 (1): 164.

Jain, S.K., Goswami, A., and Saraf, A.K. (2010). Assessment of snowmelt runoff using remote sensing and effect of climate change on runoff. *Water Resources Management* 24 (9): 1763–1777.

Khan, F., Das, B., and Mishra, R.K. (2022). An automated land surface temperature modelling tool box designed using spatial technique for ArcGIS. *Earth Science Informatics* 15 (1): 725–733.

Kulkarni, A.V., Singh, S.K., Mathur, P., and Mishra, V.D. (2006). Algorithm to monitor snow cover using AWiFS data of RESOURCESAT-1 for the Himalayan region. *International Journal of Remote Sensing* 27 (12): 2449–2457.

Kumar, V.S., Paul, P.R., Rao, C.L.R., and Haefner, H. (ed.) (1993). Snowmelt runoff forecasting studies in Himalayan basins. In: *Snow and Glacier Hydrology, Proceedings of the Kathmandu Symposium*, 85–94. Kathmandu: IAHS Publication.

Kumar, V., Singh, P., and Singh, V. (2007). Snow and glacier melt contribution in the Beas River at Pandoh dam, Himachal Pradesh, India. *Hydrological Sciences Journal* 52 (2): 376–388.

Mandal, A., Ramanathan, A., Azam, M.F. et al. (2020). Understanding the interrelationships among mass balance, meteorology, discharge and surface velocity on Chhota Shigri glacier over 2002–2019 using in situ measurements. *Journal of Glaciology* 66 (259): 727–741.

Martinec, J. (1975). Snowmelt-runoff model for stream flow forecasts. *Hydrology Research* 6 (3): 145–154.

Martinec, J. and Rango, A. (1986). Parameter values for snowmelt runoff modelling. *Journal of Hydrology* 84 (3–4): 197–219.

Martinec, J., Rango, A., and Roberts, R.T. (2008). *Snowmelt Runoff Model (SRM) User's Manual*, 175. Las Cruces, New Mexico: New Mexico State University.

Nagler, T. and Rott, H. (1997). The application of ERS-1 SAR for snowmelt runoff modelling. *Hydrological Sciences* 242: 119–128.

Nagler, T., and Rott, H. (1998). SAR tools for snowmelt modelling in the project HydAlp'. In *IGARSS'98. Sensing and Managing the Environment 1998 IEEE International Geoscience and Remote Sensing. Symposium Proceedings*, Cat. No. 98CH36174, Vol. 3, pp. 1521–1523. IEEE.

Nagler, T. and Rott, H. (2000). Retrieval of wet snow by means of multitemporal SAR data. *IEEE Transactions on Geoscience and Remote Sensing* 38 (2): 754–765.

Northern Khorasan Water Regional Company (2013). Physiography. Meteorology and Hydrology Studies Report.

Pangali Sharma, T.P., Zhang, J., Khanal, N.R. et al. (2020). Assimilation of snowmelt runoff model (SRM) using satellite remote sensing data in Budhi Gandaki River basin, Nepal. *Remote Sensing* 12 (12): 1951.

Prasad, V.H. and Mahadev, R.H. (ed.) (2006). Estimating actual evapotranspiration using RS and GIS. In: *Agriculture and Hydrology Applications of Remote Sensing*, vol. 6411, 116–126. SPIE.

Prasad, V.H. and Roy, P.S. (2005). Estimation of snowmelt runoff in Beas Basin, India. *Geocarto International* 20 (2): 41–47.

Ramahaimandimby, Z., Randriamaherisoa, A., Vanclooster, M., and Bielders, C.L. (2023). Driving factors of the hydrological response of a tropical watershed: the Ankavia River basin in Madagascar. *Water* 15 (12): 2237.

Rango, A. and Martinec, J. (1998). The snowmelt runoff model (SRM) user/s manual. Version 4. http://hydrolab.arsusda.gov/pub/srm/srm4.pdf (accessed 7 October 2024).

Singh, P., Bengtsson, L., and Berndtsson, R. (1995). Relating air temperature to the depletion of snow-covered area in a Himalayan basin. *Nordic Hydrology* 34: 267–280.

Singh, R., Archfield, S.A., and Wagener, T. (2014). Identifying dominant controls on hydrologic parameter transfer from gauged to ungauged catchments – a comparative hydrology approach. *Journal of Hydrology* 517: 985–996.

Sreenivasulu, V. and Bhaskar, P.U. (2010). Change detection in land use and land cover using remote sensing and GIS techniques. *International Journal of Engineering, Science and Technology* 2 (12): 7758–7762.

World Meteorological Organization (1986). Intercomparison of models of snowmelt runoff. WMO Operational Hydrology Report no. 23, WMO Report no. 646. Geneva, Switzerland: World Meteorological Organization.

World Meteorological Organization (1992). *Simulated Real-Time Intercomparison of Hydrological Models*. Secretariat of the World meteorological Organization.

Yuan, F., Hao, Z., Berndtsson, R. et al. (2016). The mass balance of glacier no. 1 at the headwaters of the Urumqi River in relation to northern hemisphere teleconnection patterns. *Water* 8 (3): 100.

Zhou, G., Cui, M., Wan, J., and Zhang, S. (2021). A review on snowmelt models: progress and prospect. *Sustainability* 13 (20): 11485.

14

Management of Water Resources Through the Application of Geographic Information Systems and Remote Sensing for Satpuda Mountainous Terrain, Madhya Pradesh, India

S.F.R. Khadri[1], S.F.A. Khadri[2], Mayura D. Deshmukh[3], and Vinay K. Gaddam[1]

[1] *Department of Civil Engineering, V.R. Siddhartha Engineering College, Deemed to be University, Kanuru, Vijayawada, Andhra Pradesh, India*
[2] *Department of Computer Science and Engineering, Acharya Nagarjuna University, Nagarjuna Nagar, Guntur, Andhra Pradesh, India*
[3] *Department of Geology, Shri Shivaji Science College, Amravati, Maharashtra, India*

14.1 Introduction

The research region lies between latitudes 21° 11′–21° 52′ N and longitudes 75°55′–76°30′ E found in Toposheet no. 55C/1 to C/8 (Figure 14.1). The research area Asirgarh is located in Madhya Pradesh's Burhanpur District exposing 460 m thick pile lava flows.

The Tapi and Chhota Tawa rivers drain the research region (Figure 14.1). The research location is situated between the Satpuda hills and the Chhota Tawa and Tapi River basin on level ground along the north bank of the Narmada River. Along with the Tapi River, the two other peninsular rivers in India that travel from east to west are the Mahi and Narmada Rivers. The river originates in the eastern Satpuda Mountain ranges of Maharashtra. The river's tributaries are the Boked, Sirale, Nagholi, Utavali, Sukhi, Kara, Pandhar, Masak, and Siwal Rivers. A few tributaries are the subject of research. The Chhota Tawa River is the longest tributary of the Narmada, and spans 172 km (Murthy and Mishra 1981). It originates in the Betul Satpura mountains and flows north and west until joining the Narmada near the village of Bandra Bhan in the Hoshangabad District. Founded beside the Tawa in 1865, the Bori Reserve Forest is the oldest forest preserve in India (Murthy and Mishra 1981). The Chhota Tawa River has the Wana, Abma, Sukta, Bham River, and Ladampura as tributaries.

14.2 Climate and Vegetation

Asirgarh experiences subtropical weather. It is subject to the summer, winter, and monsoon seasons as well. The study region sees temperatures above 29.4 °C from March through June, considered the summer season. It experiences a hot, dry summer (April–June), followed by monsoon rains (July–September), and a cold, generally dry winter. Quiet soil

Water Resources Management in Mountain Regions, First Edition. Edited by Virendra Bahadur Singh, Sughosh Madhav, Rakesh Kumar Gupta, Prerna Diwan, and Amit Kumar.

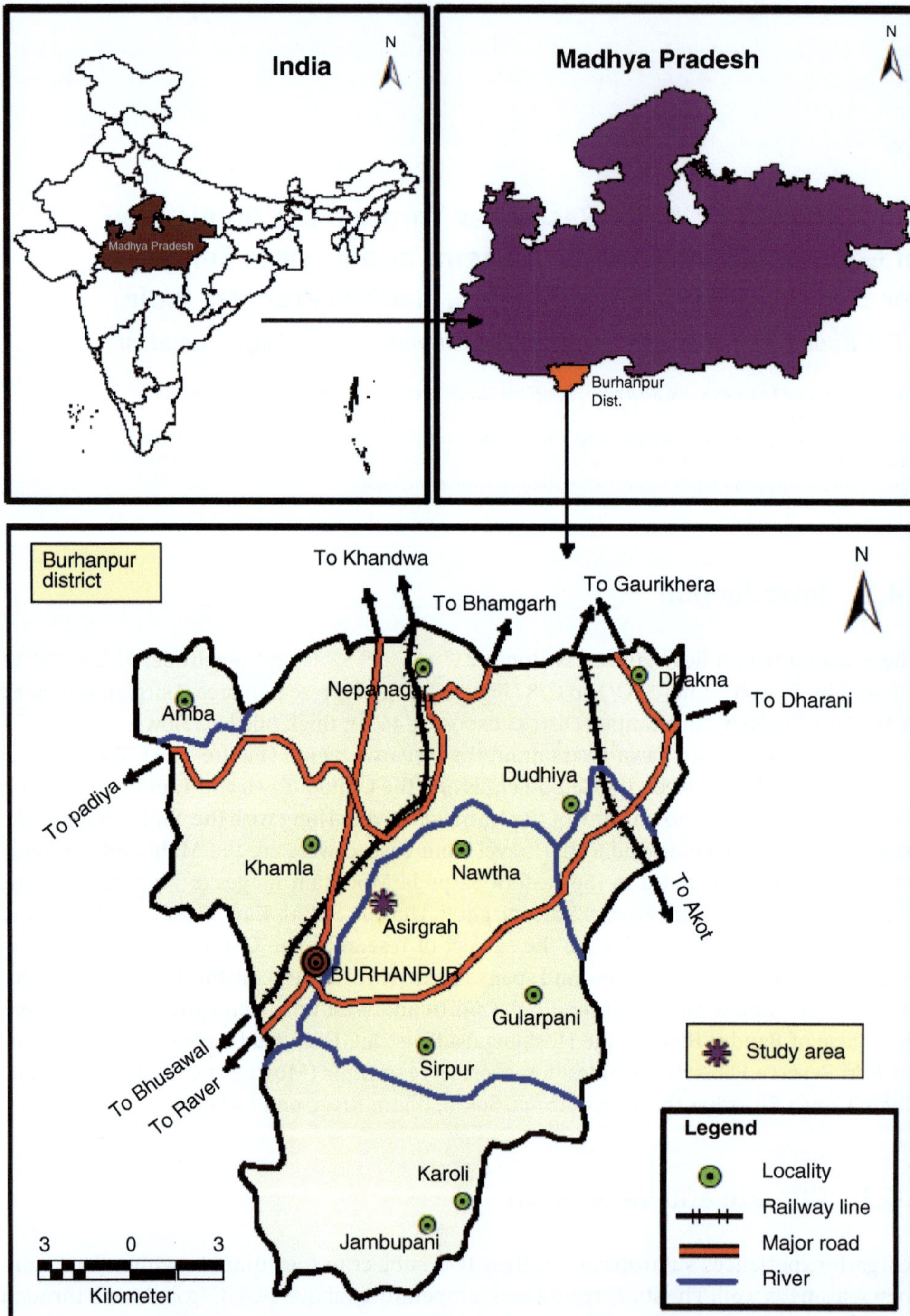

Figure 14.1 Location map of the study area.

storms are frequent in the region, which receives very little rainfall. Most of the state's rainfall occurs between June and September, with the southwest Monsoon usually arriving in the middle of June. The south and southeast of the region receive more rainfall than the north and west, which receive less. In comparison to the southern regions of the state, the northern regions of the state experience lower wintertime temperatures. Nearly every northern region of January experiences the highest temperature, which ranges from 15 to 18°C (Deshmukh 2019, PhD Thesis unpublished). In general, the weather is dry. The state can be physically split into four regions based on climate: the northern plains, the Vindhyas, the Narmada Valley, and the Malwa plateau. Rainfall falls from east to west, with 1370 mm being the average amount (Sreedevi et al. 2004).

In comparison to the southern regions of the state, the northern regions of the state experience lower wintertime temperatures. Nearly every northern region of January experience the highest temperature, which ranges from 15 to 18°C. In general, the weather is dry. The state can be physically split into four regions based on climate: the northern plains, the Vindhyas, the Narmada Valley, and the Malwa plateau. Rainfall falls from east to west, with 1370 mm being the average amount.

The state has particularly varied topography and therefore shows the difference in the soil and vegetation. Agriculturally, Malwa and Central Narmada Valley (CNV) are advanced agro-climatic zones (ACZs). The cropping patterns show a significant existence of cash crops and horticultural crops in Malwa and CNV, ACZ. The normal production of land in this area is higher than that of the state average. CNV ACZ has a higher percentage of irrigated terrain to mesh sown area at the same time as Malwa is third in the list. Region is comprised of Gwalior, and Bundelkhand in the Malwa region of the state and is wealthy in black soil. In the Narmada Valley, the composition of soil and vegetation is different. There are rich alluvial deposits in this part of the state. The soils of Madhya Pradesh range in composition from rich clayey soils to gravelly soils. Major soil groups found in the state include mixed red and black, alluvial, shallow and medium black, and medium and deep black soils.

14.2.1 Literature Survey

A brief reference to the geology of the Burhanpur district was completed by Blanford (1869) in his journal; "On the geology of the Tapti and Narmada valleys and some adjoining districts" in which the most important features of the southern parts of the Gawilgarh range and the Purna plain are outlined. Bose (1884) recognized the relationship of alkali and acidic rocks inside the Deccan Traps and the majority of the lava eruptions in the Deccan Trap likely originated from dykes. The majority of the lava eruptions likely originated from the dykes, as Oldham (1893) noted. In order to shed light on some of its essential physical, chemical, and optical properties, Washington (1922) made a statement about the unbelievable homogeneity of Deccan Basalt's quartz tholeiitic character. You may say that the present study was pioneered by Fermor (1934), who used petrographic, chemical, and field studies to map the presence of individual lava flows near Chhindwara in the N-E region of the Deccan.

Different geological epochs were assigned to the creation of the Narmada Valley by Crookshank (1936), who determined that it is a rift valley. Pascoe (1959) argued that the Western Ghat escarpment was the surface expression of a large trap-cutting fault. Krishnan

(1960) has cited the occurrence of upper, middle, and lower traps based on stratigraphy and intertrappean beds. Sukheswala and Udas (1963), Krishnamurthy (1971), and Krishnamurthy and Cox (1977) designate the presence of alkaline rocks in the western part of the Deccan volcanic province. Crawford (1978) identified the extension of the Narmada–Son lineament across northern Madagascar into Gondwana Land. Murthy and Mishra 1981 described that a zone of crustal weakness with up and down movement from Precambrian to recent times in the Narmada–Son lineament. Lowrie and Alwarez (1981) indicated that according to any polarity timescale, short reversal during the period of Deccan Trap formation was unlikely. Khadri et al. (1988a) has recognized demagnetization in a thick sequence of lava flows in Kalsubai hill in the Western Ghat. Bodas et al. (1985–1988) and Khadri et al. (1988a and 1988b) have given detailed flow-by-flow stratigraphy aided by extensive geochemical and palaeomagnetic investigation of a part of the western Deccan Basalt. Khadri et al. (1996), Beane et al. (1986), Lightfoot et al. (1990) and Lightfoot (1995) have given a detailed flow-by-flow stratigraphy aided by extensive geochemical and palaeomagnetic investigation of a part of the western Deccan basalt province between Pune and Nasik. Cameron and French (1977) defined the Basalt composition is that chemical composition that as crystallized at one atmosphere below standard situation will yield olivine, clinopyroxene, and plagioclase as the primary silicate phases.

A small section of the Asirgarh Volcanics in the Burhanpur District of Madhya Pradesh has not undergone any magnetic mineral alteration, according to preliminary palaeomagnetic research reported by Khadri et al. (2013). In the Deccan trap area near Manmad in the Indian state of Nashik, geochemists Jain and Gupta (2013) examined intrusive rocks. Tholeiitic rocks composed of lava flows and dykes make up the Manmad area's rock formations in the Deccan Trap. Two types of dykes, one include olivine and the other one without it, are described by the dolerite dykes rocks found in the Manmad region. The advent of computers and digital imagery [Geographic Information System (GIS), satellite photos] has brought about a significant shift in geomorphology. One area that is starting to reap the benefits of digitization in the Earth Sciences is geomorphological mapping. From a visual perspective, nonspecialists typically find it challenging to understand geomorphological maps due to the intricacy of the symbols used. This issue can be resolved with the use of GIS capabilities (Khadri et al. 2013).

Using uranium-lead (U-Pb) zircon geochronology to Deccan rocks, Schoene et.al. (2015) determined that the eruptions began −250 000 years before the Cretaceous–Paleogene boundary and that basalt erupted at a rate of >1.1 million cubic kilometers every -750 000 years. These findings provide credence to the theory that the Deccan Traps accelerated the rate of environmental change and biological turnover during the Late Cretaceous, which in turn triggered the mass extinctions of both marine and terrestrial species.

14.3 Geomorphological Studies

The study of landforms and landscapes, as well as the characterization, categorization, origin, evolution, and history of planetary surfaces, is known as geomorphology. A comprehensive understanding of geology and landform evolution is necessary to comprehend the occurrence of porous and permeable zones (Karanth 1999). The surface features of the research region were identified using the LISS-III satellite picture. A toposheet map combined with field observation typically displays a range of geomorphic features, including a lower dissected plateau, a highly moderately dissected plateau, ridges, valley fill, alluvial plains, and flood plains (Figure 14.2). To determine the elevation or geomorphic characteristics, geomorphologic investigations involve a thorough geomorphology and remote sensing (RS) analysis of the studied area. The LISS-III satellite imagery's visual interpretation techniques are used to create the landform (geomorphologic) map. The visual interpretation process is based on the tone, size, shape, pattern, texture, location, and related background of the image as well as previous maps.

Geomorphology is the study of landforms, their processes, formation, and deposits on Earth's surface (and, in rare cases, other planets). Geomorphology is the study of landscapes to determine how surface processes like air, water, and ice create them. Erosion or deposition produces landforms, which are subsequently transported to new sites and deposited there.

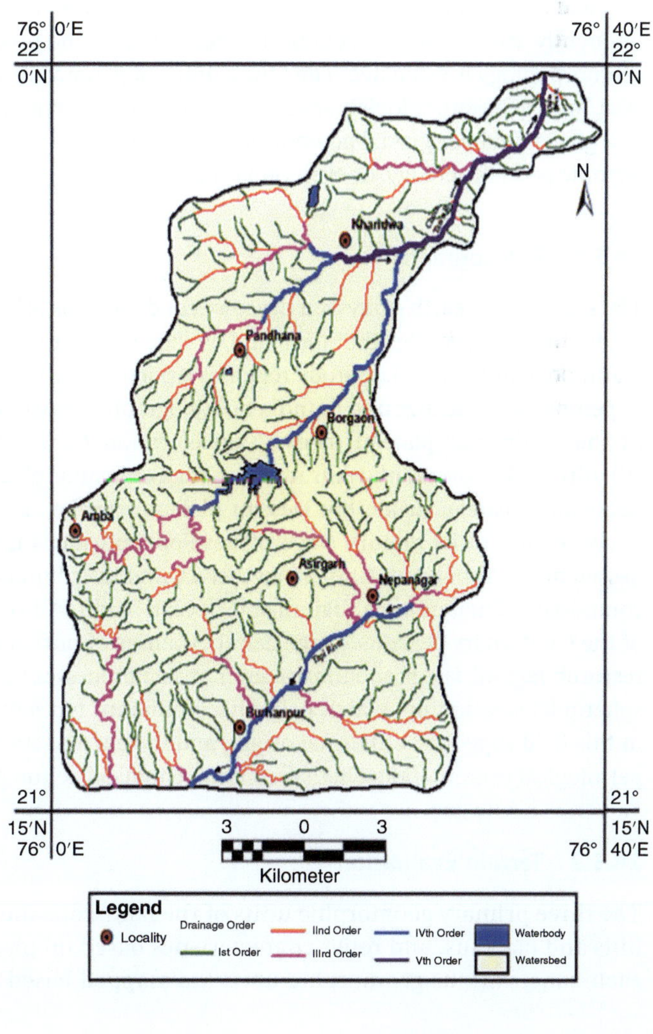

Figure 14.2 Study area drainage map.

Different landform systems result from varying climatic circumstances. Desert landforms, such as dunes and ergs, are substantially distinct from the glacial and periglacial features found in polar and subpolar regions. Recently, landforms have been created by earth-surface processes, altering the terrain, albeit frequently very slowly. The majority of geomorphological processes happen slowly; nevertheless, once in a while, a significant event like a flood or landslip happens, rapidly altering the surrounding area and occasionally endangering people. Thus, geomorphologists are interested in geological risks like landslides, earthquakes, tsunamis, and volcanic eruptions. Global distributions have been better understood by geomorphologists thanks to advances in GIS mapping and satellite RS during the past few decades.

Geomorphologists, sometimes known as "landscape detectives," delve into the past of a landscape. Many thousands of years ago, the vast majority of ecosystems, including those in Britain and Ireland, underwent glaciation. Like the steep-sided valleys of the Lake District and the drumlin fields of central Ireland, these glaciated environments have shaped the terrain. Geomorphologists can add to our knowledge of these places' historical background by studying both sedimentology and geomorphology. Particles and organic material held in lake sediments and peat, such as pollen, beetles, diatoms, and macrofossils, are frequently used as indicators of past climate change and processes. As a result, the study of geomorphology is extensive. The foundations of geomorphology apply to all settings, however, most geomorphologists specialize in one or two areas, such as planetary geomorphology, glacial and periglacial geomorphology, Aeolian (desert) geomorphology, volcanic and tectonic geomorphology, or even other topics.

14.3.1 Physiography

The study of the earth's physical processes and patterns is known as physiography. It seeks to identify the processes that form and change the weather, oceans, rocks, and patterns of world flora and fauna. The study region is a portion of the Deccan plateau, which is a form of peninsular plateau found in India's physiographic divisions. The uneven plateau, known as the peninsular plateau, rises to an elevation of 600–900 m above the river plains (Khadri 2011). Several plateaus make up the peninsular plateau. The study area, which is a peninsular plateau inside the Deccan plateau, can be classified as the central highland region owing to its mature stage of development and high concentration of lava flows passed by the Tapi River, Chotta Tawa River, and its tributaries. The study area is primarily composed of Quaternary alluvium and Deccan basalt. Alluvial plains are a defining feature of the Quaternary deposits. Large aerial expanses of horizontal lava flows characterize the research region. The predominant geomorphic processes in this area are river erosion and spheroidal weathering, which help divide different types of landforms. Previous research in this field suggested a thermal origin, and the current study, which is based on field and petrological examinations, has shown that chemical action plays a major role.

14.3.2 Terrain Evaluation

The three primary geomorphic units of the landscape include plains, piedmont zones, hills and plateaus, and relief characteristics based on physiography and relief. Within each zone, various geomorphic units are mapped based on the characteristics of the landforms, such as their thickness of deposition, depth of weathering, and areal extent. The interpreted geomorphic units/landforms are verified through field trips. Numerous elements are assessed at stream cuttings, lithology, etc., including the type of material deposited, the thickness of the deposit, the degree of weathering, etc. One phase in the transformation process is geometric rectification using ground control points (GCPs) located on the input shape file and related SOI map. The analysis of landforms and geomorphic units is based on the following fundamental concepts of landforms.

14.4 Structural Hill

The tectonic process is responsible for the formation of the structural hills. The drainage lines sever a great deal of the hills. Hills are categorized as having high, moderate, or low dissection based on drainage and joint density. This is typically deduced from plan metric satellite data, and the classification is somewhat subjective. A ridge is a type of geological feature that consists of a series of hills or mountains that continue to build a high crest over a certain length of time. In addition to being referred to as hills or mountains, ridges come in various main varieties based on their size.

14.4.1 Landforms Originating From Denudation

Where the denudation process outweighs the other process, it forms. The majority of the landform produced by this process is the product of both chemical and mechanical weathering working together. The process of removing material by weathering and erosion is called denudation. Water, ice, and wind are the agents that have a direct impact on the area's relief, particularly when it comes to reducing relief to the base level. Numerous elements, including tectonics, climate, geology, and human activity, influence denudation. Consequently, various landforms with characteristic shapes and forms are produced by weathering and erosion. Smaller pieces of the parent material, or rock-weathered material, are formed as a result of the dissolution of part of the minerals and their removal by water. Erosion is the process of removing material, and it is carried out by weathering, wind, water, and glaciers. Weathering serves as a raw material for sedimentary rock and soil. The mapped landforms affected by denudation are described below.

14.4.2 Plateau

A plateau is generally higher and more expansive than a plain, with more noticeable relief. Any relatively flat area of considerable size and elevation as well as an extensive land region significantly elevated [above 150–300 m above sea level or the neighboring country; (Khadri 1989, PhD Thesis unpublished)] and usually bounded on at least one side by an abrupt descent, with a largely smooth or nearly smooth surface that is often carved out by deep valleys or canyons and topped by ranges of tall hills or mountains, with a significant portion of its total surface at or near the summit level (Figure 14.1).

14.4.3 Mesa

Mesas are isolated, nearly level land masses that are elevated above the surrounding area. They are surrounded by abrupt or steep slope erosion scarps and are covered in layers of nearly horizontal, resistant rocks. Less precisely, a very wide, level-topped, typically solitary hill or mountain of moderate height that is bordered by a cliff or slope on at least one side and is thought to be a residue of erosion.

14.4.4 Pediment

A low-lying, broad, gently sloping erosion surface or plain created by subaerial forces (such as running water) characterizes the dry or semiarid area at the foot of an abrupt, receding mountain facade or plateau escarpment. Bedrock covers it, and older alluvial deposits may occasionally cover it as well. The bedrock could be partially exposed, but more often than not it's partially covered with a veneer of alluvium that has been transported across the surface from the upland masses.

14.4.5 Pediplain

Eventually, two or more neighboring pediments combine to generate a huge, multiconcave, rock-cut erosion surface that is the mature stage of the erosion cycle, sometimes crowned by a desert dome.

14.4.6 Basin Geometry for Drainage

The research region includes the Tapi and Chhota Tawa river drainage basins. The research location is located in the Narmada and Tapi River basin, amid the Satpura mountains, on level ground along the Tapi River's north bank. From northeast to southwest, the Tapi River flows. Two of the three peninsular routes cross the Tapi River. The Chhota Tawa River is the longest tributary of the Narmada, measuring 172 km (Sreedevi et al. 2004.). It originates in the Satpura hills of Betul and flows north and west until joining the Narmada near Bandar Bhan in the Hoshangabad District. The river basin is permeated by a dendritic to subdendritic drainage system (Figure 14.2). According to Singh (1998), the dendritic drainage pattern is one of the most common in the Himalayas and is unaffected by lithological or structural causes. A methodical explanation for the drainage basin's geometry and stream-channel system requires dimensions of its linear, areal, and relief (gradient) features as well as contributing ground slopes (Chow 1964). According to Rao et al. (2010), these morphometric measures provide information on several topographical aspects, such as the kind of bedrock, the area's infiltration capacity, and the conditions governing runoff and denudation.

14.4.7 The Basin's Linear Features

The drainage network showcases the properties of the river and its tributaries from their source to their mouth, along with the structural and lithological limits that govern the drainage basin. These linear aspects make up the drainage basin. The most important linear characteristics of a drainage basin are its order, number, bifurcation ratio, mean bifurcation ratio, mean stream length, and the ratio of mean stream length to streams length.

14.4.8 Stream Order (S_u)

This is the drainage basin's linear aspect. This marks the commencement of both the drainage basin analysis and the quantitative watershed analysis. After minor revisions, Strahler (1952) proposed the stream ordering scheme, which was initially discovered by Horton (1945). The current investigation has employed Strahler's stream ordering system to arrange streams in a particular order. Strahler (1964 and 1956) classified order 1 tributaries as the tiniest fingertip tributaries. The junction of two first-order streams creates a channel segment of order 2, while the merging of two orders 2 creates a segment of order 3, and so on. As a result, the highest-order stream portion is the stem stream, through which all water and sediment discharge occurs. Stream frequency has also been reported to decrease when stream order increases. The drainage patterns of the watershed's stream network, which range from dendritic to subdendritic in type, show a lack of structural control and texture uniformity. This pattern is distinguished by its tree- or fern-like form, with branches that primarily meet at acute angles. Table 14.1 reveals that the current study region contains five stream orders in total. Figure 14.3 depicts the stream order map of the study area.

14.4.9 Stream Number (N_u)

The stream number is the total number of stream segments ordered by definition. According to Horton (1945), the stream segment numbers for each order form an inverse geometric sequence with the order number. The current research area contains 376 streams, including 283, 71, 17, 4, and 1 (Table 14.1).

Table 14.1 Stream order (S_u), streams number (N_u), and bifurcation ratios (R_b) of the study area (Deshmukh 2019).

S_u	N_u	R_b	N_{u-r}	R_b*N_{u-r}	R_b*N_u	R_{bwm}
I	283	—	—	—	—	93.62
II	71	3.98	354	1408.92	282.580.76	
III	17	4.17	88	366.96	70.890.19	
IV	4	4.25	21	89.25	170.05	
V	1	4	5	20	40.02	
Total	376	16.4	468	1885.13	374.47	
Mean		4.1*	117	471.28	93.62	

The N_{u-r} is the number of streams utilized in the R_{bm}, which is the mean bifurcation ratio. Rates of bifurcation with weighted means (R_{bwm}).

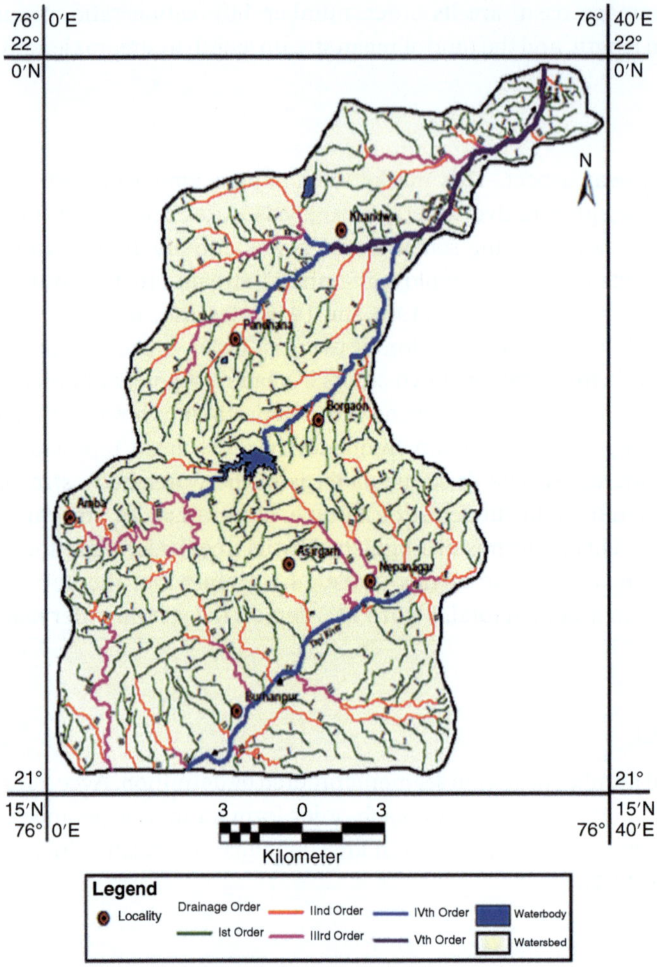

Figure 14.3 Stream order map of the study area.

14.4.10 Bifurcation Ratio (Rb)

The term "stream order ratio" refers to the proportion of stream segments in a given order (N_u) to the number of streams in the subsequent higher order (N_{u+1}), as indicated in Table 14.1. In his study, Horton (1945) employed the bifurcation ratio as a means of evaluating relief and dissertation. Strahler's (1957) work indicated that bifurcation has a limited range of variation due to the presence of considerable geological constraints in other areas. The bifurcation ratio indicates that the drainage basin's geological and lithological development influences these irregularities, which vary by order (Strahler 1964). The bifurcation ratio, which typically ranges from 3.0 to 5.0, is a dimensionless quantity. Watersheds with lower bifurcation ratio values have seen fewer structural disturbances (Strahler 1964), and Nag (1998) observes that these disturbances have not caused any distortion in the drainage pattern. A bifurcation ratio in the study area can range between 3.0 and 4.0. According to the current study, watersheds that are unaffected by structural disturbances have lower bifurcation ratio values, such as 3.98, whereas higher values, such as 4.17 and 4.27, suggest significant structural influence over drainage patterns.

14.4.11 Mean Bifurcation Ratio (R_{bm})

The average of the all-bifurcation ratio of the stream order was defined by Strahler (1953). To find the average bifurcation ratio, Schumme (1956) added up the bifurcation ratios for all the streams involved and then took the mean of the results. This was done for each pair of orders that followed. Bifurcation ratios averaging 4.1 in the study region are presented in Table 14.1.

14.4.12 Weighted Mean Bifurcation Ratio (R_{bwm})

After each set of orders, the bifurcation ratio is multiplied by the sum of all the streams that contributed to the ratio, and then the mean is determined (Strahler 1953). The weighted mean bifurcation ratio is the end result of this process. To find the mean bifurcation ratio, Schumme (1956) employed this method. The values of the weighted mean bifurcation ratio and the mean bifurcation ratio are almost the same. According to Table 14.1, the weighted mean bifurcation ratio for this study is 93.62.

14.4.13 Stream Length (L_u)

Horton (1945) and Chorley et al. (1957) used ArcGIS and SOI topographical charts to suggest legislation for the area's watershed. When the stream order grows, the total time of the segments in first-order streams tends to decrease. According to Horton's law of stream lengths, geometric resemblance is often maintained in watersheds of increasing order. Table 14.2 displays the lengths of various stream orders.

Table 14.2 Stream length and stream length ratio of study area.

S_u	L_u	L_u/S_u	L_{ur}	L_{ur-r}	L_{ur}*L_{ur-r}	L_{uwm}
I	1331.45	4.70	—	—	—	-------
II	400.84	5.64	1.2	1732.29	2078.74	532.34
III	189.51	11.14	1.97	590.35	1162.98	
IV	109.16	27.29	2.44	298.67	728.75	
V	42.38	42.38	1.55	151.54	234.88	
Total	1873.34	91.15	7.16	2772.85	2129.39	
Mean	374.66	18.23	1.79	693.21	532.34	

S_u: Sequence of streams, Two variables: stream length (L_u) and stream length ratio (L_{ur}). Mean stream length ratio (L_{urm})*, stream length (L_{ur-r}) as a ratio component L_{umm}: Ratio of weighted mean stream length.

14.4.14 Mean Stream Length (L_{um})

The average size of the drainage network's components and the watershed surfaces they contribute to are reflected in the dimensional attribute known as mean stream length, according to Strahler (1964). The surfaces of the basins and the components of the drainage network are also involved. The total number of segments divided by the total length of the order's stream is used to calculate it.

14.4.15 Mean Stream Length Ratio (L_{urm})

This measure compares the average lengths of segments in a stream from one lower order to the next. A change in the watershed is shown by the stream length ratio among streams of different orders in the study region. This shift could be accounted for by changes in slope and topography, which would place the streams in the study area in the late to early geomorphic evolution stage. The length ratio, which is found to be relatively consistent throughout successive orders of a basin (Vittala et al. 2004; Horton 1945), is defined as the ratio of the mean (L_u) of segments of order (So) to the mean (L_{u-1}) of segments of the next lower order. Each successive order of a watershed's mean stream lengths, as described by "his law of stream lengths," approaches a direct geometric sequence with the first component, "stream length," being the average length of segments in the first order (Table 14.3). As the stream length ratio changes between orders, it signifies that the geomorphic process is entering its late juvenile stage (Singh 1997).

14.4.16 Weighted Mean Stream Length Ratio

This is determined using the Horton (1945) and Chorley (1957) law, which has been computed for the area's watershed using Arc GIS software and SOI topographical sheets. The weighted mean stream length ratio in this study is 1.55.

14.4.17 Length of Main Channel (C_1)

Starting from the watershed boundary and continuing up to the highest limit of the specified sun watershed, this is the longest water course. It was determined that the main channel length is 123.61 km using the ARC GIS-10.1 program.

Table 14.3 Stream orderwise mean area in study area basin (Deshmukh 2019).

Su	Nu	Am	Ar	Arwmn
I	283	12.03	—	1.755
II	71	13.74	1.15	
III	17	23.66	1.73	
IV	4	61.25	2.59	
V	1	94.35	1.55	
Total	376	205.03	7.02	
Mean		41.006	1.755	

The variables Su, Nu, Am, Arwm, Arm, and Arc represent stream orders, number of streams, area ratio, weighted mean area, and mean area ratio*, respectively.

14.4.18 Digital Elevation Model

It summarizes the watershed's terrain feature analysis, which was generated by interpolating contour data. The contour data at 20-m intervals were derived from India's 1:50 000 scale toposheet survey. Gosain and Rao (2004) and Dabaral and Baithuri (2008) have reported on the employment of the SOI toposheet for hydrological investigations in north-east India.

The contour map is converted into a raster format by using georeferencing to specify the pixel size, number of lines and columns, and minimum and maximum coordinates of the maps in the GIS environment. The values for the pixels covered by a contour line were stored in the raster map created when the segments were converted to raster. All additional map pixels are still unknown at this time. The Digital Elevation Model (DEM) is created by applying linear interpolation to each pixel and the rasterized contour lines to provide an altitude value. Contour interpolation creates a raster map that assigns values to each pixel.

14.4.19 TIN Model

This terrain model is based on the Delaunay triangulation, which consists of randomly spaced nodes that form triangles. It employs a sheet of continuous, connected triangular facets. There are a few nonoverlapping triangles with triangulated irregular networks (TINs),which is a digital representation of a surface's morphology and surfaces. The DEM data were used to develop the TIN model for the Tapi micro watershed region (Figure 14.4). It shows how the southwest topography is mountainous and undulating, as compared to the northeast, which is comparatively vast and level. The irregular points are connected with straight-line segments to build a vector topological network with triangular facets;

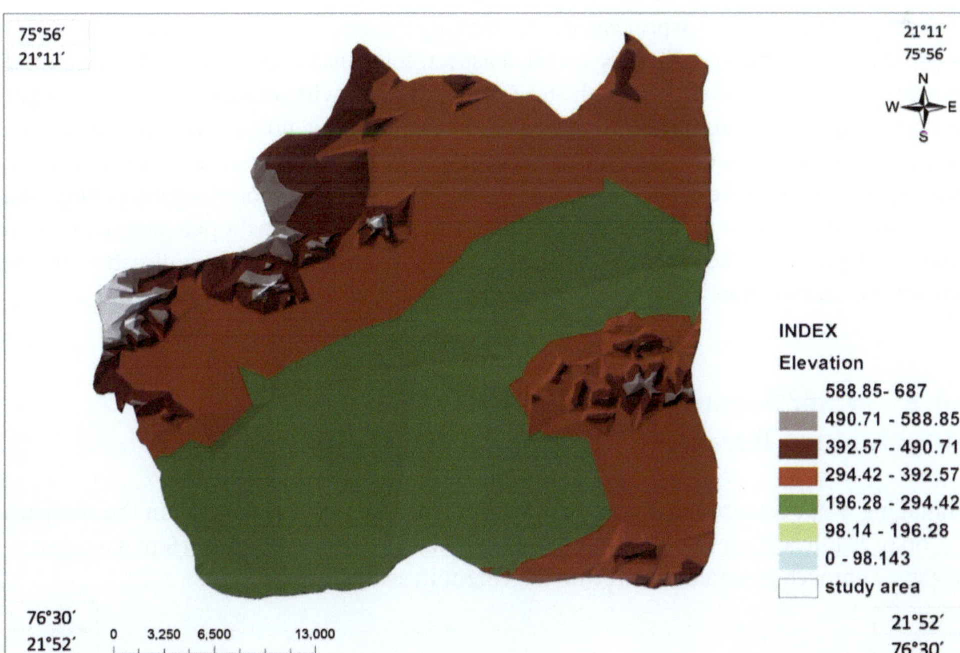

Figure 14.4 TIN map of Tapi microwatershed.

these irregularly spaced triangles serve as contiguous, nonoverlapping triangular components representing a surface. Among many other applications, a triangulated irregular direction of flow in watershed areas can be evaluated using a tree index (TIN) model. By keeping track of the topological links between the triangles, it depicts the surface as an array of nonoverlapping, continuous triangles. The terrain's roughness and the needs dictate the size of the triangles. To create triangles, the triangulated irregular network (TIN) makes use of mass points, which are a collection of points that can be considered a node set in and of itself (Deshmukh 2019).

14.4.20 Slopes and Aspect

The principles of slope and aspect are used to analyze and visualize the features of landforms. They are crucial for research on landscape units, watershed units, and morphometric measurements (Moore et al. 1991). Slope and aspect can assist with runoff computation, estimations of the forest inventory, soil erosion, site analysis, and animal habitat appropriateness when combined with other variables (Wilson and Gallant 2000). The aspect-slope map shows the slope of a landscape along with its steepness, or the direction in which it lies. Saturation, or the brightness of colors, is used to map the degree of slope classes, making steeper slopes appear brighter, while hues like red, orange, and yellow are used to portray aspect categories.

The fundamental component for deciphering and illustrating the properties of landforms is the aspect. According to Moore et al. (1991), they are significant for research on landscape units, watershed units, and morph metric measurements. Aspects can help with site analysis, animal habitat appropriateness, soil erosion, runoff computation, and estimations of forest inventories when combined with other factors (Wilson and Gallant 2000). A terrain's direction is simultaneously displayed on an aspect map. Hues (such as red, orange, yellow, etc.) are used to represent aspect categories, making the steeper slopes appear brighter.

The directional measure of slope is called aspect. It begins at the north with 0 clockwise movements and concludes there with 360. Categorical data with an aspect is often arranged in four or eight directions. The majority of the terrain is facing northward, according to an aspect map of the River (Figure 14.2). In addition, this analysis shows that there is less plain space on the western side than on the eastern side. The rate of elevation change at a surface location is known as the slope, and it can be expressed as a percentage or as an angle in degrees. At the river basin's lower end, slope westward in the direction of the northern-eastern region.

14.5 Water Resource Management of Satpuda Mountainous Terrain

Managing water resources in mountainous terrains like the Satpuda range in the Asirgarh region of Madhya Pradesh presents unique challenges and opportunities. Here is a comprehensive approach to water resource management in such an area:

1) **Watershed Management**: Implementing watershed management practices is crucial for preserving water resources in mountainous regions. This involves techniques such as contour trenching, afforestation, and soil conservation measures to prevent erosion and enhance groundwater recharge.
2) **Rainwater Harvesting**: Given the significant rainfall in the Satpuda region during the monsoon season, rainwater harvesting can be an effective way to capture and store water for future use. This can include building check dams, rooftop rainwater harvesting systems, and recharge pits.
3) **Spring Rejuvenation**: Many mountainous areas have natural springs that serve as vital water sources for local communities. Protecting and rejuvenating these springs through measures like afforestation, soil conservation, and preventing pollution can ensure their sustainability.
4) **Check Dams and Small Reservoirs**: Constructing check dams and small reservoirs along streams and rivers can help regulate water flow, control erosion, and provide water for irrigation, domestic use, and livestock during dry seasons.
5) **Community Participation**: Engaging local communities in water resource management initiatives is essential for their success and sustainability. This can involve training programs on water conservation practices, forming water user associations, and involving communities in decision-making processes.
6) **Integrated Water Resource Management (IWRM):** Adopting an IWRM approach involves considering the interconnectedness of various water sources, such as surface water, groundwater, and rainwater, and developing holistic strategies for their sustainable management.
7) **Water Quality Monitoring**: Regular monitoring of water quality is necessary to ensure that water sources remain safe for drinking, irrigation, and other purposes. Establishing water quality testing laboratories and training local personnel in water testing techniques can help in this regard.
8) **Climate Change Adaptation:** Climate change can impact water availability and exacerbate water-related challenges in mountainous regions. Implementing adaptation measures such as drought-resistant crop varieties, water-efficient irrigation techniques, and early warning systems for extreme weather events is crucial.
9) **Policy Support and Institutional Framework**: Developing supportive policies and establishing institutional frameworks for water resource management is essential. This includes clear regulations for water use, mechanisms for conflict resolution, and coordination among various government agencies and stakeholders involved in water management.
10) **Research and Innovation:** Investing in research and innovation can help identify sustainable water management solutions tailored to the specific needs of the Satpuda mountainous terrain. This can involve research on water-saving technologies, ecosystem-based approaches to water management, and climate-resilient agricultural practices.

By adopting a multifaceted approach that combines ecological, social, and technological interventions, water resource management in the Satpuda mountainous terrain can be optimized for the benefit of both people and the environment.

14.6 Summary and Conclusions

Due to its steep terrain, the Asirgarh region of Madhya Pradesh, India's Burhanpur District, has unique challenges for the management of water resources. This project aims to develop an integrated method using RS, GIS, and artificial intelligence/machine learning (AIML) technologies to increase the knowledge and sustainable use of water resources in this complex environment. Using RS and high-resolution satellite photos, the study tracks and maps the Asirgarh region's surface water bodies, vegetation, and land cover. GIS is used to analyze the spatial distribution of various land cover types to identify potential areas that may be impacted by degradation and water scarcity (Khadri and Pande 2014).

The combination of RS and GIS enables a thorough landscape assessment, which aids in identifying critical zones for targeted water resource management actions. Historical climate data, precipitation patterns, and hydrological factors are examined using AIML techniques. Predictive analytics and data-driven models are two of these techniques. The ability to predict future water availability and identify trends in the dynamics of water resources is made feasible by these models, which supports informed decision-making for sustainable water management. The research area is drained by the Tapi and Chhota Tawa rivers. Both the Tapti River and the Narmada River flow from east to west. The research region lies in the Narmada and Tapi River basin, between the Satpuda mountain ranges, on level terrain along the north bank of the Tapi River, according to Khadri and Deshmukh (2018a, 2018b).

In addition, utilizing GIS and AIML, a spatially explicit water balance model is developed to estimate water supply and demand at various geographical scales. This model considers the water balance of the Asirgarh region holistically, accounting for factors such as topography, soil types, and patterns of land use. Moreover, the GIS framework incorporates decision-support technologies to assist local authorities in optimizing the distribution and utilization of water. The study emphasizes the importance of community involvement and participatory methods in the management of water resources. Water conservation strategies work better when they integrate modern technology with traditional knowledge. The results of the study can be very helpful to policymakers, managers of water resources, and inhabitants in the hilly Asirgarh region in developing sustainable water management techniques. In summary, the integration of RS, GIS, and AIML techniques offers a solid basis for understanding, monitoring, and controlling water resources in challenging alpine settings. The outcome of this research will certainly contribute to the development of water management strategies that are tailored to the specific context of the Asirgarh region and other comparable landscapes worldwide, ensuring the sustainability and resilience of water supplies along with sustainable development.

Acknowledgments

This research would not have been possible without the substantial funding that Dr. S.F.R. Khadri received from the University Grants Commission, New Delhi.

References

Beane, J.E., Turner, C.A., Hooper, P.R. et al. (1986). Stratigraphy, composition, and form of Deccan basalts, western Ghats, India. *Bulletin of Volcanology* 48: 61–83.

Blanforord, W.T. (1869). On the geology of a portion of Cutch. *Memoirs - Geological Survey of India* 6: 17–138.

Bodas, M.S., Khadri, S.F.R., and Subbarao, K.V. (1988). Stratigraphy of the Jawhar and Igatpuri formations, Western Ghat lava pile. *Journal Geological Society of India* 10: 235–253.

Bodas, M.S., Khadri, S.F.R., Subbarao, K.V. et al. (1985). Proceedings of Indian geological congress, V session, Bombay, pp. 339–346 (19–21 December).

Bose, P.N. (1884). Geology of Narmada valley between Nimmar and Kawant. *Memoirs - Geological Survey of India* 21 (1): 332–333.

Cameron, E.P. and French, W.J. (1977). The relationship of the order of crystallization of basalt melts to their classification and to the definition of rock series. *London Mineralogical Magazine* 4: 239–251.

Chorley, R.J., Donald, E.G.M., and Pogorzelski, H.A. (1957). A new standard for estimating drainage basin shape. *American Journal of Science* 255: 138–141.

Chow, V.T. (ed.) (1964). *Handbook of Applied Hydrology*, 572. New York: McGraw Hill.

Crawford, A.R. (1978). Narmada-son lineament of India traced into Madagaskar. *Journal Geological Society of India* 19: 144–1153.

Crookshank, H. (1936). Geology of northern slop of Satpura between the Morand and shear rivers. *Memoirs - Geological Survey of India* 66: 2.

Dabaral, P.P. and Baithuri, N. (2008). Determination of surface drainage coefficient through rainfall analysis for north Lakhimpur, Assam, Soil and water conserve. *India* 7 (1): 38–42.

Deshmukh, M.M. (2019). Flow stratigraphy and geochemistry of Asirgarh Volcanics, Burhanpur District, Madhya Pradesh. PhD thesis (Unpublished). SGB Amravati University.

Fermor, L.L. (1934). On the chemical composition of the Deccan trap flow of Linga, Chindwaradistt, central province. *Records of the Geological Survey of India* 68: 344–360.

Gosain, A.K., and Rao, S. Oct 2004, "GIS based technologies for watershed management," *Special section: application of S& T rural areas, current science*, vol. 87, no. 7, pp. 948–953.

Horton, R.E. (1945). Erosional development of streams and their drainage basins: hydrophysical approach to Quantitative morphology. *Bulletin of the Geological Society of America* 56: 275–370.

Jain, P.K. and Gupta, D.C. (2013). Geochemistry of intrusive rocks of Deccan trap region around Manmad, Nasik, India. *International Journal of Earth & Environmental Sciences* 1 (1): 1–13.

Karanth, K.R. (1999). *Groundwater Assessment Development and Management*. New Delhi: Tata Mc-Graw-Hill.

Khadri, S.F.R. (1989). Flow stratigraphy, geochemistry and palaeomagnetism of a part of the Western Deccan Basalt Province, Maharashtra, India. PhD thesis (Unpublished). IIT Bombay, p. 418.

Khadri, S. F. R. (2011). Morphometric and morphotectonic analysis of Asirgah region, Burhanpur district, MP with reference to remote sensing analysis.

Proceedings of the National Seminar on "Emerging Trends in Scientific Research, India (11 October 2011), 177–188

Khadri, S.F.R. and Deshmukh, M.M. (2018a). Preliminary flow stratigraphy of Asirgarh Volcanic, Burhanpur District, Madhya Pradesh. *International Journal of Creative Research Thoughts* 6 (2): 741–745.

Khadri, S.F.R. and Deshmukh, M.M. (2018b). Petrography of Asirgarh volcanic, Burhanpur district, Madhya Pradesh. *Journal of Applied Science and Computation* 5 (5): 122–128.

Khadri, S.F.R. and Pande, C.B. (2014). Remote sensing and GIS applications of geomorphological mapping of Mahesh river basin, Akola & Buldhana districts, Maharashtra, indicating multispectral satellite data. *Indian Streams Research Journal* 4 (5): 1–7.

Khadri, S.F.R., Pande, C.B., and Mohrir, K. (2013). Geomorphological investigation of WRV-Watershed management in Wardha district of Maharashtra India; using remote sensing and geographic information system techniques. *International Journal of Pure and Applied Research in Engineering and Technology* 1 (10): 18–35.

Khadri, S.F.R., Subbarao, K.V., and Bodas, M.S. (1988a). Magnetic studies on a thick pile of Deccan Trap flows at Kalsubai. *Journal Geological Society of India* 10: 163–180.

Khadri, S.F.R., Subbarao, K.V., Hooper, P.R., and Walsh, J.N. (1988b). Stratigraphy of Thakurwadi formation, western Deccan Basalt Province, India. *Journal of the Geological Society of India* 10: 281–304.

Khadri, S.F.R., Subbarao, K.V., and Walsh, J.N. (1996). Stratigraphy from and structure of the east Pune Basalts, Western Deccan Basalt province, India. *Journal of Geological Society of India* 2: 179–202.

Krishnamurthy, P. (1971). Regional geochemical characters of the Deccan Traps lavas and their genetic implications. *Memoir-Geological Society of India* 3: 394–418.

Krishnamurthy, P. and Cox, K.G. (1977). Picrite Basalt and related lavas from the Deccan traps of Western India. *Contributions to Mineralogy and Petrology* 62: 53–75.

Krishnan, M.S. (1960). *Geology of India Burma*, 604. Madras: Highnbothams.

Lightfoot, P.C. (1995). Isotope and trace element geochemistry of the south Deccan lavas, India. PhD thesis (unpublished) Open University.

Lightfoot, P.C., Hawkesworth, C.J., Devey, C.W. et al. (1990). Source and differentiation of Deccan trap lavas: implication of geochemical and mineral chemical variations. *Journal of Petrology* 31: 1165–1200.

Lowrie, W. and Alwarez, W. (1981). 100 million years of geomagnetic polarity history. *Geology* 9: 392–397.

Moore, I.D., Grayson, R.B., and Ladson, A.R. (1991). Digital terrain modeling: a review of hydrological, geomorphologic and biological applications. *Hydrological Processes* 5 (1): 3–30.

Murty, T.V.V.G.R.K. and Mishra, S.K. (1981). The Narmada-son lineament and the structure of the Narmada rift system. *Journal Geological Society of India* 22: 112–120.

Nag, S.K. (1998). Morphometric analysis using remote sensing techniques in the Chaka sub-basin, Purulia district, West Bengal. *Journal of the Indian Society of Remote Sensing* 26 (1): 69–76.

Oldham, R.D. (1893). *Manual of the Geology of India*, 2nde. Delhi: Government of India.

Pascoe, E.H. (1959). *A Manual of the Geology of India and Burma*, 2nde, 448. Delhi: Government of India.

Rao, K.N., Latha, S., Kumar, P.A., and Krishna, M.H. (2010). Morphometric analysis of Gostani River Basin in Andhra Pradesh State, India using spatial information technology. *International Journal of Geomatics and Geosciences* 1 (2): 179–187.

Schoene, B., Samperton, K.M., Eddy, M.P. et al. (2015). U-Pb geochronology o the Deccan traps and relation to the end-cretaceous mass extinction. *Science* 347 (6218): 182–184.

Schumme, S.A. (1956). Evolution of driange systems and slopes in Badlands at Perth Amboy, New Jersey. *Bulletin of the Geological Society of America* (67): 597–646.

Singh, R.L. (1997). Morphometric analysis of terrain, Presidental address, 54 Indian Science Congress, Hyderabad.

Singh, S. (1998). *Geomorphology*, 613. Allahabad: Prayag Pustak Bhawan.

Sreedevi, P.D., Subrhmanyam, K., and Ahmad, S. (2004). The significance of Morphometric analysis for obtaining groundwater potential zones in a structurally controlled terrain. *Environmental Geology* 47: 412–420.

Strahler, A.N. (1952). Hypsometric analysis of erosional topography. *Bulletin of the Geological Society of America* 63: 1117–1142.

Strahler, A.N. (1953). Hypsometric analysis of erosional topography. *Geological Society of America Bulletin* 63: 1117–1142.

Strahler, A.N. (1956). Quantitative slope analysis. *Bulletin of the Geological Society of America* 67: 571–596.

Strahler, A.N. (1957). Quantitative analysis of watershed geomorphology. *Transactions of the American Geophysical Union* 38: 931.

Strahler, A.N. (1964). Quantitative geomorphology of drainage basin and channel network. In: *Handbook of Applied Hydrology* (ed. V. Chow), 439–476. New York: McGraw Hill.

Sukheswala, R.N. and Udas, G.R. (1963). A note on the carbonatites of Amba Dongar (Gujarat State) and its economic potentialities. *Science and Culture* 29: 563–568.

Vittala, S.R., Govindaiah, S., and Gowda, H.H. (2004). Morphometric analysis of sub watersheds in the Pavagada area of Tumkur District, South India using remote sensing and GIS technique. *Journal of the Indian Society of Remote Sensing* 32: 351–362.

Washington, H.S. (1922). Deccan traps and other plateau basalt. *Geological Society of America Bulletin* 33: 765–803.

Wilson, J.P. and Gallant, J.C. (2000). *Terrain Analysis: Principles and Applications*, 1–37. Research gate net publications.

15

Assessment and Management of Flash Floods in the India Himalayas

Challenges, Strategies, and Solutions

Tanveer Dar[1] *and Kainat Aziz*[2]

[1] *Centre for Cryosphere and Climate Change Studies, National Institute of Hydrology, Roorkee, India*
[2] *Department of Civil Engineering, Indian Institute of Technology, Guwahati, Assam, India*

15.1 Introduction

The Indian Himalayan region, characterized by its unique topography and climatic conditions, faces numerous types of natural disasters, including earthquakes, cyclones, floods, flash floods, avalanches, and glacial lake outbursts. The monsoon often triggers water-induced disasters, with cloudbursts leading to flash floods being a common consequence. A flash flood is a phenomenon marked by the sudden inundation of an area with a large volume of water over a short duration (Gaume et al. 2016). A flash flood emerges within a short timeframe following intense rainfall (Rasmussen and Houze 2012), rapid snowmelt (Wang and Russell 2016), or the sudden release of water from a glacier lake or due to infrastructure failures such as embankment breaches (Sati 2022). It may also occur with the abrupt disintegration of an ice block caused by a rapid rise in temperature (Singh et al. 2022). The impact of flash floods can be severe, resulting in significant casualties and economic losses. According to Jonkman's (2005) global analysis of flood-related fatalities, flash floods are associated with the highest mortality rates.

The Himalayas experience a wide range of climates, from subtropical to permanently snow-capped alpine summits. Because of its extreme vulnerability and propensity for disasters, the Himalayan area is often struck by weather-induced disasters (Buckley et al. 2000; Tucker 1988). Annually, significant loss of life and property damage is documented during the monsoon season. The majority of the economic losses and deaths in the Himalayas are caused by water-induced catastrophes, including flash floods and debris flows, which account for over 70% of these occurrences (Sati 2022). Environmental and human pressures have led to significant warming in the Himalayas, surpassing the world average increase of 0.74°C over the last century (Du et al. 2004; Yao et al. 2006). Due to climate change, rainfall is becoming increasingly erratic across time and space (Dar et al. 2022). Although overall trends are challenging to interpret, there are clear signs that the frequency and intensity of heavy rainfall events are rising (Goswami et al. 2006). There are also signs

Water Resources Management in Mountain Regions, First Edition. Edited by Virendra Bahadur Singh, Sughosh Madhav, Rakesh Kumar Gupta, Prerna Diwan, and Amit Kumar.

that seasonal droughts are becoming worse, and that water stress is becoming more of a concern throughout the dry season. The duration and timing of the monsoon season also seem to be shifting (Ramesh and Goswami 2007). In recent decades, the Himalayan region has seen flash floods caused by cloudbursts, mostly in the regions of Uttarakhand, Himachal Pradesh, and the Leh region of the Union Territory of Ladakh. Even though Leh is in a rain shadow zone, a review of five years of meteorological data looking at monthly rainfall, temperatures, humidity, and snowfall suggests that cloudbursts might have been caused by climate change as the region witnessed extended winters in 2010 (Thayyen et al. 2013). While the preceding hot summers and higher temperatures on the plains cause more water to evaporate, which, in turn, causes more clouds to form on the hills. This chapter discusses the historical context of flash floods, highlighting major incidents and impacts. Additionally, the chapter highlights the topographical susceptibility of various rivers in the Indian Himalayan area to flash floods produced by cloudbursts, landslides, or glacial lake outbursts, as well as the steps that may be taken to mitigate the effects of these floods.

15.2 Causes of Flash Flood

The spatial and temporal variations in rainfall patterns, landscape attributes, and the pre-existing wetness of catchments play pivotal roles in the generation of flash floods. These factors contribute to the considerable variability observed in hydrological responses across different locations and time periods (Borga et al. 2014). Flash flooding, characterized by its sudden onset and rapid escalation, can be triggered by various factors:

a) **Urbanization and Impervious Surfaces**: Rapid urbanization leads to increased areas of impervious surfaces such as pavement and concrete, which reduce infiltration and increase surface runoff. This can exacerbate flash flooding in urban areas by accelerating the flow of water into drainage systems and watercourses.
b) **Infrastructure Failures**: Failures in infrastructure such as dams, levees, or drainage systems can lead to sudden releases of water or blockages that exacerbate flash flooding. Dam breaches, in particular, can result in rapid and extensive flooding downstream.
c) **Snowmelt and Ice Jams**: In regions with cold climates, rapid snowmelt during warm periods can overwhelm rivers and streams, leading to flash flooding. Additionally, ice jams formed by melting and refreezing of river ice can block water flow, causing localized flooding when the jam releases.
d) **Debris and Sediment Accumulation**: Accumulation of debris such as fallen trees, branches, and sediment in watercourses can impede flow and increase the risk of flash flooding, particularly during intense rainfall events. Debris can also become lodged under bridges or culverts, exacerbating flooding in localized areas.
e) **Land Use Changes**: Changes in land use, such as deforestation, agricultural expansion, or urban sprawl, can alter natural drainage patterns and increase the risk of flash flooding. Clearing of forests or wetlands reduces the landscape's ability to retain water, leading to faster runoff and higher flood peaks.
f) **Climate Variability and Change**: Changes in climate patterns, including increased frequency and intensity of extreme weather events such as heavy rainfall or intense storms, can exacerbate flash flooding. Climate change may alter precipitation patterns and lead to more frequent and severe flooding events in certain regions.

g) **Anthropogenic Activities**: Human activities such as mining, construction, or improper waste disposal can disturb natural drainage systems and increase the risk of flash flooding. Alterations to the landscape, such as excavation or filling of floodplains, can disrupt natural flood defenses and exacerbate flooding.
h) **Ground Cover**: The type of ground cover, including vegetation, soil types, and land use practices, affects the infiltration of rainfall and the generation of runoff. Impervious surfaces, such as asphalt and concrete in urban areas, limit infiltration and increase surface runoff, contributing to flash flooding. Deforestation, soil compaction, and urbanization can reduce the natural storage capacity of landscapes, exacerbating flash flooding by accelerating runoff and increasing peak flows.
i) **Topography**: The topography of a region influences the flow of water and the formation of drainage pathways. Steep slopes can accelerate runoff and concentrate flow, increasing the risk of flash flooding in downstream areas. Natural features such as valleys, canyons, and gorges can channelize runoff during intense rainfall events, exacerbating flash flooding in low-lying areas.
j) **Tidal Influences**: In coastal regions, tidal influences can interact with rainfall and storm surge to exacerbate flash flooding. High tides coinciding with heavy rainfall events can impede the drainage of inland watercourses, leading to increased flood risk in coastal and estuarine areas. Storm surges associated with tropical cyclones can inundate coastal areas and exacerbate flash flooding by blocking drainage pathways and raising water levels in rivers and estuaries.

15.3 Flooding in Major River Basins of India

Flooding poses significant challenges in various river basins across India. The sudden onset of intense rainfall, coupled with geographical and climatic factors, contributes to the vulnerability of these regions. The catastrophic floods have wreaked devastation in India, resulting in the loss of irreplaceable human lives as well as their property. Table 15.1 provides an overview of the extent of the destruction that was brought about by flash floods in India.

Below is an overview of flash flooding issues in some of the major river basins in India:

a) **Brahmaputra River Basin**: The Brahmaputra River Basin, along with its tributaries, covers states such as Assam, Arunachal Pradesh, Meghalaya, Mizoram, parts of West Bengal, Manipur, Sikkim, Tripura, and Nagaland. High rainfall in the catchment areas leads to frequent and severe floods in this region. The tectonic characteristics of the Northeast region, combined with the friable nature of rocks in the hills, contribute to landslides and high silt content in rivers. Severe earthquakes further disrupt river regimes and exacerbate flooding issues. Flood problems include inundation, erosion along riverbanks, and drainage congestion due to natural and man-made factors.
b) **Ganga River Basin**: The Ganga River Basin encompasses the Ganga and its tributaries like the Yamuna, Sone, Ghaghra, Gandak, and Kosi. Flood concerns are prominent in Uttarakhand, Uttar Pradesh, Bihar, and West Bengal. While the Ganga carries substantial water discharges, flood problems are localized, primarily affecting areas on the northern bank of the river. Drainage congestion and inundation are significant issues,

Table 15.1 Summary of flash flood events in India during recent years.

Year	Villages/districts affected	State	Origin of event	River basin	Deaths	Total number of people affected
2001	Saran, Gopalganj, Purba Champaran, Madhubani districts	Bihar and Uttar Pradesh	Heavy rains		48	2 800 000
2003	Kullu district	Himachal Pradesh	Torrential rains	Parvati	150	30
2003	Solang village (Kullu district)	Himachal Pradesh			43	15
2004	Uttar Pradesh, provinces	Himachal Pradesh	Torrential rains		33	
2004	Goalpara district	Assam, Meghalaya, Arunachal Pradesh, West Bengal	Heavy rain	Krishnai, Rongjuli and Dudhnoi, Hugli	160	100 000
2005	Azamgarh district (province)	Uttar Pradesh	Heavy rains	Ghaghra, Saryu rivers	27	800 000
2006	Adilabad, Hyderabad, Karimnagar, Khammam, Mahbubnagar, Medak, Nalgonda, Nizamabad	Andhra Pradesh	Heavy rain		20	
2006	Jammu and Kashmir provinces		Monsoonal rain	Jhelum, Sutlej, Lidder, Chenab, Tavi, Tawi, Munnawar Tawi	19	15 000
2007	Kurnool, Mahbubnagar, Guntur, Prakasam, East Godavari, West Godavari, Anantapur, Cuddapah districts, Thrissur district, Mysore districts, Mumbai City, Pune districts	Andhra Pradesh, Kerala, Karnataka, Maharashtra	Monsoonal rain and tropical cyclone	Krishna, Godavari and tributaries	127	200 000
2008	Lakhimpur, Dhemaji, Sonitpur districts	Assam, West Bengal, Orissa, Bihar	Monsoonal rain	Kelegai river	1063	7 900 000

Table 15.1 (Continued)

Year	Villages/districts affected	State	Origin of event	River basin	Deaths	Total number of people affected
2008	Thiruvallur, Thoothukudi, Villupuram, Ramanathapuram, Tirunelveli Kattabo, Pudukkottai, Kanniyakumari, Nagapattinam, Cuddalore, Karaikal districts	Tamil Nadu, Puducherry	Tropical cyclone Nisha		54	803 740
2009		Delhi province	Torrential rain		11	
2010	Adilabad, East Godavari, Guntur, Hyderabad, Krishna, Kurnool, Nalgonda, Nellore, Nizamabad, Vishakhapatnam, Warangal	Andhra Pradesh	Brief torrential rain	Godavari	179	252 016
2010	Leh, Cholglamsar villages	Jammu and Kashmir	Heavy rains		196	12 500
2015	Chadoora village	Jammu and Kashmir	Heavy rains	Jhelum river	44	2122
2015	Bhadrairuppu village	Tamil Nadu	Torrential rain		8	200
2016	Chamoli, Pithoragahr districts	Uttarakhand	Torrential rains	Alaknanda and Mandakini rivers	61	
2016	Lakhimpur, Golaghat, Jorhat, Barpeta, Bongaigaon, Goalpara, Dhubri, Marigaon, Darrang, Nalbari districts	Assam	Monsoonal rain	Brahmaputra, Barak rivers	50	2 000 000
2017	Mumbai, Thane	Maharashtra	Torrential rains		14	12
2017	Chennai	Tamil Nadu	Heavy rains		21	4000
2018		Tripura	Heavy rains		20	3 000 000
2021		Himachal Pradesh, Union Territories of Jammu and Kashmir and Ladakh	Heavy rains		37	122

Source: Data from EM-DAT (2023).

particularly in Uttar Pradesh, Bihar, and West Bengal. Northern tributaries, including the Gandak and Kosi, frequently overflow, causing damage to crops and infrastructure.

c) **Indus River Basin**: This basin comprises river basins such as the Sutlej, Ravi, Beas, Tawi, Chenab, and Jhelum. Flood problems in this region are relatively less severe compared to others. Inadequate surface drainage, waterlogging in irrigated areas, and changes in river regimes due to rising groundwater levels are common issues. Drainage congestion and waterlogging are particularly prevalent in Haryana and Punjab.

d) **Central India and Peninsular Rivers Basin**: Rivers like the Narmada, Tapi, Mahanadi, Godavari, Krishna, and Cauvery are prominent in this zone, covering southern states like Andhra Pradesh, Chhattisgarh, Karnataka, Tamil Nadu, Kerala, Orissa, Maharashtra, Gujarat, and parts of Madhya Pradesh. While most rivers have stable courses, delta areas face periodic flood and drainage problems, especially during cyclonic storms. Coastal districts experience severe flooding due to heavy precipitation, silt deposition in delta areas, and overflow from smaller rivers. These regions face unique challenges related to flash flooding, influenced by their geographical, climatic, and geological characteristics. Effective flood management strategies are essential to mitigate the impact of flash floods and safeguard lives and infrastructure.

15.4 Dynamics of Flash Floods in the Himalayan Region

The Himalayas possess a distinct topography due to the arrangement of various thrust sheets caused by tectonic activity. This topography enables the southern face of the mountains to receive a wide range of rainfall. However, the fragile landscape, steep slopes, and high-relief terrain of the region make it highly vulnerable to natural disasters triggered by cloudbursts, such as debris flows, flash floods, and landslides (Schild 2008). The Himalayan region often experiences catastrophic flash floods. The occurrence of these sudden floods has notably escalated in the last ten years, showing devastating results, especially in the Uttarakhand Himalaya region. Two prominent examples include the Kedarnath flash floods of 2013, followed by the Chamoli flash floods of 2021. The occurrence of these catastrophes presents a problem in terms of reducing the sustainability of billions of people who live in the flood basins of the main rivers in the Himalaya to flash floods (Sati 2011).

According to the Emergency Events Database (EM-DAT), the number of deaths due to flash floods has seen to be escalating in the past few decades (Figure 15.1). Apart from several environmental and social pressures, many of which are already recognized as being severe, these events from the past few decades are exacerbated by the effects of climate change. A significant number of regions are seeing a larger share of total precipitation than they did in the past. The onset of snow melting occurs sooner and winter is shorter; this has an impact on river regimes, water supplies, natural hazards, and people's livelihoods and infrastructure, especially in river basins reliant upon glacier melt during the summer months.

The Himalayan region was the site of seven of the top ten natural catastrophes in terms of the number of fatalities that happened in the year 2008, as stated by the United Nations International Strategy for Disaster Reduction (UNISDR 2009). This demonstrates not just the frequency with which natural catastrophes occur in the region, but also its vulnerability as a whole to the occurrence of this kind of calamity. There is a growing consensus that the phenomenon of global warming is linked to extreme fluctuations, especially when combined with the intensification of monsoon circulations. Glacial lakes have formed

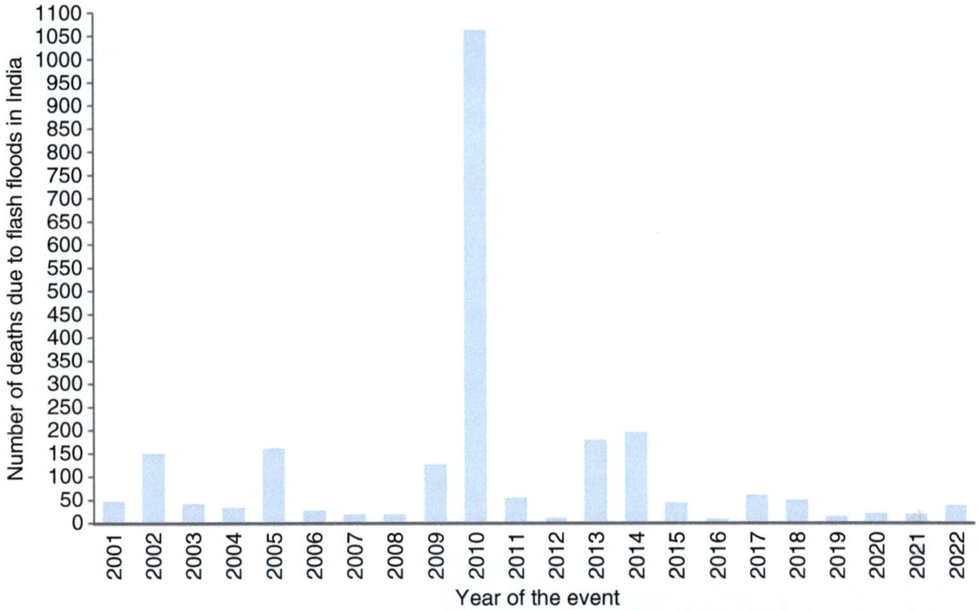

Figure 15.1 Human life loss in the past decade due to flash floods in India. *Source:* Data from EM-DAT (2023).

behind terminal moraines as a result of glacial melt that is connected with climate change. Many of these lakes at high altitudes have the potential to be hazardous. Because of their relative fragility, moraine dams are prone to rapid failure, which may result in the sudden release of enormous amounts of water and debris. As was seen in the Kedarnath Valley during the eruption of the Chorabari glacier-fed lake, the accompanying Glacial Lake Outburst Floods (GLOFs) produced catastrophic flooding in the downstream territory, causing major damage to life, agriculture, forests, property, and infrastructure.

15.4.1 Case Studies of Flash Floods in the Himalayas

The Leh District saw a cloudburst on 6 August 2010, with recorded rainfall intensities of 356 mm in two hours. However, the actual rainfall varied around the valley. Debris flows were caused by the cloudburst on the slopes of the mountains in Leh District and the communities nearby after 24 hours. In Leh and the surrounding villages, 255 people lost their lives, hundreds more were wounded, and crucial infrastructure was destroyed or seriously damaged.

On 8 July 2022, an exceptional cloudburst occurrence took place at the Amarnath Shrine, upstream, at the peak of the pilgrimage season. Consequently, a substantial quantity of debris and stones rapidly flowed downstream, causing the destruction of three temporary shelters and 40 encampments that were positioned within and beside the drainage channel. A total of 17 pilgrims lost their lives in the tragedy, with 60 pilgrims sustaining injuries and 30 pilgrims being reported as missing.

According to a study from the State Disaster Management Authority, there have been several instances of flash floods and cloudbursts in Jammu and Kashmir since 2010, resulting in the unfortunate loss of almost 300 lives. In 2022, a total number of deaths of over 40 persons was documented due to cloudbursts in the Jammu and Kashmir area. Furthermore, a considerable number of persons suffered injuries during these occurrences. The majority of cloudburst incidences have been documented during the pre-monsoon months, unlike in Himachal Pradesh and Uttarakhand where such phenomena occur from September to October.

The cloudburst incident has caused extensive damage to both public and private infrastructure and assets over the whole area of the Indian Himalayas, including internal roads, highways, tunnels, bridges, water and electricity supply lines, agriculture, crops, and orchards. The frequency and intensity of cloudburst catastrophes are increasing, but the existing mechanisms to monitor and mitigate their effects are woefully inadequate. The primary cause of the destruction is often unregulated building and encroachment on drainage systems. The mortality rate at Amarnath serves as proof of the apathy shown by the authorities. Often, the government prefers to ascribe climate change as the main factor behind many problems, suggesting a lack of ability. Although this claim cannot be ignored, it is crucial to recognize that there is much potential for taking proactive actions in the fields of cloudburst monitoring, forecasting, and management.

15.5 Factors Affecting Flash Floods

15.5.1 Influence of Climate Change and Extreme Rain Events

Flash floods pose a particularly high risk due to their rapid onset and unpredictable nature, making them responsible for the highest number of flood-related fatalities. According to the Intergovernmental Panel on Climate Change (IPCC), snowmelt and rainfall are the two factors that climate change "has detectably influenced," leading to floods. To put it another way, even if global warming may not be the primary cause of every flood, it does worsen a number of conditions that raise the danger of flooding. An atmosphere with higher temperatures may retain and subsequently release larger amounts of water. With a 1.1°C temperature rise from the preindustrial revolution period, it is expected that there would be an increase in heavy precipitation occurrences. This includes severe weather phenomena such as atmospheric rivers, in which air currents get laden with enormous quantities of water from tropical regions. Meanwhile, in areas where there is a substantial amount of snow that melts seasonally, higher temperatures might cause more instances of rain falling on the snow. This warm rain leads to quicker and earlier melting of the snow.

Climate change is amplifying the occurrence of the most intense storms, resulting in higher precipitation levels, even in regions that are not typically prone to floods (Xu et al. 2009). More water seems to be dumped by storms than it would have in the absence of climate change. The IPCC predicts that future storms would have up to 37% more water around their cores and 20% more water up to 60 miles out (IPCC 2007).

One of the most essential ways to prevent some of the worst scenarios for increasing flood risks is to mitigate climate change, and in particular, to keep the increase in the world average temperature to 1.5°C this century. The G20 nations, including governments, enterprises, industry, and the public, collectively accountable for 78% of global emissions, should establish specific goals and schedules for achieving decarbonization. We must seize the many social, economic, and political possibilities that come with moving toward a renewable energy-powered future and implementing a wide range of carbon-neutral solutions across society.

15.5.2 Flash Flooding due to Cloudbursts

There has been a significant rise in the number of recorded cloudburst incidents in recent years. Most cloudburst incidents are associated with unexpected and heavy rainfall. The Himalayas geography, with its steep and unstable slopes, provides a perfect environment for cloudburst occurrences, leading to sudden flash floods or landslides (Bhan et al. 2015; Dimri et al. 2017). This weather event has the capacity to cause circumstances similar to a flood. An example of a significant outcome of cloudbursts, in the form of floods, was seen in Jammu and Kashmir in 2014. Research experts have suggested that the combination of low-level westward moving monsoonal systems and eastward moving mid-tropospheric westerly troughs had a crucial role in causing the destructive flood that took place from 4 to 6 September 2014 (Ray et al. 2015). A total of 300 individuals tragically lost their lives in Jammu and Kashmir, with 85 of them being from the Kashmir region. An initial assessment conducted by the government indicated that the flood caused damage to more than 350 000 buildings, including 250 000 residential dwellings, and affected 1.2 million households in 5500 communities that were impacted by the flood throughout the state. Similar factors were also responsible for a two-day downpour that occurred in Uttarakhand on 16–18 June 2013. This precipitation was produced by the same conditions. As a result of the incident, several locals and pilgrims in Uttarakhand lost their lives. In 2013, the monsoon pulse rapidly moved northward throughout the Indian subcontinent, surpassing the speed of previous years. Consequently, the monsoon onset occurred across the area nearly a month earlier than usual. In recent years, there has been a rise in the frequency of a deadly confluence of humid and warm monsoon winds and chilly and dry western winds, known as western disturbance (Sikka et al. 2015). The convergence of air masses has led to the creation of a low-pressure system over Jammu and Kashmir.

Furthermore, as a result of climate change, the escalating temperatures have resulted in a significant surge in oceanic warming. This increase in temperature has made it easier for moisture-filled air to move into the Himalayan area. The warm winds cause the saturated clouds to rise vertically. However, the clouds cannot release precipitation because of the upward movement of an unusually warm air current. Raindrops, instead of falling, are carried upwards by the dominant air current. Additional droplets form when current raindrops experience dilatation. When the rains become too heavy for the cloud to carry, they fall quickly and all at once. Clouds' vertical motion provides the essential energy for the formation of a cloudburst. Usually, this phenomenon happens at altitudes between 1000 and 2500 m asl (Joshi and Maikhuri 1997).

15.5.3 Flash Floods Due to Glacier-lake Outbursts

GLOFs occur when the natural dams formed by glaciers holding back glacial meltwater or moraine-dammed lakes fail (Anacona et al. 2014). The sudden release of water leads to a rapid surge downstream, resulting in flash floods (Bhambri et al. 2019). Factors contributing to these outbursts include glacier retreat due to climate change, increased melting rates, and the formation of supraglacial or proglacial lakes (Ahmed et al. 2021). The consequences of glacier-lake outbursts can be catastrophic, with the potential for loss of life, infrastructure damage, and disruption of livelihoods (Table 15.1). Flash floods due to glacier-lake outbursts are a pressing concern in the Indian Himalayan region, where the fragile ecosystem is particularly vulnerable to the impacts of climate change (Islam and Patel 2022). The Himalayas are home to numerous glaciers and glacial lakes, which are experiencing accelerated melting rates due to rising temperatures. GLOFs have emerged as one of the most pressing hazards in the Himalayan region due to the rapid retreat of glaciers and the formation of numerous glacial lakes. The increased frequency and severity of GLOFs are attributed to climate change-induced glacier melt, which has led to the expansion of these lakes and the weakening of these lakes. The Himalayan region has experienced several devastating GLOF events over the past century, each with significant socio-economic and environmental impacts. One notable event occurred in 1929 when a GLOF originating from the Chog Kumdan ice-dammed lake in the Shyok basin of Ladakh affected approximately 48 villages downstream (Gunn 1930). In recent years, the frequency of GLOF events in the Himalayas has been increasing, posing a growing threat to communities living in the vicinity of glacial lakes. For example, the Chorabari GLOF event in 2013, caused by the breach of a glacial dam, resulted in widespread damage to infrastructure and natural habitats in the affected region (Allen et al. 2016). These events underscore the urgent need for proactive measures to mitigate the risks associated with GLOFs.

Studies by Singh et al. (2022) and Shrestha et al. (2023) have emphasized the importance of monitoring and early warning systems to mitigate the impact of GLOFs and other hydrological hazards in vulnerable mountainous areas. Furthermore, research by Pandey et al. (2022) and Gupta et al. (2022) has highlighted the role of climate change in exacerbating the risk of GLOFs by accelerating glacier melt and altering precipitation patterns. As temperatures continue to rise, the frequency and magnitude of GLOFs are expected to increase, posing significant challenges for communities living in downstream areas.

15.5.4 Flash Floods Triggered by Landslides

Flash floods triggered by landslides are a significant concern in the Indian Himalayan region, characterized by its steep slopes and fragile geology. The susceptibility of the region to heavy rainfall and seismic activity exacerbates the risk of landslides (Aziz et al. 2024), which can, in turn, lead to flash floods with devastating consequences for communities downstream.

The Indian Himalayan region has experienced numerous instances of flash floods triggered by landslides, particularly during the monsoon season when intense rainfall can destabilize slopes. Landslides deposit debris into rivers or obstruct their flow, creating temporary dams that can breach suddenly, releasing a surge of water downstream. The catastrophic flash flood in Chamoli was primarily triggered by a landslide-rock avalanche originating from the northern slope of the Trishul mountain range near the Raunthi glacier. Garg et al. (2021) conducted a detailed analysis of the geomorphological and hydrological factors contributing to the disaster, highlighting the susceptibility of the terrain to slope failures and rapid runoff generation. The detachment of a massive ice cum rock mass at

Raunthi peak, estimated at around 27 million cubic meters by Shugar et al. (2021), led to an extreme release of energy and debris flow into the Rishiganga valley. The presence of tension cracks and adverse lithological characteristics, such as migmatite gneisses and limestone, further contributed to the instability of the slope, making it susceptible to failure even during dry spells. The consequences of the Chamoli flash flood were devastating, resulting in loss of lives, infrastructure damage, and environmental degradation. The flash flood led to the destruction of two hydroelectric power projects, Rishiganga and Tapovan Vishnugad, and caused significant disruption to transportation networks and livelihoods. The event underscored the need for enhanced disaster preparedness and mitigation measures in vulnerable regions of the Himalayas, where the impact of geological and meteorological factors can lead to catastrophic events. Mitigation of flash floods triggered by landslides requires a multifaceted approach, including slope stabilization measures, early warning systems, and land-use planning to minimize risk in landslide-prone areas. Community-based disaster preparedness programs are also crucial for ensuring timely evacuation and response to landslide-related emergencies.

15.6 Mitigation Strategies for Flash Floods

The flood scenario in India is complex, making the management of floods a significant challenge. Floods in India occur frequently and show consistent patterns, with varying intensities across different regions. Among Asian countries, India has the largest number of fatalities caused by floods (Saleem Ashraf et al. 2017). Major contributors to floods in India include prolonged monsoon rains, the limited capacity of river channels to manage large water volumes, riverbank erosion, silt accumulation in riverbeds, and inadequate natural drainage in flood-prone areas. Because of the complexity of the flood situation in India, the management of floods is a laborious and time-consuming procedure. Since the country's independence, the Government of India (GoI) has implemented a variety of initiatives to lessen the likelihood of events like as the occurrence of floods and control the possible repercussions of such floods. Governments of different states and Union Territories are responsible for formulating and implementing projects for flood management and control. These projects are carried out using their own resources and according to their priorities. In order to facilitate the implementation of certain projects in crucial regions, the Central Government offers financial support to the states and UTs. An initiative known as the Flood Management Programme (FMP) has been the means by which the Central Government has been providing financial support since the eleventh plan.

To build flood-resilient communities and lessen the compounding risk, the present management system must take into account community involvement and non-structural actions. For a nation like India, there are a number of social and economic reasons that have made it difficult to actively foster community involvement. In order to save casualties and damage to property, India must establish a robust flood disaster management system. India has a two-pronged approach to managing flood disasters. Two main aspects of flood management are flood predictions and control. Predicting when and where floods will occur is the main focus of flood forecasting. The Central Flood Forecasting Centre was first created in New Delhi in 1958 by the Central Water Commission. Its technique for predicting floods was very sophisticated and scientific. The primary objective of this facility was to provide adjacent Delhi communities with early warnings of impending floods.

Mitigation strategies for flash floods in hilly areas should be comprehensive and tailored to address the specific challenges posed by the terrain and climatic conditions (Tullos et al. 2016). Before any infrastructure developments, such as roads, railways, or buildings, are constructed in hilly areas prone to flash floods, thorough environmental impact assessment studies should be conducted (Mukherjee et al. 2023). These studies will assess the potential environmental impacts, including the risk of flash floods, and propose mitigation measures to minimize these risks. This could involve measures such as avoiding construction in flood-prone areas, implementing proper drainage systems, and ensuring the structural resilience of buildings against flood damage. Additionally, tourist safety measures, such as early warning systems, evacuation plans, and education campaigns about flash flood risks, should be implemented to minimize fatalities and injuries during emergencies (Psaroudakis et al. 2021). Sustainable land management practices, such as afforestation, soil conservation, and watershed management, can help mitigate the risk of flash floods in hilly areas (Rawat et al. 2012). Vegetation acts as a natural buffer against soil erosion and runoff, reducing the likelihood of flash floods and their impacts. Implementing measures to protect natural habitats and restore degraded ecosystems can also contribute to flood risk reduction by enhancing the natural capacity of the landscape to absorb and retain rainfall. Infrastructure development in hilly areas should prioritize resilience and adaptation to flash flood risks. This includes designing roads, bridges, and other critical infrastructure to withstand extreme weather events, such as flash floods and landslides. Engineering solutions, such as slope stabilization measures, flood-resistant construction techniques, and resilient drainage systems, can help minimize the vulnerability of infrastructure to flash flood damage (Proverbs and Lamond 2017). Engaging local communities in flash flood preparedness and response efforts is essential for effective mitigation. Community-based early warning systems, disaster risk reduction training, and participatory decision-making processes can empower residents to take proactive measures to protect themselves and their properties from flash floods (Azad et al. 2019). Building local capacity for emergency response and fostering community resilience can enhance overall preparedness and reduce the impacts of flash flood disasters in hilly areas (Haque et al. 2022).

15.6.1 Early Warning Systems and Forecasting Techniques

With the process of flash flood forecasting being intricate but frequent in the era of changing climate, it necessitates the monitoring of areas with a larger spatial extent, where each small watershed of a few km^2 might potentially be impacted. Real-time observation networks and models must run at small geographical and temporal dimensions, including intervals of a few minutes and distances of a few kilometers. Moreover, there is a lack of discharge time series data for most of the potentially impacted watersheds, which presents a significant difficulty in calibrating and evaluating the models. Hydrological models are used to simulate the behavior of river systems and predict flood events. These models integrate data on precipitation, snowmelt, soil moisture, and river flow to forecast flood levels and timing. In the Indian Himalayan region, hydrological modeling is essential for assessing flood risk and developing early warning systems tailored to local conditions.

Advances in meteorological forecasting enable the prediction of extreme weather events, including heavy rainfall and thunderstorms. High-resolution weather models, coupled with observational data from weather stations and satellites, improve the accuracy of short-term forecasts, enabling authorities to issue timely warnings ahead of flash flood events. Engaging local communities in the early warning process is essential for effective disaster preparedness. Community-based early warning systems utilize indigenous knowledge, local networks, and communication channels to disseminate alerts and facilitate evacuation procedures in flood-prone areas. Effective mitigation strategies for GLOFs include the construction of protective infrastructure such as embankments and diversion channels, as well as the implementation of land-use planning measures to minimize exposure to flood risk. Additionally, community-based early warning systems and disaster preparedness training can help enhance resilience and response capabilities in vulnerable areas.

15.6.2 Satellite-based Monitoring Systems

Satellite imagery and remote sensing technologies play a vital role in observing and analyzing changes in land cover, snowmelt, and glacier dynamics. These technologies are essential for identifying potential hazards such as landslides and glacier-lake outbursts, facilitating early detection and timely mitigation efforts. High-resolution radar-based precipitation data and meteorological forecasts (with grid sizes around $1\,km^2$) are especially important for accurate flood prediction and management (Creutin and Borga 2003). Various methods are employed to forecast flash floods, with the Flash Flood Guidance (FFG) method being one of the earliest and most widely used (Georgakakos 1986). The FFG method uses rainfall-runoff modeling to determine the amount of rainfall required to produce bank-full discharge at a catchment's outlet, based on initial soil moisture conditions. Enhancements in forecasting lead times have been explored through the use of quantitative precipitation forecasts, improving the ability to predict and respond to flash flood events (Barthold and Woods 2015). Furthermore, satellite-based monitoring systems are invaluable for their ability to provide real-time data, which is crucial for dynamic flood management. This real-time data can be integrated with Geographic Information Systems (GIS) to create detailed flood maps, offering a visual representation of flood risks and aiding in the strategic planning of evacuation routes and emergency response actions.

15.6.3 Strategies to Address the Post-disaster Impact of Flash Floods

Immediately following a flash flood event, rapid emergency response and relief operations are essential for saving lives, providing medical assistance, and meeting the basic needs of affected populations. Coordination among government agencies, humanitarian organizations, and local authorities is critical for delivering timely assistance to those in need. Conducting comprehensive damage assessments is necessary to quantify the extent of infrastructure damage, loss of livelihoods, and environmental degradation caused by flash floods. Rehabilitation programs should prioritize the restoration of critical infrastructure, including roads, bridges, water supply systems, and healthcare facilities, to facilitate the recovery process and enable communities to rebuild their lives. Flash floods can have devastating impacts on local economies, particularly in rural areas dependent on agriculture and tourism. Supporting livelihood restoration initiatives, such as providing agricultural inputs, vocational training, and small business grants, helps affected communities recover economically and reduces their vulnerability to future disasters. Flash floods can have profound psychological effects on individuals and communities, causing trauma, grief, and stress. Psychosocial support services, including counseling, peer support groups, and community resilience-building activities, play a crucial role in addressing the emotional and mental health needs of flood survivors and promoting resilience in the aftermath of disasters. Flash floods can cause significant environmental damage, including soil erosion, deforestation, and water contamination. Implementing ecosystem-based approaches to flood risk reduction, such as reforestation, wetland restoration, and sustainable land management practices, helps mitigate the impacts of future flood events and enhance ecological resilience in flood-prone areas. Engaging local communities in disaster risk reduction efforts is essential for building resilience to flash floods. Community-based initiatives, such as participatory hazard mapping, early warning systems, and disaster preparedness training, empower communities to identify and address their unique vulnerabilities and strengths, thereby reducing their exposure to future flood risks.

15.7 Concluding Remarks

Throughout the chapter, we have highlighted the dynamics of flash floods in the Indian Himalayan region, exploring the various factors that contribute to their occurrence, including cloudbursts, glacier-lake outbursts, and landslides. We have examined the impact of climate change on exacerbating these events, leading to increased frequency and severity of flash floods in recent years. Additionally, we have discussed the socio-economic and environmental consequences of flash floods, highlighting the need for comprehensive mitigation strategies to minimize loss of life, infrastructure damage, and economic disruption. This chapter emphasized the importance of early warning systems and forecasting techniques in mitigating the impact of flash floods. By leveraging advancements in hydrological modeling, satellite-based monitoring, and meteorological forecasting, authorities can improve the accuracy of flood forecasts and provide timely alerts to at-risk populations, enabling them to take proactive measures to protect themselves and their communities. Furthermore, we have emphasized the significance of community engagement and participation in disaster risk reduction efforts. Community-based initiatives, such as participatory hazard mapping and disaster preparedness training, empower local communities to identify and address their unique vulnerabilities, thereby reducing their exposure to future flood risks. In conclusion, while flash floods pose significant challenges to the Indian Himalayan region, they can be mitigated through a combination of technological advancements, community-based approaches, and ecosystem-based solutions. By implementing comprehensive disaster risk reduction measures and fostering collaboration among stakeholders, it is possible to build resilience and mitigate the impact of flash floods, ultimately ensuring the safety and well-being of vulnerable populations in the region.

References

Ahmed, R., Wani, G.F., Ahmad, S.T. et al. (2021). A review of glacial lake expansion and associated glacial lake outburst floods in the Himalayan region. *Earth Systems and Environment* 5 (3): 695–708.

Allen, S.K., Rastner, P., Arora, M. et al. (2016). Lake Outburst and debris flow disaster at Kedarnath, June 2013: hydro meteorological triggering and topographic predisposition. *Landslides* 13 (6): 1479–1491.

Anacona, I.P., Norton, K.P., and Mackintosh, A. (2014). Moraine-dammed lake failures in Patagonia and assessment of outburst susceptibility in the Baker Basin. *Natural Hazards and Earth System Sciences* 14 (12): 3243–3259.

Azad, M.A.K., Uddin, M.S., Zaman, S., and Ashraf, M.A. (2019). Community-based disaster management and its salient features: a policy approach to people-centred risk reduction in Bangladesh. *Asia-Pacific Journal of Rural Development* 29 (2): 135–160.

Aziz, K., Mir, R.A., and Ansari, A. (2024). Precision modeling of slope stability for optimal landslide risk mitigation in Ramban road cut slopes, Jammu and Kashmir (J&K) India. *Modeling Earth Systems and Environment.*

Barthold, F.K. and Woods, R.A. (2015). Stormflow generation: a meta-analysis of field evidence from small, forested catchments. *Water Resources Research* 51 (5): 3730–3753.

Bhambri, R., Hewitt, K., Kawishwar, P. et al. (2019). Ice-dams, outburst floods, and movement heterogeneity of glaciers, Karakoram. *Global and Planetary Change* 180: 100–116.

Bhan, S.C., Devrani, A.K., and Sinha, V. (2015). An analysis of monthly rainfall and the meteorological conditions associated with cloudburst over the dry region of Leh (Ladakh), India. *Mausam* 66 (1): 107–122.

Borga, M., Stoffel, M., Marchi, L. et al. (2014). Hydrogeomorphic response to extreme rainfall in headwater systems: flash floods and debris flows. *Journal of Hydrology* 518: 194–205.

Buckley, R.C., Pickering, C.M., and Warnken, J. (2000). Environmental management for Alpine tourism and resorts in Australia. In: *Tourism and Development in Mountain Regions* (ed. P.M. Godde, M.F. Price, and F.M. Zimmermann), 27–45. Wallingford: CABI Publishing.

Creutin, J.D. and Borga, M. (2003). Radar hydrology modifies the monitoring of flash-flood hazard. *Hydrological Processes* 17 (7): 1453–1456.

Dar, T., Rai, N., Kumar, S., and Bhat, M.A. (2022). Climate change impact on cryosphere and streamflow in the Upper Jhelum River Basin (UJRB) of north-western Himalayas. *Environmental Monitoring and Assessment* 194 (3): 140.

Dimri, A.P., Chevuturi, A., Niyogi, D. et al. (2017). Cloudbursts in Indian Himalayas: a review. *Earth-Science Reviews* 168: 1–13.

Du, M.Y., Kawashima, S., Yonemura, S. et al. (2004). *Mutual Influence between Human Activities and Climate Change in the Tibetan*. Brussels, Belgium: EM-DAT, CRED/UCLouvain.

EM-DAT (2023). *Emergency Events Database (EM-DAT)*. Brusseks, Belgium: UCLouvain https://www.emdat.be (accessed 17 November 2023).

Garg, S., Shukla, A., Garg, P.K. et al. (2021). Revisiting the 24 year (1994–2018) record of glacier mass budget in the Suru sub-basin, western Himalaya: overall response and controlling factors. *Science of the Total Environment* 800: 149533.

Gaume, E., Borga, M., Llassat, M. C. et al. (2016). Mediterranean extreme floods and flash floods. *The Mediterranean Region under Climate Change. A Scientific Update*, 133–144.

Georgakakos, K.P. (1986). On the design of national, real-time warning systems with capability for site-specific, flash-flood forecasts. *Bulletin of the American Meteorological Society* 67 (10): 1233–1239.

Goswami, B.N., Venugopal, V., Sengupta, D. et al. (2006). Increasing trend of extreme rain events over India in a warming environment. *Science* 314: 1442–1444.

Gunn, J.P. (1930). The Shyok flood, 1929. *Himal Journal* 2: 35–47.

Gupta, A., Maheshwari, R., Guru, N. et al. (2022). Quantitative prioritization of potentially critical glacial Lakes in the Indus River basin using satellite derived parameters. *Geocarto International* 37 (25): 7508–7530.

Haque, C.E., Azad, M.A.K., and Choudhury, M.U.I. (2022). Social learning, innovative adaptation and community resilience to disasters: the case of flash floods in Bangladesh. *Disaster Prevention and Management: An International Journal* 31 (5): 601–618.

Islam, N. and Patel, P.P. (2022). Inventory and GLOF hazard assessment of glacial lakes in the Sikkim Himalayas, India. *Geocarto International* 37 (13): 3840–3876.

IPCC (2007). Climate change 2007: impacts, adaptation and vulnerability. In: *Contribution of Working Group II to the Fourth Assessment Report of the Intergovernmental Panel on Climate Change* (ed. M.L. Parry, O.F. Canziani, and J.P. Palutikof). Cambridge: Cambridge University Press.

Jonkman, S.N. (2005). Global perspectives on loss of human life caused by floods. *Natural Hazards* 34 (2): 151–175.

Joshi, V. and Maikhuri, R.K. (1997). Cloudburst: a natural calamity- a case study from Garhwal Himalaya, U.P. *Journal of Indian Builidngs Congress IV* 1: 208–219.

Proverbs, D., & Lamond, J. (2017). Flood resilient construction and adaptation of buildings. In *Oxford Research Encyclopedia of Natural Hazard Science*.

Mukherjee, M., Abhinay, K., Rahman, M.M. et al. (2023). Extent and evaluation of critical infrastructure, the status of resilience and its future dimensions in South Asia. *Progress in Disaster Science* 17: 100275.

Pandey, P., Banerjee, D., Ali, S.N. et al. (2022). Simulation and risk assessment of a possible glacial lake outburst flood (GLOF) in the Bhilangna Valley, central Himalaya, India. *Journal of Earth System Science* 131 (3): 184.

Psaroudakis, C., Xanthopoulos, G., Stavrakoudis, D. et al. (2021). Development of an early warning and incident response system for the protection of visitors from natural hazards in important outdoor sites in Greece. *Sustainability* 13 (9): 5143.

Ramesh, K.V. and Goswami, P. (2007). Reduction in temporal and spatial extent of the Indian summer monsoon. *Geophysics Research Letters* 34 (23): L23704.

Rasmussen, K.L. and Houze, R.A. (2012). A flash-flooding storm at the steep edge of high terrain: disaster in the Himalayas. *Bulletin of the American Meteorological Society* 93 (11): 1713–1724.

Rawat, P.K., Tiwari, P.C., and Pant, C.C. (2012). Climate change accelerating land use dynamic and its environmental and socio-economic risks in the Himalayas: Mitigation through sustainable land use. *International Journal of Climate Change Strategies and Management* 4 (4): 452–471.

Ray, K., Bhan, S.C., and Bandopadhyay, B.K. (2015). The catastrophe over Jammu and Kashmir in September 2014: a meteorological observational analysis. *Current Science* 580–591.

Saleem Ashraf, M.L., Iftikhar, M., Ashraf, I., and Hassan, Z.Y. (2017). Understanding flood risk management in Asia: concepts and challenges. In: *Flood Risk Management* (ed. T. Hromadka and P. Rao), 177. London: Intech.

Sati, V.P. (2022). Glacier bursts-triggered debris flow and flash flood in Rishi and Dhauli Ganga valleys: a study on its causes and consequences. *Natural Hazards Research* 2 (1): 33–40.

Sati, V.P. (2011). Climate disasters in the Himalaya: risk and vulnerability. In: *International Conference on Climate Change and Natural Hazards in Mountain Areas*, Dushanbe (19–21 September).

Schild, A. (2008). ICIMOD's position on climate change and mountain systems. *Mountain Research and Development* 28: 328–331.

Shrestha, F., Steiner, J.F., Shrestha, R. et al. (2023). A comprehensive and version-controlled database of glacial lake outburst floods in High Mountain Asia. *Earth System Science Data* 15 (9): 3941–3961.

Shugar, D.H., Jacquemart, M., Shean, D. et al. (2021). A massive rock and ice avalanche caused the 2021 disaster at Chamoli, Indian Himalaya. *Science* 373 (6552): 300–306.

Sikka, D.R., Ray, K., Chakravarthy, K. et al. (2015). Heavy rainfall in the Kedarnath valley of Uttarakhand during the advancing monsoon phase in June 2013. *Current Science* 353–361.

Singh, R., Aryan, V., and Joshi, M. (2022). Understanding the flash flood event of 7th February 2021 in Rishi Ganga basin, Central Himalaya using remote sensing technique. *Remote Sensing Applications: Society and Environment* 26: 100744.

Tucker, R.P. (1988). The British Empire and India's forest resources: the timberlands of Assam and Kumaon, 1914–1950. In: *World Deforestation in the Twentieth Century* (ed. J.F.E. Richardsa and R.P. Tucker), 91–111. Durham, N.C: Duke University Press.

Thayyen, R., Dimri, A.P., Kumar, P., and Agnihotri, P. (2013). Study of cloudburst and flash floods around Leh, India during August 4–6, 2010. *Natural Hazards* 65: 2175–2204. https://doi.org/10.1007/s11069-012-0464-2.

Tullos, D., Byron, E., Galloway, G. et al. (2016). Review of challenges of and practices for sustainable management of mountain flood hazards. *Natural Hazards* 83: 1763–1797.

UNISDR (2009). *2009 UNISDR Terminology on Disaster Risk Reduction*. Geneva: UNISDR http://www.unisdr.org/we/inform/publicaitons/7817 (accessed 15 December 2012).

Wang, S. and Russell, H.A. (2016). Forecasting snowmelt-induced flooding using GRACE satellite data: a case study for the Red River watershed. *Canadian Journal of Remote Sensing* 42 (3): 203–213.

Xu, J., Grumbine, R.E., Shrestha, A. et al. (2009). The melting Himalayas: cascading effects of climate change on water, biodiversity, and livelihoods. *Conservation Biology* 23: 520–530.

Yao, T.D., Guo, X.J., Lonnie, T. et al. (2006). δ18O record and temperature change over the past 100 years in ice cores on the Tibetan Plateau. *Science in China Series D: Earth Sciences* 49 (1): 1–9.

16

Critical Perspectives on Climate Change and Glacial Lake Outburst Floods' Impact in the Himalayas

Policy Inferences from the South Lhonak Flood of 2023

Arpan Sharma[1], *Narpati Sharma*[2], *and Mousumi Gupta*[3]

[1] *Asian Development Research Institute, Patna, Bihar, India*
[2] *Sikkim State Council of Science and Technology, Government of Sikkim, Gangtok, Sikkim, India*
[3] *Sikkim Manipal Institute of Technology, Sikkim Manipal University, Majitar, Sikkim, India*

16.1 Introduction

The Himalayan region faces a critical challenge – the accelerating expansion of glacial lakes. Climate change-induced glacier retreat fuels the formation and growth of these lakes, significantly increasing the risk of catastrophic glacial lake outburst floods (GLOFs). The recent October 2023 South Lhonak Lake GLOF event underscores these dangers, showcasing the devastating potential of these natural hazards in the heart of the Himalayas.

While GLOFs are not new to the Himalayas, their increasing frequency and intensity are alarming. Studies warn that the number of potentially dangerous glacial lakes has grown significantly in recent decades, driven by accelerated melting (Ramsankaran et al. 2018; A. Sharma et al. 2023b; Xu et al. 2019). The South Lhonak GLOF, specifically, was notable for the scale of destruction and loss of life along the Teesta River valley. This incident compels policymakers to consider the shortcomings in current risk assessments and disaster preparedness policies. Glacial lake monitoring and early warning systems are inadequate in some regions of the Himalayas, highlighting a critical gap in mitigation infrastructure. Moreover, rapid and often unplanned developments within potentially hazardous zones exacerbate communities' vulnerability to such disasters.

Dar et al. (2022) analyzed long-term geomorphic changes in the Rambiara catchment by analyzing historical imagery and surface profiles. Their study found that Pleistocene glaciations extended over 10 km further downstream than the present, indicating significantly colder and more extensive glaciation in the past. They also reported a significant warming trend over the last century based on temperature reconstructions.

This warming is consistent with other studies reporting rising temperatures across the Himalayas (Chettri et al. 2023). Such warming has serious implications for biodiversity and water resources as temperatures rise above the optimal ranges for many mountain species and ecosystems.

Water Resources Management in Mountain Regions, First Edition. Edited by Virendra Bahadur Singh, Sughosh Madhav, Rakesh Kumar Gupta, Prerna Diwan, and Amit Kumar.

Climate change is having significant impacts on the glaciers and glacial lakes in the Himalayan region. Khadka et al. (2023) provide a comprehensive review of the research on climate change and its effects in Nepal. They find that temperatures have been rising significantly throughout the country, with average increases of 0.02–0.16°C per year. Meanwhile, precipitation trends have been more complex, with some studies finding decreases in annual amounts but one finding increases from 1979 to 2016 This temperature rise has led to pronounced glacier retreat and shrinkage across Nepal. The country saw an overall 24% loss in glacier area between around 1980 and 2010. Individual basins like Koshi lost up to 26% over this period.

This glacier melt has also fed the growth of glacial lakes in Nepal. Khadka et al. (2023) observed that the total number of lakes over 0.001 km^2 expanded from 3430 to 3808 between around 1980 and 2010, even as total glacier area decreased. Individual lake surface areas have been rising rapidly as well, with an average 0.83% annual expansion rate from 1987 to 2017. This expansion poses greater hazards, as more than 45 recorded GLOF events have occurred in Nepal due to factors like moraine dam failures

A study by Nie et al. (2021) had highlighted the profound impact of climate change on the Himalayan region's glaciers and the associated hydrological consequences. Rising temperatures have led to accelerated melting of glaciers, particularly in the Himalayas, resulting in increased meltwater runoff. This initially boosts water availability during spring and early summer, but as glaciers continue to shrink, this "peak water" is expected to be followed by a significant decline in meltwater contributions, leading to potential water shortages during the drier months.

In addition to altering water availability, climate change has escalated the frequency and intensity of GLOFs. The formation and expansion of glacial lakes, caused by increased glacial melt, heightens the risk of sudden and catastrophic flooding events. GLOFs pose significant threats to communities, infrastructure, and ecosystems in the Himalayan region. Various studies (Das et al. 2024; Khadka et al. 2023; Nie et al. 2021) underscore the necessity for improved monitoring and predictive capabilities to manage these hazards and adapt to the rapidly changing glacial and hydrological landscapes. Enhanced in-situ monitoring networks for weather, hydrology, and glacier changes are crucial for developing effective strategies to mitigate the impacts of these climatic changes and ensure water and food security for the affected populations.

Hence, this chapter examines the problem of glacial lake expansion in the Himalayas, drawing insights from the South Lhonak Lake GLOF and past events. It analyzes the causes and consequences of this event, explores potential mitigation strategies, and highlights the urgent need for informed policy responses. The goal of this chapter is to inform decision-making aimed at reducing the vulnerability of Himalayan communities and critical infrastructure to future GLOF disasters, ensuring greater resilience in the face of rapid environmental change.

16.2 Outbursts and Climate Change Impact in the Last Century

The landscape of the Karakoram is characterized by extremely high relief, with altitude differences of 3000–5000 m over short distances. Certain aspects, especially northern and eastern facing slopes, promote high, surging glacier activity with potential for main valley dams. When a glacier dams a major river valley that it flows into, it can form a huge reservoir of water. If this ice dam catastrophically fails, it results in an outburst flood with discharge that far exceeds typical high flows. A GLOF is a type of outburst flood that occurs when the dam containing a glacial lake fails due to various possible triggers. These natural dams can be composed of glacier ice (ice-dammed lakes), terminal or lateral moraines (moraine-dammed lakes), or a combination of these. A GLOF involves the sudden release of water retained in a glacial lake, and it can result in the rapid discharge of a significant amount of water. However, while GLOF is a more recent term, it was more commonly known as jökulhlaup (Hewitt 1982; Tómasson 1996).

One of the earliest works related to jökulhlaup in the Himalayas can be credited to Hewitt (1982) and unpublished works by the same author. A few of the major GLOFs listed by Hewitt (1982) are as follows:

- 1926, 1929, and 1932 floods from the Chong Khumdan Glacier on the Upper Shyok River.
- Series of floods from multiple glaciers between 1926 and 1932 along over 1200 km of the Indus River, resulting from successive damming events.
- Historic flood from the late eighteenth century bursting of the Biafo Gyang Glacier dam, which formed a reservoir estimated to contain 3.3 billion cubic meters of water.
- 1907 flood from the Khurdopin Glacier dam bursting on the Virjerab tributary of the Upper Shimshall River.
- Floods attributed to the bursting of the 6 km long Kyagar Glacier lake in 1978 and subsequent years on the Upper Yarkand River. This same glacier caused flooding 50 years earlier.
- Numerous other floods dated back to the early nineteenth century involving glaciers like Khumdan, Sultan, Ishkoman, Aktash, Kichik, and others impacting the Indus, Shyok, and Yarkand river systems.

The author had documented the progression and impacts of some noteworthy outburst floods in the early twentieth century, like the series from the Chong Khumdan Glacier in the 1920s–1930s. Its monitoring had revealed that flood waves rose over 25 m and traveled hundreds of kilometers at rates up to 20 km/hr, with discharges estimated at over 800,000 cubic feet/second at the impact site. These floods posed exceptional risks and had enormous erosive power, carrying massive amounts of sediment and triggering landslides. This shows that outbursts have been happening for more than the last 10–15 decades and were well documented.

However, the early literature did not explicitly use the term “climate change impacts.” While earlier authors do not reference or use terms related to climate change in their explanation for why outburst floods occur from glacial dams in the Karakoram region, they note that dams have been observed to happen more frequently during periods associated with the Little Ice Age advances in the past. However, they were not confined only to that time period. Additionally, there was a conspicuous gap in any reported dams from the mid-1930s until 1978, which was attributed to a general period of glacier recession that was taking place in the region at that time. The focus remained more on documenting the hazard events rather than explaining their driving mechanisms (Hewitt 1982; Tómasson 1996).

Climate change has been understood as a major contributing factor in the formation of glacial lakes in the later literature. As per the study by Govindha Raj (2009), climate change during the twentieth century has had a significant impact on the glacial environments of the Himalayas. This impact is primarily evidenced by the accelerated retreat of glaciers and the subsequent formation of a large number of glacial lakes bounded by moraine dams. The warmer climates over the past 100–150 years have led to widespread glacial retreat, which in turn has resulted in the formation of these moraine-dammed lakes in many mountain ranges for common reasons.

Govindha Raj (2009) stated that the primary cause of the GLOFs could be the collapse of moraine dams, which consist of loose sediment and debris and may fail under the stress of impounded water. Glacier surges, or the rapid movement of glaciers, can also precipitate GLOFs by abruptly releasing water that has been stored within or behind the glacier. Additionally, ice avalanches, involving large chunks of ice falling into a lake, can displace significant water volumes, leading to outbursts. Some GLOF events, however, occur for reasons that have not been identified or documented. The ongoing threat of such events is exacerbated by climate change, which is contributing to the formation of more and larger glacial lakes, and thus increasing the potential for future GLOFs. The author had listed many GLOF events across the Himalayas. Two of the major reasons identified by the author were glacier surge and moraine collapse. Table 16.1 lists the common GLOF events that happened in the past.

Coming back to the early recorded outbursts, on 4 August 1985, a devastating natural disaster struck the Khumbu Himal region of Nepal when Dig Tsho, a moraine-dammed glacial lake, dramatically burst its banks. The cause of this catastrophic event was an ice avalanche that deposited a massive 150,000 m^3 of ice into the lake, generating powerful waves that ultimately breached the moraine dam. The consequences of this breach were dire: approximately five million cubic meters of water cascaded down the valleys of Bhote Kosi and Dudh Kosi, destroying critical infrastructure including a new hydroelectric power station, 14 bridges, and around 30 homes. Moreover, the flood inflicted severe damage upon local agriculture by ravaging numerous hectares of arable land and severely impacting

Table 16.1 List of common GLOF events as documented by Govindha Raj (2009).

Year	Countries	Lake	River Basin	Cause of the event
1964	China and Nepal	Gelhaipuco	Pum Qu/Arun	Surge of glacier
1977	Nepal	Nare	Dudh Koshi	Collapse of moraine
1980	Nepal	Nagma Pokhari	Tamor	Collapse of moraine
1981	China and Nepal	Zhangzangbo	Boqu / Sun Koshi	Surge of glacier
1985	Nepal	Dig Tsho	Dudh Koshi	Ice avalanche
1994	Bhutan	Lugge Tsho	Pho Chu	Collapse of moraine
1998	Nepal	Sabai Tsho	Khumbu Himal	Unknown
2008	Pakistan	Ghulkin Glacier lake	Karakoram	Collapse of moraine

Source: Govindha Raj (2009).

the area's trail network. Although human casualties were not reported in the known scientific literature, however, the flood's impact on the community's livelihood and development was profound and long-lasting (Vuichard and Zimmermann 1987).

In the Bhutan Himalaya, the catastrophic Lugge Tsho GLOF of 1994 was triggered by the failure of the moraine dam of Lugge Tsho. This failure was due to the significant hydrostatic pressure from the meltwater, which had increased the lake's depth. The collapse of the dam led to a domino effect, releasing the waters of nearby Tshopdak Tsho as well. The outburst resulted in an immense discharge of water, with the depth calculated at 23 m, causing substantial damage downstream and leading to the unfortunate loss of 21 lives. This event highlights the severe risks associated with glacial lake outbursts in the region, underscoring the need for vigilant monitoring and disaster preparedness in areas prone to such natural calamities (Watanbe and Rothacher 1996).

In the mid-2000s, the event of the breach of Parchu Lake within the Himalayas, the cause of the flood was attributed to the rapid and late snowmelt of an unusually deep and extensive snowpack, rather than heavy rainfall. This meltwater contributed to the breach of Parchu Lake, leading to a significant outburst flood. Conversely, the flood in 2000 was precipitated by intense rainfall, which increased the lake's water level and subsequently led to the dam's failure on the Tibetan plateau. The distinction between Landslide Lake Outburst Floods (LLOFs) and GLOFs is critical in this context; LLOFs are triggered by the collapse of natural dams formed by landslides or similar mass movements, while GLOFs result from the failure of glacier-dammed lakes, predominantly due to the moraine dam failure (Ruiz-Villanueva et al. 2017).

Two of the most recent flash floods in the Himalayas are those that happened at Kedarnath in 2013 and Chamoli in 2021. Both of these floods happened in the state of Uttarakhand, making it highly vulnerable to climate change impacts (Durga Rao et al. 2014; Martha et al. 2021). The 2013 Kedarnath disaster in the Indian state of Uttarakhand was one of the worst flood-related disasters in the country's history, killing over 6000 people (Sati and Gahalaut 2013). Heavy rainfall from 10–18 June triggered flash flooding and landslides across Uttarakhand, with the majority of deaths and destruction occurring as a result of a lake outburst and debris flow originating from above the village of Kedarnath on 16–17 June (Allen, Rastner, et al. 2016; Sati and Gahalaut 2013). Extreme precipitation during this period, estimated at 212 mm from 15–17 June based on TRMM remote sensing data (Allen, Rastner, et al. 2016), led to rapid snowmelt and saturation of slopes above a small seasonal lake called Chorabari Lake (Dobhal et al. 2013). The catastrophic failure and outburst of this lake then triggered large debris flows that devastated Kedarnath village and other downstream settlements (Allen, Linsbauer, et al. 2016; Dobhal et al. 2013; Sati and Gahalaut 2013). While the heavy rainfall was clearly the main triggering factor (Sati and Gahalaut 2013), analyses indicate that topographic characteristics of the lake catchment, which lacked glacial influence and favored rapid runoff, as well as recent climate-driven changes like rapid depletion of snow cover before the event and glacier wasting, also played important predisposing roles in this disaster (Allen, Rastner, et al. 2016).

A study by Durga Rao et al. (2014) analyzed the devastating 2013 Kedarnath flash floods through hydrological and hydraulic modeling using satellite data. Major findings were that unprecedented rainfall of 500–550 mm from 10–18 June across steep terrain led to saturation and heavy runoff. The sudden breach of the 3 ha Chorabari Lake on 17 June compounded issues, releasing an estimated 0.4 MCM of water that combined with rainfall runoff to produce a peak discharge of 1800 m^3/s in the Mandakini River. Hydraulic simulations found flood depths of 12 m and velocities from 2 to 8 m/s, explaining the destructive power that washed away 64 buildings in Kedarnath alone. GLOFs from such dam failures can be extremely deadly, as the rapid release of millions of cubic meters of water can overwhelm downstream areas in minutes, leaving little time for evacuation. The Kedarnath disaster demonstrated the potential hazards posed when heavy rainfall coincides with a GLOF, underscoring the need for improved flash flood forecasting.

The Chamoli disaster that occurred on 7 February 2021, was a catastrophic flood in the Rishiganga and Dhauliganga valleys in Chamoli district of Uttarakhand, India, that was triggered by a massive rockslide from the slopes of the Trishul massif near the Ronti Glacier situated at approximately 5600 m above sea level (Pandey et al. 2021). Satellite images and aerial surveys observed that approximately 0.34 square kilometers of rock mass, estimated to be around 23 million cubic meters, detached from the mountain slopes and fell from an altitude of around 1700 m (Meena, Chauhan, et al. 2021). This huge rockslide is believed to have generated substantial kinetic and thermal energy that accelerated the mobilization of water-saturated sediments through fluidization, creating devastating flood waves that severely damaged hydropower projects, bridges, and led to loss of lives in the surrounding areas (Meena, Bhuyan, et al. 2021; Pandey et al. 2021). Hydrodynamic modeling of the disaster indicated peak discharges in the range of 3500–7500 m^3/s and helped assess the impact and inundation caused by the floods in downstream locations (Pandey et al. 2021).

Another study by Martha et al. (2021) identified that a large rock avalanche triggered by wedge failure in a steep glaciated valley caused the devastating flash floods of Chamoli in February 2021. Analysis of high-resolution satellite imagery before and after the event revealed that the failure was controlled by the orientation of planar weaknesses like joints and foliation in metamorphic rocks. The avalanche volume was estimated at 29.3 million cubic meters, highlighting the tremendous potential energy released. When the mass impacted the valley floor, it generated an air blast, melted large quantities of ice through friction and fragmentation, and incorporated water and pulverized debris to form channelized mudflows. These floods inundated infrastructure down valleys.

While not explicitly mentioned as the cause, the authors note that climate change may be a contributing factor for destabilizing slopes. They observed highly crevassing glaciers near the avalanche site. With ongoing global warming leading to increased melt and reduced snowpack as precipitation declines, slopes in high-altitude glaciated regions like this are potentially experiencing enhanced fracturing. As temperatures rise further in coming decades, similar slope failures could become more common if climatic effects are weakening terrain. The study underscores the value of satellite data for characterizing landslide hazards, especially in remote areas, but also hints that climate impacts warrant careful monitoring of destabilizing conditions to help mitigate future risks from incidents of this scale.

Another study by Mir et al. (2018) had earlier focused on understanding recession of the Dalung and Padam glaciers and the consequent formation and expansion of their associated glacial lakes, which have been identified as potentially dangerous glacial lakes (PDGLs). The authors employ various datasets, including multi-temporal satellite imagery and digital elevation models, to analyze changes in glacier extents and lake sizes over time. Their findings indicate significant glacier shrinkage and lake expansion, which poses a risk for GLOF events. The use of the MIKE 11 hydraulic model to simulate peak flood hydrographs is of particular interest for hazard assessment and the development of an emergency action plan. This model helps to estimate potential peak discharges that could result from lake outbursts, enabling authorities to better understand the risks to downstream communities and infrastructure. Four key points interpreted from the above studies are as follows, from which informed policy response can be achieved.

- Glacier Recession: Mir et al. (2018) findings corroborate the broader trend of glacial retreat in the Himalayas due to climate change. This phenomenon increases the risk of GLOF events as more water is stored in expanding glacial lakes.
- GLOF Hazard Assessment: The identification of PDGLs in the Zanskar basin highlights the importance of regular monitoring and risk assessment of glacier and lakes throughout the Himalayan region.
- Hydrodynamic Modeling: The studies demonstrate the utility of hydrodynamic models like MIKE 11 in predicting the outcomes of potential GLOF events, which is crucial for disaster preparedness and mitigation strategies.
- Emergency Action Plans: The studies provide essential data that can inform emergency action plans to protect downstream communities from the destructive impacts of GLOFs.
- Regional Collaboration: GLOF risks in the Himalayas often cross-national borders, necessitating regional cooperation in terms of data sharing, early warning systems, and joint disaster response mechanisms.

While there have been many studies on informed policy response, the outburst of South Lhonak Glacier Lake in October 2023, despite many studies, underscores the complexities of predicting GLOF events and the need for continuous monitoring and updating of risk assessments. It highlights the inherent uncertainties in glacial behavior under the influence of climate change. This event is a reminder of the vulnerability of Himalayan regions to such unpredictable natural disasters. The next section of this chapter provides and overview of the South Lhonak Glacial Lake Outburst that happened recently, causing damage to the Himalayan State of Sikkim at large.

16.3 Warnings and South Lhonak Lake Outburst Flood

A recent flood that happened in the Sikkim Himalaya on 4 October 2023 was one of the biggest examples of climate-induced flash floods. The event's satellite imagery provided few critical insights into the disaster (Figure 16.1). The images revealed evidence of a landslide (lateral moraine sliding) near the northwestern part of South Lhonak Glacier Lake, which was the likely trigger for the lake breach.

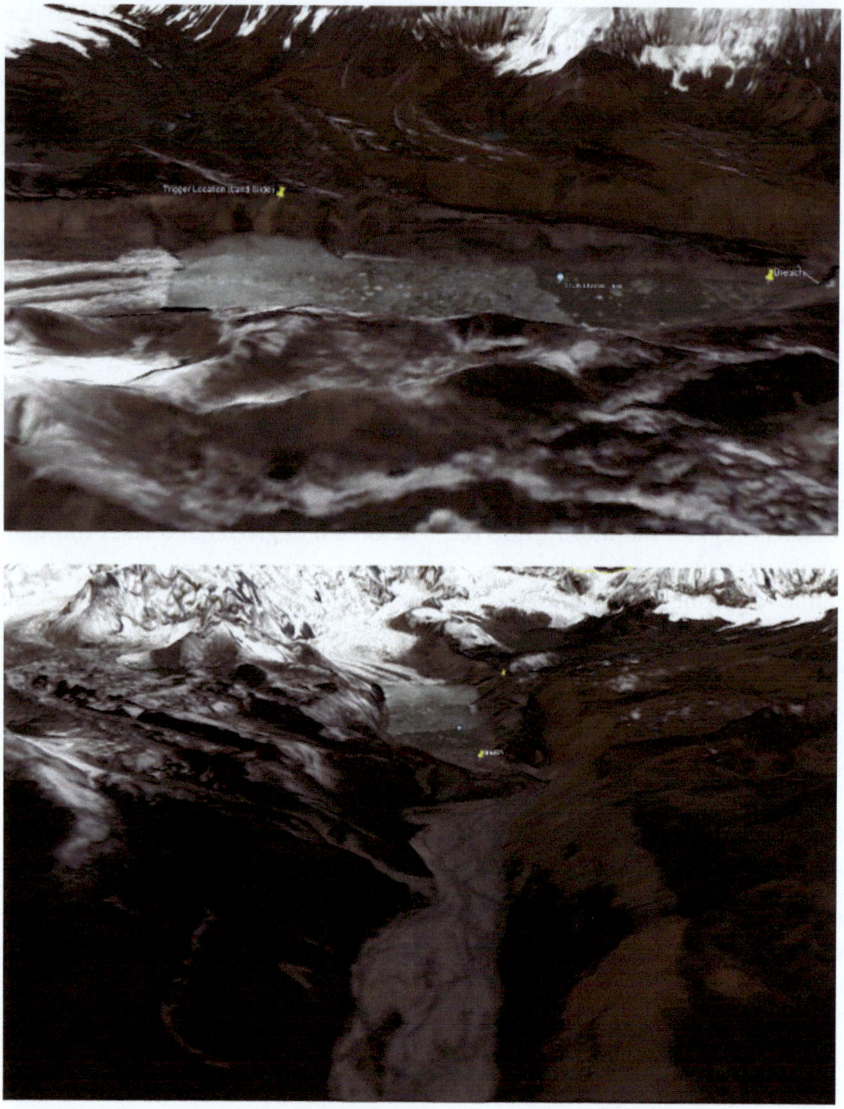

Figure 16.1 Location showing lateral moraine sliding (left yellow pin in the upper image) with breach location (right yellow pin) followed by the path of the wastewater (debris flow). *Source:* Planet Labs Imagery of 7 October 2023 rendered on Google Earth for visualization purposes.

The water flow from the breach was observed to be intense, as seen from visible sediment deposits in the initial satellite photos taken shortly after the event (Figure 16.1) by Planet Labs. Further analysis of the imagery and media reports found that while the failure of the Chungthang hydroelectric dam may have slightly intensified downstream flooding, the water flow from the lake breach itself was already quite strong based on the sedimentation evidence.

Additionally, the Dikchu dam played a key protective role by holding back excess water flows and safeguarding lives downstream during this extreme event.[1,2,3] However, the failure of the 1200-MW Chungthang Dam has raised concerns over unchecked developmental projects, especially in the Himalayan regions.[4] The GLOF event that happened in Sikkim entirely washed away the 60-meter-high dam built for Rs 13,000 crore.[4] Earlier, a study by Huber (2019) had argued that unchecked development of hydropower infrastructure in hazardous Himalayan terrains poses grave risks that have been dangerously overlooked. As this research illustrates, the rapid proliferation of dams across geologically and climatically fragile regions like Sikkim has exacerbated the threat of disasters by siting projects in risky locations prone to earthquakes, floods, and landslides.

There have been various studies that have analyzed and pointed out the dangers of the South Lhonak Glacial Lake Outburst since the past. A study by Babu Govindha Raj et al. (2013), i.e. 10 years before the disaster, had provided a detailed hazard assessment of the South Lhonak Lake in the Sikkim Himalaya, revealing that the lake posed a high risk of GLOF with a probability of 42%. It was observed that, triggered by the receding South Lhonak Glacier, which has retreated by 1.9 km from 1962 to 2008, the lake's potential outburst could result in a peak discharge of 586 m^3/s, causing devastating impacts downstream. The authors underscored the urgent need for a comprehensive glacial lake inventory, field surveys, and the implementation of early warning and mitigation strategies to safeguard communities and infrastructure. However, while early warning systems were installed multiple times, either they failed due to natural disasters like avalanches or they were stolen.[5]

A study by Begam et al. (2018), i.e. five years before the disaster, discussed the validation of an integrated model for simulating moraine dam failure by overtopping and erosion in the Himalaya, which was done through a coupled numerical simulation model. The authors applied their validated numerical simulation model to the South Lhonak Glacial Lake in the Sikkim Himalayas, India, to forecast the probable outflow hydrograph in the event of a moraine dam failure. They found that the height of the moraine dam and the volume of water in the lake upstream are critical parameters that significantly influence the peak discharge of a potential GLOF.

The burgeoning field of GLOF research has been significantly advanced through quantitative assessments of future hazards, as exemplified by the work of Sattar et al., (2019, 2021). In their comprehensive analysis of the South Lhonak Lake in the Sikkim Himalaya, the

1 What caused Sikkim floods 2023, is it a climate emergency? *Environment and Society* https://envs.scimonk.com/sikkim-glof-climate-change/, Retrieved 19th March 2024.

2 Glacial lake outburst flood kills 14 in Sikkim, 102 people missing: What is GLOF, and why does it happen? *Indian Express*, https://indianexpress.com/article/explained/explained-climate/glacial-lake-outburst-flood-glof-sikkim-8968562/, Retrieved 15th March 2024.

3 Sikkim: Threat of South Lhonak Lake bursting was forewarned by researchers two years ago, *Down to Earth*, https://www.downtoearth.org.in/news/natural-disasters/sikkim-threat-of-south-lhonak-lake-bursting-was-forewarned-by-researchers-two-years-ago-92122, Retrieved 16th March 2024.

4 Sikkim Flash Floods: Chungthang dam collapse raises concerns over safety of other hydroelectric projects, *India Times*, https://www.indiatimes.com/news/india/sikkim-flash-floods-collapse-of-chungthang-dam-concerns-over-other-hydroelectric-projects-616937.html, Retrieved 12th March 2024.

5 Sikkim: Lhonak Lake early warning system failed, says official, *Hindustan Times*, https://www.hindustantimes.com/india-news/sikkim-lhonak-lake-early-warning-system-failed-says-official-101697020649076.html, Retrieved 28th February 2024.

authors utilized glacier and hydrodynamic models to quantify the evolving threat posed by GLOFs due to climatic changes. This study is particularly salient given the geomorphological dynamics of the Teesta basin, where accelerated glacial retreat has been observed.

Through the integration of glacier dynamics and hydrodynamic simulations, the study by Sattar et al. (2021) predicted substantial growth in the volume of South Lhonak Lake. The authors identified an increased susceptibility of the lake to avalanche-induced impacts, an element critical to understanding the potential for moraine dam failure. The research presented a series of avalanche scenarios, highlighting the likelihood and consequences of overtopping events. For example, the frontal moraine was deemed susceptible to overtopping, which could trigger a catastrophic release of water. Quantitative parameters are central to GLOF prediction and were also analyzed by this study. The uncertainty in breach parameters such as breach width and the timing of failure were considered, leading to the estimation of both upper and lower hydraulic limits for potential GLOF scenarios. Dynamic flood routing models yielded projections of peak discharge rates associated with varying breach depths: 4311 m^3/s for a 20 m breach depth, 8000 m^3/s for a 30 m breach, and up to 12,487 m^3/s for a 40 m breach.

These discharge rates were then used to predict the downstream impact, with the most severe scenario indicating that the town of Chungthang could experience flow depths of 25–30 m and flow velocities of 6–9 m/s. Such parameters are critical for assessing the potential damage to infrastructure, including the major hydropower station built in 2015.

Another study by Hazra and Krishna (2021) quantitatively assessed the GLOF susceptibility of the South Lhonak Glacial Lake by examining topographical and environmental parameters. The use of AHP (Analytic Hierarchy Process) in their analysis revealed that during winter, fresh snow covers 33.14% of the glacier area, whereas in the post-monsoon season, this figure drops to 10.28%. This quantitative measure of snow cover signified a higher potential for snow avalanches in winter, which can act as a trigger for GLOFs. The rapid expansion rate of the lake at 3.64% per year between 2000 and 2018 further compounded the risk. The study's results, supported by a calculated consistency ratio (CR) of 0.0126, which is within acceptable limits, indicate that certain areas around the lake were at high risk for snow avalanches, and consequently, GLOFs.

The risks associated with these findings were significant. A GLOF event could lead to the sudden release of massive volumes of water and debris, posing a serious threat to downstream communities, infrastructure, and ecosystems. The warnings became reality when the South Lhonak Outburst Flood happened in October 2023. Various other findings had already identified that high-risk zones and the quantifiable parameters of snow cover and lake expansion are vital for authorities and policymakers. Researchers and scientists have informed the development of early warning systems, disaster risk reduction strategies, and climate adaptation measures to safeguard lives and livelihoods in the Sikkim Himalayan region.

Further, several studies have underscored specific risks associated with the South Lhonak Glacial Lake, noting its distinctive characteristics, especially in response to climate change (Basnett et al. 2013; Hazra and Krishna 2019; Remya et al. 2019; R. K. Sharma et al. 2018). Particularly, it was observed that the South Lhonak Glacier exhibits a higher average glacial velocity compared to others in the region (A. Sharma et al. 2023a). This faster movement not only contributes to the dynamic changes in the lake's size and shape but also elevates the risk profile for potential GLOF incidents, making it a critical focus for ongoing monitoring and risk assessment efforts. From the above studies and the striking of South Lhonak GLOFs, there are various policy lessons to be inferred.

16.4 Inferences from the South Lhonak Lake Outburst Flood

The South Lhonak Lake Outburst Flood of 2023 serves as a case, encapsulating the complex challenges posed by the expansion of glacial lakes in the Himalayas. The dire consequences of this particular GLOF – loss of life, environmental degradation, and infrastructure devastation – underscore the pressing need for adaptive policy frameworks capable of responding to the shifting dynamics of climate change. Through reflection on the event's antecedents, impacts, and subsequent emergency responses, critical inferences emerge to inform and refine policy decisions. The imperative for more nuanced risk assessments, alongside resilient infrastructure planning and community-based early warning systems, emerges as a high priority.

A study by R. K. Sharma et al. (2022) had called for a multifaceted policy approach to tackle the escalating threat of GLOFs, driven by climate change and other environmental factors. Key policy recommendations included in the study were the implementation of comprehensive glacial lake monitoring systems, the development of region-specific early warning and evacuation plans, and the integration of GLOF risk assessments into regional development planning. This chapter emphasized the importance of enhancing infrastructure resilience, particularly in vulnerable mountain communities, and fostering cross-border collaboration for data sharing and disaster response. Furthermore, the chapter suggested that policies should support scientific research and incorporate traditional ecological knowledge to improve understanding of glacial dynamics. The South Lhonak Lake incident exemplifies the need for informed policy interventions that can mitigate the risks and prepare the Himalayan region for future GLOF events.

However, it can be argued that the event of the GLOF at South Lhonak Lake has shown that despite major scientific warnings and policy recommendations, the damage happened, washing away lives, infrastructure, and a 1200 MW hydropower dam. Many other studies also highlight the problem of glacial lake expansion in the Himalayas (Allen, Linsbauer, et al. 2016; Khadka et al. 2018; Nie et al. 2013; Wang et al. 2015). Notwithstanding any potential biases, it can be critically inferred that there has also been a lack of "translation" of scientific research to policymakers; otherwise, the South Lhonak Outburst Flood might have been averted.

This disconnect between scientific foresight and policy implementation necessitates an immediate recalibration of how research findings are communicated to and acted upon by decision-makers. The South Lhonak disaster serves as a crucial wake-up call, indicating that current strategies and communication channels are insufficiently equipped to translate scientific understanding into proactive, preventive measures. To bridge this gap, there must be a concerted effort to establish a dialogue between scientists, local authorities, and policymakers, ensuring that complex data is rendered into actionable insights. Policies must be dynamic and science-driven, informed by the latest research and predictive models, and should be subject to regular review to adapt to the changing climate and topography. Additionally, investing in capacity building for local governments and communities to understand and apply scientific recommendations will play a pivotal role in enhancing the region's resilience. It is through such policy evolution, underpinned by a commitment to continuous learning and adaptation, that the Himalayan region can hope to avert the repeat of tragedies like the South Lhonak Lake GLOF.

16.5 Conclusion

The pervasive and escalating threat of GLOFs in the Himalayan region is a stark reminder of the tangible impacts of climate change. The South Lhonak GLOF of 2023 stands as a cautionary tale of the destruction that can be wrought when a combination of natural geological processes is intensified by human-induced climate alterations. The increasing frequency and intensity of GLOFs not only jeopardize human lives but also threaten vital infrastructure, livelihoods, and the broader ecosystem upon which the regional population depends.

In light of the inferences from the South Lhonak GLOF and other documented events, it is clear that a proactive, integrated approach to risk management is essential. This involves the improvement and implementation of early warning systems, the reinforcement of critical infrastructure, and the adoption of sustainable land-use practices that consider the heightened risks of GLOFs. Furthermore, ongoing research and monitoring of glacial lakes must be supported to better understand and predict the precursors to such catastrophic events, enabling timely evacuation and mitigation efforts.

The international community, national governments, local authorities, and scientific bodies must collaborate to develop and enforce policies that prioritize the safety of at-risk communities. Climate resilience can be significantly bolstered by investing in adaptive measures that address both immediate and long-term vulnerabilities. Education and community engagement are also crucial, empowering local populations with the knowledge and tools necessary for self-sufficiency in the face of disaster.

Ultimately, the fight against climate change is at the core of preventing future GLOFs. Global efforts to reduce greenhouse gas emissions and slow the pace of global warming are indispensable in mitigating the risks of glacial lake expansion. The Himalayas, often referred to as the "Third Pole," are a barometer of the health of our planet's climate system. Protecting this vital region not only safeguards a critical water source for billions but also serves as a commitment to preserving the Earth's delicate balance for future generations.

References

Allen, S.K., Linsbauer, A., Randhawa, S.S. et al. (2016). Glacial lake outburst flood risk in Himachal Pradesh, India: an integrative and anticipatory approach considering current and future threats. *Natural Hazards* 84 (3): 1741–1763. https://doi.org/10.1007/s11069-016-2511-x.

Allen, S.K., Rastner, P., Arora, M. et al. (2016). Lake outburst and debris flow disaster at Kedarnath, June 2013: hydrometeorological triggering and topographic predisposition. *Landslides* 13 (6): 1479–1491. https://doi.org/10.1007/s10346-015-0584-3.

Babu Govindha Raj, K., Remya, S.N., and Vinod Kumar, K. (2013). Remote sensing-based hazard assessment of glacial lakes in Sikkim Himalaya. *Current Science* 104 (3): 359–364.

Basnett, S., Kulkarni, A.V., and Bolch, T. (2013). The influence of debris cover and glacial lakes on the recession of glaciers in Sikkim Himalaya, India. *Journal of Glaciology* 59 (218): 1035–1046. https://doi.org/10.3189/2013JoG12J184.

Begam, S., Sen, D., and Dey, S. (2018). Moraine dam breach and glacial lake outburst flood generation by physical and numerical models. *Journal of Hydrology* 563: 694–710. https://doi.org/10.1016/j.jhydrol.2018.06.038.

Chettri, N., Adhikari, B., Chaudhary, S., and Wangchuk, K. (2023). Changing discourses in the third pole: a systematic review of climate change impact on biodiversity in the Hindu Kush Himalaya. *Ecological Indicators* 155: 111046. https://doi.org/10.1016/j.ecolind.2023.111046.

Dar, R.A., Murtaza, K.O., Paul, O.J. et al. (2022). River response to melting cryosphere since late quaternary in the Pir Panjal range of NW Himalaya. *Frontiers in Water* 4: 879001. https://doi.org/10.3389/frwa.2022.879001.

Das, S., Das, S., Tanu Mandal, S. et al. (2024). Inventory and GLOF susceptibility of glacial lakes in Chenab basin. *Western Himalaya* 15 (1): 2356216. https://doi.org/10.1080/19475705.2024.2356216.

Dobhal, D.P., Mehta, M., and Srivastava, D. (2013). Influence of debris cover on terminus retreat and mass changes of Chorabari Glacier, Garhwal region, central Himalaya, India. *Journal of Glaciology 59* (217): 961–971.

Durga Rao, K.H.V., Venkateshwar Rao, V., Dadhwal, V.K., and Diwakar, P.G. (2014). Kedarnath flash floods: a hydrological and hydraulic simulation study. *Current Science* 106 (4): 598–603.

Govindha Raj, B. K. (2009). Remote sensing based hazard assessment of glacial lakes – a case study from Kumaon Himalaya. *Current Science*, 2(2), 31–39. https://www.researchgate.net/publication/260268092 (accessed 26 September 2024).

Hazra, P. and Krishna, A.P. (2019). Spatio-temporal and surface elevation change assessment of glaciers of Sikkim Himalaya (India) across different size classes using geospatial techniques. *Environmental Earth Sciences* 78 (14): https://doi.org/10.1007/s12665-019-8390-1.

Hazra, P.and Krishna, A. P. (2021). Ahp based assessment of GLOF susceptibility of South Lhonak Glacial Lake, Sikkim Himalaya, India. *International Geoscience and Remote Sensing Symposium (IGARSS)*, Brussels, Belgium, (11–16 July 2021), 5489–5492. 10.1109/IGARSS47720.2021.9553439

Hewitt, K. (1982). Natural dams and outburst floods of the Karakoram Himalaya. *Hydrological Aspects of Alpine and High Mountain Areas (Proceedings of the Exeter Symposium* (July 1982). IAHS Publ. No. 138, 259–270.

Huber, A. (2019). Hydropower in the Himalayan hazardscape: strategic ignorance and the production of unequal risk. *Water* 11 (3): 414. https://doi.org/10.3390/w11030414.

Khadka, N., Chen, X., Sharma, S., and Shrestha, B. (2023). Climate change and its impacts on glaciers and glacial lakes in Nepal Himalayas. *Regional Environmental Change* 23 (4): 143. https://doi.org/10.1007/s10113-023-02142-y.

Khadka, N., Zhang, G., and Thakuri, S. (2018). Glacial lakes in the Nepal Himalaya: inventory and decadal dynamics (1977-2017). *Remote Sensing* 10 (12): 1913. https://doi.org/10.3390/rs10121913.

Martha, T.R., Roy, P., Jain, N. et al. (2021). Rock avalanche induced flash flood on 07 February 2021 in Uttarakhand, India – a photogeological reconstruction of the event. *Landslides* 18 (8): 2881–2893. https://doi.org/10.1007/s10346-021-01691-9.

Meena, S.R., Bhuyan, K., Chauhan, A., and Singh, R.P. (2021). Changes in the flood plains and water quality along the Himalayan rivers after the Chamoli disaster of 7 February 2021. *International Journal of Remote Sensing* 42 (18): 6984–7001. https://doi.org/10.1080/01431161.2021.1944696.

Meena, S.R., Chauhan, A., Bhuyan, K., and Singh, R.P. (2021). Chamoli disaster: pronounced changes in water quality and flood plains using sentinel data. *Environmental Earth Sciences* 80 (17): 601. https://doi.org/10.1007/s12665-021-09904-z.

Mir, R.A., Jain, S.K., Lohani, A.K., and Saraf, A.K. (2018). Glacier recession and glacial lake outburst flood studies in Zanskar basin, western Himalaya. *Journal of Hydrology* 564: 376–396. https://doi.org/10.1016/j.jhydrol.2018.05.031.

Nie, Y., Liu, Q., & Liu, S. (2013). Glacial lake expansion in the Central Himalayas by landsat images, 1990–2010. *PLoS ONE*, 8(12): e92654. https://doi.org/10.1371/journal.pone.0083973

Nie, Y., Pritchard, H.D., Liu, Q. et al. (2021). Glacial change and hydrological implications in the Himalaya and Karakoram. *Nature Reviews Earth and Environment* 2 (2): 91–106. https://doi.org/10.1038/s43017-020-00124-w.

Pandey, P., Chauhan, P., Bhatt, C.M. et al. (2021). Cause and process mechanism of rockslide triggered flood event in Rishiganga and Dhauliganga River Valleys, Chamoli, Uttarakhand, India using satellite remote sensing and in situ observations. *Journal of the Indian Society of Remote Sensing* 49 (5): 1011–1024. https://doi.org/10.1007/s12524-021-01360-3.

Ramsankaran, R.A.A.J., Pandit, A., and Azam, M.F. (2018). Spatially distributed ice-thickness modelling for Chhota Shigri Glacier in western Himalayas, India. *International Journal of Remote Sensing* 39 (10): 3320–3343. https://doi.org/10.1080/01431161.2018.1441563.

Remya, S.N., Kulkarni, A.V., Pradeep, S., and Shrestha, D.G. (2019). Volume estimation of existing and potential glacier lakes, Sikkim Himalaya, India. *Current Science* 116 (4): 620–627.

Ruiz-Villanueva, V., Allen, S., Arora, M. et al. (2017). Recent catastrophic landslide lake outburst floods in the Himalayan mountain range. *Progress in Physical Geography* 41 (1): 3–28. https://doi.org/10.1177/0309133316658614.

Sati, S.P. and Gahalaut, V.K. (2013). The fury of the floods in the north-west Himalayan region: the Kedarnath tragedy. *Geomatics, Natural Hazards and Risk* 4 (3): 193–201. https://doi.org/10.1080/19475705.2013.827135.

Sattar, A., Goswami, A., and Kulkarni, A.V. (2019). Hydrodynamic moraine-breach modeling and outburst flood routing - a hazard assessment of the South Lhonak lake, Sikkim. *Science of the Total Environment* 668: 362–378. https://doi.org/10.1016/j.scitotenv.2019.02.388.

Sattar, A., Goswami, A., Kulkarni, A.V. et al. (2021). Future Glacial Lake Outburst Flood (GLOF) hazard of the South Lhonak Lake, Sikkim Himalaya. *Geomorphology* 388: 107783. https://doi.org/10.1016/j.geomorph.2021.107783.

Sharma, A., Gupta, M., Sharma, N., and Gupta, S. (2023a). Line-of-sight glacier velocity estimation of transboundary glaciers in the eastern Himalayas using high-resolution TerraSAR-X data. *Proceedings* 87 (1): 25. https://doi.org/10.3390/IECG2022-13951.

Sharma, A., Gupta, M., Sharma, N., and Gupta, S. (2023b). Thickness and mass of East Rathong Glacier (ERG), West Sikkim, India, estimated using GIS. *Environment and Ecology Research* 11 (4): 554–563. https://doi.org/10.13189/eer.2023.110404.

Sharma, R.K., Kumar, R., Pradhan, P., and Sharma, A. (2022). Climate-induced glacier retreats and associated hazards: need for robust glaciers and glacial lake management policy in Sikkim Himalaya, India. In: *Climate Change: Impacts, Responses and Sustainability in the Indian Himalaya* (ed. S. Rani and R. Kumar), 161–182. Cham: Springer International Publishing https://doi.org/10.1007/978-3-030-92782-0_8.

Sharma, R.K., Pradhan, P., Sharma, N.P., and Shrestha, D.G. (2018). Remote sensing and in situ-based assessment of rapidly growing South Lhonak glacial lake in eastern Himalaya, India. *Natural Hazards* 93 (1): 393–409. https://doi.org/10.1007/s11069-018-3305-0.

Tómasson, H. (1996). The jökulhlaup from Katla in 1918. *Annals of Glaciology* 22: 249–254. https://doi.org/10.3189/1996aog22-1-249-254.

Vuichard, D. and Zimmermann, M. (1987). The 1985 catastrophic drainage of a moraine-dammed lake, Khumbu Himal, Nepal: cause and consequences. *Mountain Research & Development* 7 (2): 91–110. https://doi.org/10.2307/3673305.

Wang, W., Xiang, Y., Gao, Y. et al. (2015). Rapid expansion of glacial lakes caused by climate and glacier retreat in the Central Himalayas. *Hydrological Processes* 29 (6): 859–874. https://doi.org/10.1002/hyp.10199.

Watanbe, T., & Rothacher, D. (1996). The 1994 Lugge Tsho glacial lake outburst flood. *Mountain Research and Development*, 16(1), 77–81. http://www.jstor.orgURL:http://www.jstor.org/stable/3673897http://www.jstor.org/stable/3673897?seq=1&cid=pdf-reference#references_tab_contents (accessed 26 September 2024).

Xu, J., Grumbine, R.E., Shrestha, A. et al. (2019). The melting Himalayas: cascading effects of climate change on water, biodiversity, and livelihoods. *Conservation Biology* 23: 520–530. https://doi.org/10.1111/j.1523-1739.2009.01237.x.

17

A Comprehensive Review on Water Resource Vulnerability to Climate Change in the Mountain Region of India

Deepika Dimri[1], *Ambika Kumar*[2], *and Shalini Singh Chhetri*[1]

[1] *Himalayan Women Awareness and Livelihood (HIMWAL) Society, Dehradun, India*
[2] *Bhagalpur National College, Bhagalpur, Bihar, India*

17.1 Introduction

Climate change pertains to extensive and prolonged alterations in Earth's weather patterns and the average global temperature. These alterations arise from modifications within the climate system induced by natural and human factors (Khadka et al. 2023). Water resources are globally being highly influenced by climate change, as per the report given by the Intergovernmental Panel on Climate Change (IPCC). By the year 2050, it is anticipated that over one billion individuals in Asia will confront water scarcity due to these environmental shifts. According to the 6th Assessment Report by the Intergovernmental Panel on Climate Change, the global temperature increased by 0.85 °C from 1881 to 2012 (Prajapati et al. 2023). Climate change affects the timing, intensity, pattern, temperature, and discharge of rainfall, thereby exerting an influence on the hydrological cycle at both local and global levels (Prajapati et al. 2023).

About a quarter of the Earth's surface is occupied by mountainous regions, with roughly 26% of the global population residing in either the mountains or their foothills (Syldon et al. 2024). Mountain ecosystems, being among the Earth's most delicate, serve as crucial suppliers of water, energy, minerals, and forest products. Furthermore, these ecosystems play a pivotal role in fostering economic advancement, preserving the environment, and promoting environmental resilience (Mirjam Macchi et al. 2011; Pandey et al. 2017). Many studies have frequently documented a temperature surge in the Himalayan regions. The temperature trends in the Himalayan region indicate an average warming of (+0.28 °C) per decade from 1951 to 2020, significantly exceeding the previously reported trend of (+0.12 °C) per decade observed from 1971 to 1994 (Khan et al. 2024; Shrestha et al. 1999). Consequently, this warming trend is expected to have adverse effects on river water supply, particularly during the fall and spring seasons (Bharati et al. 2019).

Concerning climate change, vulnerability refers to the extent to which a system is prone to and incapable of managing the detrimental impacts of climate change, encompassing

Water Resources Management in Mountain Regions, First Edition. Edited by Virendra Bahadur Singh, Sughosh Madhav, Rakesh Kumar Gupta, Prerna Diwan, and Amit Kumar.

Table 17.1 Effects of climate change on water resources.

Water resources	Impact	References
Ganges River Basin	Groundwater storage decline occurring at a pace of 1.25–2.1 cm/yr	Panda and Wahr (2016)
South Lhonak Lake (Sikkim Himalaya)	In 2015, the estimated volume was approximately 60 ± 10.8 million m^3. It is anticipated that the lake may reach a maximum volume of around 90 ± 16.2 million m^3 due to the continued retreat of the glacier	Remya et al. (2019)
Lahaul Glacial Area	Between 1999 and 2011, for a 2110 km^2 glacierized area in the Lahaul and Spiti region, the mass balance was reported −(0.44) m.w.e.	Vincent et al. (2013)
Sutlej Basin	Six of the eight monitoring sites in the Sutlej basin showed an increase in the annual maximum temperatures (Tmax) between 1980 and 2010. These findings indicated a warming trend in the region	Hamid et al. (2014)

both climatic variations and extreme events (Pandey et al. 2018). The elevated occurrence of extreme climate events, along with widespread poverty, migration, and marginalization of the Himalayan population, renders mountain communities increasingly susceptible to climate variability (Gerlitz et al. 2017). In spite of the rich abundance of natural resources in the Himalayas, communities residing in these mountains find themselves among the most impoverished and vulnerable globally. The precarious situation of these mountain communities arises from factors such as geographical remoteness, unsustainable development practices, and the excessive exploitation of local natural resources (Gerlitz et al. 2017; Gupta et al. 2020; Pandey et al. 2017). However, if the current climate change scenario persists without mitigation and adaptation measures, it is anticipated that the future impacts of climate change will be severe (Khadka et al. 2023; Wester et al. 2019).

Despite receiving global attention, research on climate change in the Himalayan Mountains is still in its early stages. Consequently, there is a lack of long-term climate data in the region, and the existing data has been gathered inconsistently in terms of instrumentation and methodology. This inconsistency contributes to uncertainty regarding the accuracy and reliability of the observed data. Additionally, there are limited studies available that could illustrate the impact of climate change on mountain ecosystems (Negi and Mukherjee 2020). Hence, the present chapter provides a brief summary regarding different water resources in the Indian Himalayan region, which are being affected by climate change. Table 17.1 illustrates the impact of climate change on various water resources throughout the Himalayan region.

17.2 The Documented Consequences of Climate Change

17.2.1 Glaciers

Climate change plays a pivotal role in determining the condition and future of glaciers, influencing their size, volume, and mass equilibrium (Gardelle et al. 2013; Khadka et al. 2023). The gradual retreat of the Himalayan glaciers marks a noteworthy slow-onset phenomenon, defined as an event that unfolds progressively through incremental alterations over an extended period. Current forecasts specify that an elevation in global temperature by 1.5 °C is expected to result in a warming of approximately (2.1 ± 0.1 °C) in the Himalayan glaciers region (Talukder et al. 2021). Over the past four decades, there has been a discernible reduction in glacier area across the Himalayas, with around 20 060 out of the total 40 000 km^2 of glaciated area showing a decrease. This reduction accounts for approximately 13% of the total glacier area, encompassing all major climatic zones of the Himalayan region. The western Himalayas have seen a decrease in spring snow cover since 1993, and snow has been melting more quickly due to rising winter temperatures (Kripalani et al. 2003). Numerous investigations into the renowned Indian glacier "Gangotri," have documented a substantial retreat over the preceding decades. However, no retreat has been observed for the period between 2006 and 2010 in these studies (Kargel et al. 2011). The Karakoram Range showed a clear pattern of rising glacial reserves, in contrast to the general observation of retreating glaciers in the area (Bahuguna et al. 2014; Zhou et al. 2017). A substantial proportion of glaciers in the western Indian Himalayas, the northern central Himalayas, as well as the west Kunlun Shah, experienced a notable and swift reduction in size. Observations indicated that these regions were characterized by a notable prevalence of debris-free glaciers (Ding et al. 2006; Mir et al. 2021). Glacial melting disrupts the terrestrial carbon cycle, leading to soil carbon loss through diminished microbial communities, reduced biogeochemical and physical weathering processes, and decreased storage and cycling of organic carbon (exceeding 104 Pg C) and essential nutrients (Hood et al. 2015; Talukder et al. 2021; Wadham et al. 2019). Glacial melt plays a crucial role in regulating river discharge and groundwater levels while serving as a vital water source during dry periods. The Himalayan glaciers are key contributors to the hydrological balance of major river systems. Notably, melted glacier water constitutes 9% of the Ganges' overall annual flow, rising to 70% during the spring and autumn seasons when alternative water sources are scarce. In addition, melted glacier water accounts for 41% of the runoff in the Upper Indus basin (Biemans et al. 2019; Huss and Hock 2018; Huss et al. 2010; Radić and Hock 2014 Talukder et al. 2021). Scherler et al. (2011) conducted a study on numerous glaciers across Nepal, Bhutan, Sikkim, Uttarakhand, and Himachal Pradesh during the period of 2000–2008. As temperatures rise, the ice caps and glaciers melt, leading to the expansion of ocean water as it warms. The melting of Himalayan glaciers has led to the expansion of numerous proglacial and supraglacial lakes, along with the formation of over 5000 volatile glacier lakes, which are vulnerable to floods caused by glacial lake outbursts.

The Geological Survey of India (GSI) reported that the Bara-Shigri Glacier retreated by approximately 2755 m from 1906 to 1995, at a pace of 30.9 m annually. Saha et al. (2024) observed in their study that the Rathong glacier in the Sikkim Himalaya exhibited a maximum retreat rate of 0.11 km^2/year, along with a significant 163% expansion in the ablation zone. Tripathi et al. (2022) calculated the glacier area over a 30-year period and discovered that the glacier area is shrinking due to the increasing Land Surface Temperature (LST).

Kulkarni and Alex (2003) mapped 30 glaciers in the Baspa basin of Himachal Himalaya. Their subsequent study analyzed the area, snout, and volume retreat of 19 glaciers from

1976 to 2003, revealing a retreat in snout position and glaciers mass loss (Singh et al. 2022). Over 39 years (1980–2019), the total loss in the glaciers of the Baspa river basin has been calculated at 3.28 km^3 (Singh et al. 2022). The glaciological mass balance of the Shaune Garang glacier was analyzed in the basin from 2013 to 2016, revealing a negative trend in mass balance and a rise in temperature (Singh et al. 2022). During the investigation period from 2005 to 2017, the Gangotri and neighboring glaciers (Chaturangi, Bhagirathi-Kharak, Raktvarn, Satopanth, and Arwa) in the Central Himalayan region exhibited a retreating trend (Yousuf et al. 2024). Table 17.2 presents varying glacial retreat rates of glaciers in the Garhwal Himalayas over time.

Monitoring Himalayan glaciers using conventional methods is difficult owing to the heterogeneous terrain. However, advancements in satellite technology now enable us to monitor the extent of larger glaciers in high mountainous regions. Additionally, the application of remote sensing techniques has made this monitoring feasible.

Kulkarni et al. (2011) reported an average loss of 16% in glacier area among the 1868 glaciers examined in the Indian Himalayas. The study by Thapliyal et al. (2023) examined the dynamics of the Bhagirathi-Kharak (BKG) and Satopanth (SPG) glaciers in the Mana basin (Central Himalayan area of India), with a particular emphasis on the effects of climate change. The study involved evaluating the spatiotemporal alterations in glacier area, volume, and snout positions from 1968 to 2017, using a combination of CORONA photographs and satellite data. Their findings indicate pronounced warming across the basin, accompanied by a notable downward trend in snow cover area across various seasons. Rajat et al.'s (2022) study emphasizes the use of a fundamental semantic segmentation network (U-net) in the identification of glaciers using Indian Remote Sensing (IRS) and Landsat satellite data located in the western Indian Himalayas. Over a span of three decades (1994–2021), the researchers systematically map glaciers in the area to evaluate how climate change is affecting glacial ablation. Notably, the glaciated area in Himachal Pradesh exhibited a consistent decline, decreasing at an annual rate of 67.84 km^2 from 4020.6 km^2 in 1994 to 2198.5 km^2 in 2021. The study observes a percentage loss of approximately 1.678 per annum in glacier area from 1994 to 2021, with a decreasing decadal trend from 2.31% between 1994 and 2001 to 1.39% between 2011 and 2021. In their recent study, Saha et al. (2024) discovered that climate change is accountable for modifying the physical properties of glaciers. Their findings indicate that the Zemu glacier underwent the most significant deglaciation, with a reduction in area by 20.79 km^2, a maximum 30 m decrease in elevation, a 24 m increase in snout elevation, and a retreat of 200 m at the terminus within Kanchenjunga National Park. Kumar and Sharma (2023) conducted a comprehensive study that assessed the terminus retreat of medium-sized glaciers spanning the years 1988–2018 in Sikkim. Their findings revealed a range of glacier reduction rates, varying from 63.9 to 3.9 m/year, resulting in an overall loss of approximately 2.53% (0.08%/year) of the total glacier area in the study area. A comparative analysis of glacier retreats in Sikkim with other Himalayan regions indicates a diminishing gradient from the northwest toward the eastern Himalayas.

The literature review reveals a lack of mass balance data for glaciers. Many glacierized regions worldwide, including the Himalayas, remain unsampled, poorly sampled, and inadequately monitored (Tak and Keshari 2020). A detailed mass balance study of Indian Himalayan glaciers using the glaciological method began in 1974, conducted by the GSI

Table 17.2 Glacial retreat rate over time in the Himalayas.

S.N.	Studied glacier	Time interval	Retreat (m/yr)	References
1	Gangotri	2006–2015	9.0	Bhattacharya et al. (2016)
		2001–2015	10	Singh et al. (2017)
		2004–2010	21.3	Saraswat et al. (2013)
		2005–2014	12.01	Yousuf et al. (2024)
2	Satopanth	1968–2017	8.45	Thapliyal et al. 2023
		2005–2013	5.9	Nainwal et al. (2016)
		1962–2006	26.45	Nainwal et al. (2008)
		2005–2014	1.89	Yousuf et al. (2024)
3	Bhagirathi-Kharak	1968–2017	4	Thapliyal et al. (2023)
		2005–2013	12.2	Nainwal et al. (2016)
		2005–2017	5.99	Yousuf et al. (2024)
4	Chorabari	1990–2001	7.00	
		2001–2012	5.02	
		2012–2017	10.66	
5	Parkachik	1971–2021	4	Rana et al. (2023)
6	Chaturang	2006–2015	10.1	Bhattacharya et al. (2016)
		2006–2013	10.2	Bhattacharya et al. (2016)
		2015–2016	9.74	
		2005–2017	7.97	Yousuf et al. (2024)
7	Raktvarn	2006–2015	4.1	Bhattacharya et al. (2016)
		2005–2017	3.28	Yousuf et al. (2024)
8	Arwa	2005–2017	0.85	Yousuf et al. (2024)
9	Bara-Shigri	1863–2014	19.2	Chand et al. (2017)
10				
11	Sonapani	1971–2016	18.4	Majeed et al. (2020)
12	Sonapani	1906–2016	16.5	Majeed et al. (2020)
13	Milam Glacier	1849–1957	13.14	Cotter and Brown (1907)
14	Padam	1962–2015	50.7	Mir (2018)
14	Drang Drung	1975–2008	9.4	Kamp et al. (2011)
15	Dalung	1962–2015	20	Mir (2018)
16	Kolahoi 1	1979–2016	22	Mir (2018)
16	Kolahoi 2	1979–2016	11	Mir (2018)

(Singh et al. 2018). The study by Tak and Keshari (2020) revealed a mass deficit for the Parvati Glacier over a 19-year period from 1998 to 2016, indicating an alarming retreat of the glacier over time. They reported a mean specific mass loss of -0.49 ± 0.11 meters water equivalent (m.w.e.), and the total mass loss amounted to 3.95 gigatonnes (Gt). These findings highlight the significant impact of climate change on the glacier, affecting river flows and water availability. Over the study period, the Gara glacier in Himachal Himalaya displayed a positive mass balance in the first two years and the final year, with a negative balance for the other five years. Seasonal studies revealed that a consistent positive mass balance during summer led to a positive annual mass balance (Raina, 2009; Singh et al. 2018). The reported differences in the mass balances of the Hamtah and Chhota Shigri glaciers are likely due to the Hamtah glacier's lower maximum elevation and smaller accumulation area compared to the Chhota Shigri glacier (Singh et al. 2016). Raman et al. (2022) employed the improved accumulation area ratio (IAAR) method to estimate the mass balance of 1055 glaciers in the Alaknanda and Bhagirathi basins of the Garhwal Himalayas. For the period 2001–2013, they estimated the mean specific mass balance to be (−1.1) m.w.e.a^{-1} for the Alaknanda basin and (−1.01) m.w.e.a^{-1} for the Bhagirathi basin. Table 17.3 provides a comparative analysis of the specific mass balance among glaciers in the Himalayan region.

17.2.2 Groundwater

Climate change and variability have significant impacts on groundwater systems, both directly and indirectly. Direct effects involve changes in natural recharge processes, while climate and land cover play primary roles in governing precipitation and

Table 17.3 Analysis of specific mass balance variations among glaciers in the Himalayas.

S.N.	Glacier	Annual specific mass balance (m.w.e.)	Time period	References
1.	Chhota Shigri	−0.39	1987–1989	Dobhal et al. (1995)
2.	Chhota Shigri	0.7	2002–2005	Wagnon et al. (2007)
3.	Parvati	−0.49	1998–2016	Tak and Keshari (2020)
4.	Bara-Shigri	−0.48	1999–2011	Gardelle et al. (2013)
5.	Hamtah	−0.45	1999–2011	Vincent et al. (2013)
6.	Region-wide	−0.44	1999–2011	Vincent et al. (2013)
7.	Gangotri	−0.92	2001–2014	Aggrawal et al. (2017)
		−0.82	2006–2007	
		−0.98	1985–2005	
8.	Dokriani	−0.22	1992–1993	Dobhal et al. (2008)
		0.23	1993–1994	
		−0.38	1999–2000	
		−0.24	2010–2011	Dobhal et al. (2021)
		−0.33	2011–2012	
		−0.35	2012–2013	
9.	Mulkila	−0.66	2009–2018	Guruprasad et al. (2022)

evapotranspiration (Swain et al. 2022). Groundwater resources could be impacted by climate change as a result of anticipated shifts in rainfall patterns, evapotranspiration rates, and the spatial and temporal distribution of these crucial elements within the water balance (Guevara-Ochoa et al. 2020). Understanding how climate variables interact with groundwater is more complex than their interaction with surface water. Increased precipitation is expected to lead to higher water flow rates, thereby increasing the likelihood of flooding while simultaneously reducing the amount of groundwater recharge (Şen Z 2021). Coupled with other human-induced factors, this has led to a growing concern about the depletion of springs across the area (Rathi et al. 2020; Tambe et al. 2012)).

A substantial segment of the population residing in the Indian Himalayas relies on spring water for their livelihoods, be it for agricultural purposes or to support their livestock. The flow of rivers through deep valleys restricts their suitability for water supplies, while accessing groundwater aquifers in the challenging Himalayan terrain proves challenging and expensive. Within the remote, non-snow-fed areas and rain-shadow watershed regions, springs stand as the exclusive source of water for essential needs such as drinking, household use, and sustaining livestock (Panwar 2020). Climate change is causing a decline in the availability of water from the springs in the Indian Himalayas (Rathi et al. 2020; Tambe et al. 2012). Studies on spring discharge have conclusively indicated characteristics of the recharge region and the rainfall pattern have an impact on it (Negi and Joshi 1996; Negi et al. 2001; Rai et al. 1998). There have been reports indicating that the reduction in glacial meltwater in the central Ganges basin is resulting in a diminished groundwater recharge (Sukanya and Joseph 2023).

The spring flow in the Kashmir Himalayas is directly influenced by the melting of snow and glaciers and is inversely affected by precipitation. The higher recharge associated with higher temperatures and the melting of glaciers throughout the summer identifies upstream areas as important recharge zones (Bhat et al. 2014; Panwar et al. 2020). Agarwal et al. (2012) investigated the long-term relationship (1999–2010) between rainfall and spring discharge in the Chandrabhaga and Danda watersheds in the Garhwal Himalayas. The swift and receptive response of spring flow to rainfall indicates that aquifer recharge in the Garhwal region mostly depends on precipitation. Tambe et al. (2012) examined the springs located in the Sikkim region of the Himalayas. Their research uncovered that the flow rate of these springs peaked at 51 l/minute during the months following the monsoon season (September–November), then steadily declined to 8 l/minute in the spring months (March–May). The Kosi River basin in the Kumaun Himalayas has a significant impact on the entire distribution of rainfall. Significantly, the area has had a noticeable decrease in precipitation as a result of shifting climatic conditions. As a result, this has caused a decline in the water table, a decrease in spring flow, and a reduction in headwater flows (Grover 2015). In their study, Pant and Rawat (2015) found that there has been a drop in the number of perennial springs in the Dabka sub-catchment area of the Kosi River basin in the Kumaun Himalayas. Specifically, they observed a decline of 3 springs per year. Conversely, non-perennial and dried-up springs exhibited an increase, with a growth rate of 1 spring per year and 14 springs per decade, respectively (Panwar et al. 2020). Research integrating both factual information and human perspectives indicates a connection between declining spring flows and alterations in the climate (Kulkarni et al. 2021; Pandey et al. 2018).

17.2.3 Glacial Lakes Outburst Flood (GLOF)

Glaciers are commonly associated with multiple glacial lakes, which are particularly sensitive to warming temperatures.

GLOF refers to the rapid flow of a significant amount of water through a restricted river channel caused by the failure of a natural dam (Mohanty and Maiti 2022). The glacial lakes in the Himalayan region are experiencing rapid growth and expansion, leading to an increased presence of potentially dangerous lakes (Ahmed et al. 2021). In 1929, a GLOF caused by an ice dam occurred in Chog Kumdan Lake in the Shyok basin of the Ladakh region. This flood had a significant impact on around 48 settlements located downstream (Gunn 1930). In their study, Dubey and Goyal (2020) analyzed the possible consequences that could occur downstream from 329 lakes located in the Indian Himalayan region. They specifically identified 36 lakes that are vulnerable to being affected by avalanche impact zones. According to Shukla et al. (2018), the number of glacial lakes in the Sikkim Himalayas rose from 425 in 1975 to 466 in 2017, indicating a roughly 9% increase. Aggarwal et al. (2017) identified that 21 lakes in the Sikkim Himalaya region are under danger of experiencing outburst floods. In a recent study, Islam et al. (2022) examined 27 glacial lakes in the Chhombo Chhu Watershed in the Sikkim Himalaya that have the potential to pose a threat. Dubey and Goyal (2020) from the Indian Institute of Technology Indore, India, employed satellite data, digital elevation models (DEMs), and flood simulations to assess the potential danger posed by 329 glacial lakes in the Indian Himalayas. From 2014 to 2016, the GSI identified 486 glacial lakes in Uttarakhand. Of these, 13 lakes, situated near the Rishiganga and Dhauliganga valleys, were reported highly vulnerable to GLOFs (Nautiyal et al. 2022). A study in Uttarakhand shows that 16 out of 78 tehsils are at risk of GLOFs. The Wadia Institute of Himalayan Geology conducted a scientific evaluation, revealing that Dehradun has 1266 glacial lakes, with 10–11 moraine-dammed glacial lakes being highly susceptible to outbursts (Nautiyal et al. 2022). In the Badrinath region of the Uttarakhand Himalayas, which is already at risk from potential GLOFs originating from two valleys, a new and significant threat is expected to emerge from neighboring valleys in the future (Mal et al. 2021). Sattar et al. (2021) assessed the future GLOF risk of South Lhonak Lake, the largest proglacial lake in Sikkim. Using an ice-thickness approach, they estimated the future volume of the lake to be $114.8 \times 10^6 \, m^3$. Linsbauer et al. (2016) projected the future formation of glacial lakes in the Himalayas by modeling the topographies of glacier beds, identifying 5000 overdeepenings that could potentially become glacial lakes. Kaushik et al. (2024) conducted an inventory of glacial lakes in the Sikkim Himalaya and applied dam break flood simulations to the seven most dangerous lakes. Their results revealed that in extreme case scenarios, the highest peak discharges were $9504 \, m^3/s$ from Khanchung Lake and $8421 \, m^3/s$ from South Lhonak Lake. Their findings indicated that 10,000 people are at direct risk of GLOFs, with potential large-scale infrastructure damage impacting around 1900 settlements, 5 bridges, and 2 hydropower plants.

Studies on glacier lakes in the Himalayas have revealed a notable increase in both their surface area and volume over the last four decades (Haritashya et al. 2018). Moreover, the development of the "Himalayan Glacier Thickness Mapper" (HIGTHIM) technique has enabled the estimation of both present and future glacier lake volume and expansion. This method utilizes DEMs, surface velocity data, and moraines to derive these estimates (Kulkarni et al. 2019, 2021). These lakes serve as indicators of climate change within the vulnerable Himalayan region, posing threats to both human lives and the local economy (Mir et al. 2021). Mountain lakes are frequently regarded as harbingers of change, given their presence in delicate or harsh environments. Even minor alterations in environmental conditions or human intervention can trigger swift fluctuations in the functionality of the aquatic ecosystems they support (Tiberti et al. 2013). The emergence and growth of glacial lakes represent a substantial hazard through GLOFs. The extent and dangers associated with these events are influenced by the specific topography, geological characteristics of the region, and dynamics of the glaciers (Watanabe et al. 2009). GLOFs have occurred at an average rate of approximately 1.3 per year in the Himalayas over the past three decades. A GLOF can be incredibly destructive, with the potential to travel distances of up to 200 km, boasting a peak discharge of 30,000 m^3/s, and speeds ranging from 14 to 18 km/h (Ding et al. 2020; Sattar et al. 2020). The Chamoli district witnessed a GLOF on 7 February 2021, while the Chorabari lake outburst in Kedarnath occurred in July 2013 in the Garhwal Himalaya. These devastating events tragically resulted in the loss of a thousand lives and incurred property damages (Shugar et al. 2021). The GLOF incident in Chorabari lake was also reported by Allen et al. (2016). Rafiq et al. (2019) in their study, used bathymetry measurements to show that excessive water caused the Chorabari lake in Kedarnath to fill up to its maximum volume (3822.7 m^3). The hydrograph of the BREACH model in their study showed a peak discharge of around 1699 m^3/s on 17 June 2013 in Kedarnath.

The outburst flood from the glacier lake at South Lhonak Lake in the Sikkim Himalayas can extend as far as Chungthang town, which is located 63 km downstream, and cover an area of 0.06 km^2 in the settlement (Sattar et al. 2019). Figure 17.1 depicts the current

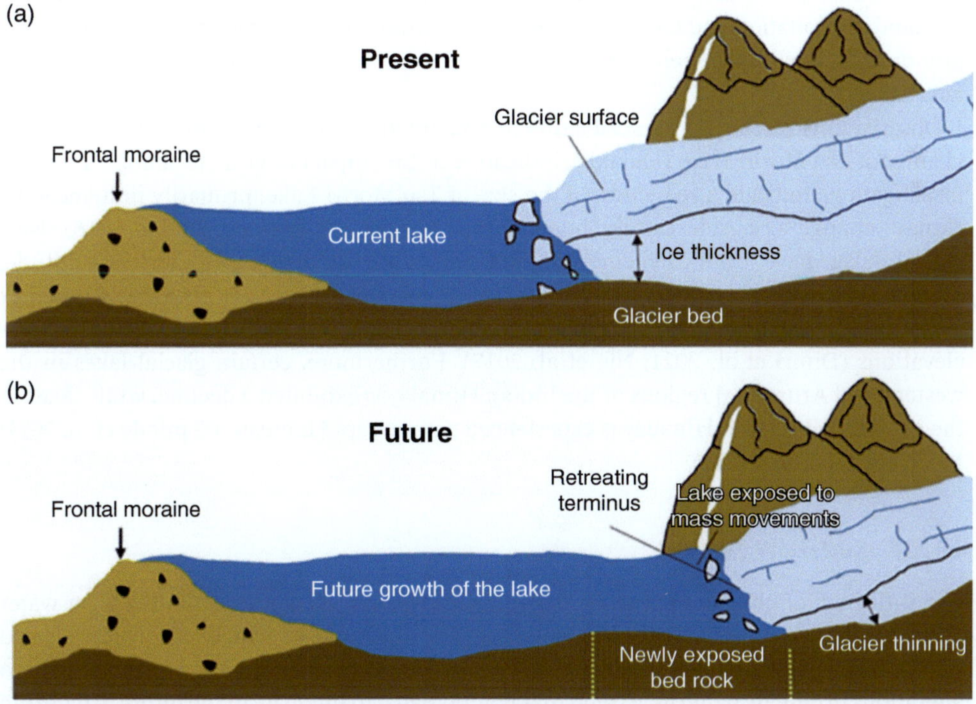

Figure 17.1 (a) Currently existing proglacial lake system of the South Lhonak Lake (Sikkim Himalaya); (b) the expansion of the lake as the over deepened glacier bed is exposed in the future due to glacier retreat, increasing its vulnerability to potential mass movements entering the lake. *Source:* Adapted from Sattar et al. (2021).

Table 17.4 Changes in the Gangabal lake area within Kashmir.

	Lake area increased					
Lake	**1972**	**1992**	**2001**	**2010**	**2020**	**Reference**
Gangabal lake (Kashmir Himalaya)	$1.42\,km^2$	$1.58\,km^2$	$1.61\,km^2$	$1.64\,km^2$	$1.66\,km^2$	Ahmed et al. 2022

proglacial lake system of South Lhonak Lake (Sikkim Himalaya), with potential expansion and increased vulnerability from glacier retreat exposing the over-deepened glacier bed.

Table 17.4 depicts changes in the Gangabal lake area within the Kashmir Himalayas across different time periods.

The phenomenon of global warming has caused an increase in the temperature of lake water, consequently fostering eutrophication and exacerbating the proliferation of algal blooms. Global warming fosters the expansion of current glacial lakes and the formation of new ones, presenting a significant hazard to human life and property. The risk level varies based on local topography, geological factors, and glaciological processes (Watanabe et al. 2009). The findings from an isotopic analysis-based mass balance study on Nainital Lake (Uttarakhand Himalaya) reveal it as a warm monomictic lake of the flow-through variety, exhibiting significant groundwater inflow and outflow (Rai et al. 2008). Research conducted on Wular Lake (Kashmir Himalaya) indicates a varying degree of decrease in seasonal precipitation and lake discharge, ranging from minor to significant, coupled with a pronounced downward trend in the extent of clear water surface area (Mushtaq and Lala 2017).

Observations reveal pronounced shifts in precipitation and temperature patterns in the vicinity of Tso Moriri Lake (Ladakh), indicative of the impact of climate change. Previous investigations indicate a reduction in the size of Tso Moriri Lake, primarily influenced by climate change alongside certain human interventions (Gupta and Shukla 2016). New glacial lakes have primarily emerged and expanded in regions of elevated altitude, predominantly ranging from 4800 to 5700 m. Some instances have even been noted at altitudes as high as 5900 m, suggesting a pronounced warming trend at these higher elevations (Dimri et al. 2021; Nie et al. 2017). Furthermore, certain glacial lakes in the western and Arunachal regions of the Indian Himalayas exhibited a decline, while lakes in the central and eastern Himalayas experienced a more rapid increase (Gardelle et al. 2011; Mohanty and Maiti 2021).

17.2.4 Surface Water

The influence of climate change on basin hydrology, and its consequent effects on water quantity and availability, is readily apparent. The streamflow, which is considered as a crucial component of the hydrological cycle, demonstrates significant sensitivity to alterations in climate patterns, a trend that is anticipated to intensify in the future. Analyzing the impact of climate change on streamflow necessitates gathering precipitation and temperature data over an extended period. This extensive timeframe is crucial for distinguishing whether the fluctuations in a specific stream result from climate change or from alterations in land use or anthropogenic activity in the region (Mir et al. 2021). Projections specific to the lower Indus, Ganges, and Brahmaputra rivers indicate a significant increase in flood frequencies throughout the twenty-first century (Wijngaard et al. 2017). The Satluj River Basin witnessed an 18% rise in runoff attributed to snowmelt and a 38% increase from glacier melt due to a 2 °C temperature rise (Mir et al. 2021). Numerous studies have illuminated the potential hydrological consequences of climate change on the future streamflow of a diverse array of river systems across the globe (Sukanya and Joseph 2023). The flow of the Kosi River in the headwater region (Uttarakhand) has also been documented to have decreased from 550 m^3/s in 2001 to 220 m^3/s in 2007 (Pant and Rawat 2015).

Pareta and Pareta (2014) reported that precipitation in Chamba town decreased from 1037.9 mm during 1990–2001 to 684 mm during 2002–2013. They also noted a reduction in the number of rainy days throughout this period. The current impact of snow and ice meltwater is recognized as significant for river runoff in the Indus and Brahmaputra basins, with respective overall annual melt contributions reaching as high as 46% and 19%. Nonetheless, within the monsoon-dominated Ganges basin, the contribution is comparatively modest, standing at just 9%. The Indus basin recorded a 19% surge in precipitation and observed a 14–90% augmentation in river flows. Projections indicate that there will be a 4% reduction in the average annual flow. The flow from glacial subbasins is predicted to reach a maximum of about 150% of the initial levels by about the year 2060. In addition, scientists observed a rise in intense precipitation occurrences in the Ganga basin, coupled with a decline in the frequency of rainy days, resulting in an estimated 18% decrease in the average annual water flow (Rees and Collins 2006). According to Immerzeel et al. (2009), there is a projected decrease in river flows in Indian rivers. The estimates suggest that the Indus basin may experience an 8.4% reduction, the Ganges basin a 17.6% reduction, and the Brahmaputra basin a 19.6% reduction between the years 2046 and 2065, compared to the levels observed in 2001–2002 (Immerzeel et al. 2009; Mir et al. 2021). Nevertheless, several research studies carried out on the Indian Himalayas concentrate on the combined meltwater from both snow and glaciers, which contributes to streamflow. This complicates the precise assessment of the specific contribution of glacier melt to the overall streamflow (Alford and Armstrong 2010; Mir et al. 2021). Lutz et al. (2014) used a detailed model to measure the amount of water flowing out of the major Himalayan river basins, including the Indus, Ganges, Brahmaputra, Salween, and Mekong rivers. They made projections for the near future based on the moderate RCP4.5 scenario. Although the nature of runoff varies throughout various basins, their research revealed a significant rise in expected runoff by 2050. This increase mostly arises from heightened precipitation in the upper Ganges, Brahmaputra, Salween, and Mekong basins, combined with increased glacier melting in the Upper Indus basin (Krishnan et al. 2020). Numerous global studies have emphasized the prevalence of elevated sediment loads in rivers due to shifts in climate patterns. These excessive sediment loads detrimentally affect river health, disrupting natural river ecosystems and imposing additional challenges on the planning of hydraulic structures (Mir et al. 2021; Rather et al. 2020; Zalaki-Badil et al. 2017). The warming climate will generally result in a rise in rainfall-runoff contribution, a prolonged decrease in snowmelt and ice melt, and an earlier onset of seasonal snow melting. These changes will modify the pattern and timing of the hydrological cycle in Himalayan Rivers (Shirsat et al. 2021).

17.3 Summary and Conclusion

The Himalayas serve as the economic lifeline and any disruption to this vital ecosystem would carry significant consequences for both the region's populace and its biodiversity. The hydrological dynamics of the broader Himalayan region exhibit significant sensitivity to climate change due to the predominant coverage of snow, glaciers, and permafrost throughout the year. The streamflow in these places is mostly determined by the runoff resulting from the melting of snow, which is crucial for the survival of millions of people living in areas downstream. Although the Himalayan range is the source of many permanent rivers, it is paradoxical that the mountain communities mostly depend on spring water for their survival. It is crucial to obtain precise data in a timely manner regarding water availability, demand, resource inventory, and socio-economic aspects. Utilizing state-of-the-art hydrological models alongside this data can aid in addressing this pressing issue effectively. The vulnerability inherent in mountain ecosystems poses significant challenges to their long-term sustainability. Ensuring sustainability requires strengthening resilience, promoting equity, and fostering social sustainability. The accelerated depletion of glacier mass carries the potential for intensified streamflow and flooding across the Himalayan river basins, consequently affecting agricultural activities downstream. An emerging consequence of climate change is the increased occurrence and severity of heavy rainfall events, leading to phenomena like flash floods, landslides, and debris flow. This trend has been notably observed in regions such as Kashmir, Himachal Pradesh, and Uttarakhand Himalayas in recent years.

If the ongoing glacial recession persists, it will greatly diminish groundwater recharge, presenting a substantial risk to water availability. The decreasing patterns of yearly precipitation within the area could impact the influx of water into the lakes, potentially leading to diminished water retention in the years to come. While advancements in remote sensing technologies have enabled the study of previously inaccessible Himalayan glaciers, there remains a need to foster a comprehensive understanding in this area.

Hence, it is imperative to consistently monitor the weather and climate in the complex and challenging terrain of the Himalayan cryosphere. Its purpose is to rectify gaps in knowledge within the field and improve the precise representation of regional cryospheric processes in climate models. To effectively address future challenges in water management, it is crucial to conduct precise assessments of how climate change affects water levels. This involves evaluating impacts at various scales, ranging from regional levels to specific basins or local areas.

References

Agarwal, A.A., Bhatnaga, N.K., Nema, R.K., and Agrawal, N.K. (2012). Rainfall dependence of springs in the midwestern Himalayan hills of Uttarakhand. *Environmental Science and Technology* 46 (7): 3557–3564.

Aggarwal, S., Rai, S.C., Thakur, P.K., and Emmer, A. (2017). 'Inventory and recently increasing GLOF susceptibility of glacial lakes in Sikkim. *Eastern Himalaya', Geomorphology* 295: 39–54.

Ahmed, R., Wani, G.F., Ahmad, S.T. et al. (2021). A review of glacial lake expansion and associated glacial lake outburst floods in the Himalayan region. *Earth Systems and Environment* 5 (3): 695–708.

Ahmed, R., Rawat, M., Wani, G.F. et al. (2022). Glacial lake outburst flood hazard and risk assessment of Gangabal Lake in the Upper Jhelum Basin of Kashmir Himalaya using geospatial technology and hydrodynamic modeling. *Remote Sensing* 14 (23): 5957.

Alford, D. and Armstrong, R. (2010). The role of glaciers in stream flow from the Nepal Himalaya. *The Cryosphere Discussions* 4 (2): 469–494.

Allen, S.K., Linsbauer, A., Randhawa, S.S. et al. (2016). Glacial lake outburst flood risk in Himachal Pradesh, India: an integrative and anticipatory approach considering current and future threats. *Natural Hazards* 84: 1741–1763.

Bahuguna, I.M., Rathore, B.P., Brahmbhatt, R. et al. (2014). Are the Himalayan glaciers retreating? *Current Science* 100 (7): 1008–1013.

Bharati, L., Bhattarai, U., Khadka, A. et al. (2019). *From the Mountains to the Plains: Impact of Climate Change on Water Resources in the Koshi River Basin*, vol. 187, 1–36. Colombo, Sri Lanka: International Water Management Institute (IWMI).

Bhat, N.A., Jeelani, G.H., and Bhat, M.Y. (2014). Hydrogeochemical assessment of groundwater in karst environments, Bringi watershed, Kashmir Himalayas, India. *Current Science* 106 (7): 1000–1007.

Bhattacharya, A., Bolch, T., Mukherjee, K. et al. (2016). Overall recession and mass budget of Gangotri Glacier, Garhwal Himalayas, from 1965 to 2015 using remote sensing data. *Journal of Glaciology* 62 (236): 1115–1133.

Biemans, H., Siderius, C., Lutz, A.F. et al. (2019). Importance of snow and glacier meltwater for agriculture on the Indo-Gangetic Plain. *Nature Sustainability* 2 (7): 594–601.

Chand, P., Sharma, M.C., Bhambri, R. et al. (2017). 'Reconstructing the pattern of the Bara Shigri Glacier fluctuation since the end of the Little Ice Age. *Chandra valley, north-western Himalaya', Progress in Physical Geography* 41: 643–675. https://doi.org/10.1177/0309133317728017.

Cotter, G.D. and Brown, J.C. (1907). Notes on certain glaciers in Kumaon. In: *Records of Geological Survey of India, 35*, 148–157. Calcutta: Government of India.

Dimri, A.P., Allen, S., Huggel, C. et al. (2021). Climate change, cryosphere and impacts in the Indian Himalayan Region. *Current Science* 120 (3): 345–356.

Ding, Y., Liu, S., Li, J., and Shangguan, D. (2006). The retreat of glaciers in response to recent climate warming in western China. *Annals of Glaciology* 43: 97–105.

Ding, M., Gao, Z., Huang, T., and Hu, X. (2020). The hazard assessment of glacial lake debris flow: a case study on Dongcuoqu, Luolong County, Tibet. *IOP Conference Series: Earth and Environmental Science* 570 (4): 042054.

Dobhal, D.P., Kumar, S., and Mundepi, A.K. (1995). Morphology and glacier dynamics studies in monsoon–arid transition zone: an example from Chhota Shigri Glacier, Himachal-Himalaya, India. *Current Science* 68: 936–944.

Dobhal, D.P., Gergan, J.T., and Thayyen, R.J. (2008). 'Mass balance studies of the Dokriani Glacier from 1992 to 2000, Garhwal Himalaya, India. *Bulletin of Glaciological Research* 25: 9–17.

Dobhal, D.P., Pratap, B., Bhambri, R., and Mehta, M. (2021). Mass balance and morphological changes of Dokriani Glacier (1992–2013), Garhwal Himalaya, India. *Quaternary Science Advances* 4: 100033.

Dubey, S. and Goyal, M.K. (2020). Glacial lake outburst flood hazard, downstream impact, and risk over the Indian Himalayas. *Water Resources Research* 56 (4): e2019WR026533.

Gardelle, J., Arnaud, Y., and Berthier, E. (2011). Contrasted evolution of glacial lakes along the Hindu Kush Himalaya mountain range between 1990 and 2009. *Global and Planetary Change* 75 (1-2): 47–55.

Gardelle, J., Berthier, E., Arnaud, Y., and Kääb, A. (2013). Region-wide glacier mass balances over the Pamir-Karakoram-Himalaya during 1999–2011. *The Cryosphere* 7 (4): 1263–1286.

Gerlitz, J.Y., Macchi, M., Brooks, N. et al. (2017). The multidimensional livelihood vulnerability index–an instrument to measure livelihood vulnerability to change in the Hindu Kush Himalayas. *Climate and Development* 9 (2): 124–140.

Guevara-Ochoa, C., Medina-Sierra, A., and Vives, L. (2020). Spatio-temporal effect of climate change on water balance and interactions between groundwater and surface water in plains. *Science of the Total Environment* 722: 137886.

Gunn, J.P. (1930). The Shyok flood. *The Himalayan Journal* 2: 35–47.

Gupta, S.K. and Shukla, D.P. (2016). Assessment of land use/land cover dynamics of Tso Moriri Lake, a Ramsar site in India. *Environmental Monitoring and Assessment* 188: 1–13.

Gupta, A.K., Negi, M., Nandy, S. et al. (2020). Mapping socio-environmental vulnerability to climate change in different altitude zones in the Indian Himalayas. *Ecological Indicators* 109: 105787.

Guruprasad, C., Gopal, D., and Devaraj, S. (2022). Mass balance estimation of Mulkila glacier, Western Himalayas, using glacier melt model. *Environmental Monitoring and Assessment* 194 (10): 761.

Hamid, A.T., Sharif, M., and Archer, D. (2014). Analysis of temperature trends in Satluj River Basin, India. *Journal Earth Science Climate Change* 5 (8): 1–9.

Haritashya, U.K., Kargel, J.S., Shugar, D.H. et al. (2018). Evolution and controls of large glacial lakes in the Nepal Himalaya. *Remote Sensing* 10 (5): 798.

Hood, E., Battin, T.J., Fellman, J. et al. (2015). Storage and release of organic carbon from glaciers and ice sheets. *Nature Geoscience* 8 (2): 91–96.

Huss, M. and Hock, R. (2018). Global-scale hydrological response to future glacier mass loss. *Nature Climate Change* 8 (2): 135–140.

Huss, M., Jouvet, G., Farinotti, D., and Bauder, A. (2010). Future high-mountain hydrology: a new parameterization of glacier retreat. *Hydrology and Earth System Sciences* 14 (5): 815–829.

Islam, A.R.M.T., Saha, A., Ghose, B. et al. (2022). Landslide susceptibility modeling in a complex mountainous region of Sikkim Himalaya using new hybrid data mining approach. *Geocarto International* 37 (25): 9021–9046.

Kamp, U., Byrne, M., and Bolch, T. (2011). Glacier fluctuations between 1975 and 2008 in the Greater Himalaya Range of Zanskar, southern Ladakh. *Journal of Mountain Science* 8: 374–389.

Kargel, J.S., Cogley, J.G., Leonard, G.J. et al. (2011). Himalayan glaciers: the big picture is a montage. *Proceedings of the National Academy of Sciences* 108 (36): 14709–14710.

Kaushik, S., Rafiq, M., Dharpure, J.K. et al. (2024). Increasing risk of glacial lake outburst flood in Sikkim, Eastern Himalaya under climate warming. *Remote Sensing Applications: Society and Environment* 36: 101286.

Khadka, N., Chen, X., Sharma, S., and Shrestha, B. (2023). Climate change and its impacts on glaciers and glacial lakes in Nepal Himalayas. *Regional Environmental Change* 23 (4): 143.

Khan, A., Potapov, P., Hansen, M.C. et al. (2024). Perennial snow and ice cover change from 2001 to 2021 in the Hindu-Kush Himalayan region derived from the Landsat analysis-ready data. *Remote Sensing Applications: Society and Environment* 34: 101192.

Kripalani, R.H., Kulkarni, A., and Sabade, S.S. (2003). Western Himalayan snow cover and Indian monsoon rainfall: a re-examination with INSAT and NCEP/NCAR data. *Theoretical and Applied Climatology* 74: 1–18.

Krishnan, R., Gnanaseelan, C., Sanjay, J. et al. (2020). 'Introduction to climate change over the Indian region', assessment of climate change over the Indian Region: a report of the Ministry of Earth Sciences (MoES). *Government of India* 1–20.

Kulkarni, A.V. and Alex, S. (2003). Estimation of recent glacial variations in Baspa Basin using remote sensing technique. *Journal of the Indian Society of Remote Sensing* 31: 81–90.

Kulkarni, A.V., Rathore, B.P., Singh, S.K., and Bahuguna, I.M. (2011). Understanding changes in the Himalayan cryosphere using remote sensing techniques. *International journal of remote sensing* 32 (3): 601–615.

Kumar, P. and Sharma, M.C. (2023). Frontal changes in medium-sized glaciers in Sikkim, India during 1988–2018: insights for glacier-climate synthesis over the Himalaya. *Iscience* 26 (10).

Linsbauer, A., Frey, H., Haeberli, W. et al. (2016). Modelling glacier-bed overdeepenings and possible future lakes for the glaciers in the Himalaya-Karakoram region. *Annals of Glaciology* 57 (71).

Majeed, Z., Mukhtar, M.A., Mir, R.A. et al. (2020). Sonapani glacier recession over a century from 1906–2016, Chandra basin, Himachal Himalaya. *Journal of the Geological Society of India* 95: 36–44.

Mir, R.A. (2018). 'Recent changes of two parts of Kolahoi Glacier and its controlling factors in Kashmir basin, western Himalaya. remote sensing applications. *Society and Environment* 11: 265–281.

Mir, B.H., Kumar, R., Lone, M.A., and Tantray, F.A. (2021). Climate change and water resources of Himalayan region—review of impacts and implication. *Arabian Journal of Geosciences* 14: 1–14.

Mirjam Macchi, M.M., Gurung, A.M., Hoermann, B., and Dhrupad Choudhury, D.C. (2011). Climate variability and change in the Himalayas: community perceptions and responses. *International Centre for Integrated Mountain Development.*

Mohanty, L. and Maiti, S. (2021). Probability of glacial lake outburst flooding in the Himalaya. Resources. *Environment and Sustainability* 5: 100031.

Mohanty, L.K. and Maiti, S. (2022). Glacial lake formation probability mapping in the Himalayan glacier: A probabilistic approach. *Journal of Earth System Science* 131 (1): 54.

Mushtaq, F. and Lala, M.G.N. (2017). 'Assessment of hydrological response as a function of LULC change and climatic variability in the catchment of the Wular Lake, J&K, using geospatial technique. *Environmental Earth Sciences* 76: 1–19.

Nautiyal, S., Goswami, M., Prakash, S. et al. (2022). Spatio-temporal variations of geo-climatic environment in a high-altitude landscape of Central Himalaya: an assessment from the perspective of vulnerability of glacial lakes. *Natural Hazards Research* 2 (4): 343–362.

Negi, G.C.S. and Joshi, V.J. (1996). Geohydrology of springs in a mountain watershed: the need for problem solving research. *Journal of the Geological Society of India* 99: 875–880.

Negi, G.C.S. and Mukherjee, S. (2020). Climate change impacts in the Himalayan Mountain ecosystems. In: *Encyclopedia of the World's biomes*, 349–354. Elsevier.

Negi, G.C.S., Kumar, K., Panda, Y.S., and Satyal, G.S. (2001). Water yield and water quality of some aquifers in the Himalaya. *International Journal of Ecology and Environmental Sciences* 27: 55–59.

Nie, Y., Sheng, Y., Liu, Q. et al. (2017). A regional-scale assessment of Himalayan glacial lake changes using satellite observations from 1990 to 2015. *Remote Sensing of Environment* 189: 1–13.

Panda, D.K. and Wahr, J. (2016). Spatiotemporal evolution of water storage changes in India from the updated GRACE-derived gravity records. *Water Resources Research* 52 (1): 135–149.

Pandey, R., Jha, S.K., Alatalo, J.M. et al. (2017). Sustainable livelihood framework-based indicators for assessing climate change vulnerability and adaptation for Himalayan communities. *Ecological Indicators* 79: 338–346.

Pandey, R., Kumar, P., Archie, K.M. et al. (2018). Climate change adaptation in the western-Himalayas: household level perspectives on impacts and barriers. *Ecological Indicators* 84: 27–37.

Pant, C.C. and Rawat, P.K. (2015). 'Declining changes in spring hydrology of non-glacial river basins in Himalaya: a case study of Dabka catchment. In: *Dynamics of Climate Change and Water Resources of Northwestern Himalaya*, Society of Earth Scientists Series (ed. R. Joshi, K. Kumar, and L.M.S. Palni), 151–179. Cham: Springer.

Panwar, S. (2020). Vulnerability of Himalayan springs to climate change and anthropogenic impact: a review. *Journal of Mountain Science* 17 (1): 117–132.

Pareta, K. and Pareta, U. (2014). Climate change impact on land and natural resource in Chamba Tehsil of Himachal Pradesh State, India. *The International Journal of Science and Technoledge* 2 (4): 38.

Prajapati, R.N., Ibrahim, N., and Thapa, B.R. (2023). Climate change impact on water availability in the Himalaya: insights from Sunkoshi River basin, Nepal. *Hydro Research* 6: 279–292.

Radić, V. and Hock, R. (2014). Glaciers in the Earth's hydrological cycle: assessments of glacier mass and runoff changes on global and regional scales. *Surveys in Geophysics* 35: 813–837.

Rafiq, M., Romshoo, S.A., Mishra, A.K., and Jalal, F. (2019). Modelling Chorabari lake outburst flood, Kedarnath, India. *Journal of Mountain Science* 16 (1): 64–76.

Rai, R.N., Singh, K.A., and Solanki, R.C. (1998). A case study of water flows of some hill springs of Sikkim. *Indian Journal of Soil Conservation* 16 (1): 52–56.

Raina, V.K. (2009). Himalayan glaciers: a state-of-art review of glacial studies, glacial retreat and climate change. Government of India.

Rajat, S., Singh, B.R., Prakash, C., and Anita, S. (2022). Glacier retreat in Himachal from 1994 to 2021 using deep learning. *Remote Sensing Applications: Society and Environment* 28: 100870.

Raman, A., Kulkarni, A.V., and Prasad, V. (2022). Glacier mass balance estimation in Garhwal Himalaya using improved accumulation area ratio method. *Environmental Monitoring and Assessment* 194 (8): 583.

Rana, A.S., Kunmar, P., Mehta, M., and Kumar, V. (2023). Glacier retreat, dynamics and bed overdeepenings of Parkachik Glacier, Ladakh Himalaya, India. *Annals of Glaciology* 1–14. https://doi.org/10.1017/aog.2023.50.

Rather, N.A., Lone, M.A., Dar, A.Q., and Mir, B.H. (2020). The influence of particle shape and gradation parameters on the permeability of filter media. *International Journal of Hydrology Science and Technology* 10 (3): 267–286.

Rathi, V.K., Ram, S., Kumar, R. et al. (2020). Hydrological classification and performance of Himalayan springs in climate change scenario–a case study. *Water Supply* 20 (2): 594–608.

Rees, H.G. and Collins, D.N. (2006). Regional differences in response of flow in glacier-fed Himalayan rivers to climatic warming. *Hydrological Processes: An International Journal* 20 (10): 2157–2169.

Remya, S.N., Kulkarni, A.V., Pradeep, S., and Shrestha, D.G. (2019). Volume estimation of existing and potential glacier lakes, Sikkim Himalaya, India. *Current Science* 116 (4): 620–627.

Saha, S., Bera, B., Bhattacharjee, S. et al. (2024). Significant changes of area, length and terminus of Sikkim Himalayan glaciers within the Kanchenjunga national park from 1990-2022. *Results in Earth Sciences* 2: 100013.

Saraswat, P., Syed, T.H., Famiglietti, J.S. et al. (2013). Recent changes in the snout position and surface velocity of Gangotri glacier observed from space. *International Journal of Remote Sensing* 34 (24): 8653–8668.

Sattar, A., Goswami, A., and Kulkarni, A.V. (2019). 'Hydrodynamic moraine-breach modeling and outburst flood routing-A hazard assessment of the South Lhonak lake, Sikkim. *Science of the Total Environment* 668: 362–378.

Sattar, A., Goswami, A., Kulkarni, A.V., and Emmer, A. (2020). Lake evolution, hydrodynamic outburst flood modeling and sensitivity analysis in the Central Himalaya: a case study. *Water* 12 (1): 237.

Sattar, A., Goswami, A., Kulkarni, A.V. et al. (2021). Future glacial lake outburst flood (GLOF) hazard of the South Lhonak Lake, Sikkim Himalaya. *Geomorphology* 388: 107783.

Scherler, D., Bookhagen, B., and Strecker, M.R. (2011). Spatially variable response of Himalayan glaciers to climate change affected by debris cover. *Nature Geoscience* 4 (3): 156–159.

Şen, Z. (2021). Reservoirs for water supply under climate change impact—a review. *Water Resources Management* 35 (11): 3827–3843.

Shirsat, T.S., Kulkarni, A.V., Momblanch, A. et al. (2021). Towards climate-adaptive development of small hydropower projects in Himalaya: a multi-model assessment in upper Beas basin. *Journal of Hydrology: Regional Studies* 34: 100797.

Shrestha, A.B., Wake, C.P., Mayewski, P.A., and Dibb, J.E. (1999). Maximum temperature trends in the Himalaya and its vicinity: an analysis based on temperature records from Nepal for the period 1971–94. *Journal of Climate* 12 (9): 2775–2786.

Shugar, D.H., Jacquemart, M., Shean, D. et al. (2021). A massive rock and ice avalanche caused the 2021 disaster at Chamoli, Indian Himalaya. *Science* 373 (6552): 300–306.

Shukla, A., Garg, P.K., and Srivastava, S. (2018). Evolution of glacial and high-altitude lakes in the Sikkim, Eastern Himalaya over the past four decades (1975–2017). *Frontiers in Environmental Science* 6: 81.

Singh, S., Kumar, R., Bhardwaj, A. et al. (2016). Changing climate and glacio-hydrology in Indian Himalayan Region: a review. *Wiley Interdisciplinary Reviews: Climate Change* 7 (3): 393–410.

Singh, D.S., Tangri, A.K., Kumar, D. et al. (2017). 'Pattern of retreat and related morphological zones of Gangotri Glacier, Garhwal Himalaya, India. *Quaternary International* 444: 172–181.

Singh, S., Kumar, R., and Dimri, A.P. (2018). Mass balance status of Indian Himalayan glaciers: a brief review. *Frontiers in Environmental Science* 6: 30.

Singh, A., Kumar, R., Kumar, R. et al. (2022). Quantification of volume loss and snout retreat from 1980 to 2019 of Baspa Basin glaciers, western Himalaya. *Materials Today: Proceedings* 49: 3331–3339.

Swain, S., Taloor, A.K., Dhal, L. et al. (2022). Impact of climate change on groundwater hydrology: a comprehensive review and current status of the Indian hydrogeology. *Applied Water Science* 12 (6): 120.

Syldon, P., Shrestha, B.B., Miyamoto, M. et al. (2024). Assessing the impact of climate change on flood inundation and agriculture in the Himalayan Mountainous Region of Bhutan. *Journal of Hydrology: Regional Studies* 52: 101687.

Tak, S. and Keshari, A.K. (2020). Investigating mass balance of Parvati glacier in Himalaya using satellite imagery based model. *Scientific Reports* 10 (1): 12211.

Talukder, B., Matthew, R., Bunch, M.J. et al. (2021). Melting of Himalayan glaciers and planetary health. *Current Opinion in Environmental Sustainability* 50: 98–108.

Tambe, S., Kharel, G., Arrawatia, M.L. et al. (2012). Reviving dying springs: climate change adaptation experiments from the Sikkim Himalaya. *Mountain Research and Development* 32 (1): 62–72.

Thapliyal, A., Kimothi, S., Taloor, A.K. et al. (2023). Glacier retreat analysis in the context of climate change impact over the Satopanth (SPG) and Bhagirathi-Kharak (BKG) glaciers in the Mana basin of the Central Himalaya, India: a geospatial approach. *Geosystems and Geoenvironment* 2 (1): 100128.

Tiberti, R., Metta, S., Austoni, M. et al. (2013). Ecological dynamics of two remote Alpine lakes during ice-free season. *Journal of Limnology* 72 (3): 401–416.

Tripathi, J.N., Sonker, I., Tripathi, S., and Singh, A.K. (2022). Climate change traces on Lhonak Glacier using geospatial tools. *Quaternary Science Advances* 8: 100065.

Vincent, C., Ramanathan, A., Wagnon, P. et al. (2013). Balanced conditions or slight mass gain of glaciers in the Lahaul and Spiti region (northern India, Himalaya) during the nineties preceded recent mass loss. *The Cryosphere* 7 (2): 569–582.

Wadham, J.L., Hawkings, J.R., Tarasov, L. et al. (2019). Ice sheets matter for the global carbon cycle. *Nature Communications* 10 (1): 3567.

Wagnon, P., Linda, A., Arnaud, Y. et al. (2007). Four years of mass balance on Chhota Shigri Glacier, Himachal Pradesh, India, a new benchmark glacier in the western Himalaya. *Journal of Glaciology* 53 (183): 603–611.

Watanabe, T., Lamsal, D., and Ives, J.D. (2009). Evaluating the growth characteristics of a glacial lake and its degree of danger of outburst flooding: Imja Glacier, Khumbu Himal, Nepal. *Norsk Geografisk Tidsskrift-Norwegian Journal of Geography* 63 (4): 255–267.

Wester, P., Mishra, A., Mukherji, A., and Shrestha, A.B. (2019). *The Hindu Kush Himalaya Assessment: Mountains, Climate Change, Sustainability and People*, 627. Berlin: Springer Nature.

Wijngaard, R.R., Lutz, A.F., Nepal, S. et al. (2017). Future changes in hydro-climatic extremes in the Upper Indus, Ganges, and Brahmaputra River basins. *PLoS ONE* 12 (12): e0190224.

Yousuf, B., Shukla, A., Ali, I. et al. (2024). Surface facies analysis of the Gangotri and neighbouring glaciers, central Himalaya. *Science of Remote Sensing* 9: 100128.

Zalaki-Badil, N., Eslamian, S., Sayyad, G.A. et al. (2017). Using SWAT model to determine runoff, sediment yield in Maroon-Dam catchment. *International Journal of Research Studies in Agricultural Sciences* 3 (12): 31–41.

Zhou, Y., Li, Z., and Li, J.I.A. (2017). Slight glacier mass loss in the Karakoram region during the 1970s to 2000 revealed by KH-9 images and SRTM DEM. *Journal of Glaciology* 63 (238): 331–342.

18

Rainfall Intensity and Return Periods of Extreme Rainfall Over the Garhwal Himalaya, Uttarakhand, India

Deen D. Khandelwal[1], *Manish Mehta*[2], *and Dwarika P. Dobhal*[1,2]

[1] *Centre for Glaciology, Wadia Institute of Himalayan Geology, Dehradun, India*
[2] *Enviromental Geomorphology, Wadia Institute of Himalayan Geology, Dehradun, India*

18.1 Introduction

The Himalaya, renowned as the youngest and largest mountain system globally, exhibits intense tectonic activity (Valdiya et al. 1999; Yin 2006). Its vast geographic expanse and intricate and rugged topography, coupled with diverse climatic conditions, render the region intrinsically susceptible to a multitude of natural hazards (Valdiya 1987; Dobhal et al. 2013). The Intergovernmental Panel on Climate Change's (IPCC) fifth assessment report (Stocker et al. 2014) unequivocally establishes robust evidence supporting a global surge in extreme events post-1950, notably an increased frequency of hot days and intense precipitation events. Goswami et al. (2006) posit that annual variability has markedly heightened the occurrence and intensity of extreme monsoonal rainfall events in India, surpassing normal rainfall levels by 50%. Khandelwal et al. (2015) further contribute to this discourse, asserting that a substantial proportion, ranging between 50 and 90%, of the total annual rainfall in the Garhwal Himalaya region transpires during the period from June to September. In light of the dynamic geological and climatic conditions of the Himalaya, these findings underscore the heightened risk of natural hazards in the region, necessitating continued research and comprehensive risk management strategies to address the multifaceted challenges posed by such environmental dynamics.

Uttarakhand experiences high to moderate rainfall, ranging from 1000 to 2000 mm/yr (Basistha et al. 2008; UNDP 2014). The primary region with high precipitation is situated in the south over the Siwalik Hills, where the mean rainfall is approximately ~1200 mm/yr. Another area of elevated precipitation is observed near the physiographic transition between the Lesser and Higher Himalayas, with an annual rainfall of around 2000 mm (Gadgil et al. 2003; Nandargi and Dhar 2011).

Extreme rainfall events associated with floods and debris flows constitute a significant hazard in Uttarakhand. Locally, such events may occur when 1-day cumulative rainfall exceeds 200 mm and 2–3 days cumulative rainfall surpasses 300–400 mm, with recurrence

Water Resources Management in Mountain Regions, First Edition. Edited by Virendra Bahadur Singh, Sughosh Madhav, Rakesh Kumar Gupta, Prerna Diwan, and Amit Kumar.

intervals on the order of decades (Starkel and Basu 2000). Rainfall data analysis from the Kumaon and Garhwal Himalaya indicates that extreme rainfall events of 200–500 mm/day occur approximately once every century (Starkel and Basu 2000). Notable instances include September 1880, with a recorded rainfall of 820 mm in a day, September 1924, with 770 mm of rainfall (Froehlich and Starkel 1987), and 15th and 17th June 2013, with a recorded 325 mm of rainfall (Dobhal et al. 2013; Mehta et al. 2017). A study by Nandargi and Dhar (2011) suggests that the maximum frequencies of extreme one-day rainfall, ranging between 70 and 90 mm, were recorded during the three monsoon months of July, August, and September. Sain et al. (2023) reported that the Maldevta area in Uttarakhand recorded 290.5 mm of rainfall on 19 August 2022 and 241.5 mm of rainfall on 20 August 2022, which was the cause of the cloudburst in this area.

The rainfall pattern is a critical parameter influencing hydrological processes and practices (Ahammed et al. 2014; Priya et al. 2016). This information holds significant relevance for applications such as flood forecasting, hydrological modeling, design of drainage structures, irrigation, and hydropower projects (Lagouvardos et al. 2013). However, in the higher elevations and steep slopes of the Himalayas, the installation and maintenance of an adequate number of stations for accurate rainfall data recording are complicated by the rugged topography and the inaccessibility of many locations. Consequently, the region suffers from a scarcity of reliable recording stations, leading to a sparse and nonuniform network of measurement points that yield limited data. This limitation makes it challenging to accurately determine the spatio-temporal distribution of rainfall (Burbank et al. 2012; Bhardwaj et al. 2017). The flow of the river of the Himalaya is highly susceptible to rainfall variations, so any change in precipitation pattern over the mountainous area may impact freshwater ability in downstream areas (Nijssen et al. 1997; Arnell 2003; Messerli et al. 2004).

In this study, a systematic analysis of rainfall frequency distribution and return period of extreme rainfall events in Garhwal Himalaya was carried out using the daily rainfall data (May to October) of four rainfall stations for the period from 1998 to 2015. Gumbel distribution and frequency distribution functions were used to estimate the return periods of extreme rainfall events and also to predict daily maximum rainfall variability.

18.2 Study Area and Environmental Setting

The study area, situated in the Garhwal Himalaya, is a part of the state of Uttarakhand, India, encompassing an approximate area of ~14 580 km^2 (Basistha et al. 2008; UNDP 2014; https://www.gktoday.in/regional-divisions-of-himalayas/). This region exhibits intricate topography, characterized by high mountain chains with glacierized basins in the north and fluvial terraces in the central and southern portions. It is geographically defined by the Indo-Tibetan international boundary to the north, the Kumaon region to the east, and the Siwalik foothills to the south, acting as a separator from the expansive plains of the Ganga River System. In addition, the Tons River to the west/northwest forms a natural boundary, demarcating the region from Himachal Pradesh (Figure 18.1).

The study area is located between latitude N 30°20'38.34" to 30°51'45" and longitude E 79°59'53.22" to 79°03'53.63." The altitudinal range of the areas extends between 600 and

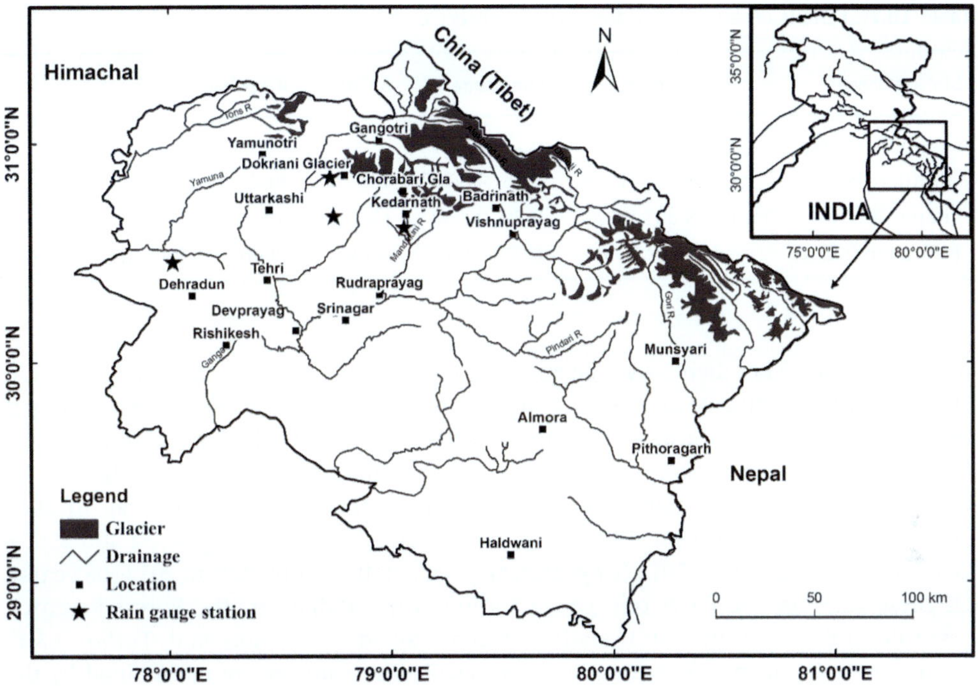

Figure 18.1 Location map of the study area showing the rain gauge stations at different altitudes (DDN—Dehradun station, GH—Ghuttu Station, DOK—Dokriani glacier station, and BS—Bhatwari Sitapur station) and drainage pattern of the area.

3700 m asl from Dehradun to the Dokriani Glacier base camp, the northwestern part of the study area. Our research endeavors to delve into the intricacies of rainfall dynamics, particularly focusing on the variability of rainfall patterns and extreme events in both the upper and lower regions of the Garhwal Himalaya. Situated on the western periphery of the central Himalayan region, our study area experiences a climatic regime dictated by two primary phenomena: the southwest monsoon during the summer months (July–September) and precipitation originating from western disturbances throughout the winter (November–February) seasons. Notably, January emerges as the chilliest month, witnessing daily high temperatures persistently below freezing, while July stands out as the warmest, with daily maximum temperatures soaring to approximately 20°C (Basistha et al. 2008; UNDP 2014).

18.3 Data and Methodology

18.3.1 Data Source

We have used rainfall data for the monsoon season (May–October) from four rain gauge stations, viz. Dehradun (DDN, 640 m asl), Ghuttu (GHT, 1836 m asl), Bhatwari-Sitapur (BHT, 1700 m asl), and Dokriani Glacier base camp (DOK, 3700 m asl) for this study (Figure 18.1). The time duration of the presented data series of these four stations are

Table 18.1 Details of rainfall stations in the study area.

Station name	Latitude	Longitude	Altitude	River basin
Dokraini (DOK)	N 30° 51' 41"	E 78° 46' 58"	3700 m	Bhagirathi
Ghuttu (GHT)	N 30° 31' 53"	E 78° 44' 52"	1836 m	Bhilangana
Sitapur-Bhatwari (BHT)	N 30° 37' 24"	E 79° 00' 02"	1700 m	Mandakini
Dehradun (DDN)	N 30° 20' 40"	E 77° 52' 12"	640 m	Plain area (Siwalik foothills)

not the same. The distinctive geographical locations of these four sites within Uttarakhand span diverse directions, contributing to a comprehensive understanding of the region's rainfall pattern. The comprehensive analysis of all four locations collectively enables a precise characterization of the rainfall dynamics in Uttarakhand. The time duration of the available rainfall data for DDN station is from 1998 to 2015 (18 years), DOK station from 1998 to 2013 (14 years, 2 years missing, 2002 and 2005), GHT station from 2007 to 2013 (07 years), and BHT station from 2006 to 2013 (08 years). Of these stations, the DDN rain gauge station is maintained by the Forest Research Institute (FRI), and the BHT station is maintained by Larson and Turbo (L&T) Hydropower Company. Whereas, GHT and DOK observatories are maintained by the Wadia Institute of Himalayan Geology (WIHG) Dehradun. We carried out several checks to ensure quality control of data (i.e. Data Collection Protocol, Instrument Calibration, Quality Assurance Checks, etc.) and check observed data (i.e. Comparison with Reference Data, Data Completeness and Consistency, Data Validation and Correction, etc.) carefully. The detail of the rain gauge station is given in Table 18.1.

18.3.2 Frequency Distribution

Observational frequencies vary across regions due to discrepancies in the number of monitoring stations and the occurrence of rainy days at each station. These disparities hinder a meaningful comparison of frequency distributions. Here, we are trying to differentiate the different frequency ranges of daily rainfall data. These rainfall frequencies are 0 mm rainfall means no rain during the past 24 hours, 0.1–10, 10.1–20, 20.1–30, 30.1–40, 40.1–50, and >50 mm, which is the nonstandard classification for mountain regions which was used by Singh et al. (2005). In this case, we are considering more than 50 mm/day rainfall as extreme rainfall or heavy rainfall during the last 24 hours. Here, we are applying the following formula to calculate the percentage of events.

$$\text{Percentage of events } \left(P_e\right) = \frac{N_f}{N_d} \times 100 \tag{18.1}$$

where N_f = Number of days for a particular frequency, and N_d = Number of days for data used

Moreover, we have calculated the total rainfall for the monsoon season and the percentage of cumulative rain for a particular frequency.

$$F_c = \left(\frac{C_r}{T_r}\right) \times 100 \tag{18.2}$$

where F_c = Percentage of cumulative rain for a particular frequency, C_r = Cumulative rainfall for a particular frequency, and T_r = Total rainfall of summer monsoon

18.3.3 Return Period (σ)

We employed descriptive statistics to analyze the rainfall data, encompassing calculations for the mean, median, mode, variability (standard deviation), symmetry (skewness), and peakedness (kurtosis). For the assessment of extreme rainfall events' return periods in the Garhwal region, we utilized the Gumbel distribution function (Gumbel 1941; Nadarshah and Choi 2007; Carvalho et al. 2013). In a comparable context, Ahammed et al. (2014) applied the Gumbel distribution function to estimate the return period of extreme rainfall events in Dhaka city. Their findings indicated that annual daily maximum rainfall equal to or exceeding 425 mm had a return period of 100 years. Moreover, Ahammed et al. (2014) employed the cumulative distribution function of the generalized extreme value (GEV) distribution as

$$F(x) = \exp\left\{-\left(1+\xi\,.\frac{x-\mu}{\sigma}\right)^{-1/\xi}\right\} \tag{18.3}$$

for $1+\xi(x-\mu)/\sigma > 0$, where μ, σ, ξ are referred to as the location, scale, and shape parameters, respectively. The equation (18.3) is referred to as a Gumbel distribution for the case of $\xi \rightarrow 0$. The GEV distribution can be expressed as an equation (18.4).

$$F(x) = \exp\left[-\exp\left\{-\left(\frac{x-\mu}{\sigma}\right)\right\}\right] \tag{18.4}$$

The probability and return period of daily maximum rainfall were estimated using equations (18.5) and (18.6), respectively,

$$P(x) = 1 - F(x) \tag{18.5}$$

$$N = \frac{1}{P(x)} \tag{18.6}$$

So finally equation (18.6) became

$$N = \left[1 - \exp\left\{-\exp\left\{-\left(x - \frac{\mu}{\sigma}\right)\right\}\right\}\right]^{-1} \tag{18.7}$$

where x = seasonal daily maximum rainfall, μ = mean of observed daily maximum rainfall, σ = standard deviation of observed seasonal daily maximum rainfall, $F(x)$ = Cumulative probability distribution, $P(x)$ = Probability distribution, N = Return period of expected seasonal daily maximum rainfall.

18.4 Result and Discussion

18.4.1 The Frequency Distribution of Rainfall

The seasonal rainfall data of these stations for different periods of monsoon months are classified to obtain the frequencies in different ranges of intensities. The average percentage of frequencies is evaluated in Figure 18.2a and given in Tables 18.2 and 18.3. The number of seasonal rainy days in these stations varies from ~55 to 65%.

The calculated data of all stations shows that the value of the higher frequency and lower frequency of seasonal rainfall is the same for all the stations. The higher frequency rainfall ranges from 0.1 to 10 mm contributed ~40% to the DOK station, ~34% to the GHT station, ~30% to the BHT station, and ~24% to the DDN station of the total seasonal rainfall. Whereas the lower frequency rainfall ranges between 40.1 and 50 mm contributed 0.6% to

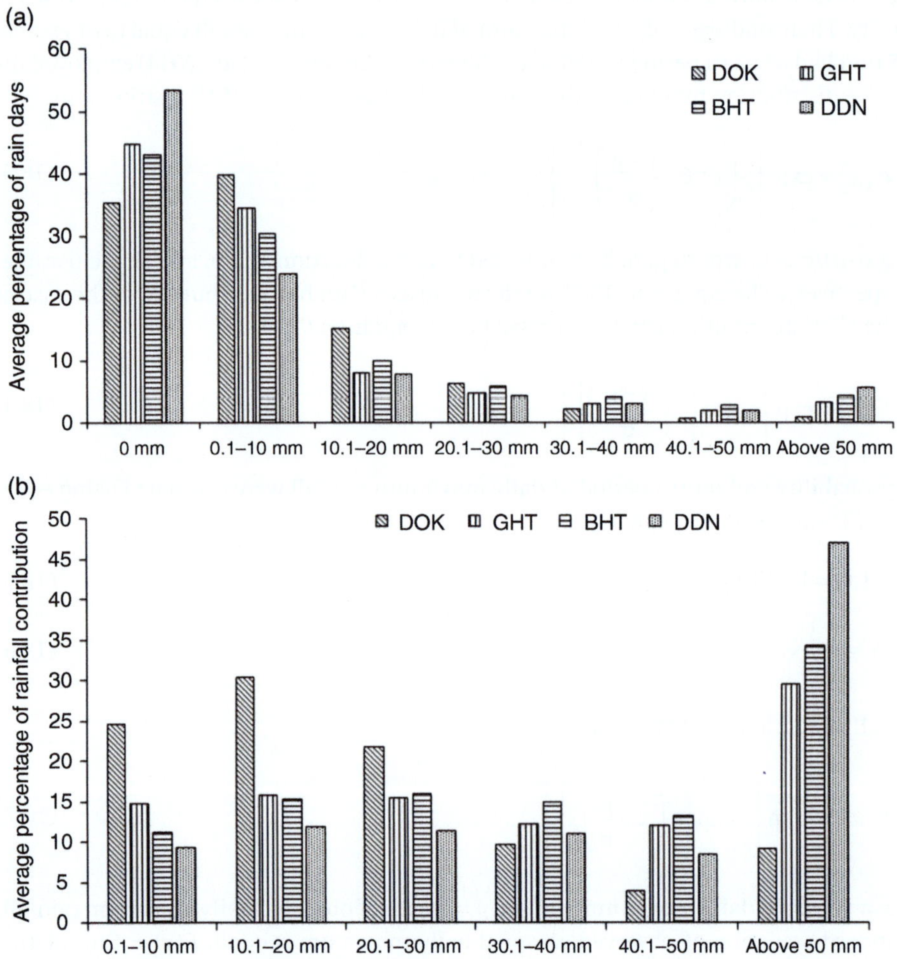

Figure 18.2 Bar diagram showing the (a) Percentage of rainfall events for different frequencies and (b) Percentage of rainfall contribution in different frequencies.

Table 18.2 Rainfall frequency distribution at the different stations.

Rainfall ranges	DOK (%)	GHT (%)	BHT (%)	DDN (%)
0 mm (No rainfall)	35.4	44.87	42.99	53.44
0.1–10 mm	39.73	34.34	30.44	24
10.1–20 mm	15.02	7.922	9.82	7.78
20.1–30 mm	6.15	4.7	5.86	4.31
30.1–40 mm	2.09	2.9	4.17	3.01
40.1–50 mm	0.6	2.03	2.79	1.84
>50 mm	0.9	3.21	4.4	5.61

Table 18.3 Rainfall contribution of different frequency.

Rainfall frequency	DOK (%)	GHT (%)	BHT (%)	DDN (%)
0.1–10 mm	24.65	14.88	11.29	9.39
10.1–20 mm	30.46	15.86	15.27	11.91
20.1–30 mm	21.79	15.44	15.94	11.4
30.1–40 mm	9.67	12.2	14.94	11.13
40.1–50 mm	3.95	12.05	13.32	8.46
>50 mm	9.26	29.52	34.35	46.99

the DOK station, 2.03% to the GHT station, 2.79% to the BHT station, and 1.84% to the DDN station of the total seasonal rainfall. (Table 18.2 and Figure 18.2a).

The contribution of rainfall for different frequencies shows a very interesting result. The high-altitude DOK station received the maximum contribution of rainfall between 0.1–10 mm (~25 %) and 10.1–20 mm (~30 %), and the minimum contribution of rainfall (~4%) pertains to 40.1–50 mm. Whereas, the lower elevated GHT, BHT, and DDN stations received the maximum contribution of rainfall, i.e. >50 mm, which contributed ~30%, ~34%, and 47%, respectively (Table 18.3 and Figure 18.2b). However, the minimum contribution of rainfall in these three stations is the same as the DOK station, i.e. 40.1–50 mm (Table 18.3). All three stations, GHT, BHT, and DDN, are in the Lesser Himalayas and Siwalik Hills, where the rainfall frequency is almost the same, whereas the Dok station is near the glacier, and here, the heavy rainfall frequency is very low.

Overall, the result suggests that the maximum frequency of rainfall recorded was 0.1–10 mm in all the stations, while the maximum frequency of extreme rainfall (>50 mm/day) was reported from a low altitude below 2000 m asl. Our study also suggests that the frequency of extreme rainfall increases with decreasing altitude and supports the theory of Nandargi and Dhar (2011). Furthermore, the contribution of rainfall was higher in lower frequency (0.1–10 to 10.1–20 mm/day) at higher altitudinal regions, whereas, with the decreased elevation, the contribution of high-intensity rainfall (>50 mm) contributed a higher amount of rainfall, like DDN station contribute 47% of extreme rainfall (>50 mm).

18.4.2 Return Period

The highest recorded 1-day rainfall values (Table 18.4) during the monsoon months in the four stations, DOK, GHT, and BHT, were between ~130 and ~170 mm/day, whereas the DDN station received the highest ~335 mm/day. On 16th and 17th June 2013, Uttarakhand received heavy rainfall and flash flood in the downstream area, which caused heavy damage to the property and livestock, and more than 5000 people lost their lives (Dobhal et al. 2013; Rautela 2013; Nirupama 2015; Mehta et al. 2017). On 16th and 17th June 2013 (2 days), Dokriani Glacier Tela camp (2500 m asl) received 269 mm rainfall, and Chorabari Glacier camp (3800 m asl) received 325 mm heavy rainfall (Mehta et al. 2017). The average seasonal maximum rainfall of these stations ranges from ~65 to ~80 mm, while the standard deviation ranges from ~23 to ~39 (Table 18.5). Considering the aforementioned event, we have calculated the return periods of an extreme event in Garhwal Himalaya. The analytic changes in return periods for rainfall events are shown in Figure 18.3 and Table 18.6. The return period of rainfall events is 50 mm/day in ~1 year for all the stations, and the return period of rainfall events 100 mm/day is ~5 years for three stations (DOK, GHT, and BHT), while the return periods of extreme rainfall events of 100 mm/day for DDN station is ~2 years (Table 18.6). The return period of extreme rainfall events 200 mm/day, which flooded Uttarakhand on 16th and 17th June 2013, is ~260 years for DOK, ~234 years for GHT, and 327 years for BHT stations. However, the return period of extreme rainfall events 200 mm/day for DDN station is ~26 years (Table 18.6 and Figure 18.3). A similar study has been carried out in different parts of the world to calculate the return period. Ahammed et al. (2014) analyzed the 57 years of rainfall data from Bangladesh, and they found that the return

Table 18.4 Highest one-day and average seasonal (May–October) rainfall of the study area.

Station	Highest 1-day rainfall (mm)	Date of occurrence	Average rainfall (in mm)
DOK	135.2	17 August 2011	1228.84
GHT	149.8	4 July 2007	1359.5
BHT	163	17 July 2010	1449.56
DDN	335.5	17 June 2013	1784.51

Table 18.5 Statistical descriptions of rainfall pattern (Seasonal maximum daily rainfall in mm).

Parameters	DOK	GHT	BHT	DDN
Mean	65.31	67.77	66.83	80.35
Maxima	135	149.8	163	335.5
Minima	44.4	45.3	45.4	45.2
Standard deviation	24.23	24.25	23	39.07
Skewness	1.65	1.73	1.99	2.53
Kurtosis	1.99	2.9	4.65	10.13

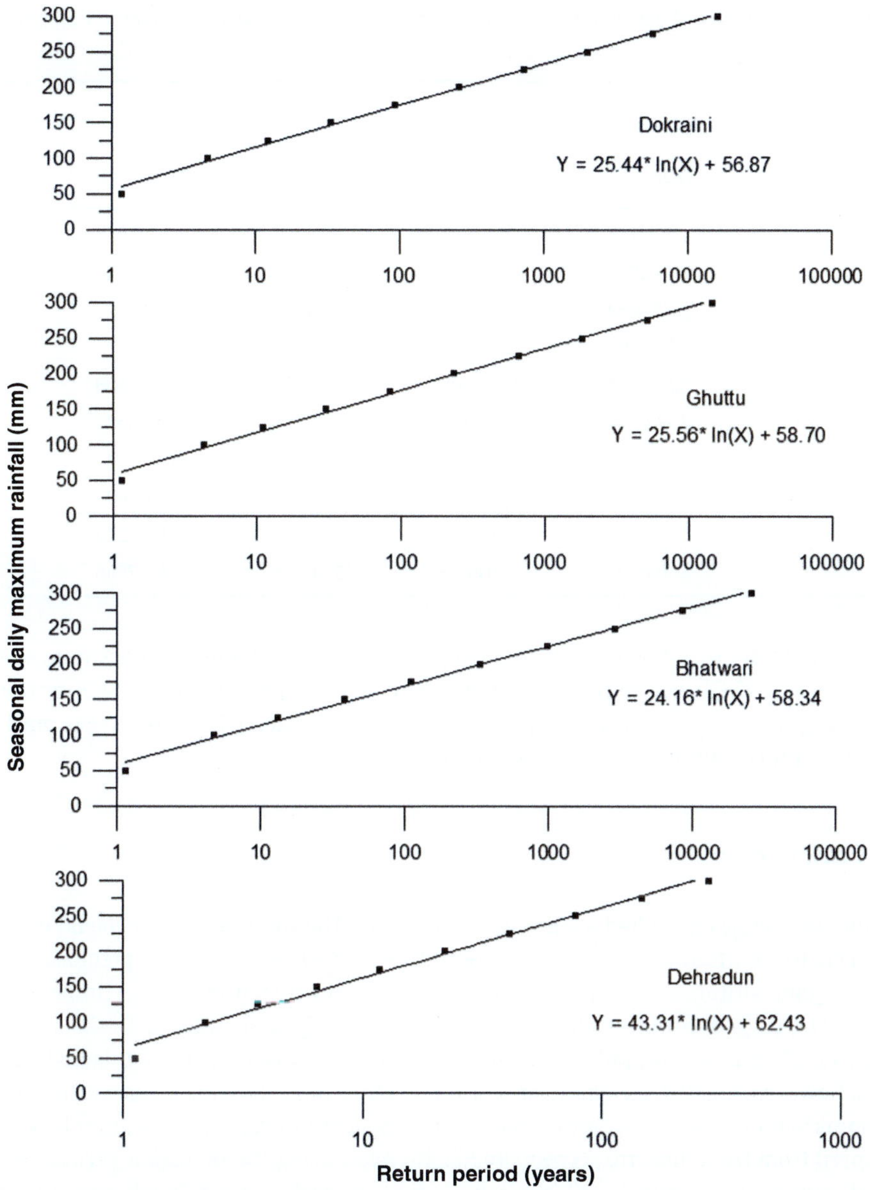

Figure 18.3 Return period of seasonal maximum daily rainfall.

period of 225 and 425 mm is 4 and 100 years, respectively. Furthermore, Nadarajah and Choi (2007) modeled the annual maxima of daily rainfall for the years 1961–2001 for five locations in South Korea. They estimate that the return periods of daily rainfall were 10, 50, 100, 1000, 50 000, 10 000, 50 000, and 100 000 years for the five stations (Nadarajah and Choi 2007). Soler et al. (2007) used 10-year precipitation data and found recurrences of extreme events using the Gumbel distribution function.

Table 18.6 Return periods of extreme rainfall events in the study area calculated by the Gumbel Distribution function.

	Return period in years			
Rainfall (mm/day)	**DOK**	**GHT**	**BHT**	**DDN**
50	1.179756	1.142521	1.143005	1.128205
100	4.702556	4.29795	4.748528	2.203479
125	12.24054	11.09359	13.04415	3.6617313
150	33.41794	30.17977	37.67293	6.4590714
175	92.83734	83.70008	110.703	11.780608
200	259.5284	233.7482	327.2234	21.880537
225	727.1425	654.4084	969.1561	41.03713
250	2038.924	1833.726	2872.333	77.365003
275	5718.818	5139.931	8514.799	146.25242
300	16041.9	14408.85	25243.36	276.87968

Our study suggests that the rainfall intensity decreases with increased elevation, the higher elevated DOK Glacier station (3700 masl) received less precipitation ~1228 mm/yr (Annual average May to October) in comparison to the low elevated DDN station (640 masl) received ~1784 mm (annual average in same months).

18.5 Conclusions

Recent climate changes have had a significant impact on the high mountain glacial environment. Due to continuously rising air temperatures all over the world, the precipitation pattern at higher altitudes is changing from solid (snow) to liquid (rain) (Immerzeel et al. 2010). The rugged terrain of the Himalayas is intimately acquainted with the severe consequences of extreme rainfall occurrences, including devastating floods, landslides, and debris flows, surpassing the challenges faced by other mountainous areas worldwide. In the Uttarakhand Himalayas, communities nestled beneath towering peaks grapple with constant peril from these hazards, especially heightened during the monsoon period. The intensified rainfall events during the monsoon season exacerbate flash floods and trigger landslides, exacerbating the risks faced by local inhabitants. Based on the aforementioned problem in the Himalaya, we have statistically analyzed the seasonal daily rainfall data of four stations of Uttarakhand Himalaya. The study can be concluded as

1. The number of seasonal rainy days of these stations varies from ~55 to 65%, and the average seasonal maximum of these stations ranges from ~65 mm to ~80 mm, while the standard deviation ranges from ~23 to ~39.

2. The return period of rainfall events 50 mm/day is ~1 year for all the stations, and the return period of rainfall events 100 mm/day is ~5 years for the stations DOK, GHT, and BHT, whereas 2 years for DDN station.

3. The return period of extreme rainfall events 200 mm/day is ~260 years for DOK, ~234 years for GHT, and ~327 years for BHT stations. However, return periods of extreme rainfall events of 200 mm/day for DDN station is ~26 years (Table 18.6 and Figure 18.3). Our result suggests that the higher elevation DOK station received less precipitation (average annual rainfall ~1228 mm) in comparison to low elevated DDN station received ~1784-mm average annual rainfall, and the further high-resolution study can be useful to future research. This study also tries to fill the gap of poor availability of rainfall data in the Himalaya.

Acknowledgments

The authors are grateful to the Director of Wadia Institute of Himalayan Geology (WIHG), Dehra Dun, India, for providing facilities and support to carry out this work. MM is especially thankful to Dr. Khayingshing Luirei, a scientist at WIHG, for their suggestions and valuable comments. MPGO Ghuttu team of WIHG for collecting continuous data through a MoES, New Delhi sponsored project (MoES/P.O. (Seismo)/NPEP(15)/2009) are also acknowledged. FRI Dehradun and L&T Sitapur –Bhatwari are duly acknowledged for providing the rainfall data. DDK and DP acknowledge and appreciate the financial support provided by DST New Delhi.

References

Ahammed, F., Hewa, G.A., and Argue, J.R. (2014). Variability of annual daily maximum rainfall of Dhaka, Bangladesh. *Atmospheric Research* 137: 176–182.

Arnell, N.W. (2003). Relative effects of multi-decadal climatic variability and change in the mean and variability of climate due to global warming, future stream flow in Britain. *Journal of Hydrology* 270: 195–213.

Basistha, A., Arya, D.S., and Goel, N.K. (2008). Spatial distribution of rainfall in Indian Himalayas – a case study of Uttarakhand region. *Water Resource Management* 22: 1325–1346.

Bhardwaj, A., Ziegler, A.D., Wasson, R.J., and Chow, W.T.L. (2017). Accuracy of rainfall estimates at high altitude in the Garhwal Himalaya (India): a comparison of secondary precipitation products and station rainfall measurements. *Atmospheric Research* 188: 30–38.

Burbank, D.W., Bookhagen, B., Gabet, E.J., and Putkonen, J. (2012). Modern climate and erosion in the Himalaya. *Comptes Rendus Geoscience* 344 (11–12): 610–626.

Carvalho, J.R.P.D., Assad, E.D., Evangelista, S.R.M., and Pinto, H.D.S. (2013). Estimations of dry spells in three Brazilian regions – analysis of extremes. *Atmospheric Research* 132–133: 12–21.

Dobhal, D.P., Gupta, A.K., Mehta, M., and Khandelwal, D.D. (2013). Kedarnath disaster: facts and plausible causes. *Current Science* 105 (2): 171–174.

Froehlich, W. and Starkel, L. (1987). Normal and extreme monsoon rains – their role in the shaping of the Darjeeling Himalaya. *Studia Geomorphologica Carpatho-Balcanica* 21: 129–160.

Gadgil, S., Shrestha, A.B., Wake, C.P. et al. (2003). The Indian monsoon and its variability. *Annual Review of Earth and Planetary Sciences* 31: 429–467.

Goswami, B.N., Venugopal, V., Sengupta, D. et al. (2006). Increasing trend of extreme rain events over India in a warming environment. *Science* 314: 1442–1445.

Gumbel, E.J. (1941). The return periods of flood flows. *Annals of Mathematical Statistics* 12 (2): 163–190.

Immerzeel, W.W., Beek, L.P.H.V., and Bierkens, M.F.P. (2010). Climate change will affected the Asian water tower. *Science* 324: 1382–1385.

Khandelwal, D.D., Gupta, A.K., and Chauhan, V. (2015). Observation of rainfall in Garhwal Himalaya, India during 2008-2013 and its correlation with TRMM data. *Current Science* 108 (6): 1146–1151.

Lagouvardos, K., Kotroni, V., Defer, E., and Bousquet, O. (2013). Study of a heavy precipitation event over southern France, in the frame of HYMEXproject: observational analysis and model results using assimilation of lightning. *Atmospheric Research* 134: 45–55.

Mehta, M., Shukla, T., Bhambri, R. et al. (2017). Terrain changes, caused by the 15–17 June 2013 heavy rainfall in the Garhwal Himalaya, India: a case study of Alaknanda and Mandakini basins. *Geomorphology* 284: 53–71.

Messerli, B., Viviroli, D., and Weingartner, R. (2004). Mountains of the world-vulnerable water towers for the 21st centure, Royal colloquium on mountain area, a global resource. *AMBIO Special Report* 13: 29–34.

Nadarajah, S. and Choi, D. (2007). Maximum daily rainfall in South Korea. *Journal of Earth System Science* 11 (4): 311–320.

Nandargi, S. and Dhar, O.N. (2011). Extreme rainfall events over the Himalayas between 1871 and 2007. *Hydrological Sciences Journal* 56 (6): 930–945.

Nijssen, B.N., Lettenmaier, D.P., Liang, X. et al. (1997). Streamflow simulation for continental-scale river basins. *Water Resources Research* 33: 711–724.

Nirupama, N. (2015). A understanding risk from floods and landslides in the Himalayan region: a discussion to enhance resilience. *Planet@Risk* 3 (2): 1–4.

Priya P, Krishnan R, Mujumdar M, Houze R A 2016 Changing monsoon and mid-latitude circulation interactions over the Western Himalayas and possible links to occurrences of extreme precipitation. *Climate Dynamics* 49: 2351–2364 doi:https://doi.org/10.1007/s00382-016-3458-z: 1-14.

Rautela, P. (2013). Lessons learnt from the Deluge of Kedarnath, Uttarakhand, India. *Asian Journal of Environment and Disaster Management* 5 (2): 43–51.

Sain, K., Mehta, M., Kumar, V. et al. (2023). A climatic surprise–slope instability triggered by heavy rain in maldevta region, Dehradun, Uttarakhand, on 20 August, 2022. *Journal Geological Society of India* 99: 317–320.

Singh, P., Haritashya, U.K., Ramasastri, K.S., and Kumar, N. (2005). Prevailing weather conditions during summer seasons around Gangotri Glacier. *Current Science* 88 (5): 753–760.

Soler, M., Regüés, D., Latron, J., and Gallart, F. (2007). Frequency–magnitude relationships for precipitation, stream flow and sediment load events in a small Mediterranean basin (Vallcebre basin, Eastern Pyrenees). *Catena* 71: 164–171.

Starkel, L. and Basu, S. (ed.) (2000). *Rains, Landslides and Floods in the Darjeeling Himalaya*, 168. New Delhi: INSA.

Stocker, T.F., Qin, D., Platner, G.K. et al. (ed.) (2014). *Climate Change 2013: The Physical Science Basis*, 1535. Cambridge: Cambridge University Press.

UNDP (2014). Uttarakhand action plan on climate change, 226.

Valdiya, K.S. (1987). *Environmental Geology: Indian Context*, 538. New Delhi: Tata McGrow Hill.

Valdiya, K.S., Paul, S.K., Tara, C. et al. (1999). Tectonic and lithological characterization of Himadri (Great Himalaya) between Kali and Yamuna rivers, Central Himalaya. *Himalayan Geology* 20: 1–17.

Yin, A. (2006). Cenozoic tectonic evolution of the Himalayan orogen as constrained by along-strike variation of structural geometry, exhumation history, and foreland sedimentation. *Earth Science Reviews* 76: 1–131.

19

Traditional Techniques of Rainwater Management and Conservation in the Mountainous Regions of North-East India

Liza G. Kiba[1], *Chitrasen Lairenjam*[1], *Arbind Kumar Verma*[1], *Prem Ranjan*[2], *and Grace Nengzouzam*[3]

[1] *Department of Agricultural Engineering, School of Agricultural Sciences, Nagaland University, Medziphema, Nagaland, India*
[2] *Department of Agricultural Engineering, North Eastern Regional Institute of Science and Technology, Nirjuli, Arunachal Pradesh, India*
[3] *Department of Agricultural Engineering, School of Engineering and Technology, Nagaland University, Kohima, Nagaland, India*

19.1 Introduction

Water, the elixir of life, is the most crucial natural resource available for all forms of life. It is mainly used for drinking and irrigation purposes, and its requirements depend upon the physical as well as socioeconomic development of the area concerned. A society can follow a desired development path with the aid of irrigation, which functions as a cultural and social tool. Due to its combination of sociocultural and socioeconomic effects, irrigation development has a unique role to play in assisting indigenous communities in achieving their cultural goals. Water from various water sources (rivers, reservoirs, canals, channels, lakes, and wells) is drawn and directed to farms or fields by irrigation systems, which are designed to support farmers and augment crop water requirements (Adamala et al. 2014). Indian traditional and indigenous irrigation methods have been around since prehistoric times. These methods are based on collecting and using rainfall as well as surface and groundwater. They are also economical and sustainable. The use of traditional irrigation systems in Indian agriculture dates back at least 5000 years, according to various archeological excavations (Sengupta 1985). These systems designed, built, and overseen by local farmers are generally intended for small-scale community applications (Srinivasan and Babu 2000).

The North Eastern Region (NER) is India's easternmost area, located between 21.5°N and 29.5°N latitude and 85.5°E and 97.5°E longitude, covering a total area of 262 000 sq. km, which is about 7.75% of India's total geographical area. It comprises eight states: Arunachal Pradesh, Assam, Manipur, Meghalaya, Mizoram, Nagaland, Sikkim, and

Water Resources Management in Mountain Regions, First Edition. Edited by Virendra Bahadur Singh, Sughosh Madhav, Rakesh Kumar Gupta, Prerna Diwan, and Amit Kumar.

Tripura. The NER drains into the Bay of Bengal primarily through the Brahmaputra and Barak rivers. The Brahmaputra basin spans Arunachal Pradesh, Assam, Meghalaya, Nagaland, and Sikkim, while the Barak basin covers Meghalaya, Manipur, Mizoram, Assam, Tripura, and Nagaland.

The region is known for its diverse social and cultural practices, challenging topography, low population density, abundant forests, and unpredictable weather patterns. Temperatures in the NER are moderate and decrease with altitude, with higher altitudes having permanent snow cover. The region experiences three distinct seasons: Winter (early November to mid-March), Summer (mid-April to mid-October), and the Rainy season, which coincides with the summer months (Dikshit and Dikshit 2014) (Figure 19.1).

The northeastern region of India exhibits a diverse climate due to its varied topography and geographical location. Tropical wet and dry climates are found in parts of Manipur, Tripura, and Mizoram. Warm temperate mesothermal climates cover much of Assam, Nagaland, northern Meghalaya, northern Manipur, and parts of Arunachal Pradesh, where mean temperatures in the coldest months range from −3 to 18°C (27–64°F). The Brahmaputra valley has a humid subtropical climate with hot summers. The eastern hills of Nagaland, Manipur, and Arunachal Pradesh, at altitudes between 500 and 1500 m (1600 and 4900 ft), experience warm summers. Areas above 1500 m (4900 ft) in Meghalaya, parts of Nagaland, and northern Arunachal Pradesh have short, cool summers. The extreme northern parts of Arunachal Pradesh have a humid continental climate, with mean winter temperatures below −3°C (27°F) (Peel et al. 2007).

The region's mountainous topography, combined with its unique geographic setting and hydro-meteorological conditions, places it among the world's highest rainfall zones. The climate ranges from subtropical to alpine. Still, most of this region experiences water shortages during the dry months, particularly from December to March, because rainfall is restricted to about 6 months a year. Thousands of people in the plain areas also experience water-related issues in the form of floods and waterlogging, in addition to the severe water shortage in the hilly regions caused by an inadequate water management system.

The tribal communities in Northeast India rely heavily on diverse agricultural practices, including shifting cultivation, fallow systems, home gardens, and wet rice cultivation (Rai 2005). About 80% of the people in the NER rely on agriculture as their primary source of income, making it an agrarian economy (Talukdar 2018). Settled farming in the plains and foothills and shifting cultivation (Jhum) on the hill slopes are the two distinct farming systems practiced in the region. The region has medium to steep slopes and challenging terrain, characterized by a wide range of soil types, erratic rainfall patterns, a lack of modern machinery and techniques, various farming methods, and other issues like slow market facilities and transportation. This region boasts one of the highest levels of ethnic diversity in the nation, with each state having its unique farming system and culture. In this region, one of the most significant traditional harvesting systems is the rainwater harvesting system. The long-standing and deeply ingrained traditions of the local population have led to the development of an environmentally sustainable rainwater harvesting system. Long channels are used to carry water from natural streams in the mountains. The water is then diverted and channeled around the fields before leaving to the river or nalas. In valley lands, people lift water by any means necessary for their daily use. In addition, people harvest and store water in tanks by inserting pipes into natural streams. These pipes carry water through gravity flow over long distances to its final destination.

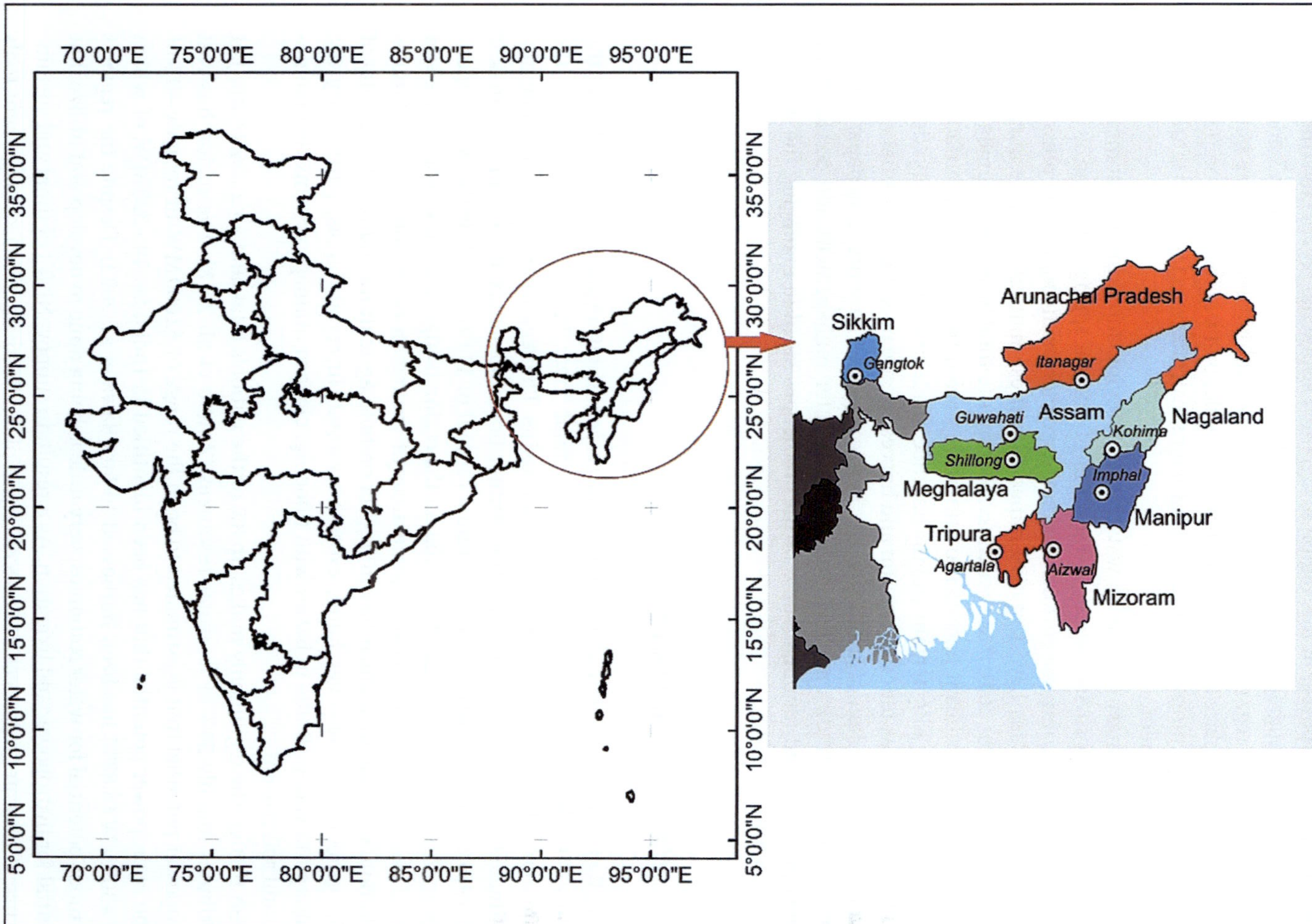

Figure 19.1 North East Region of India (Photo courtesy: Renzut, 2017).

The Northeast region of India heavily relies on natural moisture and rainfall for growing crops, resulting in instability in agricultural production. Unfortunately, the agricultural sector's overall development in this region is lagging. Despite the region covering 7.9% of the country's total geographic area, it only produces 1.5% of the country's food grain production. Since the region is primarily a rain-fed agricultural area, there are not enough irrigation facilities in place to protect it from the unpredictability of the monsoon. Therefore, irrigation is a crucial component of any program for the primary sector of Northeast India's agricultural development.

Moreover, the primary source of water in this region for drinking, domestic, and irrigation purposes is surface water in the form of natural springs, streams, and rivers. Unfortunately, indiscriminate cutting of forest trees, increasing population pressure, and developmental activities have negative impacts on these natural water resources. The monsoon rain does not contribute much to meeting the total water requirements of the region because a significant part of the rainfall is wasted in the form of surface runoff due to the undulating hilly terrain and the absence of adequate water conservation measures. As a result, rainwater conservation and harvesting are necessary remedial measures to overcome the problem of water scarcity in the region. Therefore, this paper aims to review the various traditional techniques of rainwater management and conservation systems practiced and highlight some of the challenges faced in managing the water in the hills of NER of India.

19.2 Water Resources of NER

The NER of India boasts abundant water resources, constituting approximately 46% of the nation's total water resources. Among the notable river basins in the area are the Brahmaputra, Barak, Gumti, and Imphal river basins. The region's surface water potential is estimated at 1487.2 km^3, while its groundwater potential stands at 25.3 km^3 (Borthakur et al. 1989). India receives an average annual rainfall of 1119 mm, while the NER experiences approximately 2450 mm. The rainfall of the region is high and varies with space and time on a scale of 1000–11 000 mm. Some areas of the region experience water shortages due to rainfall variability, which has an impact on both agriculture and people's quality of life in general. On the other hand, excessive rainfall in certain areas, particularly the Brahmaputra valley, results in issues with flooding, clogged drainage systems, sedimentation, and soil erosion.

Even though the region only makes up 8% of the nation's total land area (26.23 Mha), it contributes roughly 34% of its water resources (Zhimo et al. 2018). The region has vast hydropower potential that is estimated to be in the range of 62 000 MW, or roughly 40% of all the hydropower potential that has been identified in India, i.e. 145 320 MW, of which only 4309 MW (6.88%) has been harnessed so far (Bawa et al. 2022). Despite the region's enormous potential for water resources, very little is being done to develop and utilize that potential for hydropower and irrigation. As a result, to support NER's optimal and sustainable use of its water resources, it is necessary to plan effectively using management tools and collaborating with institutions and then to start taking appropriate action for water control, conservation, and utilization throughout the region.

19.3 Traditional Water Management Techniques in NER

The NER experiences uneven distribution of rainfall. From October to February, the region receives little precipitation, and the average evaporation is higher than the average precipitation by 8–357 mm. Therefore, irrigation becomes necessary to conserve and supplement soil moisture. The region is home to numerous tribes, and each state of the region has its unique environment, local conditions, socioeconomic situations, festivals, and rituals associated with agricultural practices. There are a total of 145 communities in this region (Sen et al. 2015). The ethnic communities of Northeast India, with rich histories and traditions, have pioneered water management techniques that integrate eco-friendly systems for the conservation, preservation, and sustainable utilization of natural resources. Some popular water management system is given in Table 19.1 and discussed briefly below:

19.3.1 Bamboo Drip Irrigation

Meghalaya, renowned for receiving the highest rainfall globally, has farmers in the Jaintia and Khasi hills who have devised a distinctive bamboo drip irrigation system. This system harnesses springs and stream water to irrigate crops such as betel leaf or black pepper, commonly planted in areca nut orchards or mixed orchards (Sanjay-Swami 2021). Due to the hilly and undulating terrain and the poor water-holding capacity of the soil, it is an excellent irrigation system for crops grown in dry regions. With this system, a hilltop stream's water is directed along a split bamboo line, acting as an open channel. These bamboo channels allow gravity flow to carry water to the field, where it is subsequently distributed among the plants surrounding the orchard. The channel sections, constructed from bamboo, divert and transport water to the plot site, where it is distributed without leakage into branches made from various forms of bamboo pipes. Installing secondary, tertiary, or even multiple bamboo pipelines is how water is distributed to the orchard. Each branch has an unconventional water circulation system installed. At the end of each plant's branch, water is finally dripped onto the base of the

Table 19.1 Various traditional water management techniques of NER.

Sl. No.	Name	State mainly practiced
1	Bamboo-Drip irrigation	Meghalaya
2	Dong	Assam
3	Kulo	Sikkim
4	Paddy-cum-fish farming	Arunachal Pradesh
5	Pukhris	Manipur
6	Roof-top rainwater harvesting	Mizoram
7	Thuikhir	Mizoram
8	Zabo farming system	Nagaland
9	Panikheti	Nagaland

plant at a rate of 10–100 drops per minute. Occasionally, this system is used to carry water to far-off mountains for residential use.

Bamboo of different diameters is used for constructing channels (Sanjay-Swami 2021). When constructing the channels, approximately one-third of the outer casing's length and internodes of bamboo pieces must be removed. The system is laid out in a way that the ground clearance of the channels gradually decreases from a few meters to 10–15 cm. This is achieved by reducing the height of the channel supports (Borthakur 1992). This system's benefit is that it runs without fuel or electricity. It might be put into practice in areas where bamboo is inexpensive or free to obtain. One significant disadvantage is that, after 2–3 years, the entire network must be rebuilt because bamboo rots when it rains. Since bamboo is available freely in this region, the cost of building the system is minimal (Sanjay-Swami 2021). Overall, it appears to be a low-cost, environmentally friendly irrigation system that can be installed for both urban permaculture gardens and agricultural needs.

19.3.2 Dong

The Bodo and the Madahis community, who settled on the foothills of Assam, have used the Dong (canal) water management system for decades now. It is referred to as Doisa in the Bodo dialect. As population growth escalated, so did the demand for food, necessitating the adoption of wet rice cultivation. In addition, the region faced a scarcity of perennial drinking water sources, exacerbating the water shortage. In response to these challenges, local inhabitants developed their water management system. Initially, this system involved diverting water from the Pagladiya River to agricultural fields through community engagement (Saha et al. 2020).

Natural streams are the sources of dongs, and canals are cut to direct water to the fields. Water is redirected and stored in ponds known as Pukhuri in the area via dongs made of earthen canals. Occasionally, water for irrigation is taken straight out of the earthen canals. To protect the earthen canals from excessive runoff during the rainy season, it is necessary to build these dongs annually. Lahoni, a traditional water-lifting device, is used to raise the water in the pond. It can harvest about 10 l of water, which is sufficient to water a plot of two bighas within a single day (Sarma and Goswami 2015). Koon, a leg-operated wooden boat-like water lifting device that can collect up to 25 l of water at a time, is also used in the field. Another type of doisa or dong is practiced by cultivators in the Kokrajhar district, where rainwater is used as the primary source of water. They trap rain through construction ponds, and the accumulated water is distributed by canal systems to agricultural fields in the villages (Talukdar 2018). Dongs are ancient water channels used to convey water from streams or ponds to irrigate fields. Water from tube wells can carry iron, which can be harmful to the crops when used for irrigation purposes, but dongs carry surface water and rainwater, which are free from impurities, to the irrigation field and, therefore, are sustainable irrigation systems.

19.3.3 Kulo—A Water-storing Structure

Kulo is the traditional spring diverted channel adopted for irrigating crops in Sikkim (Patle 2013). This practice is commonly used by the people of Sikkim, who are skilled in water conservation and management in hilly regions. Springs have been the primary source of water for drinking and other uses, including agriculture, in this area for many years. The region receives little to no rainfall from December to March, resulting in a severe shortage of drinking water during winter due to drying or a drastic reduction in spring discharge. Although water is available in the valleys, the terrain constraints force a significant portion of agricultural fields to rely solely on rain. A long channel is used to transport water from the natural streams and direct it around the field, eventually leaving it for the river or nalas. For regular use, water is collected in the valley land through various methods. Plastic pipes can also be inserted into natural streams by people to harvest and store water in tanks. These pipes carry water by gravity and flow long distances to the destination. This practice can also be found in other states in the Northeast, such as Arunachal Pradesh, Meghalaya, Manipur, and Nagaland.

19.3.4 Paddy-cum-fish Farming

The Apatani tribes residing in the Ziro Valley of Arunachal Pradesh employ the rice cum fish farming method. This agricultural system combines paddy and fish farming along with millet cultivation on the bunds separating each plot. Depending on the slope, size, and shape of the terraces, the bunds can range in width from 0.6 to 1.4 m and height from 0.2 to 0.6 m (Rai 2005). The skill of integrating rice and fish farming, known as Aji-ngyii, was introduced in the 1980s with great success and has been continuously evolving since then (Lahkar 2019). It is believed to be one of the most productive and efficient agricultural systems in the region. (Singh and Gupta 2002). This integrated rice and fish farming is very organic because no inorganic fertilizers are used; instead, wastes from households, livestock, and crops are used. After the crops from the previous season are harvested, the wastes are dumped in the field, and the preparation of the field for the next season starts immediately, which continues until the end of spring. Weeds are manually removed from the field three to five times during the cultivation cycle and then deposited back onto the field for decomposition. The irrigation channel thus carries natural fertilizers from decaying materials to the fields (Rai 2005). In this system of farming, a pit/channel is dug at appropriate intervals on the paddy field, and fingerlings are reared there. During the monsoon season, when water levels are high, farmers practice shallow submergence of their paddy fields, typically ranging from 5 to 10 cm. This inundation allows fish to emerge from their pits and move freely within the submerged area. Even in the dry season, adequate water remains in the pits and channels, ensuring the survival of the fish. This integrated system benefits the fish by providing better nutrition through the manuring of the paddy fields, and their growth is enhanced due to the increased surface area available for movement during the submergence period. Effective rainwater management enables the simultaneous production of both paddy and fish in these fields.

19.3.5 Pukhris

Pukhris or ponds are traditional water harvesting structures constructed by the people of Manipur to store rainwater. The name is believed to be borrowed from the Bengali word "pukur" meaning pond (Kshetrimayum 2013). Earlier, one pond was shared by two or more households, but nowadays, community ponds are commonly found in the settlements

(Singh 2004). Due to the pathetic water supply system in the settlement area, community ponds are mostly dug and maintained. The water harvested in this structure is primarily used for irrigating private kitchen gardens and household uses, but over the years, people have bred fish in these ponds for their consumption and additional income. Such structures are also constructed at the lower end of the plot in a paddy field to provide water to the field during dry spells. The majority of the population still relies on pukhris to meet their water needs due to the scarcity of water.

19.3.6 Rooftop Rainwater Harvesting

Rooftop rainwater harvesting involves utilizing the roof of a house or building as a catchment area to collect rainwater, which is then stored in tanks, either underground or above-ground, for future use. The roof surface can be composed of various materials such as tiles, metal sheets, or plastics, which intercept the flow of rainwater in conjunction with gutters and pipes. These systems can range from large tanks serving entire communities to simple setups like a bucket positioned beneath a roof with a gutter. Rooftop rainwater harvesting is primarily practiced in Mizoram and Nagaland.

In Mizoram, the terrain is characterized by steep hills with gradients exceeding 50%, interspersed with deep river gorges. The geological composition of the region limits water retention, leading to the drying up of springs and small streams during the winter season. In response to this challenge, local communities have adopted rooftop harvesting structures, which have become widespread across the state.

The typical approach to rainwater storage involves the installation of horizontal rain gutters along the edges of sloping roofs, often made of corrugated iron sheets. Rainwater collected in these gutters is directed into pipes connected to storage tanks, traditionally constructed using galvanized plain sheets or Galvanized Corrugated Iron (GCI) sheets. However, there is a growing trend toward utilizing reinforced cement concrete tanks, which can be situated in courtyards or beneath houses.

The government of Mizoram has initiated policies to promote the adoption of GCI sheet roofs over thatched roofs in village homes, and financial assistance in the form of loans is provided to encourage rooftop water harvesting (Satapathy 2001). Harvested rainwater is stored in tanks or earthen ponds with compacted floors to minimize seepage. To reduce evaporation losses, these ponds are covered with opaque plastics, straw, or locally available materials.

19.3.7 Tuikhur

In Mizoram, water is harvested from small hilly streams through the construction of tanks known as Tuikhur or village spring sources at the foothills. These water sources are vital for the daily needs of villagers, particularly for low-income families, and nearly every village has access to them. Tuikhur water is perennial, providing a reliable source throughout the year. However, recent fluctuations in rainfall patterns and various human activities have led to the drying up or significant reduction in the volume of many of these water sources. The harvested water is transported or transferred by different means for use in different purposes. Bamboo pipes and rubber/iron pipes are some of the means through which water from Tuikhur is transported to the household. During the dry season, villagers are restricted from using Tuikhur water for cleaning and washing purposes, although it is utilized for irrigation, drinking, and other essential needs throughout the year (Biswas and Azyu 2021).

19.3.8 Zabo System

Zabo, a practice rooted in indigenous wisdom, blends traditional methods with modern techniques to harness rainwater for crops, horticulture, and livestock. This holistic approach focuses on natural resource conservation, creating a self-sustaining agricultural ecosystem. Known as Zabo in the Chakhesang area and Ruza in other regions, which means “impounding water,” this system combines irrigation, agriculture, forestry, and animal husbandry while also addressing soil erosion, water resource development, and the preservation of natural vegetation. The distinct system consists of rice fields below the harvesting tank/pond, a livestock farm adjacent to the water collection tank in the middle, and a small area of forest land atop the hill (Maibam and Ritse 2018).

Indigenous people in Nagaland, India, use the Zabo farming system, a sustainable farming method based on traditional knowledge for managing natural resources. Along the hillside’s slope, the system includes integrated agricultural practices like animal husbandry, paddy cultivation, water harvesting, and forest land for silviculture (Murry and Das 2021). The Zabo system (Figure 19.2) of farming utilizes protected forestland situated atop a hill. Water-harvesting tanks are positioned in the central area, while cattle and paddy fields are situated at the lower elevation. Water from the harvesting pond traverses through the cattle yard before reaching the paddy field for irrigation. This water carries with it dung and urine from the cattle, serving as a valuable source of nutrients for the crops cultivated at the lower elevation (Singh et al. 2018).

19.3.9 Panikheti

Weed infestation poses a significant challenge in rice cultivation. Due to hilly terrains, most of the rainfall is lost as surface runoff. Panikheti is a terrace-based farming method developed by local farmers mainly to harness the runoff water in the terraces (Das et al. 2012).

Figure 19.2 Zabo farming system in Kikruma Village.

In the upper reaches of the hills, natural forests remain primarily undisturbed and are predominantly utilized for paddy cultivation. This farming method involves the construction of terraces of varying sizes, supported by bunds, depending on the slope of the hill. Terraces are leveled using the cut-and-fill technique. Runoff water from hilltops and slopes is utilized for irrigation in the terraced fields, while excess water is efficiently channeled through well-designed drainage channels to storage tanks in lower areas. The decomposition of tree leaves, forest litter, crop residues, weeds, and cattle dung, either through burning or incorporation into the terraced fields, contributes to the maintenance of soil fertility for rice cultivation. This farming system is particularly prevalent in Nagaland, Manipur, and Sikkim.

19.4 Challenges in the Water Management System

The majority of the NER's villages are perched on steeply sloping hilltops, meaning that surface runoff loses most of the rainfall that falls in those areas. Even today, rainwater is the primary source of agricultural water (Singh 2004). In addition, since its establishment, the population has grown, making it impossible to meet the current water demand for drinking and sanitation (Sangtam 2018). Along with population growth, the use of fertilizers, pesticides, insecticides, and herbicides has resulted in a reduction in water availability and deterioration in water quality. It is therefore necessary to expand the villages' additional water supply sources. The water crisis in the region is anticipated to escalate due to significant temporal and spatial fluctuations in water availability, exacerbated by various factors such as the impacts of climate change, resulting in heightened occurrences of water-related disasters like flash floods, river meandering, and soil fertility loss. The natural water infrastructure, which is primarily composed of springs, streams, and other bodies of water, is becoming more and more contaminated. Owing to extensive temporal and spatial fluctuations in water availability, which are anticipated to escalate due to various factors, including the impacts of climate change, the incidence of water-related disasters such as soil fertility loss, flash floods, and river meandering is expected to exacerbate, deepening the water crisis in the state. Furthermore, natural water bodies, streams, and springs, which constitute the fundamental components of the state's natural water infrastructure, are experiencing escalating pollution levels.

Typically, urban plain regions are situated downstream, compared with rural hilly areas situated upstream within hydrological systems. Winter months see little to no rain, and most rains occur during the monsoon season. When monsoon rains fall heavily, the upland rural areas in the hills above experience severe soil erosion, and the urban areas downstream experience flash floods. In both upstream hills and downstream plains, winter is characterized by drought and the ensuing drop in groundwater levels. In the rural upstream regions, the water shortage causes less cultivation to occur during the winter. The primary challenge in the study area is the imbalance between the population size and the per capita water availability. The region's topography complicates water transportation, and the existing systems are outdated and insufficient for the current population.

19.5 Conclusion

In most NERs, due to unpredictable climatic patterns and the lowering of groundwater, there is a huge gap between the supply and demand for water. The effective management of spring sheds and rainwater harvesting systems has provided numerous rural villages self-sufficient. It is imperative to undertake the scientific capacity building of institutions for integrated water resource management in the region. Traditionally, urban and rural boards and authorities have addressed these issues separately. However, this approach lacks scientific rigor, as rural upstream and urban downstream areas are hydrologically interconnected, necessitating a comprehensive and integrated management practice.

Effective watershed management is the key to bridging the gap between urban and rural areas and achieving comprehensive regional development that takes the North East as a whole. Although the NER's states are divided politically, they are all connected hydrologically. All state authorities must present a unified front to tackle the problems associated with water resources and devise strategies for the comprehensive development of the area. This will not only increase farming operations but will also lessen flooding, open more work opportunities in both urban and rural areas, foster a sense of fraternity and interdependence, and, most importantly, promote peace in the area. When there is a shortage of water, these water harvesting techniques and storage buildings can help meet household water demands. The field crops may benefit from additional irrigation thanks to harvested rainfall. Modern irrigation techniques like drip and sprinkler irrigation can also help ensure the wise use and management of rainfall that has been harvested. The NER's way forward is to make the most of and plan for its current resource – rainfall, in order to improve springshed management and recharge, impound rainfall by building different water ponds, and install appropriate rainwater harvesting systems in both homes and communities.

Acknowledgment

The authors would like to thank all the farmers from Northeastern India who provided information on the indigenous techniques they have practiced for generations.

References

Adamala, S., Raghuwanshi, N., and Mishra, A. (2014). Development of surface irrigation systems design and evaluation software (SIDES). *Computers and Electronics in Agriculture* 100C: 100–109. https://doi.org/10.1016/j.compag.2013.11.004.

Bawa, J.S., Roy, A., and Awasthi, V. (2022). Identify and harness hydropower potential in North Eastern Region. *International Conference on Hydropower and Dam Development for Water and Energy Security – Under Changing Climate*, Rishikesh, India (7–9 April 2022).

Biswas, B. and Azyu, A. (2021). Water resources and management system of the Himalayan regions: case study of Mizoram, India. *Nature Environment and Pollution Technology* 20 (1): 193–201.

Borthakur, D.N. (1992). *Agriculture of the North-Eastern Region*. Guwahati, India: Bee Cee Prakashan.

Borthakur, H.P., Bora, C.K., Phukan, U. et al. (1989). *Agro-climatic Regional Planning of the Eastern Himalayan Region – Profile and Strategy*. Jorhat: Assam Agricultural University.

Das, A., Ramkrushna, G.I., Choudhury, B.U. et al. (2012). Natural resource conservation through indigenous farming systems: wisdom alive in North East India. *Indian Journal of Traditional Knowledge* 11 (3): 505–513.

Dikshit, K.R. and Dikshit, J.K. (2014). Weather and climate of North-East India. In: *North-East India: Land, People and Economy. Advances in Asian Human-Environmental Research*, 149–173. Dordrecht: Springer https://doi.org/10.1007/978-94-007-7055-3_6.

Kshetrimayum, K. (2013). An ogle on our Pukhri(s). *E-Pao* (6 May). https://e-pao.net/epSubPageExtractor.asp?src=education.Science_and_Technology.An_ogle_on_our_Pukhris (accessed 28 September 2024).

Lahkar, B. (2019). Practice of agro-pisciculture among the Apatani tribe of Ziro Valley, Arunachal Pradesh. *Restaurant Business* 118 (9): 213–225.

Maibam, M. and Ritse, N. (2018). Rüza: sustainable development practice in Kikruma Village, Nagaland, India. *International Journal of Humanities and Social Science Invention* 7 (1): 37–41.

Murry, N. and Das, S. (2021). Zabo farming system-a sustainable farming based on traditional knowledge for natural resource management practiced by tribal in Nagaland, India. *International Journal of Agriculture, Environment and Biotechnology* 14 (2): 203–205.

Patle, G.T. (2013). Water harvesting methods and storage structures in Sikkim. *Indian Farming* 63 (3): 37–40.

Peel, M.C., Finlayson, B.L., and McMahon, T.A. (2007). Updated world map of the Koppen-Geiger climate classification. *Hydrology and Earth System Sciences* 11 (5): 1633–1644.

Rai, S.C. (2005). Apatani paddy-cum-fish cultivation: an indigenous hill farming system of North East India. *Indian Journal of Traditional Knowledge* 4 (1): 65–71.

Renzut (2017). North East India. https://en.m.wikipedia.org/wiki/File:Northeast_india.png (accessed 19 November 2017).

Saha, S., Deka, N., and Bhagabati, A.K. (2020). Traditional water management system and agricultural stainability in a Himalayan foothill village of Assam, India. In: *Habitat, Ecology and Ekistics, Case Studies of Human-Environment Interactions in India* (ed. Rukhsana, A. Haldar, A. Alam, and L. Satpati), 279–292.

Sangtam, T. (2018). *Rooftop Rain Water Harvesting: A Case Study*. Munich: GRIN Verlag https://www.grin.com/document/427542 (accessed 28 September 2024).

Sanjay-Swami (2021). Bamboo drip irrigation: traditional way of irrigating crops in Meghalaya. *Just Agriculture: e-Magazine* 1 (8): 18–21.

Satapathy, K.K. (2001). Water harvesting and recycling in hills. In: *Lecture Notes: Integrated Watershed Development and Management* (ed. S.K. Srivastava and A.K. Sahu). Nirjuli, India: Winter School held at Department of Agricultural Engineering, NERIST.

Sarma, A. and Goswami, D.C. (2015). Sustainable agricultural practices and the methods of traditional water harvesting in North East Region of India. *Scholars Research Library* 7 (4): 23–30.

Sen, P., Devi, N.L., and Singha, D. (2015). Understanding indigenous irrigation systems in North East India. *Eco Whisper* 2 (1): 8–15.

Sengupta, N. (1985). Irrigation: traditional vs modern. *Economic and Political Weekly* 20 (45/47): 1919–1938.
Singh, L.D. (2004). Water harvesting and water conservation in imphal East I Block, Imphal East District, Manipur, 90.
Singh, R.A. and Gupta, R.C. (2002). Traditional land and water management systems of North East hill region. *Indian Journal of Traditional Knowledge* 1: 32–39.
Singh, R.K., Hannah, K., Asangla et al. (2018). Zabo: a time-tested integrated farming system practiced by Chakesang tribe of Nagaland. *Indian Journal of Hill Farming* 31 (1): 188–192.
Srinivasan, R.K. and Babu, S.S.V. (2000). *A Water Harvesting Manual for Urban Areas*. New Delhi, India: Centre for Science and Environment.
Talukdar, R. (2018). Achievement towards sustainability: a study in the North-East India with special reference to eco-friendly harvest. *International Journal of Creative Research Thoughts* 6 (1): 989–1005.
Zhimo, V.V., Imchen, N., and Rawat, M.S. (2018). Water resources management for sustainable development in Nagaland, North East India: a review. *International Journal of Environmental Sciences* 8 (1): 6–7.

20

Sustainable Management of Groundwater Along the Foothills Focusing on Mountain Front Recharge

A Global Review and Status with Regard to Indian Context

Banajarani Panda[1], *Sabarathinam Chidambaram*[2], *Nagappan Ganesh*[3], *Subhashree Biswal*[1], *and Bedour Alsabti*[2]

[1] *Department of Geology, Ravenshaw University, Cuttack, India*
[2] *Water Research Centre, Kuwait Institute for Scientific Research (KISR), Kuwait City, Kuwait*
[3] *Department of Earth Sciences, Annamalai University, Chidambaram, Tamil Nadu, India*

20.1 Introduction

Groundwater is an essential and important commodity that could provide enough water for industries farms and households, particularly in semiarid regions, because of its excellent quality. Therefore, it is necessary to accurately measure the rates at which natural groundwater is replenished to effectively manage this valuable resource. The mountainous areas contribute approximately four times more to the annual discharge of the river basin in comparison to the flat areas (Viviroli et al. 2003). Mountain front recharge (MFR) is a term that refers to the study of hydrological processes that take place from the high elevation to the deeper depth of flowing sub-surface water. The majority of the water in the area stems from the primary mountain block (MB) and the surrounding river systems. This includes all forms of rainfall, including melting snow, runoff from the surface, transpiration, faults, and fractures. Depending on the quantity as well as distribution of superficial and subterranean components, the water from MB can flow into the adjacent watershed by means of MFR (Viviroli et al. 2003).

Multiple authors have made attempts to establish a precise meaning for the word "mountain front" (MF), although a definitive and easily understandable definition for MF is still absent. The primary issue lies in determining if the MF is a clearly defined line or a limited area or similar to the areas impacted by the intrusion of seawater into the coastal aquifer.

If it is considered as a line, there are various, such as borders between different types of vegetation, boundaries between different types of soil, boundaries between different slopes, faults that define the edges of mountains, or even the line where snow begins. Ruxton (1961) provides three possible boundaries for the MF, as shown in Figure 20.1.

Water Resources Management in Mountain Regions, First Edition. Edited by Virendra Bahadur Singh, Sughosh Madhav, Rakesh Kumar Gupta, Prerna Diwan, and Amit Kumar.

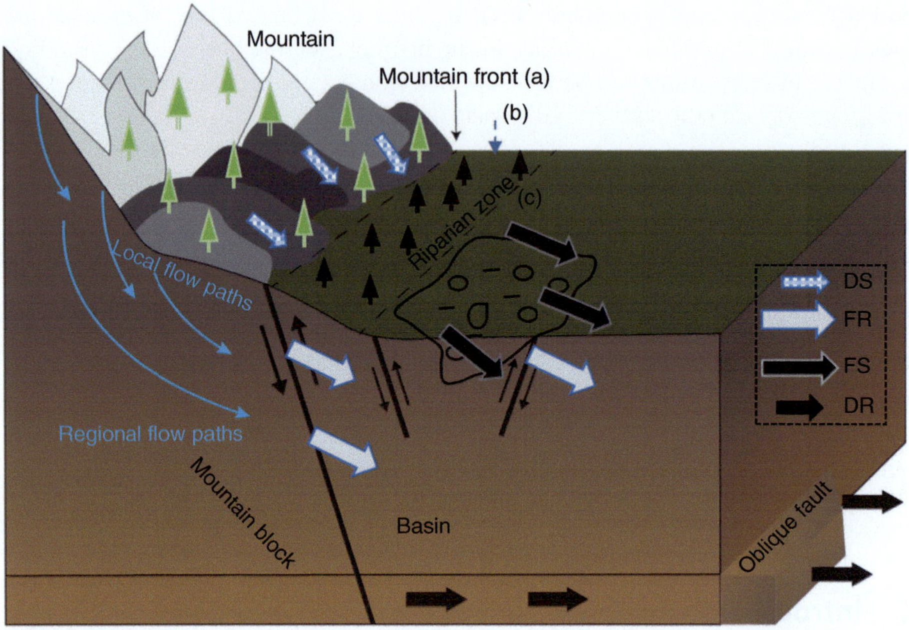

Figure 20.1 Diagram shows possible mountain front limits depicting the zones of mountain block (MB), mountain front (MF) and the riparian block; the pathways that groundwater in MB travels through; the basic components of the MFR. *Source:* Adapted from Wilson and Guan 2004; Toth 1963; Keith 1980.

The area exhibits a noticeable alteration in vegetation, as seen in Figure 20.1 (a).
Intermediate zone between mountain and piedmont (b).
The plinth angle is the intersection of the piedmont and the basin floor periphery (c).

However, if we define the MF as a limited area located between a mountain and a basin floor, then any borders depicted in the aforementioned Figure 20.1 can be regarded as part of the MF zone. However, when studying and considering the objective of defining MF, it is most appropriate to use the term MFR in semiarid and dry regions, by considering the piedmont zone, the area between the mountain edge and riparian zone (RZ), which is referred to as the region where the land meets a river or stream (Figure 20.1). Thus, "MFR" can be referred to as the total volume of water that infiltrates into the MB and MF basin aquifers. Direct MFR entails recharging the MF, while indirect MFR requires recharging the groundwater of MF to the aquifer. These two types of MFR are defined by their respective definitions. There are focused and diffused channels due to the direct and indirect recharge processes, which take place near the surface and below the earth's surface, respectively. These pathways give rise to four distinct components of MFR, as illustrated in Figure 20.1 (Wilson and Guan 2004).

The Near Surface Focused Recharge (FS): Surface stream runoff (FS1, clearly quantifiable) and sub-surface water that is shallow moving through the stream bed deposits (FS2, hard to quantify) are the two forms of recharge that occur along the main flow path (MF).

The Near Surface Diffuse Recharge (DS): It is the runoff from steep slopes and water flowing through the thin soil layers.

The Subsurface Focused Recharge (FR): It refers to the movement of groundwater through natural or manmade openings in the bedrock, such as faults, fractures, joints, and pipes. These openings can be a result of tectonic activity or human actions. The water then flows into the aquifer in the basin and the MB.

Diffuse Recharge (DR): Groundwater that flows across the interface between the MB bedrock and the sediments of the basin aquifer.

According to the component descriptions mentioned above, the formula for MFR is calculated by adding FS1, FS2, DS, FR, and DR together.

20.2 Estimation Methods

The amount of groundwater in the region with arid and semiarid climates has been estimated by adopting various methods, each with notable levels of uncertainty (Allison et al. 1994; Dettinger 1989; Scanlon et al. 2002; Wilson and Guan 2004). These techniques encompass the use of environmental tracers, physical and empirical approaches, as well as numerical models. The study has emphasized literature on various methodologies used in groundwater recharge studies.

20.2.1 Geophysical Approach

The hydrogeological structures of an aquifer impacting groundwater recharge have been identified by using resistivity survey (Beresnev et al. 2002; Danielsen et al. 2003, 2007; Tizro et al. 2012). A resistivity study using the Wenner configuration was conducted in Seethanagaram, India by Mondal et al. (2008), to identify the prospectivity of groundwater recharge. The survey revealed that the potential water resource is located at depths of 50–70 m and has a high yield. Amigun et al. (2012) adopted the vertical electrical sounding approach to acquire the structural and hydrogeological nature of the aquifer for groundwater resource development in Nigeria. Nejad et al. (2012) conducted a resistivity survey in the Curin basin to assess the hydrogeology. Various authors have utilized the ER techniques to assess the quality, to delineate potential groundwater zones (McGlynn et al. 2002), detecting saline water intrusion (Swartz 1937), identifying recharge areas, determining aquifer parameters (Akhter and Hasan 2016), and evaluating the properties of the sediments in the unsaturated zone (Braga 2008; Hussain et al. 2016b).

The integration of resistivity data with the Remote Sensing (RS) & Geographical Information System (GIS) method is used to determine the groundwater prospects in many regions worldwide, including both hard rock and sedimentary aquifers (Israil et al. 2006; Srivastava and Bhattacharya 2006; Srinivasa et al. 2000).

20.2.2 Remote Sensing

Multiple investigations conducted by Edet et al. (1998), Teeuw (1994), and Travaglia and Ammar (1998) inferred that the source and amount of recharge varies with respect to different areas. All of the studies mentioned above utilize the digital elevation models (DEMs) to obtain the regional slope which is crucial to determine the migratory pathway of groundwater and identifying suitable locations for artificial recharge (Saraf and Choudhury 1997, 1998; Saraf et al. 1999).

Integrating several thematic layers such as land use, lithology, geomorphology, rainfall, and water level spatially, zones with higher groundwater recharge prospects can be demarcated. Various studies have employed these techniques to map the regions of groundwater prospects (Krishnamurthy et al. 1992; Obi Reddy et al. 1994), identify a suitable location for recharge (Balachandar et al. 2010; Sarup et al. 2011), and analyze the spatial distribution of high groundwater potential zones (Khadri and Moharir 2014).

Several studies conducted by Cheng et al. (2008), Das et al. (1997), Kumar and Tomas (1998), Machiwal et al. (2011), Pratap et al. (2000), Prabir et al. (2011), and Vaidyanathan (1964) have focused on assessing the groundwater prospects by analyzing geomorphological features obtained from satellite imagery. Identifying earth surface features like lineaments and drainage using RS and GIS methods and also in analyzing groundwater replenishment using the same has been carried out by Leblanc et al. (2003); Sener et al. (2005).

20.2.3 Hydrometeorology

The increase in the water level due to rainfall has been observed in numerous fractured rock areas worldwide. Rodhe and Bockgård (2006) observed a quick response of rainfall to water level fluctuation in granitic aquifers. This aquifer is covered by a 10-m thick layer of till and shows rapid movement of water through the permeable strata. A notable and varying response to rainfall events in the aquifers of Ontario was observed. The water level at one well increases at a pace of 1.0–3.0 m during a few hours, while in surrounding wells, it is <1.0 m (Milloy 2007). The quick response to rainfall suggests good groundwater recharge (increase in water level). Among various methodologies for estimating recharge, the connection between specific yield and water table fluctuations (WTFs) was advocated by Healy and Cook (2002) and later by Healy (2010). In their study, WTF was used in identifying the groundwater recharge of the aquifer by using comprehensive time series data. It is also suggested that fluctuation in groundwater recharge may be due to change in land use pattern or climatic conditions (de Vries and Simmers 2002).

20.2.4 Hydrogeochemical Aspects

Groundwater geochemistry is crucial in identifying the migratory pathways of groundwater. The source and recharge of groundwater relies on the signatures of major ions and heavy metals. These geochemical signatures also help to determine the geochemical variation, the residence time, and the demarcation of recharge zones (Rivera and Guerrero 2007). Therefore, the spatial distribution of major ions in an area in both time and space helps to identify the heterogeneity and connectedness of the aquifer. It also helps in explaining the variations in hydrochemistry and the recharge origin (Chenini and Khemiri 2009; Chidambaram et al. 2005; Güler et al. 2002) Hydrogeochemical approaches have been employed to ascertain the origin or recharge source of groundwater through various techniques as described below.

20.2.4.1 Statistics

Cluster analysis is a powerful tool for dividing large datasets into more manageable subsets according to shared characteristics, which may be used for purposes such as identifying their origin or finding correlations between them (Helstrup et al. 2007; Mahlknecht et al. 2004). Groundwater characteristics, including its quantity and quality, can be effectively analyzed and predicted using multivariate statistical data. These data are commonly used to study groundwater dynamics and deviations induced by natural and anthropogenic variables in recharge. Several studies (Chidambaram 2000; Chidambaram et al. 2013; De Ceballos et al. 1998; Reisenhofer et al. 1995; Srinivasamoorthy et al. 2008) have utilized multivariate statistical data to depict and assess groundwater recharge and its quality.

20.2.4.2 Stable Isotopes

The water isotopes ($\delta^{18}O$, $\delta^{2}H$) have been utilized since the pioneering research conducted by Craig (1961) and Dansgaard (1964) to track the flow of water in the hydrological cycle. To better comprehend hydrogeological processes like basin hydrology, groundwater recharge, interactions between surface and sub-surface water, and precipitation, stable isotopes serve as markers. (Vandenschrick et al. 2002). Stable isotopes are useful for detecting the recharge source, its origin, migratory pathways, and the transmission of pollutants (Gibson et al. 2005). The oxygen-18 ($\delta^{18}O$) and deuterium ($\delta^{2}H$) isotope composition in rainwater and groundwater can shed light on the recharge process (Blasch and Bryson 2007; Li et al. 2008).

The aquifers in dry regions primarily contain aged groundwater as the aquifers are less frequently recharged. Therefore, numerous researchers have utilized environmental isotopes such as ^{14}C (Bretzler et al. 2011; Cartwright and Morgenstern 2012) noble gases (Kulongoski et al. 2008) or ^{36}Cl (Guendouz and Michelot 2006), to estimate groundwater recharge rates of both historical and present conditions, along with their age. D-excess, as a component of isotopic studies in groundwater research, helps in identifying meteoric sources of water, confirm the temporal variation and the fractionation process that takes place before being recharged. (Deshpande et al. 2003; Gibson et al. 2005).

The recent groundwater recharge has been generally determined by tritium (^{3}H) (Begemann and Libby 1957). However, the ^{3}H indicates that the age of groundwater is relatively recent (less than about 50–60 years old). However, the lack of thermonuclear tritium should not be mistaken for a lack of recent recharge. The absence of tritium indicates that the recharge is >~ 60 years (Clark and Fritz 1997).

20.2.4.3 Heavy Metals

Many groundwater studies on the occurrence of heavy metals (occurring in traces and are even referred to as trace metals), their relation to their migratory pathways and its sources of recharge have been conducted worldwide (Table 20.1). These trace elements are discharged through a range of natural and anthropogenic sources, which have a detrimental impact on the environment (Romic and Romic 2003). Their origin is typically from the dispersed components found in the sediment on the bed, which occurs when water flows (Bradley and Cox 1986). According to Siepak et al. (2004), approximately ~90% of these elements are absorbed by suspended particles and subsequently settle at the bottom as sediment in surface-water bodies. Consequently, as the river/lake water infiltrates into the

Table 20.1 Research targeting heavy metals as potential recharge sources.

Serial number	Study focus	Authors and Year
1	An evaluation of the concentration of heavy metals in the subsurface water and surface soil	Landmeyer (1994)
2	Estimation of subsurface recharge through the trace element signatures	Clare (1996)
3	Assessing the extent to which stream water contributes to groundwater pollution by determining the concentration of heavy metals.	Land et al. (2000)
4	The comprehensive observation of the process of urban groundwater recharge	Lerner (2002)
5	The contamination of groundwater as a result of agricultural activities and the identification of the source of recharge	Bohlke (2002)
6	Role of heavy metals in addressing recharge and coastal saline water ingression	Hanson (2003)
7	Trace elements and their relation to the evolution of groundwater chemistry	Edmunds et al. (2006)
8	Relationship of heavy metals to ascertain the migratory pathway of groundwater	Rivera and Guerrero (2007)
9	Trace elements to predict the surface–subsurface interactions	Brenot et al. (2008)
10	Recent groundwater studies adopting trace elements for multiple approaches	Shotyk et al. (2010)
11	Source of recharge identification to assess groundwater pollution	Ayotte et al. (2011)
12	The variation in the chemical status of shallow groundwater in agricultural and forest areas after one hundred years of changes in land use	Kellner et al. (2015)

capillary zone and replenishes the zone of saturation, it results in an increased concentration of these elements. The periodic buildup of these metals can result in an increase in concentration levels (Walna and Siepak 2012; Yu et al. 2003; Wu et al. 2010).

While there are several environmental tracers used for the purpose of identifying and estimating the origin of recharge, as they are discharged into the subsurface through varied origins (Lerner 2002).

Heavy metal contamination of water is an important issue for environmental protection agencies in both developed and developing countries (Khan et al. 2013a). Heavy metals seep into water sources as a consequence of crustal substances due to the process of weathering (Biney and Christopher 1991). Leaching can cause the infiltration of various heavy metals from fertilizers, effluents, residential and municipal sewage into the groundwater (Alloway and Jackson 1991; Biney and Christopher 1991). Groundwater contamination mostly occurs as a result of infiltration and recharge, leading to significant impacts on the availability of drinking water in both urban and rural regions (Thuyet et al. 2016). Polluted water containing trace elements in higher concentrations can lead to negative health impacts when consumed by humans (Karim 2011).

20.2.5 Groundwater Modelling

Several locations within the dry to semiarid basin were investigated by Flint et al. (2004) using a basin characterization model to compute potential recharge rates. The computed rates ranged from several hundred millimeters per year. They determined that recharge takes place in <5% of basins and is not present every year. The estimates derived from the regions adjoining Nevada and eastern Utah using a similar approach indicate that the annual average recharge is 7% of total rainfall, which is >60% of the estimates by the Maxey–Eakin model (Flint and Flint 2007). According to the Maxey–Eakin model, recharge is scanty when the amount of rainfall is ~200 mm per year. Dettinger (1989) and Avon and Durbin (1994) conducted studies in other Nevada basins using different independent methodologies. Their findings were broadly consistent with those of Maxey and Eakin.

20.3 Estimation of MFR

The amount of rainfall received is more in mountainous areas because of orographic effects. The amount of water received by basin aquifers from mountain aquifers is termed as MFR. Quantification of the MFR is accomplished by quantifying the water supplied by mountains to the adjacent aquifer that is adjacent to the mountains. According to Keith (1980), MFR is a process of recharge of an aquifer located at the foothills of a mountainous region. The system consists of two main elements: (i) underground water flow from the nearby mountains and (ii) the process of water seeping into the streams close to the MF (Anderson et al. 1992; Chavez et al. 1994a)

There have been numerous researches conducted that specifically examine how stream flow responds to precipitation in humid environments, and they aid in improved understanding of the hydrology of the mountain areas (Peter et al. 1995). Studies have also reported the relationship of precipitation to recharge in arid and semiarid areas (Puigdefabregas et al. 1998).

To gain a comprehensive understanding of MFR, certain hydrologists employed major ions as environmental tracers (Belkhiri et al. 2010; Grace and Bollen 2008). The presence of chlorides in semiarid environments offers valuable insights into historical climatic change and land use changes (Allison and Hughes 1983; Jolly et al. 1989; Rose et al. 1979). The rise in MFR within a specific area is regarded as the outcome of alterations in land use and cover, as documented by Rose et al. (1979) and Thorburn and Rose (1990).

The researchers used pre-monsoon data to detect the recharge in the MF and RZ of Courtallam, Tamil Nadu, India (Banaja et al. 2018). Banaja et al. (2017a) conducted a study to examine the impact of rainfall on the water level of the northeast monsoon and the aquifer response in selected areas in the same region. They also analyzed stable isotopes as part of their investigation. Banaja et al. (2017c) conducted a resistivity survey to identify the prospective areas for groundwater along the MF and RZ of Couratallam foothills, with the aim of determining suitable locations for the construction of wells. Banaja et al. (2017b) conducted a hydrochemical analysis using PRM data in the Courtallam foothills. The purpose of these studies was to gain insight into the dominant processes occurring along the MF and RZ.

20.4 "MFR" Perspectives

Typically, MFR is examined from two perspectives: (i) the conventional basin-centered perspective and (ii) the mountain-centered perspective.

To establish boundary conditions for the aquifers in the basin, it is necessary to adopt a basin-centered perspective. To simplify the hydrological complications above the MF, the MF is commonly employed as a boundary condition. Hydrochemistry is a significant method for analyzing and understanding the origin and composition of groundwater in a basin.

The mountain-centered approaches primarily focus on the rainfall intensity over the mountains and its impact on MFR rates. However, these methods do consider the aquifer properties or the aquifer continuity beneath the MB. An illustration of this approach involves (i) comparing the hydrochemical characteristics and stable isotopes of rainwater to MF groundwater, (ii) establishing MFR linked to the correlation between precipitation and water level fluctuation, and (iii) calculating the difference between estimated evapotranspiration and precipitation, etc.

20.5 Mountain Groundwater Under Climate Change

The role of mountain hydrology, which includes groundwater, is of utmost importance to society in various intricate ways. Mountains serve as crucial sources of fresh water for both nearby residents and those located downstream (Viviroli et al. 2020). Mountainous communities are currently facing the effects of climate change and are especially susceptible to its effects (Gurgiser et al. 2016; Heikkinen 2017). Groundwater offers a certain level of resistance to the impact of high mountain water supplies, but it is equally susceptible to the effects of long-term climate change (Somers and McKenzie 2020).

Scopus data search on mountain recharge, groundwater, and climate change yielded 211 results; however, data search on mountain recharge and climate change yielded 282 results and data search on MFR, groundwater and climate change yielded 15 results worldwide. These demonstrate the prevalence of the keywords groundwater, climate change, mountain recharge, and MFR, all of which have been investigated by multiple authors from this standpoint. Research on mountain recharge was first carried out by Heinl and Brinkmann (1989) by adopting a groundwater model integrating hydrogeological and geological data of the Nubian aquifer in Egypt. Figure 20.2a,b displays a chronological progression of research on groundwater recharge, mountain recharge, and its linkage to climate change.

Figure 20.2 Interlinkages of the author and index keywords on Scopus literature search for mountain recharge and groundwater based on 493 articles, with a minimum of 25 occurrences. The shades of color show the chronology of the keywords used. The number of occurrences is linked to the size of the bubble. The linkage of key words related to (a) Mountain Recharge, Groundwater & Climate change, (b) Mountain Recharge & Climate change.

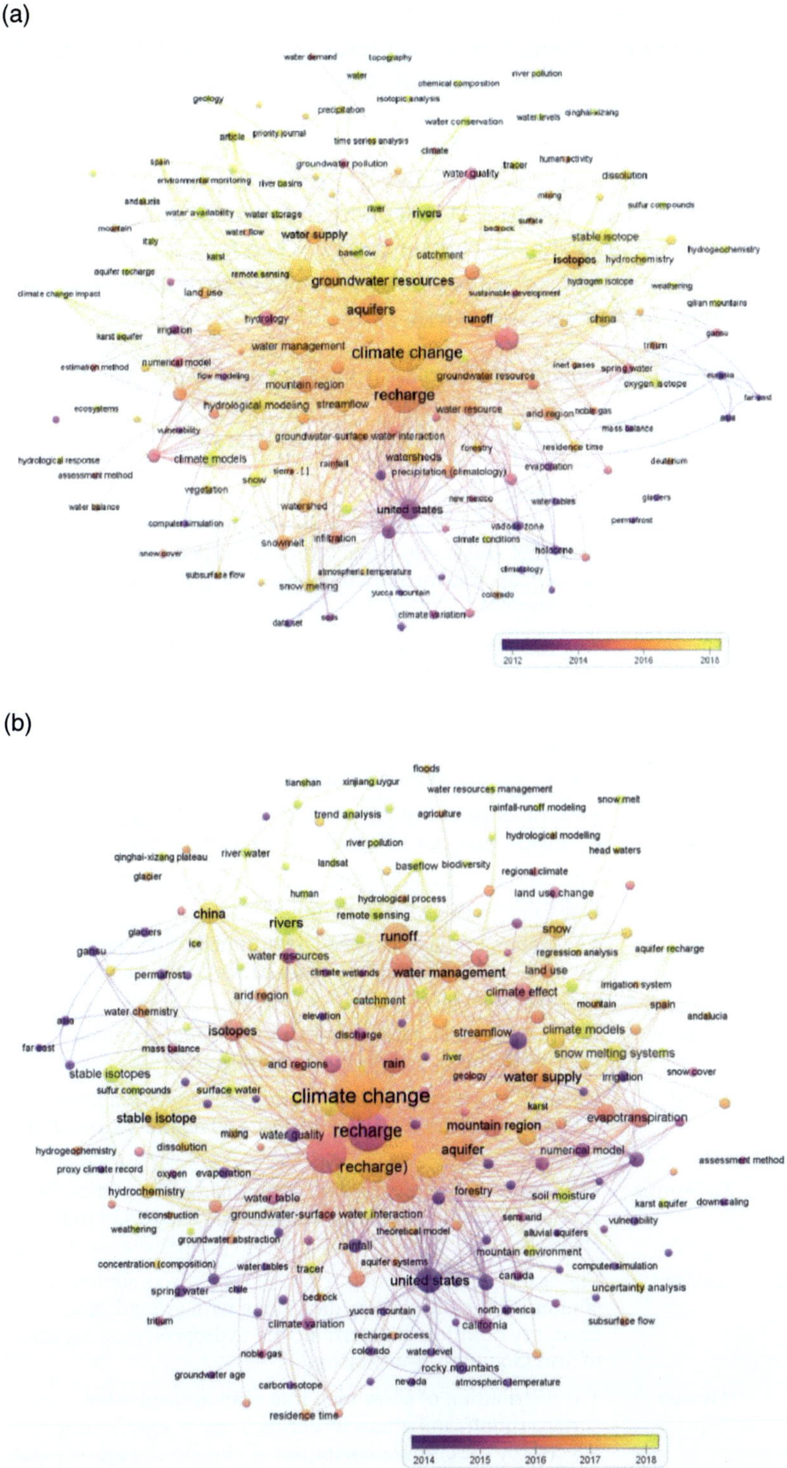
(a)
groundwater resources
aquifers
climate change
recharge
water supply
rivers
isotopes
runoff
china
hydrology
water management
mountain region
hydrological modeling
streamflow
groundwater-surface water interaction
climate models
watersheds
united states
2012
2014
2016
2018
(b)
china
rivers
runoff
water management
climate effect
isotopes
stable isotopes
stable isotope
climate change
recharge
recharge)
aquifer
mountain region
water supply
evapotranspiration
hydrochemistry
groundwater-surface water interaction
united states
2014
2015
2016
2017
2018

There are studies showcasing the peer-reviewed research conducted on groundwater processes in high mountain environments reporting research on this issue, which commenced in 1972 and has steadily increased over time. In addition, it should be mentioned that the research reached its highest point in 2014.

Other than that, below Table 20.2 lists the research carried out relating to MFR, groundwater, and climate change worldwide. There are only 15 articles (extracted from scopous data) noted worldwide as listed in Table 20.2. Mountain hydrology and hydrogeology have become crucial research fields in relation to climate change. Therefore, it is imperative to conduct thorough and extensive research at the current moment to ensure the practicality of high mountain groundwater research for society.

Table 20.2 Research carried out on mountain front recharge linked to groundwater and climate change worldwide.

Authors and publication year	Region	Inferences
Markovitch et al. (2019)	Arizona	Recharge and energy transport models are developed using numerical experiments, components of MFR, and well age distributions.
Morway et al. (2023)	Nevada	Using the groundwater modeling tools MODFLOW-NWT and MODFLOW, hydrogeological research was conducted.
Gómez-Alday et al. (2022)	United Arabs emirates	Stable isotopic fingerprinting using $\delta^{18}O_{H2O}$, $\delta^{2}H_{H2O}$, $\delta^{18}O_{NO3}$, $\delta^{15}N_{NO3}$, $\delta^{18}O_{SO4}$, $\delta^{34}S_{SO4}$, $\delta^{11}B$ was used to address the climate change impact on groundwater in MF.
Warix et al. (2024)	USA	Field observations of groundwater age, discharge, and chemistry were used to analyze C-Q relationships.
Sailo et al. (2022)	India	The various hydrogeological studies imply that MFR is the major source of recharge in Mizoram
Gleason et al. (2020)	USA	The recharge source has been determined using stable isotopic data and a chloride mass-balance approach, in MFR and mountain block (MB).
Anning and Leenhouts (2010)	Arizona	A combination of natural and anthropogenic variables impacts the quality of groundwater resources.
Markovich et al. (2019)	World	Various theoretical data suggest that recharge from MBs is inferred to contribute 5–50% of basin-fill aquifers around the world.
Faunt et al. (2009)	California	The natural water distribution differs spatio-temporally for agriculture and urban areas, and this has resulted in the creation of a vast surface water delivery and diversion system.
Meyers et al. (2021)	California	The severe climate event during 2011–2017 was studied by combining remote sensing data along with groundwater residence duration to determine the hydrogeological aspects of arid mountain-front springs.
Molotch et al. (2009)	Mexico	The distribution of snow melt into evapotranspiration, surface runoff, and groundwater recharge is significantly affected by the cumulative impact of climate change on plant and vegetation structures.

Table 20.2 (Continued)

Authors and publication year	Region	Inferences
Thiros et al. (2010)	California	A significant amount of treated wastewater, which now makes up a significant portion of the river's base flow, as well as imported water penetration are factors affecting the groundwater quality due to the increased recharge of the Santa Ana River.
Hutchison and Hibbs (2008)	Texas	The Rio Grande Alluvium's groundwater has long provided the Hueco Bolson's Chihuahua section with recharging power. The rate of recharge in the metropolitan region has increased as a result of extensive groundwater extraction leading to a deeper water table prior to recharge.
Jourde et al. (2012)	Chile	The ENSO and LNSO events in the semiarid region of North Chico (Chile) greatly affect groundwater recharge.

20.6 Major MF Aquifers in India and its Significance in Recharging Groundwater

The major MF aquifers in India play a crucial role in replenishing groundwater.

India is divided into four primary hydrogeologic provinces.

1) The alluvium found in the Indo-Gangetic plain in north India is of Quaternary age.
2) Basalt, specifically the Deccan trap in western and central India, is of Cretaceous age.
3) The region of southern and south-eastern India is known for its abundance of granites and gneisses, which are of the Archaean age.
4) Sandstones of various ages (proterozoic, paleozoic, and mesozoic) are found throughout the country in a semiconsolidated state (Rangarajan and Athavale 2000).

Within hard-rock regions, particularly in granitic landscapes, the degree and distribution of weathering can significantly differ, even within limited areas. This variation is primarily influenced by the presence of cracks below the surface and the surface features of the terrain. Therefore, in the context of groundwater investigations, it is essential to identify and analyze underground fractures and buried lineaments in hard-rock terrains. Traditionally, apparent resistivity values are employed for the detection of subsurface structures (Anbazhagan 1993; Aswathanarayana and Srinivas 1991; Gerhard Kunze 1998; Muralidharan 1996; Ravi Prakash and Krishna Rao 1991; Suresh et al. 1992; Verma et al. 1980). Mondal et al. (2008) conducted a study on the use of electrical resistivity imaging to identify hidden features in granitic terrain. Chidambaram et al. (2010) sought to estimate recharge using geochemical criteria.

In the study, Ravishankar and Mohan (2005) used a GIS-based hydrogeomorphic approach to analyze the Deccan Volcanic Province in an effort to identify site-specific artificial recharge strategies. Previous research by Anbazhagan (1997) and Saraf and Choudhury (1998) additionally examined recharge using a combined GIS technique, which was also considered in this study. Anbazhagan (1997) covered the procedures and standards for identifying artificial recharge mechanisms, particularly in the Ayyar basin. In an effort to find recharge basins or reservoirs for groundwater in hard rock terrain, Saraf and Choudhury (1998) used an integrated method.

The recharge was determined by injecting tritium. Tritium was typically injected at depths ranging from 0.6 to 0.8 m prior to the monsoon season, which occurs from June to September. Recharge studies were often carried out in arid grassland areas that rely on rainfall for water supply. Tritium was typically injected at 25–30 specific places at each site, and a mean recharge rate was computed for each site. The average rates of recharge at all locations varied from 24 to 198 mm/year, which accounted for 4–20% of the area's average seasonal precipitation (Rangarajan and Athavale 2000). The typical rates of recharging at nine (semi) dry locations varied between 46 and 161 mm/year, which accounted for 9–20% of the normal seasonal precipitation in those areas. The recharge at each site exhibited significant local variability, with a coefficient of variation ranging from 40 to 90%. This variability was mostly related to the heterogeneity of the soil. The results obtained from the tritium injection approach were found to be comparable to those obtained from the water table fluctuation method, as reported by Athavale et al. (1983) and Rangarajan and Athavale (2000).

The research on MFR has been conducted in the vicinity of Courtallam, utilizing hydrogeochemistry and geostatistical techniques (Banaja et al. 2017). Isotopic experiments were conducted in the same location as well. The isotopic data indicates that the rainfall in the foothills of Courtallam is the primary source of groundwater recharge in the basin block regions (Panda et al. 2018).

20.7 A Case Study on Hydrochemical Approach to Estimate MFR in Tamil Nadu, India

20.7.1 Hydrogeochemical Facies

To classify water types and establish the hydrochemical trends, this concept was evolved, which helps in identifying the groundwater flow direction and condition and its chemical evolution (Piper 1944). In the current study, four water types were identified, as discussed further, and also the pathway trend has been identified as given in Figure 20.3. The most common water type Na–Ca–Cl–HCO_3 is along the foothills (underlying litho-unit hornblende, biotite, quartz, and plagioclase feldspar) (Figure 20.3), thus believed to have evolved from the reaction between groundwater and these minerals. On the other hand, this water type was also noted in the location, with agricultural activities, thus may have been influenced by this as well. Hence, there could be two reasons behind this water type: natural rock-water interaction and agricultural activities.

The northeast and southeast parts of the area are dominated by Na–Ca–HCO_3–Cl water type, which might originate from the weathering process (Ophori and Toth 1989). Considering the flow pathways of dominant water types, divided into (i) ion exchange process attributed to increases in Ca or Na ion (ii) weathering processes (Figure 20.4). The prevalence of weathering processes suggests a substantial influx of water along the foothills (Srinivasamoorthy et al. 2008).

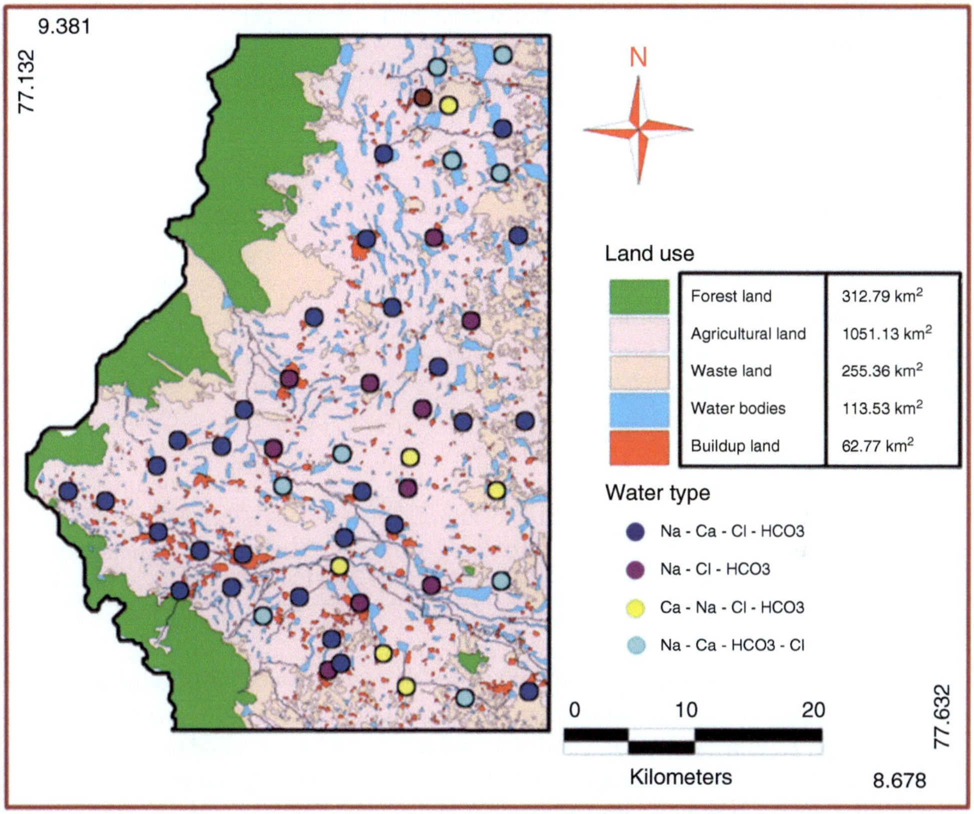

Figure 20.3 Distribution of water types along with land use pattern of the study area (Banaja et al. 2018).

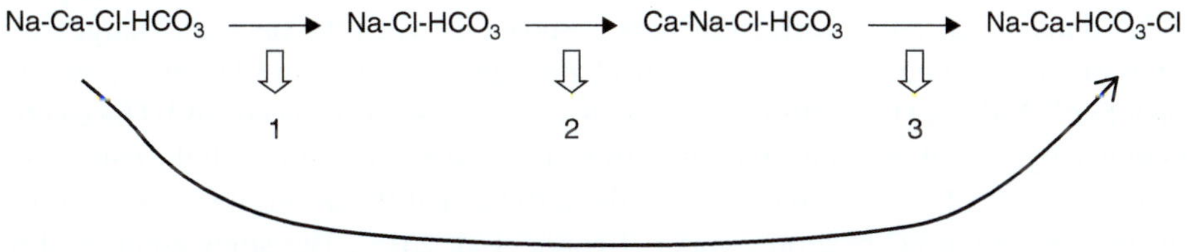

Figure 20.4 Three evolutionary pathways of dominant facies (Banaja et al. 2018).

20.7.2 Correlation Matrix

This provides insights into the interdependencies among the variables in the dataset. The Pearson correlation coefficient is considered the most reliable approach for assessing the correlation between ion pairs, as it is based on the covariance method. The correlation matrix table of the groundwater data obtained from this study is given in Table 20.3, the inferences drawn are given in the following.

There is a strong correlation exists between Mg^{2+} and Na^{+}, Ca^{2+} and Na^{+}, and Cl^{-} and SO_4^{2-}. The association between Ca^{2+} and K^{+} is statistically negligible ($r = 0.24$), as is the correlation between Mg^{2+} and K^{+} ($r = 0.25$).

Table 20.3 The correlation between ions and the numbers in bold shows a strong correlation among variables (Banaja et al. 2018).

	pH	EC	Ca^{2+}	Mg^{2+}	Na^{+}	K^{+}	Cl^{-}	HCO_3^{-}	NO_3^{-}	PO_4^{3-}	SO_4^{2-}	SiO_2
pH	1.00											
EC	−0.10	1.00										
Ca^{2+}	−0.18	**0.72**	1.00									
Mg^{2+}	−0.10	**0.71**	**0.68**	1.00								
Na^{+}	0.10	**0.76**	**0.52**	**0.59**	1.00							
K^{+}	0.30	0.28	0.24	0.25	0.33	1.00						
Cl^{-}	−0.09	**0.86**	**0.81**	**0.79**	**0.80**	0.34	1.00					
HCO_3^{-}	0.36	0.50	0.19	0.35	**0.69**	0.47	0.37	1.00				
NO_3^{-}	0.13	0.09	0.14	0.02	0.10	0.11	0.06	0.17	1.00			
PO_4^{3-}	0.13	−0.04	0.08	−0.15	−0.07	0.13	−0.07	0.14	0.36	1.00		
SO_4^{2-}	0	**0.67**	**0.56**	**0.60**	**0.65**	0.22	**0.68**	0.40	0.24	0.05	1.00	
SiO_2	−0.15	0.03	0.05	0.09	−0.04	0.01	0.08	−0.13	0.07	−0.09	0.08	1.00

A correlation of 0.56 exists between Ca^{2+} and SO_4^{2-}, 0.60 between Mg^{2+} and SO_4^{2-}, 0.80 between Na^{+} and Cl^{-}, and 0.69 between Na^{+} and HCO_3^{-}. There are weaker correlations exists between K^{+} and Cl^{-}, HCO_3^{-}, and NO_3^{-} with "r" of 0.34, 0.47, and 0.11, respectively.

The correlation exists between Ca^{2+} and Mg^{2+} ions, with a correlation coefficient of 0.68. In addition, there is a minor correlation exists between Cl and HCO_3^{-} ($r = 0.37$), Cl^{-} and NO_3^{-} ($r = 0.06$), and HCO_3^{-} and NO_3^{-} ($r = 0.17$).

There is a strong correlation ($r = 0.81$) between Ca^{2+} and Cl^{-} ions, a strong correlation ($r = 0.79$) between Mg^{2+} and Cl^{-} ions, and a moderate correlation ($r = 0.65$) between Na^{+} and SO_4^{2-} ions noted. These correlations are observed between the opposite charges and differing valence numbers of the ions. A negligible and negative correlation is observed between Mg^{2+}, Na^{+}, and Cl^{-} with PO_4^{3-} ($r = -0.15, -0.07, -0.07$ respectively). The electrical conductivity (EC) of groundwater has a strong positive relationship with the concentrations of Ca^{2+}, Mg^{2+}, Na^{+}, Cl^{-}, and SO_4^{2-}. This suggests that these ions have a substantial role in increasing the EC of groundwater (George et al. 2015). The study conducted by Subramani et al. (2010) found a strong link between Cl and the presence of Ca^{2+}, Mg^{2+}, and Na^{+} ions. This suggests that leaching and ion exchange processes have a significant impact on the study area's groundwater. A strong correlation is observed between ions such as SO_4^{2-}, Ca^{2+}, Mg^{2+}, Na^{+}, and Cl^{-}. In addition, there is a correlation between Na^{+}, Cl^{-}, and HCO_3^{-}, indicating the influence of septic tanks and the weathering process as well.

20.7.3 Factor Analysis

Three distinct components were derived from the dataset, which accounted for 67.69% of the overall variability in the original dataset (Table 20.4). Factor 1 explained 40.34% of the overall variability, showing positive associations of EC, Ca, Mg, Na, Cl, and SO_4 (Table 20.4).

Table 20.4 Factor analysis table extracting three major factors from the study area (Banaja et al. 2018).

	Factor		
Parameter	**1**	**2**	**3**
pH	−0.14	**0.77**	0.15
EC	**0.91**	0.07	0
Ca	**0.82**	−0.18	0.14
Mg	**0.85**	−0.03	−0.11
Na	**0.82**	0.36	−0.04
K	0.33	0.55	0.16
Cl	**0.95**	0.01	−0.03
HCO_3	0.47	**0.71**	0.12
NO_3	0.11	0	**0.82**
PO_4	−0.08	0.11	**0.80**
SO_4	**0.78**	0.05	0.20
SiO_2	0.12	0.44	0.12

Bold – Strong positive loadings.

The presence of these ions suggests human activities, such as the release of sodium and chloride from cesspits; however, others are from sewage. Hence, the infiltration of waste from human activities into certain aquifers, with sewer leaks, faulty septic tank, and leachates from landfills, result in elevated levels of EC and chloride (Cl) in groundwater (Panno et al. 1999). The increased concentration of calcium (Ca) and sulfate (SO_4) can be ascribed to the application of pesticides in farmlands and the discharge of residential sewage from densely populated urban regions (Fianko et al. 2009).

Factor 2 explains 15.02% of the total variance and shows significant positive correlations with pH, K, and HCO_3 (Table 20.4), suggesting weathering (Drever 1988). The significant positive loadings of potassium (K) and bicarbonate (HCO_3) indicate that the weathering process is primarily driven by the breakdown of potassium feldspar.

Factor 3 exhibits significant positive correlations between PO_4 and NO_3, accounting for 12.32% of the total variance (Table 20.4). This factor reflects the process of NO_3 and PO_4 leaching from fertilizers used in farmlands into groundwater (Chidambaram et al. 2013). Agriculture often involves the use of chemical fertilizers, resulting in a higher concentration of NO_3 in the area. Excessive farming practices, the dispersal of animal excrement, sewage sludge, and waste materials can all result in the leaching of nitrate (NO_3) (Chidambaram et al. 2013). NO_3 leaching is mostly caused by the accumulation of excessive nitrate in the soil during one season, which is subsequently washed into the groundwater during the following season. The amount of rainfall and the rate at which denitrification occurs also have an impact on the concentration of NO_3 in water that has been leached (Fianko et al. 2009).

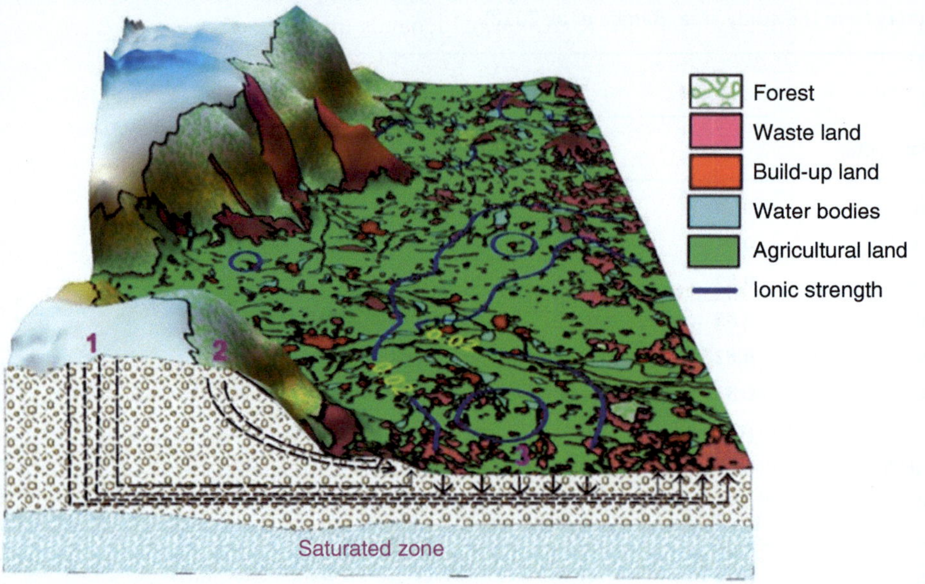

Ionic strength	Water type	Inference
1 Higher	Ca-Na-Cl-HCO3	More residence time and recharged from higher elevation.
2 Lesser	Na-Ca-Cl-HCO3, Na-Ca-HCO3-Cl	Less residence time, Foot hill recharge
3 Highest (>0.019)	Na-Cl-HCO3	Recharge along flood plain.

Figure 20.5 The diagram is drawn from the inferences drawn from the present study (Banaja et al. 2018).

The main conclusions drawn from the study are that the samples collected at the foothills exhibit varying levels of ionic strength, ranging from 0.008 to 0.013 and 0.013 to 0.018. In addition, there are three distinct water types found in the foothills, as shown in Figure 20.5. Ca–Na–Cl–HCO_3 water-type samples with higher ionic strength are believed to have originated from the MB and high residence time. On the other hand, the Na–Ca–Cl–HCO_3 and Na–Ca–HCO_3–Cl water types with low ionic strength are getting recharged in the foothills and have a lesser host rock residence time. The Chittar River is believed to facilitate the process of leaching ions and the replenishment of river water.

20.8 Conclusion and Future Work

MFR plays a crucial role in effectively managing groundwater resources in basin aquifers located in arid and semiarid regions. Therefore, it is imperative to improve and update the methods used to estimate MFR to efficiently manage water resources. As previously mentioned, the estimation of MFR has been done using basin-centered approaches. The mountain-centered view of MFR estimation depends upon the equation derived from the installed instrument in the watershed. However, this approach may not be applicable on a global scale due to potential inaccuracies caused by changes in climate, vegetation, or land use. Therefore, in addition to the conventional basin-centered and mountain-centered methods, using advanced techniques such as remote sensing to measure evapotranspiration, snow cover, and land use/cover, as well as using DEM and GIS, is crucial for improving our comprehension and description of MFR. To estimate the MFR of the MB hydrologic systems, it is necessary to employ geophysical techniques to illustrate the geology of the MB and MF, as well as to study the variation of subsurface formation. However, the hydrochemical approach considering environmental isotopes is mainly important for identifying groundwater flow direction, its source of recharge, and groundwater residence time with the host rock. Similarly, water sampling and real-time observation of water level fluctuation in accordance with rainfall variation are important to unravel the complexity of the MB hydrologic systems.

Acknowledgment

The authors would like to thank The Science and Engineering Research Board (SERB), New Delhi (No: SB/S4/ES-699/2013) for providing the necessary financial support for this study.

References

Akhter, G. and Hasan, M. (2016). Determination of aquifer parameters using geoelectrical sounding and pumping test data in Khanewal District, Pakistan. *Open Geosciences* 8 (1): 630–638.

Allison, G.B. and Hughes, M.W. (1983). The use of natural tracers as indicators of soil-water movement in a temperate semi-arid region. *Journal of Hydrology* 60: 157–173.

Allison, G.B., Gee, G.W., and Tyler, S.W. (1994). Vadose-zone techniques for estimating groundwater recharge in arid and semiarid regions. *Soil Science Society of America Journal* 58 (1): 6–14.

Alloway, B.J. and Jackson, A.P. (1991). The behavior of heavy metals in sewage sludge-amended soils. *Science of the Total Environment* 100: 151–176.

Amigun, J.O., Owolabi, S.T., and Adiat, K.A.N. (2012). Geophysical investigation at a proposed residential estate in Ibadan of Nigeria: hydrological and geotechnical implication. *Journal of Emerging Trends in Engineering and Applied Sciences* 3 (2): 368–374.

Anderson, T.W., Freethey, G.W. and Tucci, P. (1992). Geohydrology and water resources of alluvial basins in south central Arizona and parts of adjacent states. U.S Geological Survey Professional Paper, 1406-B.

Anning, D.W. and Leenhouts, J.M. (2010). Section 8.-Conceptual understanding and groundwater quality of the basin-fill aquifer in the Upper Santa Cruz Basin, Arizona.

Aswathanarayana, L.G. and Srinivas, G. (1991). Resistivity survey of Ghataprabha river basin in parts of Karnataka and Maharashtra, South India. *Proceedings of the First International*

Seminar and Exhibition on Exploration Geophysics in Nineteen Nineties, Hyderabad, India, pp. 262–269 (25–30 November 1991). Hyderabad: Association of Exploration Geophysicists, 1.

Athavale, R.N., Singh, V.S., and Subrahmanyam, K. (1983). A constant discharge device for aquifer tests on large-diameter wells. *Groundwater* 21 (6): 752–755.

Avon, L. and Durbin, T.J. (1994). Evaluation of the Maxey–Eakin method for estimating recharge to ground-water basins in Nevada. *JAWRA Journal of the American Water Resources Association* 30 (1): 99–111.

Ayotte, J.D., Szabo, Z., Focazio, M.J., and Eberts, S.M. (2011). Effects of human-induced alteration of groundwater flow on concentrations of naturally-occurring trace elements at water-supply wells. *Applied Geochemistry* 26 (5): 747–762.

Balachandar, D., Sundararaj, P., Rutharvel, M.K., and Kumaraswamy, K. (2010). An investigation of groundwater quality and its suitability to irrigated agriculture in Coimbatore District, Tamil Nadu, India – a GIS approach. *International Journal of Environmental Sciences* 1 (2): 176190.

Banaja, P., Chidambaram, S., Ganesh, N. et al. (2018). A hydrochemical approach to the estimation of mountain front recharge in an aquifer system in Tamil Nadu, India. *Acta Geochimica* 37: 465–488. https://doi.org/10.1007/s11631-017-0229-4.

Banaja, P., Chidambaram, S., Ganesh, N. et al. (2017a). A study on mountain front recharge by using integrated techniques in the hard rock aquifers of southern India. *Environment, Development and Sustainability* 15: https://doi.org/10.1007/s10668-017-9987-8.

Banaja, P., Chidambaram, S., Thilagavathi, R. et al. (2017b). Geochemical signatures of groundwater along mountain front and riparian zone of Courtallam, Tamil Nadu. *Groundwater for Sustainable Development* 7: 372–386. https://doi.org/10.1016/j.gsd.2017.10.003.

Banaja, P., Chidambaram, S., and Ganesh, N. (2017c). An attempt to understand the subsurface variation along the mountain front and riparian region through geophysics technique in South India. *Modeling Earth Systems and Environment* 3: 783–797. https://doi.org/10.1007/s40808-017-0334-8.

Begemann, F. and Libby, W.F. (1957). Continental water balance, ground water inventory and storage times, surface ocean mixing rates and world-wide water circulation patterns from cosmic-ray and bomb tritium. *Geochimica et Cosmochimica Acta* 12 (4): 277–296.

Belkhiri, L., Boudoukha, A., Mouni, L. et al. (2010). Application of multivariate statistical methods and inverse geochemical modeling for characterization of groundwater – a case study: Ain Azel plain (Algeria). *Geoderma* 159, 390–398.

Beresnev, I.A., Hruby, C.E., and Davis, C.A. (2002). The use of multi-electrode resistivity imaging in gravel prospecting. *Journal of Applied Geophysics* 49 (4): 245–254.

Biney, C.A. and Christopher, A.B. (1991). Trace metal concentrations in fish and sediments from the WIWI: a small urban river in Kumasi, Ghana. *Tropical Ecology* 32 (2): 197–206.

Blasch, K.W. and Bryson, J.R. (2007). Distinguishing sources of ground water recharge by using δ^2H and $\delta^{18}O$. *Groundwater* 45 (3): 294–308.

Böhlke, J.K. (2002). Groundwater recharge and agricultural contamination. *Hydrogeology Journal* 10: 153–179.

Bradley, S.B. and Cox, J.J. (1986). Heavy metals in the Hamps and Manifold valleys, North Staffordshire, U.K.: distribution in floodplain soils. *Science of The Total Environment* 65: 105–128.

Braga, A.C.O. (2008). Estimation of the natural vulnerability of aquifers: a contribution from the resistivity and longitudinal conductance. *Brazilian Journal of Geophysics* 26: 61–68.

Brenot, A., Baran, N., Petelet-Giraud, E., and Négrel, P. (2008). Interaction between different water bodies in a small catchment in the Paris basin (Brevilles, France): tracing of multiple Sr sources through Sr isotopes coupled with Mg/Sr and Ca/Sr ratios. *Applied Geochemistry* 23 (1): 58–75.

Bretzler, A., Osenbrück, K., Gloaguen, R. et al. (2011). Groundwater origin and flow dynamics in active rift systems – a multi-isotope approach in the Main Ethiopian Rift. *Journal of Hydrology* 402 (3–4): 274–289.

Cartwright, I. and Morgenstern, U. (2012). Constraining groundwater recharge and the rate of geochemical processes using tritium and major ion geochemistry: oven's catchment, southeast Australia. *Journal of Hydrology* 475: 137–149.

Chavez, A., Davis, S.N., and Sorooshian, S. (1994a). Estimation of mountain-front recharge to regional aquifers 1: development of an analytical hydroclimatic model. *Water Resources Research* 30 (7): 2157–2167.

Cheng, C.H., Lehmann, J., and Engelhard, M.H. (2008). Natural oxidation of black carbon in soils: changes in molecular form and surface charge along a climosequence. *Geochimica et Cosmochimica Acta* 72: 1598–1610. https://doi.org/10.1016/j.gca.2008.01.010.

Chenini, I. and Khemiri, S. (2009). Evaluation of ground water quality using multiple linear regression and structural equation modeling. *International Journal of Environmental Science and Technology (IJEST)* 6 (3): 509.

Chidambaram, S. (2000). *Hydrogeochemical Studies of Groundwater in Periyar District, Tamilnadu, India*. Department of Geology, Annamalai University.

Chidambaram, S., Ramanathan, A.L., Anandhan, P. et al. (2005). A study of the coastal groundwaters from Puduchattiram to Coleroon, Tamilnadu, India. *International of Ecology and Environmental Sciences* 31 (3): 299–306.

Chidambaram, S., Ramanathan, A.L., Prasanna, M.V. et al. (2010). Study on the hydrogeochemical characteristics in groundwater, post- and pre-tsunami scenario, from Portnova to Pumpuhar, southeast coast of India. *Environmental Monitoring and Assessment* 169: 553–568.

Chidambaram, S., Bala Krishna Prasad, M., Manivannan, R. et al. (2013). Environmental hydrogeochemistry and genesis of fluoride in groundwaters of Dindigul district, Tamilnadu (India). *Environmental Earth Sciences* 68: 1–10.

Clare, N. (1996). Ground water recharge using waters of impaired quality. *Environmental Geology* 27: 376–377.

Clark, I.D. and Fritz, P. (1997). *Environmental Isotopes in Hydrogeology*. Boca Raton, FL: CRC Press.

Craig, H. (1961). Standard for reporting concentrations of deuterium and oxygen-18 in natural waters. *Science* 133 (3467): 1833–1834.

Danielsen, J.E., Auken, E., Jørgensen, F. et al. (2003). The application of the transient electromagnetic method in hydro-geophysical surveys. *Journal of Applied Geophysics* 53 (4): 181–198.

Danielsen, J.E., Dahlin, T., Owen, R. et al. (2007). Geophysical and hydrogeologic investigation of groundwater in the Karoo stratigraphic sequence at Sawmills in northern Matabeleland, Zimbabwe: a case history. *Hydrogeology Journal* 15 (5): 945.

Dansgaard, W. (1964). Stable isotopes in precipitation. *Tellus* 16 (4): 436–468.

Das, S., Behera, S.C., Kar, A. et al. (1997). Hydro geomorphological mapping in ground water exploration using remotely sensed data – a case study in Keonjhar district, Orissa. *Journal of the Indian Society of Remote Sensing* 25 (4): 247–259.

De Ceballos, B.S.O., Koning, A., and De Olivera, J.F. (1998). Dam reservoir eutrophication: a simplified technique for a fast diagnosis of an environmental degradation. *Water Research* 32 (11): 3477–3483.

De Vries, J.J. and Simmers, I. (2002). Groundwater recharge: an overview of processes and challenges. *Hydrogeology Journal* 10 (1): 5–17.

Deshpande, R.D., Bhattacharya, S.K., Jani, R.A., and Gupta, S.K. (2003). Distribution of oxygen and hydrogen isotopes in shallow groundwaters from Southern India: influence of a dual monsoon system. *Journal of Hydrology* 271 (1): 226–239.

Dettinger, M.D. (1989). Reconnaissance estimates of natural recharge to desert basins in Nevada, U.S.A., by using chloride balance calculations. *Journal of Hydrology* 106: 55–78.

Drever, J.I. (1988). *The Geochemistry of Natural Waters*, vol. 437. Englewood Cliffs, NJ: Prentice Hall.

Edet, A.E., Okereke, C.S., Teme, S.C., and Esu, E.O. (1998). Application of remote sensing data to groundwater exploration: a case study of the Cross River State, southeastern Nigeria. *Hydrogeology* 6: 714–722.

Edmunds, W.M., Ma, J., Aeschbach-Hertig, W. et al. (2006). Groundwater recharge history and hydrogeochemical evolution in the Minqin Basin, North West China. *Applied Geochemistry* 21 (12): 2148–2170.

Faunt, C.C., Hanson, R.T., Belitz, K. et al. (2009). Chapter C. Numerical model of the hydrologic landscape and groundwater flow in California's Central Valley. In: *Groundwater availability of the Central Valley aquifer of California*, 121–212.

Fianko, J.R., Osae, S., Adomako, D., and Achel, D.G. (2009). Relationship between land use and groundwater quality in six districts in the eastern region of Ghana. *Environmental Monitoring and Assessment* 153: 139–146.

Flint, A.L. and Flint, L.E. (2007). Application of the basin characterization model to estimate in-place recharge and runoff potential in the Basin and Range carbonate-rock aquifer system, White Pine County, Nevada and adjacent areas in Nevada and Utah, U.S. Geological Survey Scientific Investigations Report 2007–5099, 20 p.

Flint, A.L., Flint, L.E., Hevesi, J.A., and Blainey, J.M. (2004). Fundamental concepts of recharge in the Desert Southwest: a regional modeling perspective. In: *Groundwater Recharge in a Desert Environment: The Southwestern United States*, Water Science and Applications Series, vol. Vol. 9 (ed. J.F. Hogan, F.M. Phillips, and B.R. Scanlon), 159–184. Washington, D.C.: American Geophysical Union.

George, N.J., Ibanga, J.I., and Ubom, A.I. (2015). Geoelectrohydrogeological indices of evidence of ingress of saline water into freshwater in parts of coastal aquifers of Ikot Abasi, southern Nigeria. *Journal of African Earth Sciences* 109: 37–46.

Gerhard Kunze, A.W. (1998). Implications of electrical resistivity data regarding ground-water lenses on San Salvador island, Bahamas. *Environmental & Engineering Geoscience* 4 (1): 55–76.

Gibson, R.N., Atkinson, R.J.A., and Gordon, J.D.M. (2005). Ecology of cold seep sediments: interactions of fauna with flow, chemistry and microbes. *Oceanography and Marine Biology: An Annual Review* 43: 1–46.

Gleason, C.L., Frisbee, M.D., Rademacher, L.K. et al. (2020). Hydrogeology of desert springs in the Panamint range, California, USA: identifying the sources and amount of recharge that support spring flow. *Hydrological Processes* 34 (3): 730–748.

Gómez-Alday, J.J., Hussein, S., Arman, H. et al. (2022). A multi-isotopic evaluation of groundwater in a rapidly developing area and implications for water management in hyper-arid regions. *Science of the Total Environment* 805: 150245.

Grace, J.B. and Bollen, K.A. (2008). Representing general theoretical concepts in structural equation models: the role of composite variables. *Environmental and Ecological Statistics* 15: 191–213.

Guendouz, A. and Michelot, J.L. (2006). Chlorine-36 dating of deep groundwater from northern Sahara. *Journal of Hydrology* 328 (3): 572–580.

Güler, C., Thyne, G.D., McCray, J.E., and Turner, K.A. (2002). Evaluation of graphical and multivariate statistical methods for classification of water chemistry data. *Hydrogeology Journal* 10 (4): 455–474.

Gurgiser, W., Juen, I., Singer, K. et al. (2016). Comparing peasants' perceptions of precipitation change with precipitation records in the tropical Callejón de Huaylas, Peru. *Earth System Dynamics* 7 (2): 499–515.

Hanson, R.T. (2003). Geohydrologic Framework of Recharge and Seawater Intrusion in the Pajaro Valley, Santa Cruz and Monterey Counties, California. US Department of the Interior, US Geological Survey.

Healy, R.W. (2010). *Estimating Groundwater Recharge*. Cambridge: Cambridge University Press.

Healy, R.W. and Cook, P.G. (2002). Using groundwater levels to estimate recharge. *Hydrogeology Journal* 10 (1): 91–109.

Heikkinen, A, 2017, 'Climate change in the Peruvian Andes: a case study on small-scale farmers' vulnerability in the Quillcay River basin', *Iberoamericana–Nordic Journal of Latin American and Caribbean Studies*, 46(1), 77–88.

Heinl, M. and Brinkmann, P.J. (1989). A groundwater model of the Nubian aquifer system. *Hydrological Sciences Journal* 34 (4): 425–447.

Helstrup, T., Jorgensen, N.O., and Banoeng-Yakubo, B. (2007). Investigation of hydrochemical characteristics of groundwater from the Cretaceous-Eocene limestone aquifer in southern Ghana and southern Togo using hierarchical cluster analysis. *Hydrogeology Journal* 15 (5): 977.

Hussain, I., Sulatan, A., Shinwari, Z.K. et al. (2016b). Genetic diversity based on morphological traits in Walnut (Juglans regia L.) landraces from Karakoram region-I. *Pakistan Journal of Botany* 48 (2): 653–659.

Hutchison, W.R. and Hibbs, B.J. (2008). Ground water budget analysis and cross-formational leakage in an arid basin. *Groundwater* 46 (3): 384–395.

Israil, M., Al-Hadithi, M., and Singhal, D.C. (2006). Application of a resistivity survey and geographical information system (GIS) analysis for hydrogeological zoning of a piedmont area, Himalayan foothill region, India. *Hydrogeology Journal* 14 (5): 753–759.

Jolly, D., Schutz, H., Diez-Knauf, K., and Johal, J. (1989). Organic foods: consumer attitudes and use. *Food Technology* 43 (11): 61–66.

Jourde, H., Blanc, M., Rochette, R. et al. (2012). Characterization of the interactions between subterranean and superficial fluxes within an Andean catchment as a function of the spatio-temporal variability of climate. *Houille Blanche-Revue Internationale De L Eau* 2: 18–25.

Karim, Z. (2011). Risk assessment of dissolved trace metals in drinking water of Karachi, Pakistan. *Bulletin of Environmental Contamination and Toxicology* 86 (6): 676–678.

Keith, S.J. (1980). Chapter 4: Mountain front recharge. In: *Regional Recharge Research for Southwest Alluvial Basins* (ed. L.G. Wilson et al.), 4-1–4-44. Tucson: University of Arizona.

Kellner, E., Hubbart, J.A., and Ikem, A. (2015). A comparison of forest and agricultural shallow groundwater chemical status a century after land use change. *Science of the Total Environment* 529: 82–90.

Khadri, S.F.R. and Moharir, N.K. (2014). Spatial analysis of groundwater quality for Man River basin in Akola and Buldhana District of Maharashtra using GIS. *Golden Research Thoughts* 3: 12.

Khan, K., Lu, Y., Khan, H. et al. (2013a). Heavy metals in agricultural soils and crops and their health risks in Swat District, northern Pakistan. *Food and Chemical Toxicology* 58: 449–458.

Krishnamurthy, J., Manavalan, P., and Saivasan, V. (1992). Application of digital enhancement techniques for ground water exploration in a hard rock terrain. *International Journal of Remote Sensing* 13: 2925–2942.

Kulongoski, J.T., Hilton, D.R., Cresswell, R.G. et al. (2008). Helium-4 characteristics of groundwaters from Central Australia: comparative chronology with chlorine-36 and carbon-14 dating techniques. *Journal of Hydrology* 348 (1): 176–194.

Kumar, A. and Tomas, S. (1998). Groundwater assessment through hydrogeomorphological and geophysical survey – a case study in Godavari sub-watershed, Giridih, Bihar. *Journal of the Indian Society of Remote Sensing* 26 (4): 177–184.

Land, M., Ingri, J., Andersson, P.S., and Ohlander, B. (2000). Ba/Sr, Ca/Sr and 87 Sr/86 Sr ratios in soil water and groundwater: implications for relative contributions to stream water discharge. *Applied Geochemistry* 15 (3): 311–325.

Landmeyer, J.E. (1994) Assessment of concentrations of trace elements in ground water and soil at the small-arms firing range, Shaw air force base, South Carolina. US Department of the Interior, US Geological Survey.

Leblanc, M., Leduc, C., Razack, M. et al. (2003). Application of remote sensing and GIS for groundwater modelling of large semiarid areas: example of the Lake Chad Basin, Africa. In: *Hydrology of Mediterranean and Semiarid Regions Conference, Montpieller, France*, Red Books Series, vol. 278, 186–192. Wallingford: IAHS.

Lerner, D.N. (2002). Identifying and quantifying urban recharge: a review. *Hydrogeology Journal* 10 (1): 143–152.

Li, J.W., Zhao, X.F., Zhou, M.F. et al. (2008). Origin of the Tongshankou porphyry–skarn Cu–Mo deposit, eastern Yangtze craton, Eastern China: geochronological, geochemical, and Sr–Nd–Hf isotopic constraints. *Mineralium Deposita* 43 (3): 315–336.

Machiwal, D., Jha, M.K., and Mal, B.C. (2011). Assessment of groundwater potential in a semi-arid region of India using remote sensing, GIS and MCDM techniques. *Water Resources Management* 25: 1359–1386.

Mahlknecht, J., Schneider, J.F., Merkel, B.J. et al. (2004). Groundwater recharge and geochemical evolution in semi-arid central Mexico. *Hydrogeology Journal* 12: 511–530.

Markovich, K.H., Manning, A.H., Condon, L.E., and McIntosh, J.C. (2019). Mountain-block recharge: a review of current understanding. *Water Resources Research* 55 (11): 8278–8304.

McGlynn, B.L., McDonnel, J.J., and Brammer, D.D. (2002). A review of the evolving Perceptual model of hillslope flow paths at the Maimai catchments, New Zealand. *Journal of Hydrology* 257: 1–26.

Meyers, Z.P., Frisbee, M.D., Rademacher, L.K., and Stewart-Maddox, N.S. (2021). Old groundwater buffers the effects of a major drought in groundwater-dependent ecosystems of the eastern Sierra Nevada (CA). *Environmental Research Letters* 16 (4): 044044.

Milloy, C. (2007). Measurement of hydraulic head in the evaluation of groundwater recharge to discrete fracture zones in a crystalline bedrock aquifer. Unpublished M. Sc. Thesis. Queen's University. Kingston.

Molotch, N.P., Brooks, P.D., Burns, S.P. et al. (2009). Ecohydrological controls on snowmelt partitioning in mixed-conifer sub-alpine forests. *Ecohydrology: Ecosystems, Land and Water Processes Interactions, Ecohydrogeomorphology* 2 (2): 129–142.

Mondal, N.C., Singh, V.S., Saxena, V.K., and Prasad, R.K. (2008). Improvement of groundwater quality due to fresh water ingress in Potharlanka Island, Krishna delta, India. *Environmental Geology* 55 (3): 595–603.

Morway, E.D., Buto, S.G., Niswonger, R.G., and Huntington, J.L. (2023) Assessing potential effects of changes in water use in the middle carson river basin with a numerical groundwater-flow model, eagle, Dayton, and Churchill valleys, West-central Nevada (No. 2023-5008). US Geological Survey.

Muralidharan, D. (1996). A semi-quantitative approach to detect aquifers in hard rocks from apparent resistivity data. *Journal of the Geological Society of India* 47: 237–242.

Nejad, H.T., Hoseini, F.Z., Mumipour, M. et al. (2012). Delineation of the Aquifer in the Curin Basin, south of Zahedan city, Iran. *The Open Geology Journal* 6: 1–6.

Obi Reddy, G.P., Suresh, B.R., and Rao Sambasiva, M. (1994). Hydrology and hydro-geomorphological condition of Anantapur district using remote sensing data. *Indian Geographical Journal* 69 (2): 128–235.

Ophori, D.U. and Toth, J. (1989). Patterns of ground-water chemistry, Ross Creek Basin, Alberta, Canada. *Groundwater* 27 (1): 20–26.

Panda, B., Chidambaram, S., Thilagavathi, R. et al. (2018). 'Geochemical signatures of groundwater along mountain front and riparian zone of Courtallam, Tamil Nadu. *Groundwater for Sustainable Development* 7: 372–386.

Panno, S.V., Nuzzo, V.A., Cartwright, K. et al. (1999). Impact of urban development on the chemical composition of ground water in a fen-wetland complex. *Wetlands* 19: 236–245.

Peter, D.L., Buttle, J.M., Taylor, C.H., and LaZerte, B.D. (1995). Runoff production in a forested, shallow soil, Canadian Shield basin. *Water Resources Research* 31: 1291–1304.

Piper, A.M. (1944). A graphic procedure in the geochemical interpretation of water-analyses. *Eos, Transactions American Geophysical Union* 25 (6): 914–928.

Prabir, M., Singh, C.K., and Mukherjee, S. (2011). Delineation of groundwater potential zones in arid region of India, a remote sensing and GIS approach. *Water Resources Management* 26: 2643–2672.

Pratap, K., Ravindran, K.V., and Prabakaran, B. (2000). Groundwater prospect zoning using remote sensing and geographical information system: a case study in Dala-Renukoot Area, Sonbhadra District Uttar Pradesh. *Journal of the Indian Society of Remote Sensing* 28 (4): 249–263.

Puigdefabregas, J., del Barrio, G., Bore, M.M. et al. (1998). Differential responses of hillslope and channel elements to rainfall events in a semi-arid area. *Geomorphology* 23: 337–351.

Rangarajan, R. and Athavale, R.N. (2000). Annual replenishable ground water potential of India – an estimate based on injected tritium studies. *Journal of Hydrology* 234 (1–2): 38–53.

Ravi Prakash, S. and Krishna Rao, G. (1991). Geo-electrical resistivity studies of Paravada area, Visakhapatnam District, Andhra Pradesh. *Hydrology J IAH* XIV: 1–15.

Ravi Shankar, M.N. and Mohan, G. (2005). A GIS based hydrogeomorphic approach for identification of site-specific artificial-recharge techniques in the Deccan Volcanic Province. *Journal of Earth System Science* 114 (5): 505–514.

Reisenhofer, E., Picciotto, A., and Li, D. (1995). A factor analysis approach to the study of the eutrophication of a shallow, temperate lake (San Daniele, North Eastern Italy). *Analytica Chimica Acta* 306 (1): 99–106.

Rivera, J.J.C. and Guerrero, M.A.O. (2007). *Groundwater Flow Understanding: From Local to Regional Scale*. Boca Raton, FL: CRC Press.

Rodhe, A. and Bockgård, N. (2006). Groundwater recharge in a hard rock aquifer: a conceptual model including surface-loading effects. *Journal of Hydrology* 330 (3): 389–401.

Romic, M. and Romic, D. (2003). Heavy metals distribution in agricultural topsoils in urban area. *Environmental Geology* 43: 795–805.

Rose, A.W., Hawkes, H.E., and Webb, J.S. (1979). *Geochemistry in Mineral Exploration*, 490–517. New York, NY: Academic Press.

Ruxton, B.P. (1961). Weathering profiles and geomorphic position on granite in two tropical regions. *Revue de Géomorphologie Dynamique* 12: 16–31.

Sailo, L., Das, M., and Vanlalhruaia, H. (2022). Geochemical evolution of spring water sources in West Phaileng, Mizoram. In: *International Conference on Emerging Global Trends in Engineering and Technology*, 79–88. Singapore: Springer Nature.

Saraf, A.K. and Choudhury, P.R. (1997). Integrated Application of Remote Sensing and GIS Groundwater exploration in hard rock terrain. *Proceedings: International Symposium on Emerging Trends in Hydrology*, Roorkee, India (25–27 September). Roorkee, India: Department of Hydrology, pp. 435–442.

Saraf, A.K. and Choudhury, P.R. (1998). Integrated remote sensing and GIS for groundwater exploration and identification of artificial recharge sites. *International Journal of Remote Sensing* 19 (10): 1825–1841.

Saraf, A.K., Foster, J.L., Singh, P., and Tarafdar, S. (1999). Passive microwave data for snow-depth and snow-extent estimations in the Himalayan Mountains. *International Journal of Remote Sensing* 20 (1): 83–95.

Sarup, M.K., Tiwari, V., and Khatediya (2011). Delineate groundwater prospect zones and identification of artificial recharge sites using geospatial technique. *International Journal of Advance Technology & Engineering Research* 1: 6–20.

Scanlon, B.R., Healy, R.W., and Cook, P.G. (2002). Choosing appropriate techniques for quantifying groundwater recharge. *Hydrogeology Journal* 10 (1): 18–39.

Sener, E., Davraz, A., and Ozcelik, M. (2005). An integration of GIS and remote sensing in groundwater investigations: a case study in Burdur, Turkey. *Hydrogeology Journal* 13 (5–6): 826–834.

Shotyk, W., Krachler, M., Aeschbach-Hertig, W. et al. (2010). Trace elements in recent groundwater of an artesian flow system and comparison with snow: enrichments, depletions, and chemical evolution of the water. *Journal of Environmental Monitoring* 12 (1): 208–217.

Siepak, M., Niedzielski, P., and Bierła, K. (2004). Determination of inorganic speciation forms of arsenic, antimony and selenium in water from a grate ashes dumping ground as an element of hydrogeochemical monitoring of pollution spread. *Polish Journal of Environmental Studies* 13 (6): 709–713. http://www.pjoes.com/pdf/13.6/709-713.pdf.

Somers, L.D. and McKenzie, J.M. (2020). A review of groundwater in high mountain environment. *Wiley Interdisciplinary Reviews: Water* 7 (6): e1475.

Srinivasa, R., Reddy, T.V.K., and Nayudu, P.T. (2000). Ground water targeting in a hard rock terrain using fracture pattern modelling, Niva River basin, Andhra Pradesh, India. *Hydrogeology Journal* 8: 494–502.

Srinivasamoorthy, K., Chidambaram, S., Prasanna, M.V. et al. (2008). Identification of major sources controlling groundwater chemistry from a hard rock terrain – a case study from Metturtaluk, Salem district, Tamilnadu India. *Journal of Earth System Science* 117 (1): 49–58.

Srivastava, P.K. and Bhattacharya, A.K. (2006). Groundwater assessment through an integrated approach using remote sensing, GIS and resistivity techniques: a case study from a hard rock terrain. *International Journal of Remote Sensing* 27 (20): 4599–4620.

Subramani, T., Rajmohan, N., and Elango, L. (2010). Groundwater geochemistry and identification of hydrogeochemical processes in a hard rock region, Southern India. *Environmental Monitoring and Assessment* 162: 123–137.

Suresh, T.S., Naganna, C., and Srinivas, G. (1992). Exploration for groundwater in Hemavathy river basin, Karnataka: using resistivity techniques. *Journal of the Indian Academy of Geoscience* 35: 25–30.

Swartz, J.H. (1937). Resistivity-studies of some salt-water boundaries in the Hawaiian Islands. *Transactions - American Geophysical Union* 18 (2): 387–393.

Teeuw, R.M. (1994). Groundwater exploration using remote sensing and a low-cost geographic information system. *Hydrogeology Journal* 3 (3): 21–30.

Thiros, S.A., Bexfield, L.M., Anning, D.W., and Huntington, J.M. (2010). Conceptual understanding and groundwater quality of selected basin-fill aquifers in the Southwestern United States (No. 1781). US Geological Survey.

Thorburn, P.J. and Rose, C.W. (1990). Interpretation of solute profile dynamics in irrigated soils. III. A simple model of bypass flow. *Irrigation Science* 11: 219–225.

Thuyet, D.Q., Saito, H., Saito, T. et al. (2016). Multivariate analysis of trace elements in shallow groundwater in Fuchu in western Tokyo Metropolis, Japan. *Environmental Earth Sciences* 75: 1–14.

Tizro, A.T., Voudouris, K., and Basami, Y. (2012). Estimation of porosity and specific yield by application of geoelectrical method–a case study in western Iran. *Journal of Hydrology* 454: 160–172.

Toth, J. (1963). A theoretical analysis of groundwater flow in small drainage basins. *Journal of Geophysical Research* 68 (16): 4795–4812.

Travaglia, C. and Ammar, O.1998). Groundwater exploration by satellite remote sensing in the Syrain Arab Republic, Technical Report, FAO.TCP/SYR/6611, 33.

Vaidyanathan, R. (1964). Geomorphology of Cuddapah basin. *Journal of Indian Geosciences Association* 4: 29–36.

Vandenschrick, G., Van Wesemael, B., Frot, E. et al. (2002). Using stable isotope analysis ($\delta D-\delta^{18}O$) to characterize the regional hydrology of the Sierra de Gador, south east Spain. *Journal of Hydrology* 265 (1): 43–55.

Verma, R.K., Rao, M.K., and Rao, C.V. (1980). Resistivity investigations for groundwater in metamorphic areas near Dhanbad, India. *Ground Water* 18: 46–55.

Viviroli, D., Weingartner, R., and Messerli, B. (2003). Assessing the hydrological significance of the world's mountains. *Mountain Research and Development* 23 (1): 32–40.

Viviroli, D., Kummu, M., Meybeck, M. et al. (2020). Increasing dependence of lowland populations on mountain water resources. *Nature Sustainability* 3 (11): 917–928.

Walna, B. and Siepak, M. (2012). Heavy metals: their pathway from the ground, groundwater and springs to Lake Góreckie (Poland). *Environmental monitoring and assessment* 184: 3315–3340.

Warix, S., Navarre-Sitchler, A., and Singha, K. (2024). Water-rock interactions drive chemostasis. *Hydrological Processes* 38 (2): e15078.

Wilson, J.L. and Guan, H. (2004). Mountain block hydrology and mountain front recharge. In: *Groundwater Recharge in a Desert Environment: The Southwestern United States*, Water Science and Applications Series, vol. 9 (ed. J.F. Hogan, F.M. Phillips, and B.R. Scanlon), 113–137. Washington, DC: American Geophysical Union.

Wu, B., Zhang, Y., Zhang, X., and Cheng, S. (2010). Health risk from exposure of organic pollutants through drinking water consumption in Nanjing, China. *Bulletin of Environmental Contamination and Toxicology* 84: 46–50.

Yu, S., Shang, J., Zhao, J., and Guo, H. (2003). Factor analysis and dynamics of water quality of the Songhua River, Northeast China. *Water, Air, and Soil Pollution* 144: 159–169.

Index

Water Resources Management in Mountain Regions, First Edition. Edited by Virendra Bahadur Singh, Sughosh Madhav, Rakesh Kumar Gupta, Prerna Diwan, and Amit Kumar.

t

u

v

w